Possible Scenarios for Homochirality on Earth

Possible Scenarios for Homochirality on Earth

Special Issue Editor

Michiya Fujiki

MDPI • Basel • Beijing • Wuhan • Barcelona • Belgrade

Special Issue Editor
Michiya Fujiki
Nara Institute of Science and Technology (NAIST)
Japan

Editorial Office
MDPI
St. Alban-Anlage 66
4052 Basel, Switzerland

This is a reprint of articles from the Special Issue published online in the open access journal *Symmetry* (ISSN 2073-8994) from 2018 to 2019 (available at: https://www.mdpi.com/journal/symmetry/special_issues/Possible_Scenarios_Homochirality_Earth).

For citation purposes, cite each article independently as indicated on the article page online and as indicated below:

LastName, A.A.; LastName, B.B.; LastName, C.C. Article Title. *Journal Name* **Year**, *Article Number*, Page Range.

ISBN 978-3-03921-722-9 (Pbk)
ISBN 978-3-03921-723-6 (PDF)

Contents

About the Special Issue Editor

Michiya Fujiki, Dr., was born in Fukuoka, Japan, in 1954. He received his BS and MS degrees (supervisor Prof. Toyoki Kunitake and co-supervisor Prof. Seiji Shinkai) in Chemistry of Organic Synthesis from Kyushu University, Fukuoka, Japan, in 1976 and 1978, respectively, and was awarded his PhD degree (supervisor, Prof. Emeritus, Toyoki Kunitake) from Kyushu University, Fukuoka, Japan, in 1993. Since 1978, he has worked for Electrical Telecommunication Laboratories (ETL) of Nippon Telegraph and Telephone Cooperation (NTT). From 1978 to 1982, he studied low-loss optical plastic fibers, and from 1983 to 1987, the preparation and thin film properties of semiconducting phthalocyanine derivatives. In 1987, he joined the Basic Research Laboratories (BRL) of NTT. From 1987 to 2002, he studied the structure–photophysical property relationship of newly designed and synthesized helical-conjugating polysilanes. He served as Principal Investigator (representative leader) of the Japan Science and Technology Agency (JST)—Core Research for Evolutional Science and Technology (CREST) program, directed by Emeritus Prof. Hideki Sakurai (Tohoku University), from 1998 to 2003. In May 2002, he started as Full Professor at the Graduate School of Materials Science, now the Division of Materials Science, Nara Institute of Science and Technology (NAIST), in Japan. His interests, which are the foundation of his life's work, include the rational design and functionality of hierarchical architecture using group 14 element polymers, π-conjugated molecules, and phthalocyanine supramolecular polymers, and the most fundamental question of the origin of homochirality on Earth. His final goal is to unify four fundamental physical forces (gravity, strong, weak, and electromagnetic) since the Big Bang and to understand the roles of several elemental particles, including Higgs boson, neutrino, and photon particles. He has approximately 330 publications corresponding to an h-index of 59 and 10,000 citations as of October 2019. Although he retired from NAIST in March 2019 at the official retirement age (65 years old), he has continued his association as an Emeritus, keeping up-to-date in his main areas of interest.

Preface to "Possible Scenarios for Homochirality on Earth"

As evidenced in several myths, humankind has long been thinking of life: Where did life come from? And where will life go? A more apt question is whether life exists only on Earth. In 1903, Svante Arrhenius proposed the radiopanspermia hypothesis for the origin of life; specifically, that certain seeds of life in the size of 200–300 nm travelled slowly by solar radiation pressure and landed on Earth. This scenario led to the lithopanspermia and ballistic panspermia hypotheses. In 1978, Fred Hoyle proposed that interstellar comets carrying several viruses landed on Earth in accordance with these panspermia hypotheses. With respect to life on Earth, the origin of homochirality has been the greatest mystery because life cannot exist without molecular asymmetry. Living organisms on Earth are always temporospatial and metastable as a consequence of a far-from-equilibrium open system. Life can survive only under open flows of energy and chemical sources because life eats low-entropy foods and/or harvests solar/thermal energy. In recent years, several discoveries led by modern spacecraft and telescopes have brought about new horizons regarding the origins of life and chirality. For example, the Hubble Space Telescope (NASA) captured geysers of liquid water from Europa, one of Jupiter's moons. Rosetta (ESA) detected prebiotic constituents and water on the comet 67P/Churyumov–Gerasimenko that is part of the Jupiter family. Using a quadrupole gravity measurement system, Cassini (NASA) discovered water under the surface of Enceladus, one of Saturn's moons. The Jet Propulsion Laboratory (NASA) showed that Titan, the largest Saturnian moon, has water containing inorganic salts. Kepler (NASA) discovered over 2000 Earth-like exoplanets located in habitable zones. More recently, in 2016, radio telescope astronomers detected chiral propylene oxide in a giant molecular cloud called Sagittarius B2 in the Milky Way galaxy by analyzing absorption bands at mm wavelengths, although the L-D preference is yet to be characterized. Thus far, many scientists have proposed several possible hypotheses to answer this long-standing L-D question. Previously, Martin Gardner raised the question about mirror symmetry and broken mirror symmetry in terms of the homochirality question in his monographs (1964 and 1990). Possible scenarios for the L-D issue can be categorized into (i) Earth and exoterrestrial origins, (ii) by-chance and necessity mechanisms, and (iii) mirror-symmetrical and non-mirror-symmetrical forces as physical and chemical origins. These scenarios should involve further great amplification mechanisms, enabling a pure L- or D-world. Recent studies have demonstrated that even nearly racemic substances as low as 0.00001% enantiomeric excess (ee) can be significantly amplified to nearly 100% ee. The present Special Issue encouraged researchers from a broad range of disciplines to publish original papers, account papers, and reviews which describe possible scenarios of bimolecular handedness, followed by L-D amplification on Earth, as well as the possibility of life under exoterrestrial cosmological environments.

Michiya Fujiki
Special Issue Editor

Review

Origin of Terrestrial Bioorganic Homochirality and Symmetry Breaking in the Universe

Jun-ichi Takahashi * and Kensei Kobayashi

Faculty of Engineering, Yokohama National University, Yokohama 240-8501, Japan
* Correspondence: takahashi-junichi-bd@ynu.ac.jp; Tel.: +81-45-339-3938

Received: 5 June 2019; Accepted: 5 July 2019; Published: 15 July 2019

Abstract: The origin of terrestrial bioorganic homochirality is one of the most important and unresolved problems in the study of chemical evolution prior to the origin of terrestrial life. One hypothesis advocated in the context of astrobiology is that polarized quantum radiation in space, such as circularly polarized photons or spin-polarized leptons, induced asymmetric chemical and physical conditions in the primitive interstellar media (the cosmic scenario). Another advocated hypothesis in the context of symmetry breaking in the universe is that the bioorganic asymmetry is intrinsically derived from the chiral asymmetric properties of elementary particles, that is, parity violation in the weak interaction (the intrinsic scenario). In this paper, the features of these two scenarios are discussed and approaches to validate them are reviewed.

Keywords: bioorganic homochirality; circularly polarized photon; spin-polarized lepton; parity violation in the weak interaction

---

1. Introduction

A full explanation of the maximally broken symmetry in terrestrial bioorganic homochirality (the enantiomeric domination of L-form amino acids in proteins and D-form sugars in DNA/RNA) has not been sufficiently achieved despite their significance in resolving the problems concerning the origin of terrestrial life. One of the hypotheses for the origin of terrestrial bioorganic homochirality has been advocated to geological points of view based on prebiotic terrestrial circumstances (the terrestrial scenario), such as the preferential adsorption of prebiotic organic molecules on chiral mineral surfaces (e.g., left- and right-handed crystal quartz).

However, extraterrestrial origins of bioorganic homochirality have also been advocated by the researchers, even in the fields of biology and chemistry. One of the most attractive recent hypotheses in the context of astrobiology advocates that polarized quantum radiation in space, such as circularly polarized photons [1–3] or spin-polarized particles, can induce asymmetric conditions in the primitive interstellar media resulting in terrestrial bioorganic homochirality (the cosmic scenario). In particular, nuclear-decay- or cosmic-ray-derived leptons (i.e., electrons, muons, and neutrinos) in nature have a specified helicity, that is, they have a spin angular momentum polarized parallel or antiparallel to their kinetic momentum due to parity violations (PV) in the weak interaction as a result of intrinsic symmetry breaking in nature [4–14].

A second hypothesis has also been advocated in the context of intrinsic symmetry breaking in nature (the intrinsic scenario). This scenario advocates that the bioorganic asymmetry is universally derived from chiral asymmetry breaking at the level of elementary particle interactions, such as PV in the weak interaction, as well as at the level of astrophysics, such as asymmetry in cosmic constructions [15,16]. In this case, serious problems related to the considerable discrepancies in the hierarchical structures between the evolution of matter and the chemical evolution of bioorganic compounds needs to be universally resolved.

From the standpoint of asymmetry in nature, spin-polarized leptons in the above-mentioned cosmic scenario should be included in the intrinsic scenario due to the generation of spin-polarized leptons derived from PV in the weak interaction. In this article, the features of the cosmic and intrinsic scenarios are discussed and experimental research approaches to validate these scenarios are comparatively reviewed. Results of laboratory verification experiments are shown to explain the relationship between bioorganic homochirality and astrophysical asymmetric radiation sources, which are derived from the intrinsic asymmetry of the universe.

2. Typical Scenarios for the Origin of Homochirality

2.1. The Cosmic Scenario

Because bioorganic compounds synthesized under abiotic circumstances are intrinsically racemic mixtures including equal amounts of L- and D-bodies, it is hypothesized that asymmetric products originated from "chiral radiation," that is, physically asymmetric excitation sources in space. As a result, "asymmetric seeds" were transported to primitive Earth resulting in terrestrial homochirality via several types of amplification mechanisms (the cosmic Scenario) [1]. One of the most attractive cosmic scenario hypotheses can be summarized in the context of astrobiology as shown below.

(i) Prebiotic simple molecules were densely accumulated on interstellar dust surfaces in dense molecular clouds [17];
(ii) "Chiral radiation" in space, for example, circularly polarized photons or spin-polarized particles, induced asymmetric reactions and produced non-racemic mixtures of chiral complex organic materials including bioorganic precursors as "asymmetric seeds";
(iii) The "asymmetric seeds" were transported with meteorites or asteroids to the primitive Earth resulting in terrestrial bioorganic homochirality via some types of "asymmetric amplification" mechanism.

As evidence supporting this scenario, L-enantiomeric excesses have been observed in isovaline, an α-methyl amino acid, and several other α-methyl amino acids in the Murchison meteorite [18]. In general, the racemization rate of α-methyl amino acids is smaller than that of α-hydrogen amino acids. The observed enantiomeric excesses of α-methyl amino acids in meteorites suggest that asymmetric reactions on the surface or in the interior of space materials actually occurred and that the consequent enantiomeric excesses were preserved due to their reduced racemization rate. As for "asymmetric amplification" mechanism, the Soai self-autocatalytic reaction with positive feedback amplification can support the almost perfect asymmetry in terrestrial biomolecular homochirality [19].

2.1.1. Circularly Polarized Photons

- Observations of circularly polarized radiation in space

Of the known chiral radiation, circularly polarized photons have been generally considered to be the most typical polarized quantum radiation in space. In fact, high-energy polarized photons have been observed in the X-ray to gamma-ray region. In this high photon energy region, linearly polarized X-rays, which are principally generated by synchrotron radiation (SR) from relativistic kinetic electrons captured by the strong magnetic fields around neutron stars in supernova remnant, have been observed emanating from the Crab Nebula [19–21]. Gamma-ray bursts (GRBs) from supernova explosion areas have been successfully observed with highly linear polarization (up to 70–80%) by the ICAROS/GAP satellite project [22]. In addition, faint circularly polarized light (0.6% in fraction) has been observed in the visible region of GRB afterglow, which is supposed to be derived from SR due to anisotropic electron motion in plasma. In general, long bursts are generated by the gravitational collapse of high mass stars and slow bursts are generated by binary neutron star mergers [23].

Recently, polarized photons in near infrared (NIR) regions have also been observed in scattered light from interstellar dust clouds in star formation areas by terrestrial ground-level observatories.

In particular, increases in both the photon flux and the ratio of radiated circular polarization with increasing star formation area have been reported [24,25]. In the middle infrared (MIR) region, the AKARI satellite has detected a polarization pattern at wavelengths of 3–11 μm from protostars [26].

Circularly and linearly polarized light derived from absorption lines of solar elements due to Zeeman splitting have been observed (anti-Zeeman effects). These effects depend on the angle correlation between the magnetic field and the line of sight. Spectro-polarimetric observations of sunspot magnetic fields have been observed by using near-infrared cameras (e.g., large-format infrared array detectors) [27].

From the far infrared (FIR) to terahertz (THz) wave region, theoretical calculations for the evolution of elliptical galaxies have predicted that, prior to 4 Gyr, star formation areas radiated intense FIR to THz photons and gradually darkened after the star formation events. This prediction means that, in the era of prebiotic molecule formation, FIR to THz photons were more intense than visible or UV photons. This suggests that polarized photons in the FIR to THz region are also important candidates for "chiral radiation."

In contrast to the above-mentioned area- and time-localized radiation sources in space, cosmic microwave background radiation is thought to be equally distributed in space; however, localized polarization patterns spreading in a specific direction (B-mode/E-mode polarization) have recently been observed. The direct relationship between polarized microwave radiation and chiral asymmetric reactions in bioorganic molecules remains uncertain. The observed polarization is likely relevance to the non-uniform structure of space resulting in the above-mentioned generation of the polarized radiation sources in space [28].

- Chiral reaction scheme via circularly polarized photons

Asymmetric reaction schemes due to circularly polarized photons can be classified corresponding to the photon energy region. Gamma-rays (the photon-nuclear reaction physics region) indicate nuclear excitation with an intense electromagnetic field. X-rays (the intermediate region between radiation chemistry and photochemistry) indicate core electron excitation to optical-active excited states via single photon absorption. VUV-UV-VIS (the photochemistry region) indicates valence electron excitation to optical-active excited states via single photon absorption. NIR-MIR (the intermediate region between photochemistry and thermochemistry) indicates vibrational or rotational excitation of molecular bonds to optical-active excited states. FIR-THz-MW (radiochemistry region) indicates perturbation to intermolecular interactions via multi-photon absorption.

- Circular dichroism

Interactions of circularly polarized photons with chiral molecules can be estimated by using circular dichroism (CD), which is defined as the photo-absorption difference between left- and right-circularly polarized photons at specific wavelengths. Chiral bioorganic molecules generally have intense CD chromophores derived from characteristic electronic transitions corresponding to photon energies in the visible to UV region. Because the electronic transitions related to CD chromophores reflect the bond structures and base conformations of molecules, CD spectra sensitively reflect the steric structures of chiral molecules with a high degree of accuracy [29,30]. From this point of view, it is suggested that the expected characteristics of asymmetric photochemical reactions strongly depend on the irradiated circularly polarized light (CPL) wavelength. For example, theoretical calculations of CD spectrum of alanine molecules have revealed that the chromophores derived from electronic transitions correspond to several different wavelengths in the 120–230 nm region [31]. Chiral molecules also have CD chromophores derived from characteristic vibrational and rotational transitions corresponding to photon energies in the IR region. In terahertz wave region, absorption spectra for chiral molecule crystals can discriminate between chiral crystals (L-alanine and D-alanine) and achiral crystals (DL-alanine) because terahertz region spectra reflect minute intermolecular interactions [31].

2.1.2. Spin-Polarized Leptons

Another type of polarized source that is consistent with the proposed cosmic scenario hypothesis is spin-polarized leptons, i.e., electrons/positrons, muons, and neutrinos, produced in space. Leptons are classified as basic elementary particles, constructed in the standard model of particle physics via quarks, gauge bosons, and Higgs bosons. Lepton and quark interactions are dominated by the parity-violating weak interaction.

- Electrons/Positrons

Of the leptons, electrons are one of the most universally present particles in ordinary materials. Spin-polarized electrons in nature are emitted with β^--decay from radioactive nuclear particles derived from PV involving the weak nuclear interaction ($n \rightarrow p + e^- + \overline{\nu}_e$) and spin-polarized positrons (the anti-particle of electrons) from β^+-decay ($p \rightarrow n + e^+ + \nu_e$). In β^-/β^+-decay, with the weak interaction, the spin angular momentum vectors of electron/positron are perfectly polarized as antiparallel/parallel to the vector direction of the kinetic momentum. In this meaning, spin-polarized electrons/positrons are "chiral radiation" as well as muons and neutrinos which will be mentioned at the paragraphs below. It is expected that the spin-polarized leptons will induce the different type of chemical reactions form CPL induced reactions.

Radiation of spin-polarized electrons and positrons are likely frequent events from the nuclear β^-/β^+-decay of radioactive nuclei in space. In particular, chiral chemical reactions should be induced by negative helicity electrons in β^--decay from radioactive nuclei such as ^{60}Co and ^{90}Sr in space [4–9]. It has been well known that β^-/β^+-decays heat of ^{60}Fe and ^{26}Al in meteorite parent body can operate as a heat source for the formation and differentiation of its organic materials. From the standpoint of astrochemistry, spin-polarized electrons in β^-/β^+-decay potentially operate as chiral radiation to induce chiral reaction processes in organic materials in the meteorites.

Interaction schemes of spin-polarized positrons with chiral molecules should be recognizable as being very different from those of electrons. It has been suggested that the electron-positron annihilation phenomena induced by spin-polarized positrons present a dependence on the spin distribution conditions of the target bioorganic molecules.

Accompanied by electron/positron radiation in β^-/β^+-decay from radioactive nuclei, gamma-rays can be emitted from the disintegrated nuclei via the electromagnetic transition from excited states to lower-level states. The polarization of the emitted gamma-rays depends on the conservation lows of the spin and orbital angular momentums between the nuclei, electron/positron, anti-neutrino/neutrino, and gamma-photon, which is limited according to conservation laws by PV in the weak interaction. For example, β^--decay from ^{60}Co is accompanied by circularly polarized gamma-rays whereas β^- decay from ^{90}Sr is not accompanied by any gamma-rays due to its direct transition to the ground state of the disintegrated nucleus.

- Muons

In the atmosphere of the Earth, muons are generated as a secondary cosmic ray via pion-decay following the impact of a cosmic ray proton with an atmospheric atom. The spin angular momentum of positive (μ^+) and negative (μ^-) muons in nature is intrinsically polarized with negative and positive helicity (left- and right-handed polarization), respectively, by PV in the weak interaction. Muons are lepton particles with the same value of negative (μ^-) or positive (μ^+) charge as electrons or positrons, respectively, and the same value of spin (1/2) as electrons and positrons. Because the muon mass is much larger than the electron mass (m_μ/m_e ~200) and close to the proton mass (m_μ/m_p ~1/9), interactions of spin-polarized muons with organic molecules, including asymmetric reactions, are different from those of spin-polarized electrons or positrons.

In the positive muon irradiation case, muonium (a positive muon-electron pair (Mu)) and a non-bonding lone electron can be formed on a double-bond of an organic molecule resulting in the formation of "Mu radicals." As a result, radical-mediated asymmetric reactions can be expected in organic molecules. Double bonds of carbon–carbon can be reduced to a single bond resulting in rotation around the single-bond axis. In addition, hot Mu generated by a muon beam can induce hydrogen extraction from organic molecule (Mu* + RH → MuH + R). Via these types of reactions, racemization, that is, the enantiomer type conversion of chiral molecules) such as amino acids, is strongly expected. In addition, Mu can attack non-bonding electrons of the carboxyl bases of amino acids, resulting in asymmetric electronic excitation depending on the direction of spin polarization of the muon irradiation via several types of processes. Mu can be thought of a type of isotope of the H atom whose atomic total mass is 1/9 that of hydrogen. Conversely, the reduced mass of Mu (positive muon-electron) and an H atom (proton-electron) with a 1 s orbital electron are approximately equivalent values. From the standpoint of the electronic state, Mu and an H atom can be regarded as being chemically equivalent.

Interactions of positive and negative muons with organic molecules are very different from each other due to the relatively heavy mass of a muon. In the negative muon beam case, a negative muon combined with an atomic nucleus makes a muonic atom (atomic nucleus-negative muons pair). Muonic atom formation can induce a charge shielding effect due to heavy negative muons captured in small orbit radii, resulting in the formation of apparently Z-1 nucleus and negative ion. Further, atomic conversion ($Z \rightarrow Z - 1$) via β-decay ($\mu^- + p \rightarrow n + \nu_\mu$) can occur in large atomic number nucleus. It is also expected that the chemical activity of muonic atoms depends on the direction of spin polarization of the muon irradiation.

From these viewpoints, spin-polarized muon irradiation has the potential to induce novel types of optical activities different from those of polarized photon and spin-polarized electron irradiation.

- Neutrinos

In general, the interaction of neutrinos with materials is very weak on the Earth. However, reactions including neutrinos accompanied by the weak interaction cannot be neglected in high density nuclear materials (>10^{11} g/cm^3) in space.

As for high-energy astrophysical objects, neutrons released in supernova explosions can cause chiral asymmetry in molecules in interstellar gas-dust clouds. The specific physical mechanism of a relativistic neutron fireball, which is a relativistic chiral electron–proton plasma, has been advocated, in which the electrons carrying their helicity to the cloud show high chiral efficiency [5,6,9].

Neutrinos with a specified spin angular momentum due to the PV weak interaction are also radiated in supernova explosions. Recently, it has been argued that selective destruction of one amino acid chirality by electron anti-neutrinos and the magnetic field from a supernova resulted in the production of meteorites including left-handed amino acids [11–14]. In this hypothesis, spin-spin interactions between neutrinos and nitrogen nuclei are emphasized.

2.2. The Intrinsic Scenario

Another hypothesis has been advocated in the context of intrinsic symmetry breaking in nature, that is, the bioorganic asymmetry is universally derived from chiral asymmetry breaking at the level of elementary particles, such as PV in the weak interaction [15]. It has been argued that PV leads to energy differences in enantiomers in the 10^{-12} joule/mole range and that theoretical calculations of the energy differences in enantiomers indicate that the intrinsic energies of L-amino acids and D-ribose are slightly lower than those of their corresponding enantiomers (D-amino acids and L-ribose). Quack considered that electroweak quantum chemistry methods predict such energy differences to be one to two orders of magnitude larger than previously accepted, but still very small, and discussed the current status of theory and some current experimental approaches [16].

3. Experimental Approaches to Examine the Scenarios

3.1. The Cosmic Scenario

3.1.1. Circularly Polarized Photons

The first asymmetric photolysis experiment on a racemic organic molecule (2-bromopropanoic acid ethyl ester) using CPL was reported by Kuhn [32] in 1929. In the 1970's, several asymmetric photolysis experiments were conducted on racemic molecules, i.e., metal oxalate ions [33] and amino acid ((RS)-leucine) [34], using CPL from high-pressure mercury lamps.

After the 1990's, SR facilities were used as light sources for photochemical reactions, spectroscopic analyses, and micro-fabrication. In particular, the asymmetric photolysis of racemic amino acids in an aqueous solution using CPL from SR was reported as a validation of the "cosmic scenario" in the context of astrobiology. Takano et al. reported the asymmetric photolysis of racemic isovaline [35], and Nishino et al. reported the pH dependence of leucine asymmetric photolysis [36]. In these experiments, they were able to selectively photolyze one optical isomer in an optically active compound depending on its CD in the wavelength region of the irradiated CPL.

Further, Takano et al. reported the CPL irradiation of an aqueous solution of complex organic compounds with high molecular weight, which involved amino-acid precursors with molecular weights of several thousands. After CPL irradiation followed by the acid-hydrolysis of the solution, positive and negative enantiomeric excesses in alanine were successfully detected [37,38].

Conversely, in molecular clouds, organic molecules on interstellar dust surfaces should be in their solid phase, for example, the condensed ice mantle around silica dust core (Greenberg Model [17]). From this viewpoint, solid-phase experiments can contribute to verifying reaction models of on-surface or surface-catalytic reactions on materials such as interstellar dust. Meierhenrich et al. [39] reported solid-phase leucine photolysis and Takahashi reported solid-phase phenylalanine photolysis using polarized SR in the VUV region [40].

Recent verification experiments have been performed to examine the relationship between optical activity emergence into several target materials and asymmetric radiation, typically CPL. We irradiated racemic D L -amino acid films with monochromatic CPL with a wavelength of 215 nm from the free electron laser (FEL) of UVSOR-II at Institute for Molecular Science (IMS), Okazaki, Japan [31]. This result suggest that the racemic mixture of chiral amino acid was differentially photolyzed or transformed depending on the CD at the irradiated CPL wavelength.

We are using CD spectroscopy to evaluate optical activity emergence because the CD spectrum sensitively reflects the steric structures of chiral molecules with a high degree of accuracy. Theoretical calculations of the CD spectrum of the alanine molecule have revealed that the circular dichroism chromophores derived from characteristic electronic transitions (n–π^*, π–π^* for the carboxyl group and n–σ^*, π–σ^* for the amino group) correspond to several different wavelengths in the 120–230 nm region [30]. From this standpoint, the expected characteristics of optical activity emergence strongly depend on the irradiated CPL wavelength. In further experiments to validate the scenario, we are investigating the dependence on the irradiation wavelength using CPL with wavelengths of 230 nm, 203 nm, 180 nm, and 155 nm from the undulator beam line BL1U of UVSOR-III at IMS. To examine optical activity emergence, we measured the CD and photo absorption spectra of the deposited films using a commercial CD spectrometer (JASCO J-725) and the SR CD beam line BL-12 of Hiroshima Synchrotron Radiation Center. A detailed analysis of the CD spectra is in progress to determine the complete mechanism for optical activity emergence [41,42].

Similar irradiation experiments for solid-phase films of an intrinsically optically non-active achiral molecule, hydantoin (glycine precursor molecules), and glycine have also been performed using CPL with a wavelength of 215 nm from FEL of UVSOR-II and BL1U of UVSOR-III, respectively [43]. The CD spectra of irradiated hydantoin and glycine films have many common characteristics, that is, clear peaks newly appear in the CD spectra at 215 nm and 180 nm and nearly completely symmetric spectra were observed with L- and R-CPL irradiation. Further, the intensity ratio of the two peaks and the intensity and sign of the CD peaks changed with the irradiation dose. These results suggest that several new chiral structures were introduced into the achiral molecule film by the CPL irradiation. The absorbance of hydantoin after the CPL irradiation increased on the low-energy side but decreased on the high-energy side, suggesting that a molecular structure change occurred in the film.

As for the hydantoin case, first principal simulation calculations of the asymmetric optical response of hydantoin have also been performed. The calculation results suggest that some types of chiral structures derived from a distortion of the five-ring construction were introduced into the racemic film by the CPL irradiation. Analogous to hydantoin, it is suggested that some types of chiral structures were introduced into the racemic glycine film by the CPL irradiation [44]. It is known that crystalized glycine contains several crystal polymorphs including left- or right-handed spiral constructions, similar to crystalline quartz. The experimental results suggest that the CPL irradiation induced construction distortion including left- or right-handed spiral crystal polymorphs depending on the irradiated CPL chirality. Similar types of mechanisms including chiral structure formation are potentially applicable to racemic mixtures of optically active amino acids (i.e., DL-alanine), which is potentially relevant to the origin of the terrestrial bioorganic homochirality stimulated by "chiral radiation."

In addition to solid films of amino acids, ribose and related sugars following ultraviolet irradiation of interstellar ice fixed in ice analogs have also been reported with high-resolution mass spectroscopy [44–46].

3.1.2. Spin-Polarized Leptons

- Electrons/Positrons

The first pioneering experiment examining asymmetric radical formation in D- and L-alanines using β^--irradiation was conducted by Akaboshi, et al. [47]. Tsarev et al. reported that, after amino acid/metal complexes were irradiated with high flux β^--rays from an ^{90}Sr-^{90}Y source (50 Ci), at the Russian Federal Nuclear Center in Snezhinsk, and small amounts of enantiomeric excesses were detected in the irradiated products using CD spectroscopy [7–9]. As for interactions of spin-polarized electrons with chiral molecules, Kessler, et al. reported on a low energy (<10 eV) spin-polarized electron beam with a chiral bromocamphor molecule, that is, so-called "spin-polarization dichroism" [48]. Recently, Rosenberg et al. reported interactions of spin-polarized X-ray photoelectron emitted from solid surfaces with chiral molecules adsorbed on them [49,50].

We also recently carried out the same irradiation experiments with amino-acid precursors or solid films of amino acids and also detected enantiomeric excesses using CD spectroscopy on an irradiated isovaline film [51]. Because the helicity of the spin-polarized electron beam in a β^--ray is negative due to PV in the weak interaction, verification experiments are required using a spin-polarized electron beam with positive helicity to detect enantiomeric excesses with the opposite sign with respect to the negative helicity case. Our present research focuses on simulated experiments using a spin-polarized electron beam with a helicity-controlled irradiating amino acid film in the achiral state (optically non-active) and observing the emergence of optical activity.

- Muons

No significant difference has been reported for the asymmetric reactions in enantiomers of organic compounds using muon beams, except that Lemmon et al. who reported reactions of spin-polarized positive muons with alanines and octanols [52]. We conducted a trial of spin-polarized muons to examine the optical activity of organic compounds such as amino acids and their precursor molecule in achiral states. In the experiments, a racemic amino acid (DL-alanine) and a precursor molecule of glycine (hydantoin) were irradiated by positive muons with negative helicity and negative muons with positive helicity using the Muon S beam line of the Materials and Life Science Experimental Facility (MLF) in the Japan Proton Accelerator Research Complex (J-PARC). Because the penetrating depth of the muon beam through the polyethylene pellet is estimated to be less than 1 mm, nearly all the muon beam should be captured with the target materials packed in the polyethylene or potassium bromide circular cylinder pellet. Optical activity measurements of the irradiated samples are in progress. We expect to reveal a mechanism for optical activity emergence whose sign depends on the direction of the spin polarization and the charge of the muon. We are also planning to observe the disturbance of the muon spin polarization after chiral chemical reactions due to muonium or muonic atom formation in the target chiral molecules. The disturbance of the muon spin polarization can be observed using measurements of the distribution of positrons decayed from muons.

3.2. The Intrinsic Scenario

Recently, the first set of observations of PV energy differences between two enantiomers of chiral molecules was presented using high-resolution laser spectroscopy [53]. However, at present, no successful experiments have been reported. The role of PV in the weak interaction in the origin of terrestrial homochirality is a subject that cuts across research fields, including fundamental physics, nuclear cosmology, theoretical chemistry, and quantum spectroscopy.

4. Conclusions

Various scenarios for the origin of the homochirality of terrestrial bioorganic molecules in the context of astrobiology (the cosmic scenario) have been proposed. Asymmetric radiation in space, such as circularly polarized photons or spin-polarized leptons, induced asymmetric conditions in the primitive interstellar media. We are conducting cooperative investigations using the following observations, experiments and computational calculations to examine the cosmic scenario [54,55].

(1) Astronomical observations of polarized radiation from various star-forming and high-energy phenomena burst regions in space using highly sensitive polarization detecting systems settled in astronomical observatories;
(2) Experiments with polarized quantum beams from high-energy particle accelerators irradiating amino acids or sugars and their precursor molecules, followed by chemical and optical measurements of enantiomeric excesses;
(3) First principal calculations of asymmetric optical responses and the subsequent asymmetric chemical reactions for amino acids or sugars and their precursor molecules, including the intrinsic energy difference between the enantiomers derived from PV.

The synergetic use of these three approaches is expected to help solve the problems of the origin of terrestrial biological homochirality.

Funding: This research was funded by Astrobiology Center, National Institutes of Natural Sciences (NINS-ABC) Project Research-2015 and -2017.

Acknowledgments: The author would like to thank Vladimir A. Tsarev, Takeshi Saito, Kazumichi Nakagawa, Uwe J. Meierhenrich, Masahiro Katoh, Motohide Tamura, Masayuki Umemura, Yoshinori Takano, Koichi Matsuo, Yudai Izumi, Masateru Fujimoto, Yoko Kebukawa, Hiromi Shibata, M. Kenya Kubo, Yasuhiro Miyake and Michiya Fujiki for their collaborations and useful discussions.

Conflicts of Interest: The authors declare no conflicts of interest.

References

1. Bonner, W.A. The origin and amplification of bioorganic chirality. *Orig. Life Evol. Biosph.* **1991**, *21*, 59–111. [CrossRef] [PubMed]
2. Bailey, J.; Chrysostomou, A.; Hough, J.; Gledhill, T.; McCall, A.; Clark, S.; Menard, F.; Tamura, M. Circular polarization in star-formation regions: Implications for bioorganic homochirality. *Science* **1998**, *281*, 672–674. [CrossRef] [PubMed]
3. Meierhenrich, U.J. Amino acids and the Asymmetry of Life. In *Advances in Astrobiology and Biogeophysics*; Brack, A., Horneck, G., McKay, C.P., Stan-Lotter, H., Eds.; Springer: Berlin/Heidelberg, Germany, 2008.

4. Garay, A.S.; Keszthelyi, L.; Demeter, I.; Hrasko, P. Origin of asymmetry in biomolecules. *Nature* **1974**, *250*, 332–333. [CrossRef] [PubMed]
5. Cline, D.B.; Liu, Y.; Wang, H. Effect of a chiral impulse on the weak interaction induced handedness in a prebiotic medium. *Orig. Life Evol. Biosph.* **1995**, *25*, 201–209. [CrossRef] [PubMed]
6. Cline, D.B. Supernova antineutrino interactions cause chiral symmetry breaking and possibly homochiral biomaterials for life. *Chirality* **2005**, *17*, S234–S239. [CrossRef] [PubMed]
7. Gusev, G.A.; Saito, T.; Tsarev, V.A.; Uryson, A.V. A relativistic neutron fireball from a supernova explosion as a possible source of chiral influence. *Orig. Life Evol. Biosph.* **2007**, *37*, 259–266. [CrossRef] [PubMed]
8. Gusev, G.A.; Kobayashi, K.; Moiseenko, E.V.; Poluhina, N.G.; Saito, T.; Ye, T.; Tsarev, V.A.; Xu, J.; Huang, Y.; Zhang, G. Results of the second stage of the investigation of the radiation mechanism of chiral influence (RAMBAS-2 experiment). *Orig. Life Evol. Biosph.* **2008**, *38*, 509–515. [CrossRef] [PubMed]
9. Tsarev, V.A. Physical and astrophysical aspects of the problem of origin of chiral asymmetry of the biosphere. *Phys. Part Nucl.* **2009**, *40*, 998–1029. [CrossRef]
10. Cline, D.B. Possible physical mechanisms in the galaxy to cause homochiral biomaterials for life. *Symmetry* **2010**, *2*, 1450–1460. [CrossRef]
11. Boyd, R.N.; Kajino, T.; Onaka, T. Supernovae and the chirality of the amino acids. *Astrobiology* **2010**, *10*, 561–568. [CrossRef]
12. Boyd, R.N.; Kajino, T.; Onaka, T. Supernovae, neutrinos, and the chirality of the amino acids. *Int. J. Mol. Sci.* **2011**, *12*, 3432–3444. [CrossRef] [PubMed]
13. Famiano, M.; Boyd, R.N.; Kajino, T.; Onaka, T.; Koehler, K.; Hulbert, S. Determining amino acid chirality in the supernova neutrino processing model. *Symmetry* **2014**, *6*, 909–925. [CrossRef]
14. Famiano, M.; Boyd, R.N.; Kajino, T.; Onaka, T. Selection of amino acid chirality via neutrino interactions with ^{14}N in crossed electric and magnetic fields. *Astrobiology* **2018**, *18*, 190–206. [CrossRef] [PubMed]
15. Mason, S.F. Origin of bioorganic handedness. *Nature* **1984**, *311*, 19–23. [CrossRef] [PubMed]
16. Quack, M. How important is parity violation for molecular and bioorganic chirality? *Angew. Chem. Int. Ed.* **2002**, *41*, 4618–4630. [CrossRef] [PubMed]
17. Greenberg, J.M.; Kouchi, A.; Niesson, W.; Irth, H.; van Paradijs, J.; de Groot, M.; Hermsen, W. Interstellar dust, chirality, comets, and the origins of life: Life from dead stars? *J. Biol. Phys.* **1994**, *20*, 61–70. [CrossRef]
18. Cronin, J.R.; Pizzarello, S. Enantiomeric excesses in meteoritic amino acids. *Science* **1997**, *275*, 951–955. [CrossRef]
19. Soai, K.; Kawasaki, T. Asymmetric Autocalysis with Amplification of Chirality. *Top. Curr. Chem.* **2008**, *284*, 1–33.
20. Chauvin, M.; Florén, H.-G.; Jackson, M.; Kamae, T.; Kawano, T.; Kiss, M.; Kole, M.; Mikhalev, V.; Moretti, E.; Olofsson, G.; et al. Observation of polarized hard X-ray emission from the Crab by the PoGOLite Pathfinder. *Month. Not. R. Astron. Soc. Lett.* **2016**, *456*, L84–L88. [CrossRef]
21. Chauvin, M.; Florén, H.-G.; Friis, M.; Jackson, M.; Kamae, T.; Kataoka, J.; Kawano, T.; Kiss, M.; Mikhalev, V.; Mizuno, T.; et al. Shedding new light on the Crab with polarized X-rays. *Sci. Rep.* **2017**, *7*, 7816. [CrossRef]
22. Yonetoku, D.; Murakami, T.; Sakashita, T.; Morihara, Y.; Kikuchi, Y.; Takahashi, T.; Gunji, S.; Mihara, T.; Kubo, S. Detection of Gamma-Ray Polarization in Prompt Emission of GRB 100826A. *Astrophys. J.* **2011**, *743*, L30. [CrossRef]
23. Wiersema, K.; Covino, S.; Toma, K.; van der Horst, A.J.; Varela, K.; Min, M.; Greiner, J.; Starling, R.L.C.; Tanvir, N.R.; Wijers, R.A.M.J.; et al. Circular polarization in the optical afterglow of GRB 121024A. *Nature* **2014**, *509*, 201–204. [CrossRef] [PubMed]
24. Fukue, T.; Tamura, M.; Kandori, R.; Kusakabe, N.; Hough, J.H.; Lucas, P.W.; Bailey, J.; Whittet, D.C.B.; Nakajima, Y.; Hashimoto, J.; et al. Near-infrared circular polarimetry and correlation diagrams in the Orion Becklin-Neugebauer/Kleinman-Low region: Contribution of dichroic extinction. *Astrophys. J. Lett.* **2009**, *692*, L88–L91. [CrossRef]
25. Kwon, J.; Tamura, M.; Lucas, P.W.; Hashimoto, J.; Kusakabe, N.; Kandori, R.; Nakajima, Y.; Nagayama, T.; Nagata, T.; Hough, J.H. Near infrared circular polarization images of NGC 6334-V. *Astrophys. J. Lett.* **2013**, *765*, L6. [CrossRef]
26. Neha, S.; Maheswar, G.; Soam, A.; Lee, C.W. Polarization of seven MBM clouds at high galactic latitude. *Month. Not. R. Astron. Soc.* **2018**, *476*, 4442–4458. [CrossRef]

27. Sakurai, T.; Yanagisawa, K.; Kobiki, T.; Kasahara, S.; Nakakubo, K. Sunspot magnetic fields observed with a large-format infrared array. *Publ. Astron. Soc. Jpn.* **2001**, *53*, 923–930. [CrossRef]
28. Barkats, D.; Bischoff, C.; Farese, P.; Fitzpatrick, L.; Gaier, T.; Gundersen, J.O.; Hedman, M.M.; Hyatt, L.; Mcmahon, J.J.; Samtleben, D.; et al. First measurements of the polarization of the cosmic microwave background radiation at small angular scales from CAPMAP. *Astrophys. J.* **2005**, *619*, L127–L130. [CrossRef]
29. Matsuo, K.; Matsushima, Y.; Fukuyama, T.; Senba, S.; Gekko, K. Vacuum-ultravilet circulardichroism of amino acids as revealed by synchrotron radiation spectrophotometer. *Chem. Lett.* **2002**, *31*, 826–827. [CrossRef]
30. Tanaka, M.; Kodama, Y.; Nakagawa, K. Circular dichroism of amino acid films in UV-VUV region. *Enantiomer* **2002**, *7*, 185–190. [CrossRef]
31. Kaneko, F.; Yagi-Watanabe, K.; Tanaka, M.; Nakagawa, K. Natural circular dichroism spectra of alanine and valine films in vacuum ultraviolet region. *J. Phys. Soc. Jpn.* **2009**, *78*, 013001. [CrossRef]
32. Kuhn, W.; Braun, E. Photochemische erzeugung optisch aktiver stoffe. *Naturwissenschaften* **1929**, *17*, 227–228. [CrossRef]
33. Norden, B. Optical activity developed by preferential racemization of one enantiomer in Racemic Cr(III) $(ox)_3{}^{3-}$ induced by irradiation with circularly polarized light. *Acta Chem. Scand.* **1970**, *24*, 349–351. [CrossRef]
34. Flores, J.J.; Bonner, W.A.; Massey, G.A. Asymmetric photolysis of (RS)-leucine with circularly polarized ultraviolet light. *J. Am. Chem. Soc.* **1977**, *99*, 3622–3625. [CrossRef] [PubMed]
35. Takano, Y.; Kaneko, T.; Kobayashi, K.; Takahashi, J. Asymmetric photolysis of (DL)-isovaline by synchrotron radiation. *Orig. Life Evol. Biosph.* **2002**, *32*, 447–448.
36. Nishino, H.; Kosaka, A.; Hembury, G.A.; Shitomi, H.; Onuki, H.; Inoue, Y. Mechanism of pH dependent photolysis of aliphatic amino acids and enantiomeric enrichment of racemic leucine by circularly polarized light. *Org. Lett.* **2001**, *3*, 921–924. [CrossRef] [PubMed]
37. Takano, Y.; Takahashi, J.; Kaneko, T.; Marumo, K.; Kobayashi, K. Asymmetric synthesis of amino acid precursors in interstellar complex organics by circularly polarized light. *Earth Planet. Sci. Lett.* **2007**, *254*, 106–114. [CrossRef]
38. Kobayashi, K.; Kaneko, K.; Takahashi, J.; Takano, Y. High molecular weight complex organics in interstellar space and their relevance to origins of life. In *Astrobiology: From Simple Molecules to Primitive Life*; Basiuk, V., Ed.; American Scientific Publisher: Valencia, CA, USA, 2008.
39. Meierhenrich, U.J.; Nahon, L.; Alcaraz, C.; Bredehöft, J.H.; Hoffmann, S.V.; Barbier, B.; Brack, A. Asymmetric vacuum UV photolysis of the amino acid leucine in the solid state. *Angew. Chem. Int. Ed.* **2005**, *44*, 5630–5634. [CrossRef] [PubMed]
40. Takahashi, J. Asymmetric photolysis of thin solid film of aromatic amino acid with circularly polarized light. *Orig. Life Evol. Biosph.* **2006**, *36*, 280–282.
41. Takahashi, J.; Shinojima, H.; Seyama, M.; Ueno, Y.; Kaneko, T.; Kobayashi, K.; Mita, H.; Adachi, M.; Hosaka, M.; Katoh, M. Chirality emergence in thin solid fFilms of amino acids by polarized light from synchrotron radiation and free electron laser. *Int. J. Mol. Sci.* **2009**, *10*, 3044–3064. [CrossRef]
42. Matsuo, K.; Izumi, Y.; Takahashi, J.; Fujimoto, M.; Katoh, M. Emergence of biological homochirality by irradiation of polarized quantum beams. *UVSOR Act. Rep. 2016* **2017**, *44*, 157.
43. Takahashi, J.; Sakamoto, T.; Izumi, Y.; Matsuo, K.; Fujimoto, M.; Katoh, M.; Kebukawa, Y.; Kobayashi, K. Circular dichroism analysis of optical activity emergence in amino-acid thin films irradiated by vacuum-ultraviolet circularly-polarized light. In Proceedings of the 22nd Hiroshima International Symposium on Synchrotron Radiation, Higashi-Hiroshima, Japan, 7–8 March 2019.
44. Takahashi, J.; Suzuki, N.; Kebukawa, Y.; Kobayashi, K.; Izumi, Y.; Matsuo, K.; Fujimoto, M.; Katoh, M. Optical activity emergence in glycine by circularly polarized light. *UVSOR Act. Rep. 2016* **2017**, *44*, 156.
45. Meinert, C.; Myrgorodska, I.; Marcellus, P.; Buhse, T.; Nahon, L.; Hoffmann, S.V.; d'Hendecourt, L.L.S.; Meierhenrich, U.J. Ribose and related sugars from ultraviolet irradiation of interstellar ice analogs. *Science* **2016**, *352*, 208–212. [CrossRef] [PubMed]
46. Garcia, A.D.; Meinert, C.; Sugahara, H.; Jones, N.C.; Hoffmann, S.V.; Meierhenrich, U.J. The Astrophysical Formation of Asymmetric Molecules and the Emergence of a Chiral Bias. *Life* **2019**, *9*, 29. [CrossRef] [PubMed]
47. Akaboshi, M.; Noda, M.; Kawai, K.; Maki, H.; Kawamoto, K. Asymmetrical radical formation in D- and L-alanines irradiated with yttrium-90β-rays. *Orig. Life* **1979**, *9*, 181–186. [CrossRef] [PubMed]

48. Kessler, J. Electron Dichroism: Interaction of Polarized Electrons with Chiral Molecules. *Phys. Essays* **2000**, *13*, 421–426. [CrossRef]
49. Rosenberg, R.A.; Abu Haija, M.; Ryan, P.J. Chiral-selective chemistry induced by spin-polarized secondary electrons from a magnetic substrate. *Phys. Rev. Lett.* **2008**, *101*, 178301. [CrossRef]
50. Rosenberg, R.A.; Mishra, D.; Naaman, R. Chiral selective chemistry induced by natural selection of spin-solarized electrons. *Angew. Chem. Int. Ed.* **2015**, *54*, 7295–7298. [CrossRef]
51. Burkov, V.I.; Goncharova, L.A.; Gusev, G.A.; Hashimoto, H.; Kaneko, F.; Kaneko, T.; Kobayashi, K.; Mita, H.; Moiseenko, E.V.; Ogawa, T.; et al. Asymmetric reactions of amino-acid-related compounds by polarized electrons from beta-decay radiation. *Orig. Life Evol. Biosph.* **2009**, *39*, 295–296.
52. Lemmon, R.M.; Crowe, K.M.; Gygax, F.N.; Johnson, R.F.; Patterson, B.D.; Brewer, J.H.; Fleming, D.G. Search for selectivity between optical isomers in reactions of polarized positive muons with alanines and octanols. *Nature* **1974**, *252*, 692–694. [CrossRef]
53. Darquié, B.; Stoeffler, C.; Shelkovnikov, A.; Daussy, C.; Amy-Klein, A.; Chardonnet, C.; Zrig, S.; Guy, L.; Crassous, J.; Soulard, P.; et al. Progress toward a first observation of parity violation in chiral molecules by high-resolution laser spectroscopy. *Chirality* **2010**, *22*, 870–884. [CrossRef]
54. Takahashi, J.; Katoh, M.; Tamura, M.; Umemura, M.; Kusakabe, N.; Kwon, J.; Kobayashi, K.; Takashima, Y.; Hosaka, M.; Zen, H.; et al. NINS Astrobiology Center project: The origin of terrestrial bioorganic homochirality relevance to asymmetry of the universe—Approaches with synergy effects of observations, experiments and computations. In Proceedings of the 26th Goldschmidt Conference (Goldschmidt2016), Yokohama, Japan, 26 June–1 July 2016.
55. Takahashi, J. Biological homochirality and symmertry breaking of the universe. In Proceedings of the XVIII International Conference on the Origin of Life (ISSOL2017), San Diego, CA, USA, 16–21 July 2017.

Article

Astrophysical Sites that Can Produce Enantiomeric Amino Acids

Michael Famiano [1,2,*], Richard Boyd [3], Toshitaka Kajino [2,4,5], Takashi Onaka [6] and Yirong Mo [7]

1. Department of Physics, Western Michigan University, Kalamazoo, MI 49008, USA
2. National Astronomical Observatory of Japan, Division of Science, Mitaka, Tokyo 181-8588, Japan; kajino@nao.ac.jp
3. Department of Physics, Department of Astronomy, The Ohio State University, Columbus, OH 43210, USA; richard11boyde@comcast.net
4. Department of Physics, Graduate School of Science, University of Tokyo, Tokyo 113-8654, Japan
5. School of Physics, Beihang University, Beijing 100083, China
6. Department of Astronomy, Graduate School of Science, University of Tokyo, Tokyo 113-8654, Japan; onaka@astron.s.u-tokyo.ac.jp
7. Department of Chemistry, Western Michigan University, Kalamazoo, MI 49008, USA; yirong.mo@wmich.edu

* Correspondence: michael.famiano@wmich.edu; Tel.: +1-269-387-4931

Received: 28 November 2018; Accepted: 23 December 2018; Published: 28 December 2018

Abstract: Recent work has produced theoretical evidence for two sites, colliding neutron stars and neutron-star–Wolf–Rayet binary systems, which might produce amino acids with the left-handed chirality preference found in meteorites. The Supernova Neutrino Amino Acid Processing (SNAAP) model uses electron antineutrinos and the magnetic field from source objects such as neutron stars to preferentially destroy one enantiomer over another. Large enantiomeric excesses are predicted for isovaline and alanine; although based on an earlier study, similar results are expected for the others. Isotopic abundances of ^{13}C and ^{15}O in meteorites provide a new test of the SNAAP model. This presents implications for the origins of life.

Keywords: origin of life; amino acid handedness; nucleus–molecular coupling; chirality

---

1. Introduction

Recent studies [1–4] have identified two astrophysical sites in which the combination of intense magnetic fields and high fluxes of electron antineutrinos could produce amino acids with a high probability of favoring left-handed chirality.

Enantiomeric excess *ee* is defined as $ee = (N_L - N_D)/(N_L + N_D)$, where $N_{L(D)}$ is the number of left- (right-)handed molecules in a population. An ensemble of amino acids with $ee = 1$ are purely left-handed, while those with $ee = -1$ are purely right-handed. The ensemble is racemic if $ee = 0$, having equal left- and right-handed molecule parts.

Our studies have shown that *ee*s of the order of a percent or higher could be produced from two-neutron-star coalescence, and somewhat lower, but still appreciable *ee*s could arise from a close binary system consisting of a massive star and a neutron star. Note, though, that this latter site might ultimately evolve into a two-neutron-star coalescence site.

Amino acids are essential for the continuing existence of life, and may have even enabled its formation. The basic steps by which life might have developed from these basic molecules are not fully known, though much progress has recently been made for understanding them [5–7]. All amino acids (except for achiral glycine) used by Earth's living creatures are homochiral ($ee = 1$), to the near exclusion of right-handed forms. Molecular chirality was originally studied by Pasteur [8], and the homochirality of the amino acids was subsequently deduced. However, the origin of amino acid chirality is still a mystery.

Mid-20th century experiments [9,10] suggested that organic molecules, including amino acids, may have been assembled in an early earthly lighting storm. However, this result does not explain how amino acids achieved homochirality. Possible mechanisms to explain the conversion of racemic amino acids to purely left-handed ones in terrestrial processes have been reviewed by Bonner [11]. It was concluded that these processes would likely not result in amino acids of a single chirality. Additional evaluations have been made by Mason [12] and Barron [13].

Furthermore, it was concluded by Goldanskii and Kuzmin [14], and by Bonner [11], that amino acid homochirality is essential for the perpetuation of life.

Amino acids have been found in meteorites, and are therefore produced in outer space [15–20]. Further, some of these have been found to have nonzero *ee*s, of typically a few percent, and to mostly be left-handed, suggesting a possible cosmic origin of amino acids. Earth may, therefore, have been seeded with chirally selected amino acids. The observed *ee*s, however, require amplification, presumably via autocatalysis [14,21,22], to achieve earthly homochirality. Autocatalysis is thought to have converted small *ee*s to the homochirality observed today. Autocatalytic processes have been demonstrated in laboratory experiments [23–26] to produce homochiral populations from very small *ee*s. Although attempts have been made to sample *ee*s of amino acids on objects in outer space [27,28], those missions have not yet provided definitive results.

The production of enantiomeric amino acids in space could possibly be explained by multiple models. One possible model explains a production of chiral amino acids with ultraviolet circularly polarized light (CPL). Mie scattering from an extremely hot star was suggested by Flores et al. [29] and Norden [30]. This has been extensively studied [31–36]. This model has the advantage that its chiral selectivity can be experimentally demonstrated with beams of polarized photons from an accelerator. However, significant molecular selectivity also requires the destruction of most pre-existing amino acids of both chiralities. The CPL model can produce either positive or negative *ee*s, whereas the Supernova Neutrino Amino Acid Processing (SNAAP) model is capable of producing only one enantiomeric excess in a single site.

Another model, Magneto-Chiral Anisotropy (MCA), has been developed by Wagniere and Meier [37], explored experimentally by Rikken and Raupach [38], and extended by Barron [39]. In this model, the interaction between photons from an intense light source, for example, a supernova, and molecules in a magnetic field, possibly from the supernova's nascent neutron star, would produce a chirality-dependent destruction effect on the amino acids. The dielectric constant of a medium depends on $\mathbf{k} \cdot \mathbf{B}$, where $\mathbf{k}$ is the wavevector in the direction of travel of the incident light, and $\mathbf{B}$ is an external magnetic field. The dielectric constant increases (decreases) when incident light travels in the same (opposite) direction as the external magnetic field. This effect has opposite signs for L- and D-enantiomers. The net result is that one enantiomer absorbs more of the incident light (and is thus preferentially destroyed) than the other. Experimental studies on this effect have resulted in *ee*s on the order of 10^{-4} for chiral molecules [40]. The MCA model can also produce *ee*s of either sign.

Other models exist, but these are generally less developed than the MCA or CPL models [40,41]. The SNAAP model [3,4,42], described below, utilizes intense magnetic fields and electron antineutrinos to create enantiomeric amino acids. Quantum molecular calculations indicate that the SNAAP model can produce enantiomeric amino acids with significant positive *ee*s [4]. A particularly viable astrophysical scenario for this model was suggested [2] to be a neutron-star–Wolf–Rayet-star binary system.

In this work, we discuss the SNAAP model as it applies to three systems. The first is a core-collapse supernova that, albeit problematic, does illustrate some of the basic features of the model. The second is a neutron star and massive star binary system. The third is one with two neutron stars that merge. We estimated the *ee*s produced in each site for the alanine and isovaline amino acids over a range of parameters within the sites, applying the quantum molecular calculations developed for previous studies.

SNAAP model basics are described in Section 2. Application of the SNAAP model to various sites is discussed in Section 3, followed by a discussion of the simulations relevant to their development in Section 4. Section 5 presents the results for each site. Section 6 discusses how the SNAAP model might explain the isotopic abundances observed in the meteorites, and Section 7 gives our conclusions.

2. SNAAP Model

In this model [1–4,42], amino acids contained within meteoroids in the vicinity of an intense magnetic field and electron antineutrino (hereafter denoted 'antineutrino') flux are processed. As a result of the parity-violating weak interaction induced by antineutrino interactions with the ^{14}N nuclei of the amino acids, one enantiomer is destroyed over another.

The relevant nuclear reaction is:

$$\bar{\nu}_e +^{14} N \rightarrow e^+ +^{14} C \tag{1}$$

where $\bar{\nu}_e$ is an electron antineutrino, and e^+ is an antielectron—a positron. If the ^{14}N spin (1, in units of $\hbar$, where $\hbar$ is Planck's constant divided by 2π) is antiparallel to the $\bar{\nu}_e$ spin (spin 1/2), the total quantum mechanical angular momentum of the reaction can be 1/2. The must be equal to the sum of the spins of ^{14}C (spin 0) and the positron (spin 1/2) in order to conserve angular momentum so that this reaction can proceed. Alternatively, if the ^{14}N spin is aligned parallel to the $\bar{\nu}_e$ spin, the total angular momentum can only be 3/2, and, for angular momentum to be conserved, one unit of angular momentum must be provided by the wavefunction of either the incoming $\bar{\nu}_e$ or the outgoing positron. This process is known from basic nuclear physics [43] to occur at a much smaller rate (roughly one order of magnitude) than the antialigned case.

Although this transition must be between nuclear states of opposite parity, for $^{14}N \rightarrow ^{14}$C, both ground states have positive parity. Thus, two units of angular momentum must come from the antineutrino or positron wave functions for that transition to occur. Thus, the inhibition may be closer to two orders of magnitude. This is the origin of the preferential destruction postulated by the SNAAP model.

The energies of the antineutrinos from some of the sources we considered are sufficiently large that it might be possible for a transition to occur between the ^{14}N ground state and the negative-parity ^{14}C first excited state. However, transitions to the negative-parity states in ^{14}C all require not only the weak interaction, but also excitation of one of the nucleons in ^{14}C from the p-shell to the higher lying sd-shell. Such reactions would thus have their own additional level of inhibition, and would therefore probably not be more likely than the ground state to ground-state transition.

Molecular interactions with the external magnetic field have been studied via quantum molecular calculations [1,4]. Such a field can come from a neutron star. In the molecular rest frame, an electric field is produced by the motion of the meteoroids in the magnetic field. This produces a truly chiral environment [13]. In this situation, more right-handed amino acids are destroyed by the interaction of the ^{14}N nuclei with the $\bar{\nu}_e$s than left-handed ones [4].

The destruction mechanism is nuclear, but amino acid chirality is molecular, so it must be shown how the nucleus and molecule are coupled. The external magnetic field aligns the ^{14}N nuclei via their nuclear magnetic moments, whereas the effective electric field aligns the molecular electric dipole moments, which depend on chirality. The external magnetic field, however, is modified at the nucleus by the effects of the orbital electrons, known as shielding—a phenomenon central to nuclear magnetic resonance. The rank-2 shielding tensor (i.e., a two-dimensional relationship, expressed in the form of a matrix, which relates each vector component of the external field to each vector component of the shift in the local field) depends on the electron orbital configuration. Because the electron orbital configuration depends on the molecular geometry, the shielding tensor depends on the chirality of the molecule. As a result, off-diagonal elements in the shielding tensor are asymmetric under parity transformation, meaning that they are asymmetric under a change in chiralty [1,3]. A chirality-dependent magnetization (a bulk property) is created.

The vectors associated with this scenario are illustrated in Figure 1. Here, a scenario for a meteoroid at or near the equatorial plane of the magnetic dipole is studied. There, it can be seen that the external electric field vector $\mathbf{E}_{TS}$ (which is coming out of the page in this diagram) is induced by the meteoroid's velocity vector $\mathbf{v}_m$ through external magnetic flux $\mathbf{B}$. Here, an antineutrino velocity vector $\mathbf{v}_{\bar{\nu}}$ makes an angle θ with respect to the meteoroid velocity vector. The bulk magnetization vector is $\mathbf{M}$ for a meteoroid at rest. For moving meteoroids, the induced electric field creates additional transverse magnetization components $\Delta\mathbf{M}_{\chi}$, where χ represents the chiral state. This induced magnetization is chirality-dependent, and results in net-positive and -negative spin components aligned along the magnetization vectors. The population of nuclei with spins along these components is labeled as $N_{+,-}$ in the figure. The angle 2ϕ is the separation of net magnetization vectors $\mathbf{M}_{\chi}$, where $\phi = \tan^{-1}(\Delta M/M)$. The difference in angle between net magnetization and neutrino velocity results in a different reaction rate for each chiral state [3,4].

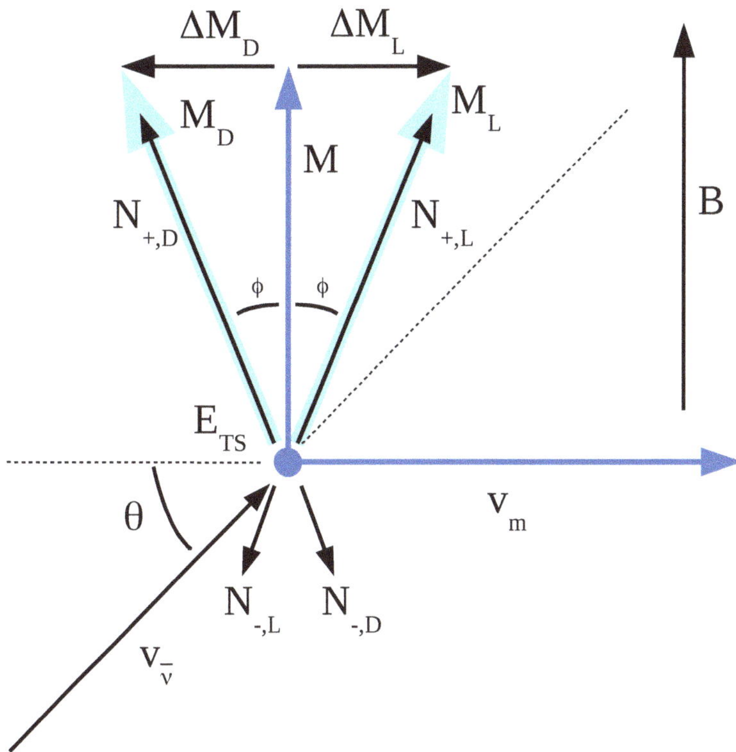

Figure 1. Vectors relevant to the processing of amino acids in this model. The vectors and labels are explained in the text [1]. Used with the permission of Astrobiology.

The external electric field enhances this asymmetry by aligning the molecule along its electric dipole moment [44], thus coupling the nuclear spin and the molecular chirality to result in the selective destruction of one enantiomer [4]. The destruction rate of individual enantiomers and spin states can be used to determine the evolution of the *ee* in time.

Figure 1 shows that the net magnetization, which is an average of the ^{14}N spins, is not necessarily parallel or antiparallel to the antineutrino velocity (and spin) vector. Thus, the antineutrino/nucleus wavefunction consists of a mixture of antialigned and aligned states. The interactions are a result of the projection of the nitrogen spin onto the antineutrino spin vector. However, because of the splitting of the magnetization vector, the components of this mixture are chirality-dependent. In Figure 1, it can be seen that the nuclei of the D enantiomers have a net magnetization that is more antialigned with the antineutrino spin than those of the L enantiomers. This is because the magnetic field for D enantiomers shifts in the direction toward the incoming neutrino and shifts in the opposite direction for the L enantiomer. This results in a higher number of spins of the D enantiomer pointing toward the neutrino. Thus, on average, the nuclei of the D enantiomers are subject to larger destruction cross-sections than the nuclei of the L enantiomers. To take into account this splitting, we adopted the factor in which the value of the reaction rate is proportional to $1 - \cos\Theta$, where Θ is the angle between the antineutrino spin vector and the magnetization vector [45], a good approximation for the considered geometries.

It is also possible that the electron neutrinos could interact with ^{14}N to produce ^{14}O. However, the energy threshold of this reaction is much higher (greater than 5 MeV compared to the order of 1 MeV). Because the cross-section for neutrino-capture processes increases as the square of the energy above threshold, and the antineutrino energies expected [46] from one of the sites, we consider, that coalescing neutron stars are predicted to be much larger for antineutrinos than neutrinos (16 MeV versus 10 MeV), this reaction does not have a significant result in producing a negative enantiometrism from the combined flux from antineutrinos and neutrinos—at least in the two-neutron-star coalescence model.

The size of the meteoroid or planetoid in this model is not constrained, as the antineutrinos are able to pass through the intervening material. They simply must be large enough to survive the passage through space and atmospheric entry. Possible candidates that may provide the necessary conditions for this model include massive-star-neutron-star binaries, magnetars, Wolf–Rayet stars, or "silent supernovae". These are stars massive enough to collapse into black holes. In this process, they create magnetic fields large enough for this model to work. They also create a very large antineutrino flux, but very few photons. Another possible candidate is the site of merging neutron stars, which also produces these same conditions, but with greater magnetic field strength and antineutrino flux at the location of the meteoroid than in the other sites, primarily due to the meteoroid's closer proximity to the event [47,48].

3. Amino Acid Processing in Various Possible Sites

3.1. Processing from a Core-Collapse Supernova

The most obvious candidate for enabling the SNAAP model is a core-collapse supernova from a single massive star. The magnetic field produced by the nascent neutron-star remnant and the antineutrino flux are both enormous. However, this site has a fatal flaw: when the massive star goes into its red-giant phase of stellar evolution, its periphery expands out to about 1 AU. However, our calculations have shown that, because the magnetic field decreases significantly with radius, amino acids would not be significantly affected beyond about 0.01 AU [43,49]. The amino acids that could be processed are therefore inside the star, and would surely be incinerated. Note that the same problem exists for the MCA model; a single isolated supernova cannot work there either.

3.2. Processing from a Close Neutron-Star–Massive-Star Binary

A system such as this would presumably have been born from two massive stars in close proximity. When the more massive one completed its stages of stellar evolution and became a supernova, it would create a neutron star, assuming it wasn't so massive so as to collapse into a black hole. If those two stars remained together, the neutron star would attract the outer one or two layers from the remaining massive star, creating a Wolf–Rayet (WR) star. It would continue its stages of stellar evolution, and would eventually become a core-collapse supernova, known as a Type Ib or Ic.

The material that had been attracted to the neutron star would form an accretion disk around the neutron star. These have been studied in detail, and have been found to enable the formation of dust grains, meteoroids, and even planets [50]. The cooler outer regions could presumably also permit the creation of amino acids, possibly on the dust grains that have formed. These may be shielded from radiation from the WR star, for the portions of their trajectory that are closest to the WR star, by more outward facing regions of the disk. Amino acids have been shown to form on dust grains [51], although under different conditions than would be expected in the outer regions of an accretion disk.

An important consideration is the region in which amino acids might be formed in the accretion disk around one of the neutron stars. Most disk simulations do not extend to temperatures at which molecules might form, but they agree that the temperature falls off roughly as $r^{-3/4}$. D'Alessio et al. [50] found that the midplane temperature depends on the assumed grain size, but was typically several hundred K at 1 AU for the system they considered. Thus, molecules might begin to form around an AU from the central object. However, this is outside the region in which the nuclei could be oriented by the magnetic field from the neutron star.

Although there may be considerations that would allow amino acids to exist closer to their parent neutron star, this might be a long-term prospect that would result in the destruction of the formed amino acids in all but the largest meteoroids. As smaller objects with amino acids agglomerated into larger ones, the material in the disk would impede their velocity, reducing it to less than that required for them to maintain a stable orbit. Thus, they would gradually sink toward the neutron star. Although it might be difficult to have very many amino acids existing within 0.01 AU, the maximum distance within which they would experience a sufficiently large magnetic field to sustain a selection between chiral states, some might survive their trek to that radius if they were in large meteoroids.

3.3. Processing from the Merger of Two Neutron Stars

The two neutron stars in close orbit might have begun their existence as two massive stars. When the first exploded as a supernova, it became a neutron star that drew the outer one or two shells off of the remaining massive star, creating an accretion disk around the neutron star, leaving the other as a WR star. However, when the second star exploded, the result might well have been a two-neutron-star system, with the two in close proximity. One of them would also probably retain some semblance to the accretion disk it had before the WR star exploded.

What is known from GW170817 [52], simultaneously observed by the FERMI [53] and INTEGRAL [54] gamma-ray detectors, and by the CHANDRA [55] X-ray detector, as well as by many optical telescopes, is that a lot of heavy nuclides were synthesized via the rapid-neutron-capture process resulting from the merger of the two neutron stars. While the details of this depend on the masses of the two stars prior to the merger, the final-state neutron star or black hole may not be of great consequence to our considerations. The actual amount of created heavy nuclides has been estimated [56] to be several tenths of a solar mass. Since this was made largely from neutron matter, a huge flux of electron antineutrinos must have been produced by the process that converted the essentially pure neutron matter to the neutron-rich progenitors of the r-process nuclei. (Here, "r-process nuclei" are isotopes thought to be produced via neutron-capture reactions in explosive environments. The r-process is responsible for production of nearly all of the elements heavier than iron [57].)

Fortunately, enough theoretical work [46,58] has been done on two-neutron star mergers that good estimates of the needed parameters to perform SNAAP model calculations exist. In particular, the maximum magnetic field generated at the composite neutron star is around 10^{17} G [46]. That permits the magnetic-field orientation region to extend nearly an order of magnitude beyond what it would be for the supernova from a neutron-star–WR-star binary system.

Perego et al. [58], and Rosswog and Liebendorf [46] calculated both the expected fluxes for electron neutrinos and antineutrinos, and their energies. The electron antineutrinos are the dominant species, and their total flux is expected to exceed 10^{53} ergs in the fraction of a second during which they would be emitted. Their mean energy is predicted to be 16 MeV.

However, for creating enantiomeric amino acids, it is important to consider the disk. When the second neutron star began to converge on its companion to a distance where it would begin to intercept the outer regions of the disk, the disk would be disrupted by the gravitational field of the second neutron star. Amino acid-laden meteoroids might be pulled into a close orbit with the second star, or might be deflected into elongated orbits. As the second star continued to plow through increasingly dense regions of the disk, it would thoroughly mix the disk material, dragging some of the meteoroids from the outer disk regions into regions closer to one or the other neutron star and sending others into orbits that would allow them to pass by the central objects after longer, and highly variable, times. As the two stars grew even closer, this disk material would begin to orbit both stars, and would be compressed under the increased gravitational pull of the two stars.

Presumably, many of the amino acids, especially those enclosed in larger meteoroids with highly elongated orbits, could survive the higher temperatures in which they would find themselves for some time, in some cases long enough for the two stars to complete their spiral into an object. The huge magnetic field of the resulting neutron star, along with the enormous flux of electron antineutrinos, would also surely produce *ee*s in many of these amino acids. The expanding "butterfly" inner disk of matter that is predicted to occur [46,59] would presumably eventually push the outer disk into outer space, there to seed the surrounding volume with enantiomeric amino acids.

4. Simulations

Approximations have been performed of the level of conversion of amino acids that might have existed in the accretion disk from racemic to enantiomeric. Important factors include the gravitational field, the magnetic field, the meteoroid orbital characteristics, and the electron antineutrino flux. Many of these factors are dependent on the meteoroid distance from the neutron star. This includes the meteoroid velocity, which is closely linked to the distance from the star. The details of the calculation are described in prior work [1], so we only summarize them here.

We have computed the shielding tensors and electric dipole moments in each enantiomer of isovaline and alanine using the `Gaussian16` [60] quantum chemistry code [4]. The environmental conditions that approximate the expected environment in the space surrounding a neutron-star merger were simulated. For a typical merger event, the two stars of 1 solar mass each, with a net surface field of 10^{11} T, are assumed. Because the dynamics of the fields (gravitational and magnetic) and the neutrino flux can be complicated in a typical event, we assumed a spherical mass of 2 $M_{\odot}$ with a dipole field. A constant antineutrino flux of 10^{57} $cm^{-2}s^{-1}$ was assumed at the surface of the merger event for one second with an average cross-section for the $^{14}N(\bar{\nu}, e^{+})^{14}C$ of 10^{-40} cm^2. The net antineutrino interaction rate, f, relative to half the ^{14}N relaxation time, T_1, is defined by a unitless fraction [1]:

$$f \equiv 2\frac{\lambda_{\bar{\nu}}}{\lambda_R} = 2\lambda_{\bar{\nu}}T_1 \tag{2}$$

Using the computed shielding tensor, the nuclear magnetic polarizabilities for cationic isovaline and alanine were computed with a density functional theory (DFT) calculation using a `pcS-2` basis set [61]. Prior to this, the initial electronic wavefunctions were optimized using an HF + MP2 [62] computation with the `aug-cc-pVDZ` basis.

With the resultant shielding-tensor asymmetries, the difference in magnetic field vector for each enantiomer was determined, $\mathbf{B}_{\chi} = \mathbf{B}_{\circ} + \Delta\mathbf{B}_{\chi}$, where χ represents the chirality of a particular enantiomer. The antineutrino-interaction rates vary as $\sigma \cdot \mathbf{B}_\chi \rightarrow \mathbf{v} \cdot \mathbf{B}_\chi$[63], where σ and $\mathbf{v}$ are the antineutrino spin and velocity vectors, respectively. Assuming a massless antineutrino, its spin vector points in the same direction as its momentum vector.

5. Results

Several simulations were run, in which the *ee*s of multiple amino acids were computed as a function of time in the vicinity if a neutron-star merger.

The net *ee* as a function of time and orbital radius is shown in Figure 2 for isovaline. Because the merger event is so fast, it can be assumed that the orbital radius changes little over the course of an event. Disk viscosity, however, is an unknown parameter in this model, so a constant meteoroid velocity of 1% of the vacuum orbital velocity was assumed for all orbital radii. Larger velocities commensurate with close radii would result in a higher computed *ee*.

In Figure 2, a plot of the *ee* up to 1 s, the assumed duration of the antineutrino pulse, is also shown for various assumptions of the meteoroid velocity at a constant orbital radius of 1500 km from the merger event. It can be seen that the achieved *ee*s can become large, comparable to those observed in meteorite analyses, at least at that assumed radius. For lower velocities or greater distances from the resultant neutron star, *ees* of the order of a percent can still be achieved.

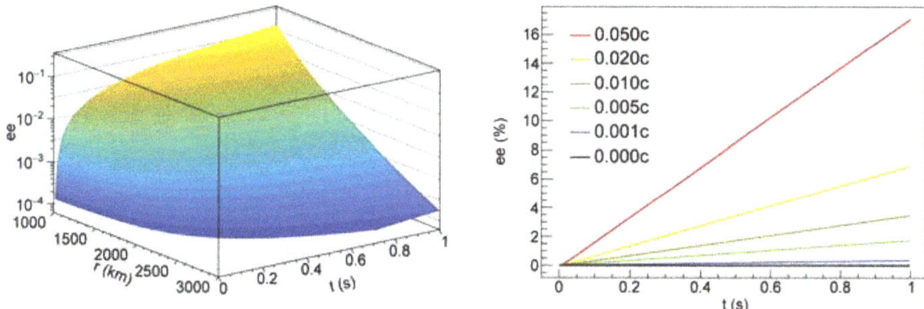

Figure 2. Left: *ee* as a function of time and radius from a neutron binary merger for isovaline. Velocity for a circular orbit is assumed to be damped to 1% of the orbital velocity of a circular orbit. **Right**: *ee* as a function of time for various meteoroid velocities at a fixed distance of 1500 km from the merger, where *c* is the velocity of light.

The conditions in Figure 2 are a result of large magnetic fields. This model also produces large electric fields; these may possibly exceed the dielectric strengths of the amino acids. However, this effect could be mitigated in several ways. A strong magnetic field and a weak electric field can result in the same chiral selection. Additionally, this effect could work if the amino acids were contained in crystalline structures for which a much smaller electric field may suffice.

In prior work, it was shown that, if the antineutrino flux continues for too long, a total destruction of both L- and D-enantiomers could result, and a sudden drop in *ee* occurs as all amino acids are destroyed. This is partially due to their thermalization. In the case of the NN merger event, the antineutrino pulse is so short that the *ee* rapidly increases and the antineutrino flux stops well before all of the amino acids are destroyed.

The results for alanine are shown in Figure 3. Here, it is seen that the *ee* as a function of time and distance is similar to that of isovaline. It is noted that, after 1 s, the *ee* of isovaline is slightly larger than that of alanine. As explained in previous work [4], the product of the asymmetric components of the shielding tensor and the electric dipole moment is larger for isovaline than for alanine. It is seen, however, that a sizable *ee* can still result for alanine. Although the present results were only achieved for isovaline and alanine, our previous work [4] strongly suggests that similar results would be obtained for most other amino acids.

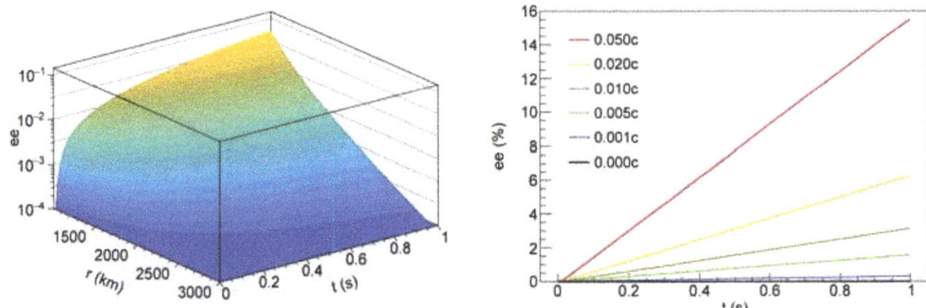

Figure 3. Left: ee as a function of time and radius from a neutron binary merger for alanine. Velocity for a circular orbit is assumed to be damped to 1% of the orbital velocity of a circular orbit. **Right:** ee as a function of time for various meteoroid velocities at a fixed distance of 1500 km from the merger, where c is the velocity of light.

The results for the neutron-star–WR-star binary system are not as impressive as those from the two-neutron-star merger model, but appear to produce viable *ee*s nonetheless. We estimate those to be roughly four orders of magnitude less than those for the two-neutron-star merger case. In this case, the amino acids would obviously require amplification via autocatalysis to reach the levels found in meteorites, but their *ee*s are still competitive with those from other models of chiral selection. Note, though, that these two sites are sufficiently different that other issues may affect the ratio of achievable *ee*s. For example, the magnetic field for the two-neutron-star merger may be so high that processing of the amino acids cannot occur as close to the merged object (because the resulting electric field might destroy the molecules) as it could to the neutron star in the neutron-star–WR-star site. As noted above, though, that site might well evolve to the two-neutron-star case when the WR star explodes, in which case considerably higher *ee*s would be achieved.

The single supernova is problematic, though, because its red-giant phase would envelop any amino acids that could be processed by its electron antineutrinos. We estimate that the maximum *ee* it could achieve would be 10^{-10} or even less. This is probably not a viable site for providing Earth's enantiomeric amino acids.

6. Test of the SNAAP Model

Recent analyses of amino acids in meteorites [64,65] have presented an interesting test of the SNAAP model. The relative abundances of ^{15}N and ^{13}C, both stable nuclides with low normal relative abundances (^{15}N is 0.366 percent of natural N and ^{13}C is 1.11 percent of natural C) were observed to be enhanced in meteoritic amino acids. The abundance of ^{15}N is considerably larger, about a factor of 10, than that of ^{13}C, although the actual enhancement factor considerably varies from one amino acid to another.

We developed a simple model to see if the SNAAP model is consistent with these isotopic enrichments. Both charged-current and neutral-current reactions can change the mass numbers of nuclei if they go to sufficiently highly excited states in the residual nucleus that they can decay by proton or neutron emission. We assume in the following discussion that neutral-current reactions perform the transitions, although the results would not be different if charged-current reactions were included. Then, the relative abundance of ^{15}N to ^{14}N is enhanced both by ^{15}N production from antineutrino reactions on ^{16}O and by ^{14}N destruction by antineutrino reactions. Similarly, the ^{13}C to ^{12}C abundance ratio is enhanced by production from antineutrino reactions on ^{14}N and destruction of ^{12}C by antineutrinos.

In order to simulate the neutral-current reactions, we approximated the antineutrino energy distribution, which has a mean energy of 16 MeV [46,58], with a flat distribution from 8 to 24 MeV, and assumed the reactions varied with the square of the energy. (The results from our model are rather insensitive to the details of the distribution.) This model produced relative destruction rates for ^{16}O:^{14}N:^{12}C of 8.3:22.7:2.3, with the destruction rate of ^{14}N being the largest because it has the lowest energy threshold of all the possible reactions. In each case, the yields from the two possible reactions were added to produce the result, since the two reactions produce the same final nucleus following the beta decay of the unstable member. For example, the interaction of an antineutrino with ^{16}O can produce either ^{15}O plus a neutron or ^{15}N plus a proton, but ^{15}O decays to ^{15}N.

To determine the enhancement of ^{15}N to ^{14}N from the destruction rates on ^{16}O and ^{14}N, one must take into account the relative abundances of ^{16}O and ^{14}N, and similarly for determining the production of ^{13}C to ^{12}C. The natural relative abundances are, roughly, for O:N:C = 10:1:4 [66] (these depend on whether they are galactic or solar abundances, so the numbers given are a rough average of those from the two sets of values). However, these may not be the abundances in the meteoroids when they get processed. The abundances of the three elements are more likely to be comparable in amino acids, but those are not the only molecules formed in the meteoroids. Indeed, there would most likely be some water, and that would enhance oxygen abundance. In the absence of better information, we have just assumed the natural abundances. When these are included, they give a relative enhancement factor of ^{15}N to ^{13}C of about 8.5.

This calculated ratio of the rare isotope enhancements is similar to the observed enhancements in meteorites of rarer isotopes ($\sim$ 2–10) to well within the observed fluctuations for different amino acids [64,65]. Since both the charged-current and neutral-current interactions would affect most of the nuclides other than hydrogen in the amino acids, they produce atomic detritus that have insufficient recoil energy to move very far from where it is produced. Thus, after the neutrino burst ends, it would be expected that these atoms would recombine into new molecules, some of which would be amino acids. It should also be noted that the dust grains on which much of the interstellar chemistry is thought to occur have both carbon and water, that is, oxygen. The neutrinos certainly produce more nitrogen from the oxygen that was not part of the original amino acid inventory, but might well be when recombination occurs. If antineutrino exposure results in a production of nitrogen, then it might be possible to produce more amino acids afterward, as more nitrogen is available for their production, assuming that available nitrogen is the limiting factor in amino acid production. Given the possibility of autocatalysis, that might well increase the *ee*s.

7. Conclusions

The present results suggest that NN-star mergers are a potentially ideal site for producing appreciable enantiomeric excesses in meteoroids that were nearby when the merger occurred. Furthermore, the ability of the SNAAP model to reproduce the ratio of the isotopic enhancements of ^{13}C and ^{15}N does lend confidence in the basic features of the SNAAP model. Finally, the SNAAP model may well explain the enantiomeric excesses observed in meteorites and even, possibly, the origin of life on Earth.

Author Contributions: M.A.F. performed the calculations and wrote the descriptions thereof; R.N.B. did the calculations of the isotopic abundances and wrote their description; M.A.F. and R.N.B. wrote the rest of the paper jointly; T.K., T.O., and Y.M. edited the draft of the manuscript.

Funding: This research was funded by Moore Foundation grant #7799, a Western Michigan University Faculty Research and Creative Activities Award, and by Grants-in-Aid for Scientific Research of JSPS (15H03665, 17K05459). The APC was funded by Moore Foundation grant #7799.

Acknowledgments: M.A.F. is supported by a Moore Foundation, grant #7799, and by a Western Michigan University Faculty Research and Creative Activities Award. T.K. is supported by Grants-in-Aid for Scientific Research of JSPS (15H03665, 17K05459).

Conflicts of Interest: The authors declare no conflict of interest.

Abbreviations

The following abbreviations are used in this manuscript:

AU	Astronomical Unit
CPL	Circularly Polarized Light
ee	enantiomeric excess
MCA	Magneto-Chiral Anisotropy
SNAAP	Supernova Neutrino Amino Acid Processing
WR	Wolf–Rayet (star)

References

1. Famiano, M.A.; Boyd, R.N.; Kajino, T.; Onaka, T. Selection of Amino Acid Chirality via Neutrino Interactions with 14N in Crossed Electric and Magnetic Fields. *Astrobiology* **2018**, *18*, 190–206. [CrossRef] [PubMed]
2. Boyd, R.N.; Famiano, M.A.; Onaka, T.; Kajino, T. Sites that Can Produce Left-handed Amino Acids in the Supernova Neutrino Amino Acid Processing Model. *Astrophys. J.* **2018**, *856*, 26–30. [CrossRef]
3. Boyd, R.; Famiano, M. *Creating the Molecules of Life*; IOP: London, UK, 2018.
4. Famiano, M.; Boyd, R.; Kajino, T.; Onaka, T.; Mo, Y. Amino Acid Chiral Selection Via Weak Interactions in Stellar Environments: Implications for the Origin of Life. *Sci. Rep.* **2018**, *8*, 8833. [CrossRef] [PubMed]
5. Rode, B.; Fitz, D.; Jakschitz, T. The First Steps of Chemical Evolution towards the Origin of Life. *Chem. Biodivers.* **2007**, *4*, 2674–2702. [CrossRef] [PubMed]
6. Pennisi, E. The power of many. *Science* **2018**, *360*, 1388–1391. [CrossRef] [PubMed]
7. Sadownik, J.; Mattia, E.; Nowak, P.; Otto, S. Diversification of self-replicating molecules. *Nat. Chem.* **2016**, *8*, 264–269. [CrossRef] [PubMed]
8. Pasteur, L. Mémoire sur la relation qui peut exister entre la forme crystalline et la composition chimique, et sur la cause de la polarisation rotatoire. *C. R. Acad. Sci. Paris* **1848**, *26*, 535–538.
9. Miller, S. The Production of Amino AcidsUnder Possible Primitive Earth Conditions. *Science* **1953**, *117*, 528. [CrossRef]
10. Miller, S.; Urey, H. Organic Compound Synthes on the Primitive Eart. *Science* **1959**, *130*, 245–251. [CrossRef]
11. Bonner, W. The origin and amplification of biomolecular chirality. *Orig. Life Evolut. Biosph.* **1991**, *21*, 59–111. [CrossRef]
12. Mason, S.F. Origins of biomolecular handedness. *Nature* **1984**, *311*, 19–23. [CrossRef] [PubMed]
13. Barron, L. Chirality and Life. *Space Sci. Ser.* **2008**, *135*, 187–201. [CrossRef]
14. Goldanskii, V. Spontaneous mirror symmetry breaking in nature and the origin of life. *Orig. Life Evolut. Biosph.* **1989**, *19*, 269–272. [CrossRef]
15. Kvenvolden, K.; Lawless, J.; Pering, K.; Peterson, E.; Flores, J.; Ponnamperuma, C. Evidence for extraterrestrial amino-acids and hydrocarbons in the Murchison meteorite. *Nature* **1970**, *228*, 923–926. [CrossRef] [PubMed]
16. Bada, J.; Cronin, J.; Ho, M.S.; Kvenvolden, K.; Lawless, J.; Miller, S.; Oro, J.; Steinberg, S. On the reported optical activity of amino acids in the Murchison meteorite. *Nature* **1983**, *301*, 494–496. [CrossRef]
17. Cronin, J.; Pizzarello, S. Enantiomeric excesses in meteoritic amino acids. *Science* **1997**, *275*, 951–955. [CrossRef] [PubMed]
18. Cronin, J.; Pizzarello, S.; Cruikshank, D. Meteorites and the Early Solar System. In *Meteorites and the Early Solar System*; University of Arizona Press: Tucson, Arizona, 1998; pp. 819–857.
19. Glavin, D.; Dworkin, J. Enrichment of the amino acid L-isovaline by aqueous alteration on CI and CM meteorite parent bodies. *Proc. Natl. Acad. Sci. USA* **2009**, *106*, 5487–5492. [CrossRef]
20. Herd, C.; Blinova, A.; Simkus, D.; Huang, Y.; Tarozo, R.; Alexander, C.; Gyngard, F.; Nittler, L.; Cody, G.; Fogel, M.; et al. Origin and evolution of prebiotic organic matter as inferred from the Tagish Lake meteorite. *Science* **2011**, *332*, 1304–1307. [CrossRef]
21. Frank, F. On spontaneous asymmetric synthesis. *Biochim. Biophys. Acta* **1953**, *11*, 459–463. [CrossRef]
22. Kondepudi, D.K.; Nelson, G.W. Weak neutral currents and the origin of biomolecular chirality. *Nature* **1985**, *314*, 438–441. [CrossRef]

23. Klussmann, M.; Iwamura, H.; Mathew, S.P.; Wells, D.H.; Pandya, U.; Armstrong, A.; Blackmond, D.G. Thermodynamic Control of Asymmetric Amplification in Amino Acid Catalysis. *Nature* **2006**, *441*, 621–623. [CrossRef] [PubMed]
24. Breslow, R.; Levine, M.S. Amplification of enantiomeric concentrations under credible prebiotic conditions. *Proc. Natl. Acad. Sci. USA* **2006**, *103*, 12979–12980. [CrossRef] [PubMed]
25. Arseniyadis, S.; Valleix, A.; Wagner, A.; Mioskowski, C. Kinetic Resolution of Amines: A Highly Enantioselective and Chemoselective Acetylating Agent with a Unique Solvent-Induced Reversal of Stereoselectivity. *Angew. Chem. Int. Ed.* **2004**, *43*, 3314–3317. [CrossRef]
26. Soai, K.; Kawasaki, T.; Matsumoto, A. The Origins of Homochirality Examined by Using Asymmetric Autocatalysis. *Chem. Rec.* **2014**, *14*, 70–83. [CrossRef] [PubMed]
27. Goesmann, F.; Rosenbauer, H.; Bredehöft, J.H.; Cabane, M.; Ehrenfreund, P.; Gautier, T.; Giri, C.; Krüger, H.; Le Roy, L.; MacDermott, A.J.; et al. Organic compounds on comet 67P/Churyumov-Gerasimenko revealed by COSAC mass spectrometry. *Science* **2015**, *349*. [CrossRef] [PubMed]
28. Kitajima, F.; Uesugi, M.; Karouji, Y.; Ishibashi, Y.; Yada, T.; Naraoka, H.; Abe, M.; Fujimura, A.; Ito, M.; Yabuta, H.; et al. A micro-Raman and infrared study of several Hayabusa category 3 (organic) particles. *Earth Planets Space* **2015**, *67*, 20. [CrossRef]
29. Flores, J.J.; Bonner, W.A.; Massey, G.A. Asymmetric photolysis of (RS)-leucine with circularly polarized ultraviolet light. *J. Am. Chem. Soc.* **1977**, *99*, 3622–3625. [CrossRef]
30. Norden, B. Was photoresolution of amino acids the origin of optical activity in life? *Nature* **1977**, *266*, 567–568. [CrossRef]
31. Bailey, J.; Chrysostomou, A.; Hough, J.H.; Gledhill, T.M.; McCall, A.; Clark, S.; Ménard, F.; Tamura, M. Circular Polarization in Star- Formation Regions: Implications for Biomolecular Homochirality. *Science* **1998**, *281*, 672–674. [CrossRef]
32. Takano, Y.; Takahashi, J.i.; Kaneko, T.; Marumo, K.; Kobayashi, K. Asymmetric synthesis of amino acid precursors in interstellar complex organics by circularly polarized light. *Earth Planet. Sci. Lett.* **2007**, *254*, 106–114. [CrossRef]
33. Takahashi, J.I.; Shinojima, H.; Seyama, M.; Ueno, Y.; Kaneko, T.; Kobayashi, K.; Mita, H.; Adachi, M.; Hosaka, M.; Katoh, M. Chirality Emergence in Thin Solid Films of Amino Acids by Polarized Light from Synchrotron Radiation and Free Electron Laser. *Int. J. Mol. Sci.* **2009**, *10*, 3044–3064. [CrossRef] [PubMed]
34. Meierhenrich, U.J.; Filippi, J.J.; Meinert, C.; Bredehöft, J.H.; Takahashi, J.I.; Nahon, L.; Jones, N.C.; Hoffmann, S.V. Circular Dichroism of Amino Acids in the Vacuum-Ultraviolet Region. *Angew. Chem. Int. Ed.* **2010**, *49*, 7799–7802. [CrossRef] [PubMed]
35. de Marcellus, P.; Meinert, C.; Nuevo, M.; Filippi, J.J.; Danger, G.; Deboffle, D.; Nahon, L.; Le Sergeant d'Hendecourt, L.; Meierhenrich, U. Non-racemic Amino Acid Production by Ultraviolet Irradiation of Achiral Interstellar Ice Analogs with Circularly Polarized Light. *Astrophys. J. Lett.* **2011**, *727*, L27–L32. [CrossRef]
36. Meinert, C.; Hoffmann, S.V.; Cassam-Chenaï, P.; Evans, A.C.; Giri, C.; Nahon, L.; Meierhenrich, U.J. Photonenergy-Controlled Symmetry Breaking with Circularly Polarized Light. *Angew. Chem. Int. Ed.* **2014**, *53*, 210–214. [CrossRef] [PubMed]
37. Wagnière, G.; Meier, A. The influence of a static magnetic field on the absorption coefficient of a chiral molecule. *Chem. Phys. Lett.* **1982**, *93*, 78–81. [CrossRef]
38. Rikken, G.; Raupach, E. Enantioselective magnetochiral photochemistry. *Nature* **2000**, *405*, 932–935. [CrossRef] [PubMed]
39. Barron, L.D. Chirality, Magnetism, and Light. *Nature* **2000**, *405*, 895. [CrossRef]
40. Guijarro, A.; Yus, M. *The Origin of Chirality in the Molecules of Life*; RSC Publishing: Cambridge, UK, 2009.
41. Meierhenrich, U. *Amino Acids and the Asymmetry of Life: Caught in the Act of Formation*; Springer-Verlag: Berlin/Heidelberg, Germany, 2008.
42. Boyd, R.; Kajino, T.; Onaka, T. Supernovae and the chirality of the amino acids. *Astrobiology* **2010**, *10*, 561–568. [CrossRef]
43. Boyd, R.N. *An Introduction to Nuclear Astrophysics*; The University of Chicago Press: Chicago, IL, USA, 2008; pp. 126–132.
44. Buckingham, A.; Fischer, P. Direct chiral discrimination in NMR spectroscopy. *Chem. Phys.* **2006**, *324*, 111–116. [CrossRef]

45. de Shalit, A.; Feshbach, H. *Theoretical Nuclear Physics I: Nuclear Structure*; John Wiley & Sons, Inc.: New York, NY, USA, 1974.
46. Rosswog, S.; Liebendörfer, M. High-resolution calculations of merging neutron stars - II. Neutrino emission. *Mon. Not. R. Astron. Soc.* **2003**, *342*, 673–689. [CrossRef]
47. Tian, J.Y.; Patwardhan, A.V.; Fuller, G.M. Neutrino flavor evolution in neutron star mergers. *Phys. Rev. D* **2017**, *96*, 043001. [CrossRef]
48. Price, D.J.; Rosswog, S. Producing Ultrastrong Magnetic Fields in Neutron Star Mergers. *Science* **2006**, *312*, 719–722. [CrossRef] [PubMed]
49. Famiano, M.; Boyd, R.; Kajino, T.; Onaka, T.; Koehler, K.; Hulbert, S. Determining Amino Acid Chirality in the Supernova Neutrino Processing Model. *Symmetry* **2014**, *6*, 909–925. [CrossRef]
50. D'Alessio, P.; Calvet, N.; Hartmann, L. Accretion Disks around Young Objects. III. Grain Growth. *Astrophys. J.* **2001**, *553*, 321–334. [CrossRef]
51. Muñoz Caro, G.M.; Meierhenrich, U.J.; Schutte, W.A.; Barbier, B.; Arcones Segovia, A.; Rosenbauer, H.; Thiemann, W.H.P.; Brack, A.; Greenberg, J.M. Amino Acids From Ultraviolet Irradiation of Interstellar Ice Analogues. *Nature* **2002**, *416*, 403–406. [CrossRef] [PubMed]
52. Abbott, B.; Abbott, R.; Abbott, T.; Acernese, F.; Ackley, K.; Adams, C.; Adams, T.; Addesso, P.; Adhikari, R.; Adya, V.; et al. GW170817: Observation of Gravitational Waves from a Binary Neutron Star Inspiral. *Phys. Rev. Lett.* **2017**, *119*, 161101. [CrossRef] [PubMed]
53. Goldstein, A.; Veres, P.; Burns, E.; Briggs, M.; Hamburg, R.; Kocevski, D.; Wilson-Hodge, C.; Preece, R.; Poolakkil, S.; Roberts, O.; et al. An Ordinary Short Gamma-Ray Burst with Extraordinary Implications: Fermi-GBM Detection of GRB 170817A. *Astrophys. J. Lett.* **2017**, *848*, L14. [CrossRef]
54. Savchenko, V.; Ferrigno, C.; Kuulkers, E.; Bazzano, A.; Bozzo, E.; Brandt, S.; Chenevez, J.; Courvoisier, T.J.L.; Diehl, R.; Domingo, A.; et al. INTEGRAL Detection of the First Prompt Gamma-Ray Signal Coincident with the Gravitational-wave Event GW170817. *Astrophys. J. Lett.* **2017**, *848*, L15. [CrossRef]
55. Troja, E.; Piro, L.; van Eerten, H.; Wollaeger, R.; Im, M.; Fox, O.; Butler, N.; Cenko, S.; Sakamoto, T.; Fryer, C.; et al. The X-ray counterpart to the gravitational-wave event GW170817. *Nature* **2017**, *551*, 71–74. [CrossRef]
56. Kasen, D.; Metzger, B.; Barnes, J.; Quataert, E.; Ramirez-Ruiz, E. Origin of the heavy elements in binary neutron-star mergers from a gravitational-wave event. *Nature* **2017**, *551*, 80–84. [CrossRef]
57. Arnould, M.; Goriely, S.; Takahashi, K. The r-process of stellar nucleosynthesis: Astrophysics and nuclear physics achievements and mysteries. *Phys. Rep.* **2007**, *450*, 97–213. [CrossRef]
58. Perego, A.; Rosswog, S.; Cabezón, R.; Korobkin, O.; Käppeli, R.; Arcones, A.; Liebendörfer, M. Neutrino-driven winds from neutron star merger remnants. *Mon. Not. R. Astron. Soc.* **2014**, *443*, 3134–3156. [CrossRef]
59. Fernández, R.; Kasen, D.; Metzger, B.; Quataert, E. Outflows from accretion discs formed in neutron star mergers: effect of black hole spin. *Mon. Not. R. Astron. Soc.* **2015**, *446*, 750–758. [CrossRef]
60. Frisch, M.J.; Trucks, G.W.; Schlegel, H.B.; Scuseria, G.E.; Robb, M.A.; Cheeseman, J.R.; Scalmani, G.; Barone, V.; Petersson, G.A.; Nakatsuji, H.; et al. *Gaussian 16 Revision A.03*; Gaussian Inc.: Wallingford, CT, USA, 2016.
61. Jensen, F. Basis Set Convergence of Nuclear Magnetic Shielding Constants Calculated by Density Functional Methods. *J. Chem. Theory Comput.* **2008**, *4*, 719–727. [CrossRef] [PubMed]
62. Jensen, H.; Aa, J.; Jørgensen, P.; Ågren, H.; Olsen, J. Second-order Møller-Plesset perturbation theory as a configuration and orbital generator in multiconfiguration self-consistent field calculations. *J. Chem. Phys.* **1988**, *88*, 3834–3839. [CrossRef]
63. Morita, M. *Beta Decay and Muon Capture*; Benjamin: Reading, MA, USA, 1973.
64. Chan, Q.H.S.; Chikaraishi, Y.; Takano, Y.; Ogawa, N.O.; Ohkouchi, N. Amino acid compositions in heated carbonaceous chondrites and their compound-specific nitrogen isotopic ratios. *Earth Planets Space* **2016**, *68*, 7. [CrossRef]
65. Glavin, D.P.; Elsila, J.E.; Burton, A.S.; Callahan, M.P.; Dworkin, J.P.; Hilts, R.W.; Herd, C.D.K. Unusual nonterrestrial L-proteinogenic amino acid excesses in the Tagish Lake meteorite. *Meteorit. Planet. Sci.* **2012**, *47*, 1347–1364. [CrossRef]
66. Anders, E.; Ebihara, M. Solar-System Abundances of the Elements: A New Table. *Meteoritics* **1982**, *17*, 180.

Review

The Chirality Induction and Modulation of Polymers by Circularly Polarized Light

Guang Yang [1], Siyu Zhang [1], Jingang Hu [2], Michiya Fujiki [3,*] and Gang Zou [2,*]

[1] College of Science, Northeast Forestry University, Harbin 150040, China; guangyang@nefu.edu.cn (G.Y.); siyuzhang@nefu.edu.cn (S.Z.)

[2] CAS Key Laboratory of Soft Matter Chemistry, Department of Polymer Science and Engineering, Key Laboratory of Optoelectronic Science and Technology in Anhui Province, University of Science and Technology of China, Hefei 230026, China; kiko@mail.ustc.edu.cn

[3] Graduate School of Materials Science, Nara Institute of Science and Technology, 8916-5 Takayama, Ikoma, Nara 630-0192, Japan

* Correspondence: fujikim@ms.naist.jp (M.F.); gangzou@ustc.edu.cn (G.Z.)

Received: 12 February 2019; Accepted: 26 March 2019; Published: 3 April 2019

Abstract: Chirality is a natural attribute nature of living matter and plays an important role in maintaining the metabolism, evolution and functional activities of living organisms. Asymmetric conformation represents the chiral structure of biomacromolecules in living organisms on earth, such as the L-amino acids of proteins and enzymes, and the D-sugars of DNA or RNA, which exist preferentially as one enantiomer. Circularly polarized light (CPL), observed in the formation regions of the Orion constellation, has long been proposed as one of the origins of single chirality. Herein, the CPL triggered asymmetric polymerization, photo-modulation of chirality based on polymers are described. The mechanisms between CPL and polymers (including polydiacetylene, azobenzene polymers, chiral coordination polymers, and polyfluorene) are described in detail. This minireview provides a promising flexible asymmetric synthesis method for the fabrication of chiral polymer via CPL irradiation, with the hope of obtaining a better understanding of the origin of homochirality on earth.

Keywords: homochirality; circularly polarized light; asymmetric reaction; polymer

1. Introduction

Homochirality is one of the most valuable aspects of science and is an essential molecular characteristic of terrestrial life [1–4]. Asymmetric conformation represents the chiral structure of biomacromolecules in living organisms on earth. Homochirality in biomolecular building blocks almost exclusively results in the use of only one enantiomer for the molecular architecture, such as the D-sugars of nucleic acids and L-amino acids of proteins [5–11]. In the achiral environment, the chemical and physical properties of D- and L-enantiomers were not different except for tiny energy differences which can be attributed to the parity violation of the weak interactions. However, these small energy differences have been theoretically proposed, but they are hard to be detected by conventional experiments [12]. During the process of building up biopolymers such as proteins, enzymes and nucleic acids, an important requirement is the selection of one enantiomer. Therefore, the origin of homochirality in nature has been widely often exploited, and the homochirality of life remains an important subject to be researched.

The asymmetric structure of polymers plays an important role in the maintenance of life processes, metabolism and evolution, and chiral polymers have been widely used in asymmetric synthesis, chiral recognition, and enantiomeric separation [13–23]. Moreover, chiral polymers also hold potential for application in chiral catalysts, liquid crystals, nonlinear optical materials, and the biomedical industry [24–33]. In recent years, scientists have reported several methods to synthesize chiral materials, such as the use of chiral solvents or templates, polymerization with chiral monomers, substituted achiral polymers with a chiral center, supramolecular self-assembly and circularly polarized light (CPL) irradiation. Compared with these methods, circularly polarized light (CPL), proposed to be one of the origins of homochirality in nature, is regarded as an important tool to prepare photo-active chiral materials [34].

As shown in Figure 1, CPL is one kind of electromagnetic waves with a spiral arrangement of the electric field vector along the propagation direction [35]. Right and left-handed CPL are considered to be plausible candidates to introduce the initial chiral asymmetry into biomolecular building blocks. In 1929, Kuhn conducted an experiment by irradiating a racemic organic molecule solution with CPL for the first time, and the results demonstrated that CPL could successfully induce the asymmetric photolysis of organic molecules [36]. Moreover, Kagan also prepared the helical structure of an olefin molecule with a redox reaction between diarylethene and iodine accompanied by the CPL irradiation. Interestingly, the results showed that the final obtained products behaved with chiral signals opposite to that of the external circularly polarized light [37]. In recent years, with the development of CPL technology, CPL has been widely utilized in areas such as asymmetric photolysis, asymmetric polymerization, and deracemization reaction [38–42]. Meinert and co-workers reported that the enantiomeric excess (ee) value of molecules was not only dependent on the polarization state of the CPL but also deeply affected by the wavelength of the CPL [43–45]. Although the CPL irradiation method is considered to be an important tool to synthesize helical biomolecules, the enantiomeric excess is quite a small (<4%). Therefore, how to improve the ee value based on CPL technology remains a question to be answered. In 2009, Vlieg et al. reported that with CPL irradiation, a racemic solution of amino acid derivatives could induce a small amount of chiral bias that was then amplified to give a pure chiral solid phase. The solution of racemic amino acids could be converted to single-handedness through an abrasive grinding process, and the final chirality of the solid phase could be totally controlled by the external handedness of the CPL [38]. Meanwhile, Kotov et al. demonstrated that left- (right-) handed CPL illumination of racemic CdTe nanoparticles in a dispersion state could induce the same direction twisted nanoribbons. Moreover, the chiral nanoribbons were generated in an enantiomeric excess exceeding approximately 30%, and this ee value was substantially higher than that obtained from traditional CPL-induced reactions [46]. Their results opened a pathway for the preparation of helical photonic structures and provided a scenario for the plausible origin of homochirality in biomolecules during the evolution of early earth. Kim et al prepared helical structure based on triphenylamine derivatives, and the helicity of the aggregation was totally controlled by the handedness of the CPL [47]. Therefore, the utilization of CPL irradiation technology in areas including the chirality induction, transference and amplification of liquid crystal or polymers would be of great value especially in asymmetric reaction.

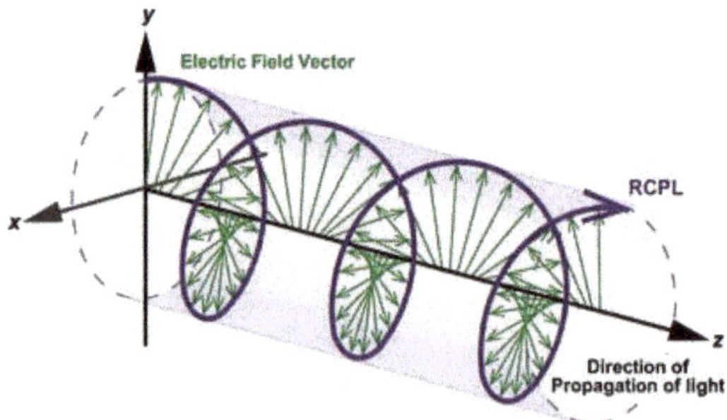

Figure 1. The schematic of right-handed circularly polarized light (R-CPL), reprinted with permission from Reference [34]. Copyright 2018 Elsevier.

In this review, the recent advances in the preparation and modulation of chiral polymers based on CPL irradiation technology are outlined, and the asymmetric mechanism and influence factors are also discussed. Our main purpose is to offer a comprehensive understanding of helical structure construction based on CPL technology which mainly focuses on polydiacetylene, polyfluorene, azobenzene, chiral coordination polymers and so on. This review provides a promising flexible asymmetric synthesis method for the preparation of chiral polymers based on CPL technology, hoping to help us obtain a deeper understanding of the plausible origin of homochirality on earth.

2. The Asymmetric Synthesis of Chiral Polymers Based on Circularly Polarized Light

2.1. The Asymmetric Polymerization of Diacetylene

Polydiacetylene (PDA) is a novel photosensitive material, which possesses conjugated backbone chains, and can be easily formed in different structures by self-assembled systems. Diacetylene (DA) monomers can be polymerized with the irradiation of UV light or γ-rays. With external stimuli (ions, pH, temperature, etc.), PDA exhibits an apparent color and fluorescence change, thereby making it an ideal material for sensing in different forms such as liposomes, vesicles, films or microtubes [48–56]. In 2006, Iwamoto et al. demonstrated the enantioselective synthesis of chiral PDA triggered by ultraviolet circularly polarized light (Figure 2). The DA monomer without any chiral centers was used to prepare the monomer film and ultra-violet CPL was the only chiral source. Interestingly, the formed films irradiated with different-handed (left or right) CPL distinguishably induced the opposite handedness of PDA films [57].

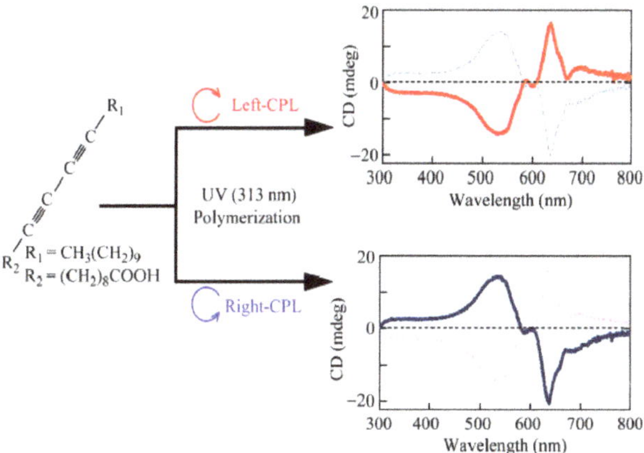

Figure 2. The preparation of chiral polydiacetylene films upon irradiation with left- or right-handed circularly polarized light (reprinted with permission from Reference [57]. Copyright 2006 the Chemical Society of Japan).

Owing to the C-C bond being able to rotate along with the direction of the CPL, a disturbance could be generated in the PDA backbone chains. Thus, the formed PDA, which was irradiated with L- or R-CPL, definitely yielded the opposite chiral polymer. However, upon irradiation with unpolarized ultraviolet light alone, no CD signals could be noticed at the corresponding absorption band for the PDA film. It was reported that visible light could also maintain the polymerization of DA monomers when the number of repeat units of PDA oligomer was more than five. In this case, the enantio-selective polymerization of diacetylene monomers triggered by circularly polarized visible light (CPVL) was realized for the first time (Figure 3). The 532 nm CPVL could effectively offer the chiral information and controlled the handedness of the final PDA chains. This work offered a new method for the synthesis of chiral optical polymers by the visible light region [58].

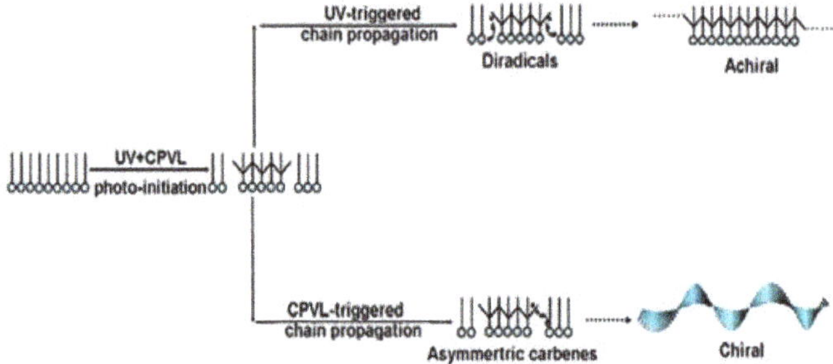

Figure 3. The mechanism of Enantio-selective polymerization of diacetylene (DA) films triggered by circularly polarized visible light (reprinted with permission from Reference [58]. Copyright 2014 The Royal Society of Chemistry).

CPL is one kind of electromagnetic waves and is considered to be a chiral form of light, which exhibits a helical trajectory propagation of the electric field vector [59–61]. However, natural CPL emitted from star formation is usually located in the long-wavelength light such as the IR region, while most chiroptical reactions were induced by ultraviolet light or visible light. Therefore, it was reasonable to extend the chirality induction to the IR region (Figure 4). It was demonstrated that by incorporating $NaYF_4$ up conversion particles, the enantioselective photoinduced polymerization of achiral benzaldehyde-functionalized DA monomers could be realized with the irradiation of 980 nm CPL, which based on the multiphoton up conversion mechanism. The 980 nm CPL acted as the only chiral source and the screw direction of the chiral polymer chain followed the handedness of 980 nm near CPL. This work paved the way for a deeper understanding of the possible origins of homochirality in living systems [62].

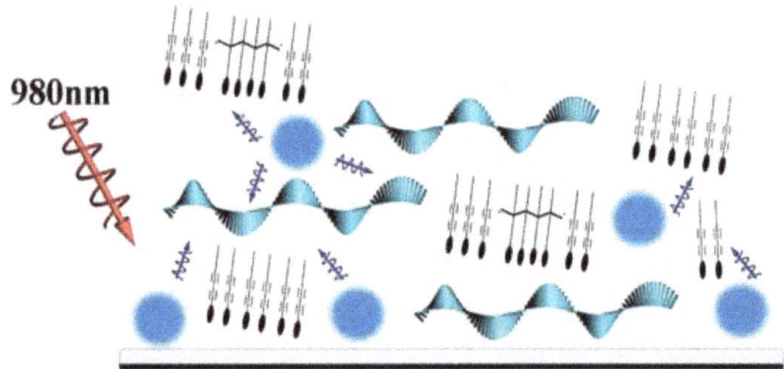

Figure 4. The mechanism of Enantio-selective polymerization of DA films triggered by 980nm circularly polarized light by doping $NaYF_4$ nanoparticles (reprinted with permission from Reference [62]. Copyright 2017 Wiley).

Moreover, according to the novel method for the synthesis of helical conjugated polymers in a nematic liquid crystal phase reported by Akagi et al., helical PDA structures in the liquid crystal (LC) phase could also be synthesized successfully. In 2014, Xu et al. synthesized the 1,3,5-tris(1-alkyl-1H-1,2,3-triazol-4-yl) benzene (TTB) molecule, and the HB complex was obtained by mixing DA with TTB in a 3:1 molar ratio through the self-assembly method (Figure 5). Interestingly, The DA units in the crystal phase could not form helical chains due to the restriction of the crystal lattice, while in the lamellar columnar mesophase, the molecular motion of the hydrogen-bonded complex was relatively free, and the external CPL could effectively direct the screw orientation of the final PDA chains [63]. This work offered a new method for the fabrication of helical polymers in the liquid crystal phase.

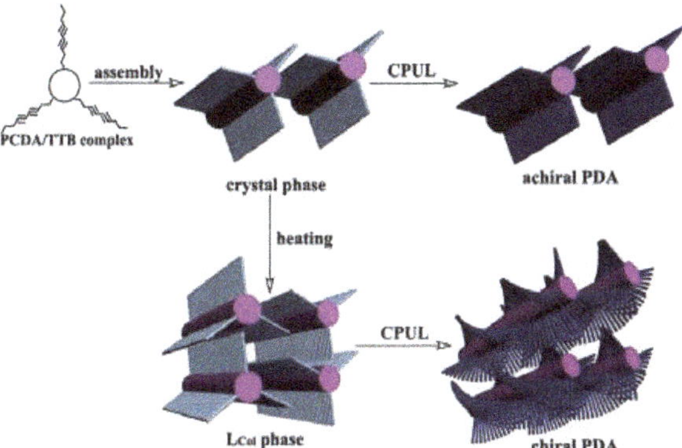

Figure 5. The mechanism of formation helical PDA films in the Lcol liquid crystal state (reprinted with permission from Reference [63]. Copyright 2014 The Royal Society of Chemistry).

In 2018, Zou and co-workers synthesized benzaldehyde-functionalized diacetylene (BSDA) monomers and demonstrated that these DA monomers could be polymerized by the visible light region. Interestingly, super-chiral light (SCL) was introduced to achieve an enhancement in the dissymmetry of BSDA molecules in this work (Figure 6). The super-chiral light was generated by the interference of two circularly polarized lights with the same wavelength, yielding opposite handedness but with a different intensity. It should be noted here that the SCL could generate a greater chiral transfer and amplification than that of the traditional circularly polarized light during the asymmetric photo-polymerization reaction of BSDA monomers. Moreover, the formed helical PDA films irradiated with SCL exhibited an excellent chiral recognition ability and could be utilized to construct a visual sensor for the discrimination of several specific enantiomers. This work offered a new method and might open a pathway for other asymmetric photochemical systems [64].

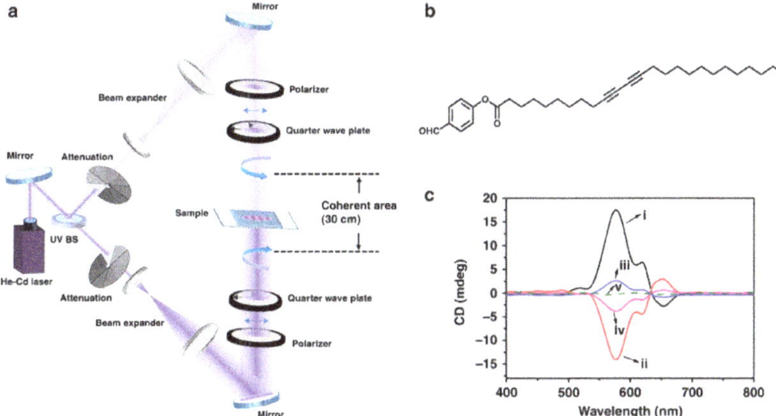

Figure 6. Experimental set-up, molecular structure and CD spectra. (**a**) Experimental set-up for SCL generated by two counter-propagating CPL waves with the same frequency and opposite handedness. The coherent length of the laser is 30 cm and the optical path difference of the two counter-propagating CPL waves is less than 2 cm; (**b**) The molecular structure of BSDA monomer; (**c**) CD spectra of thus-formed PDA films by application of (i) left-handed or (ii) right-handed SCL; (iii) left-handed or (iv) right-handed CPL; (v) LPL, respectively. The wavelength of SCL, CPL and LPL were all 325 nm. The irradiation time was 40 min (reprinted with permission from Reference [64]. Copyright 2018 Springer Nature).

2.2. Enantioselective Synthesis of Chiral Coordination Polymers with CPL

Helical coordination polymers have been widely used in asymmetric catalysis, ferroelectrics, nonlinear optical effects, and chiral resolution [65–67]. Scientists have developed a series of chiral coordination polymers (CCPs) from achiral materials without the doping of any chiral agents. However, it is impossible to predict the absolute configuration of CCPs and the asymmetric synthesis of CCPs remains challenging. With the inspiration of a CPL-triggered solid-liquid mixture of a racemic amino-acid derivative reported by Vlieg et al., Wu and co-workers synthesized the chiral copper (II) CCP [{P or M-Cu(succinate)(4,4′-bipyridine)}$_n$]·$(4H_2O)_n$, which adopts a three-dimensional helical configuration (Figure 7), and the enantioselective synthesis of chiral coordination polymers by the visible-light region was successfully obtained (Figure 8). To discovered the influence of CPL, final products were separated into two zones, bath in the CPL zone (Light-R, L) or bath in dark zone (Dark-R, L). Moreover, nearly 92 samples were selected during the experiments. In the CPL light zone, the fragment [Cu(succinate)]$_x$ acted as the chiral center, which could exist in one preferential configuration and directed the helical assembly direction upon irradiating with CPL. The value of enantioselectivity was at most 80% and the helical structure of the crystalline product was the same as that obtained by the irradiation of circularly polarized light. In the visible-light region, circular dichroism effect was relatively weak, and the size of light spot was small than the cuvette, this size mismatch could not effectively offer enough light bath area, all that lead the enantioselectivity could not reach 100%. Interestingly, a similar CD signals were also discovered in the dark zone, but CD spectra demonstrated dissonance, this was attribute to the sharing of one cuvette compared with Light-zone. This work provided a good avenue to control the enantioselective synthesis of coordination polymers [66].

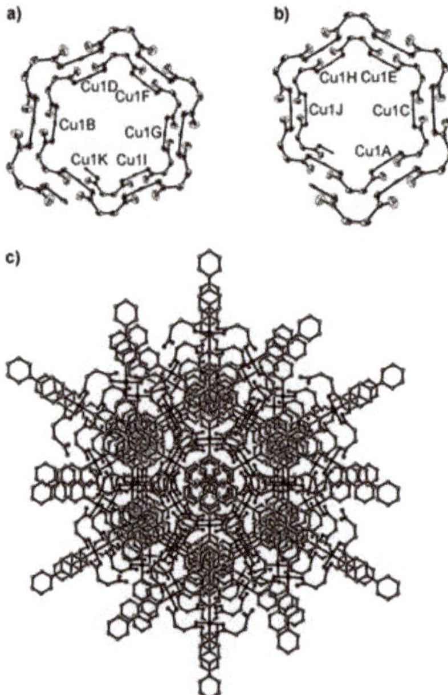

Figure 7. ORTEP plots with thermal ellipsoids set at the 30% probability level showing (**a**) the right-handed helix of polymeric [{Cu(succinate)}$_n$]; (**b**) the left-handed helix of polymeric [{Cu(succinate)}n], and (**c**) the overall structure of [{Cu(succinate)(4,4′-bipyridine)}$_n$]·$(4H_2O)_n$ viewed along the c axis (reprinted with permission from Reference [65]. Copyright 2007 Wiley).

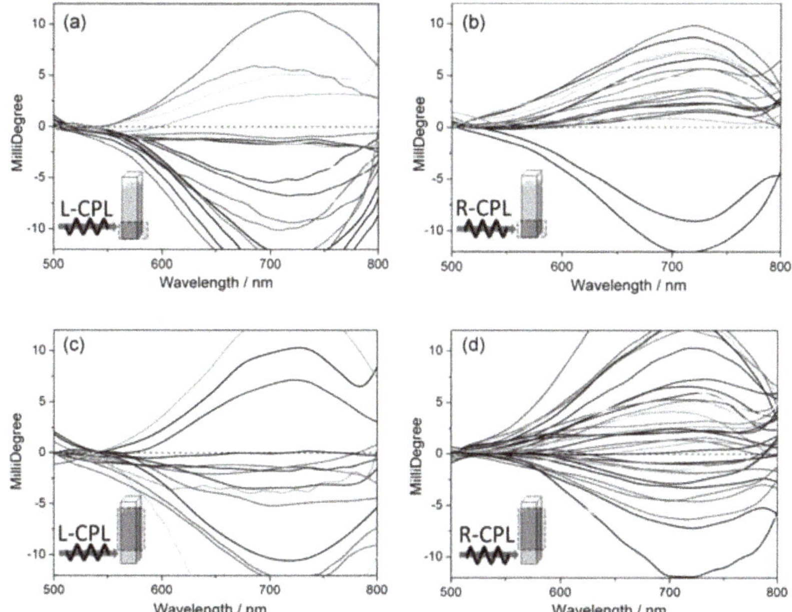

Figure 8. The CD spectra of (**a**) light-L; (**b**) light-R; (**c**) dark-L and (**d**) dark-R CCPs (reprinted with permission from Reference [66]. Copyright 2014 Wiley).

2.3. Enantioselective Thiol-ene Polymerization Reaction Triggered by CPL

The thiol-ene reaction, which given its high yield, rapid rate, and mild reaction conditions, has been widely used in the synthesis of novel organic compounds and smart functional polymers, especially in surface modification, drug-controlled release and advanced optical materials synthesis. However, how to realize the asymmetric click reaction from a racemic mixture remains a question to be solved. In 2017, the allyl-(1-((3-(dimethylamino)propyl)amino)-4-mercapto-1-oxobutan-2-yl)carbamate (DPAMOC) enantiomers were synthesized, and circularly polarized light was utilized to trigger an asymmetric polymerization reaction by our group. The results demonstrated that without any chiral dopant or catalyst, the chiral optically active polymer could also be obtained from racemic monomers with the irradiation of CPL (Figure 9). Via the CPL-triggered enantioselective polymerization click reaction, chiral linear and hyperbranched polymers were easily synthesized with the CPL acting as the only chiral source. Interestingly, the inducible chiral signals of the final polymers could be flexibly controlled by the handedness of the external CPL as well as the irradiation time. This work paved the way for expanding to other common asymmetric click reactions for the preparation of chiral polymers with controllable enantioselectivity [68].

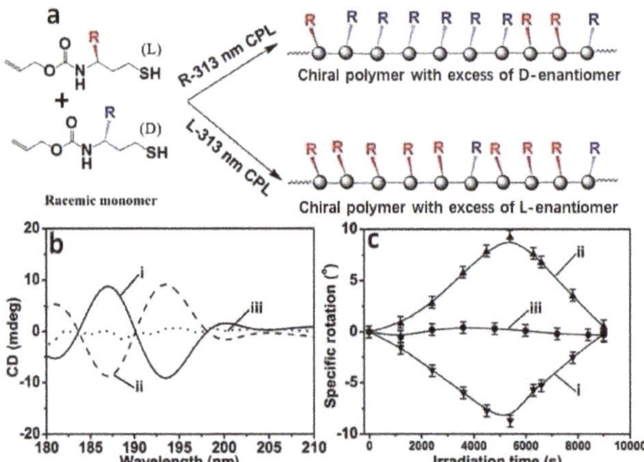

Figure 9. (**a**) Schematic illustration of preparing chiral polymer from racemic monomers through the asymmetric click polymerization process irradiated by 313 nm CPL irradiation. (**b**) CD signals and (**c**) Time-resolved development of the specific rotation values of the final polymer triggered by (i) L-CPL, (ii) R-CPL or (iii) normal UV light, respectively (reprinted with permission from Reference [68]. Copyright 2017 The Royal Society of Chemistry).

3. The Asymmetric Photo-Modulation of Chirality Polymers Based on Circularly Polarized Light

The photo-modulation of chiroptical properties based on functional materials has gained research interest, and the approach could lead to the rapid development of smart materials or devices for reversible information storage. It is believed that chiroptical polymers originate from the properties of natural polymers, which have a specific one-handed helical structure in living matters. However, the single-helical configuration (right- or left-handed) of chiroptical polymers was not stable. With an external physical or chemical stimuli such as light, heating, ions, pH or solvents, the helical configuration could be changed and may be reversed to opposite handedness. Recently, the circularly polarized light-triggered photo-modulation of chiroptical properties has been widely researched in many kinds of photochromophores [69–72]. Herein, the photo-modulation of the chiroptical properties of polymers based on CPL is described.

3.1. Enantioselective Photo-Modulations of Azobenzene Polymers

Polymers that contain azobenzene chromophores have been widely investigated due to their fortunate optical storage properties. Nikolova et al. first reported that with the illumination of circularly polarized light, side-chain azobenzene liquid crystalline polyesters could exhibit a very large circular anisotropy, and the CPL was the only chiral center [73]. An amorphous achiral azobenzene (Azo) liquid crystalline polymer (p4MAN) was synthesized by Iftime et al. With the irradiation of 514 nm CPL, the opposite handedness of CPL produced enantiomeric structures (Figure 10). However, upon switching the handedness of the external CPL, a reversible chiral signal switching between two enantiomeric superstructures of the azobenzene liquid crystalline polymer could be achieved successfully. While after several cycles of switching the handedness of the external CPL, the circular dichroism (CD) signals of the polymers tended to decrease, and this phenomenon was attributed to the orientation of the several azobenzene units which were perpendicular to the Azo film plane. After the photoisomerization, the azobenzene chromophores would exhibit a cis-to-trans transfer, whereby the Azo film plane underwent an angular reorientation. However, several Azo chromophores were also out of the polymer film's plane, and the final numbers of Azo units in the polymer film plane would decrease so that the Azo liquid crystalline polymers exhibited some fatigue, leading to the CD signal decrease [74].

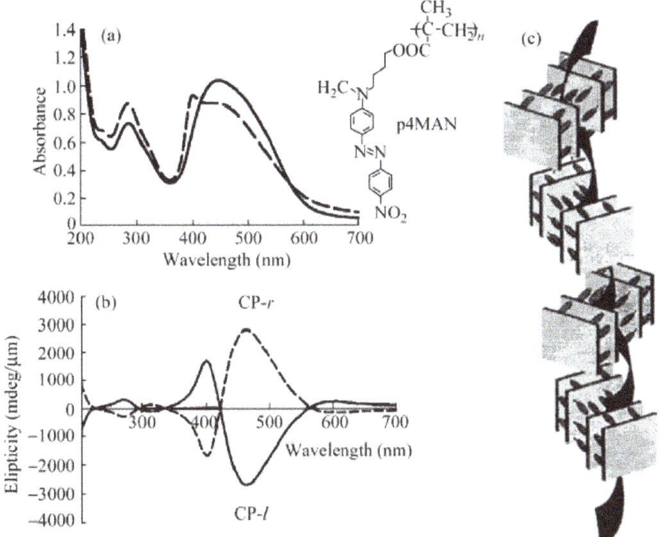

Figure 10. (**a**) UV-vis absorbance and (**b**) CD spectra of the p4MAN thin film after irradiating with circularly polarized light; (**c**) The model of p4MAN to demonstrate the helical arrangement after irradiation with circularly polarized light (reprinted with permission from Reference [74]. Copyright 2000 American Chemical Society).

Ivanov et al. demonstrated that not only the liquid crystalline phase but also the amorphous phase of achiral Azo polymers could form a helical structure upon illumination with CPL (Figure 11). The liquid crystalline orientation represented one of the important factors in the fabrication of chiral superstructure, as the circular momentum could transfer from the CPL to the azobenzene moiety in the polymers [75].

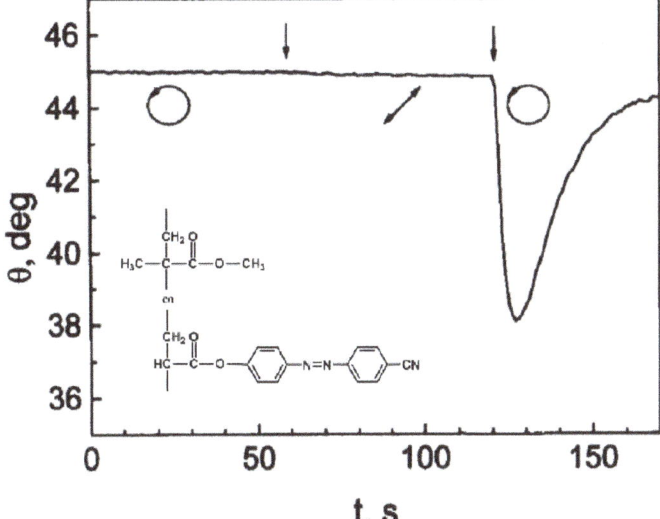

Figure 11. The structure and the rotation of the probe beam azimuth of the Azo polymers (reprinted with permission from Reference [75]. Copyright 2017 Taylor & Francis).

Kim et al. reported the CPL-driven chiral formation based on amorphous azobenzene polymer films, and achiral epoxy-substituted azobenzene polymer PDO3 was synthesized in this work [76]. The amorphous azo polymer chains in the film were in a state of several layers, upon irradiation with left circularly polarized Ar^+ laser light, the linear polarized beam of the incident light would lead to the azobenzene chromophores orienting in one way, which was perpendicular to the main axis of the incident light (Figure 12). After passing across the first layer, the major axis of the incident light could rotate following the counterclockwise direction. Therefore, the major axis of incident light could rotate the same angle in the same direction after passing through the successive layers and the final optical rotation of the polymer would be extended, thus generating the same handedness helical structure in the amorphous azopolymer. This work not only broadened the asymmetric modulation of Azo containing materials but is of great value to the deep understanding of the mechanisms behind chirality photoinduction.

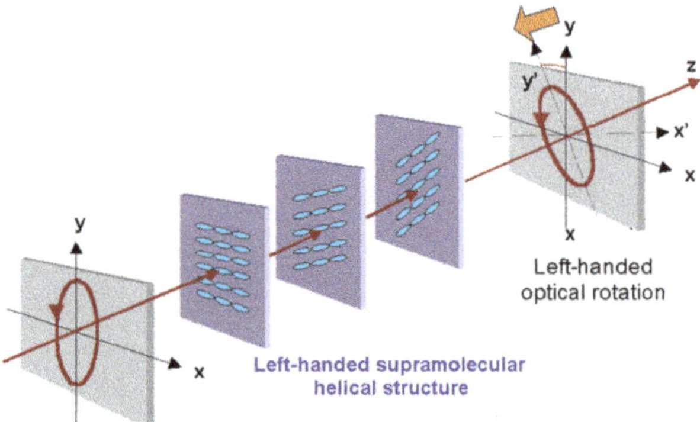

Figure 12. Schematic illumination of a proposed mechanism for the CPL induced chiral helical structure based on an amorphous azopolymer film (reprinted with permission from Reference [76]. Copyright 2000 American Chemical Society).

To further study the influences of the azobenzene chromophores structure and the spacer length during the CPL-triggered helical structure formation, several chiral azobenzene-containing homopolymers were synthesized by Zheng et al (Figure 13). The results indicated that the above samples with a short spacer length (0 or 2) did not generate any CD signals, while those with a longer spacer length (6 or 11) produced clear CD signals. Interestingly, all the films irradiated with 442 nm linearly polarized light displayed a CD signal enhancement in the azobenzene moieties absorbance region between 260 nm and 360 nm. However, the homopolymers with six methylene units demonstrated the largest level of enhancement. During the modulation of the chirality process, the cooperative dipolar interaction with the chiral side chains acted as a key factor in the arrangement of the main chains of the polymers. With longer spacers (chiral side chains), the aggregation level of chiral side chains was higher, which was convenient for the chirality transference and formation of the helical backbones. This work will be of benefit to the design of more sensitive chiral polymers for information storage and chiroptical switching [77].

The chiroptical properties of the azobenzene-substituted diacetylene (NADA) were also researched by Zou et al (Figure 14). With the irradiation of 313 nm circularly polarized ultraviolet light (CPUL), the LB films displayed supramolecular chirality, and evident CD signals were measured by the CD spectra. Consequently, the handedness of the obtained LB films was consistent with that of the CPUL. In this system, the chirality could transfer from the azobenzene units in the side chains to the PDA backbone and could determine the helical direction of the PDA chains. During the chirality transfer process, the stereoregular packing of the azobenzenes was believed to play an important role in the determination of the enantiomeric helical PDA chains. Moreover, the above chiral LB film (irradiated with right-handed CPUL) was also irradiated with left-handed 442 nm CPL. The azobenzene chromophores first changed their stereoregular packing to the opposite helical manner (L-handed), then lead the PDA chains to form the opposite helical structure. Therefore, the modulation of poly-azobenzene-substituted diacetylene (PNADA) LB film chirality could be achieved easily by the CPL treatment. However, another finding should be noted, by changing the stereoregular packing manner of the azobenzene chromophores, a partial inhomogeneous perturbation force was generated against the chiral arrangement of the Azo chromophores, which was accompanied by generating variations in the weakness of CD signals. In this way, the CD signals would decrease after a few cycles. This research offered a novel model system for the deep understanding of the chirality transfer and modulation based on azobenzene polymers [78].

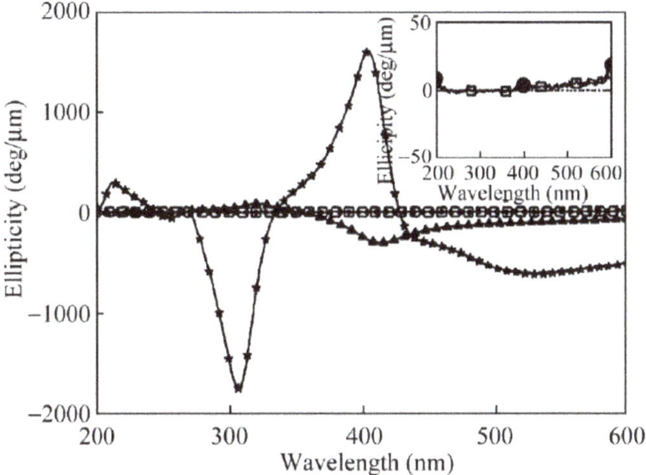

Figure 13. The CD spectra of PMxAP (x = 0, 2, 6, 11) film (reprinted with permission from Reference [77]. Copyright 2000 Elsevier).

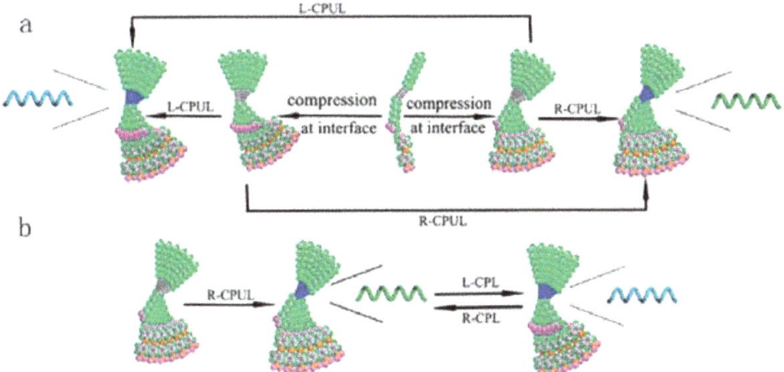

Figure 14. Schematic mechanisms of (**a**) chirality induced and (**b**) chirality modulation for PNADA films with CPL treatment (reprinted with permission from Reference [78]. Copyright 2010 The Royal Society of Chemistry).

3.2. Enantioselective Photo-Modulations of Ketone-Containing Polymers

To further study the influence of CPL during the modulation chirality process, ketone-containing polymers have also been researched by several groups. In 1999, Schuster and co-workers synthesized a racemic acrylic-substituted bicyclic ketone, and the cholesteric phase of liquid crystals based on a mixture of 4-cyano-4′-*n*-alkylbicyclohexanes could be obtained upon irradiation with CPL (Figure 15). Owing to the different absorption properties of the isomers in the presence of left- or right-handed CPL, the photoisomerization reaction of racemic acrylic-substituted bicyclic ketones could be triggered and lead to an enantiomeric excess. The chirality could be transferred through the polymer chains and induced the polymer to form the same helical structure. The enantiomeric excess (ee) value of the bicyclic ketone could be easily modulated by controlling the handedness of the CPL and the irradiation time, which could control the screw pitch and the switch from nematic to cholesteric forms of the liquid crystalline materials [79].

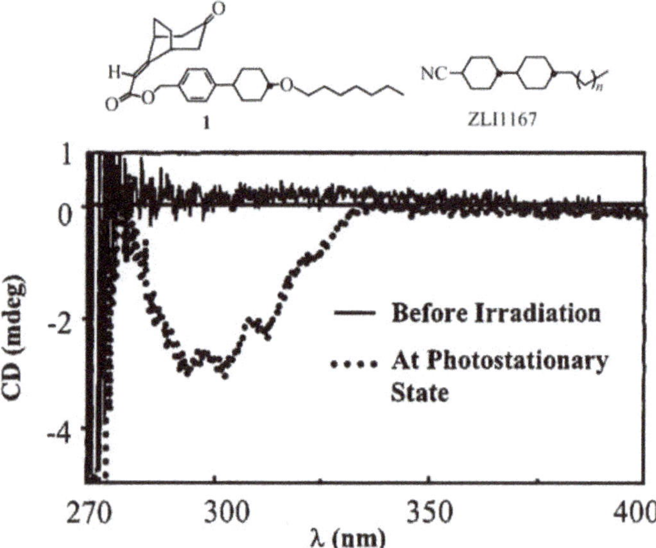

Figure 15. The molecular structure and CD spectra of the racemic samples before and after CPL irradiation (reprinted with permission from Reference [79]. Copyright 1999 American Chemical Society).

With the same idea, Selinger et Al. switched the photoresolvable polymers between mirror images with the tool of CPL in 2000 (Figure 16). The racemic mixture of the ketone-containing group was induced to the polyisocyanate matrix. The polymers first produced no CD signals due to the equal amount of isomers, however, after irradiation with CPL, the polymers could generate a small enantiomeric excess, and noticeable CD signals could be measured in the region of the ketone chromophore. This result demonstrated that circularly polarized light could enforce a disproportionate excess to form a fixed helical structure in the polymers, even with the influence of large proportions of other achiral pendants. Therefore, the helicity of the obtained polymers could be switched reversibly with the alternation of the handedness of the CPL and could also easily return to the original states without a CD signal upon irradiation by plane polarized light [80].

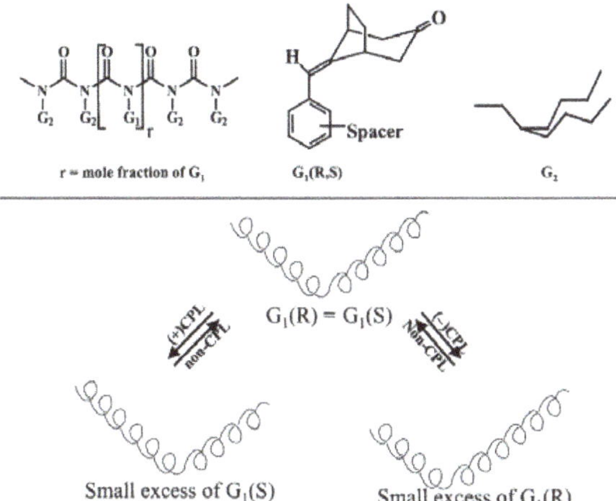

Figure 16. Schematic illumination of the switch between two-handed backbone helices for the polymers irradiated with (+) or (−) circularly polarized light (CPL) or non-circularly polarized light (non-CPL) (reprinted with permission from Reference [80]. Copyright 2000 American Chemical Society).

3.3. Enantioselective Photo-Modulations of Fluorene-Based Polymers

The preparation of optical helical polymers and asymmetric synthesis based on CPL as the only chiral source have been contributing great value and have been gaining increasing attention from researchers. Herein, the CPL-triggered optical chirality induction and asymmetric synthesis of chiral fluorene-based polymers (PDOF) have been outlined.

Similar to diacetylene, the structure of PDOF polymers also has no chiral center, due to the single bonds of the fluorene units could be rotated. Therefore, the exchange between the two conformations (P- and M-twists) can be achieved, and the enrichment of enantiomeric excess can also be obtained by external stimuli, such as CPL. In 2012, Nakano et Al. prepared an achiral polymer film based on the poly(9,9-di-n-octylfluoren-2,7-diyl). After irradiating with R-CPL for 6min, the optically active PODF film displayed intense negative CD signals (CD-1, π–π* transition, approximately 400 nm) (Figure 17). Interestingly, upon irradiating with L-CPL for 6 min, the CD signal of the PDOF film (CD-2) disappeared completely. After additional irradiation with L-CPL for 6 min, an intense positive CD signal could be noted at the same CD bond (CD-3), the spectra of CD-1 and CD-3 were almost symmetrical. Therefore, the helical structure could be reversibly modulated with CPL [81].

In order to obtain a better understanding the mechanism during chirality induction and the switching process of PDOF when irradiated by CPL, simulations of the chirality-switching free-energies based on poly(9,9-dioctylfluoren-2,7-diyl) (PDOF) were calculated not only on an amorphous silica surface but also in the vacuum phase by Nakano and co-workers (Figure 18). Based on the free-energy landscape analysis, the achiral-to-chiral switching of PDOF occurred easily only on the matrix of amorphous silica, where the activation free-energy was calculated to be 35 kcal mol^{-1}. The interactions between PDOF and amorphous silica played an important role during the chirality switching. Compared with PDOF in the solution state or in a suspension, the fluorene-fluorene dihedral of a PDOF film which was deposited on quartz glass could be twisted in a stepwise manner with the irradiation of external CPL [82].

In 2013, for the first time, Fujiki et al. achieved the mirror symmetry breaking of achiral azobenzene-alt-fluorene copolymer particles under the condition of optofluidic organic solvents as well as with CPL irradiation (Figure 19). It was demonstrated that the medium of optofluidic organic solvent, the wavelength, irradiation time and the ellipticity of the external CPL played important roles in the chirality generation, switching, racemization and retention of the copolymer particles. The CPL could trigger the two conformations (P- and M-twists) with asymmetric broken, and the enrichment of one enantiomeric excess could be finally obtained. With continual switching of the CPL handedness, the reversible modulation of chirality in the fluorene-alt-azobenzene copolymer particles could be achieved successfully [83]. This research would be helpful for the design of smart memory devices with the use of the nanosized supramolecular assembly.

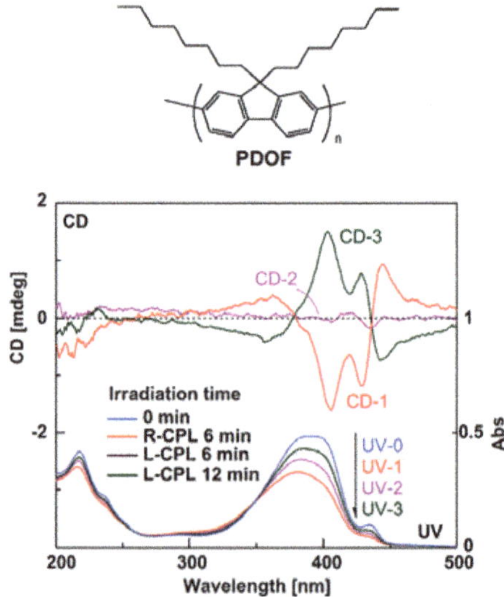

Figure 17. Molecular structure, CD and UV spectra switches of PDOF film upon irradiation with different handedness CPL for different times (reprinted with permission from Reference [81]. Copyright 2012 The Royal Society of Chemistry).

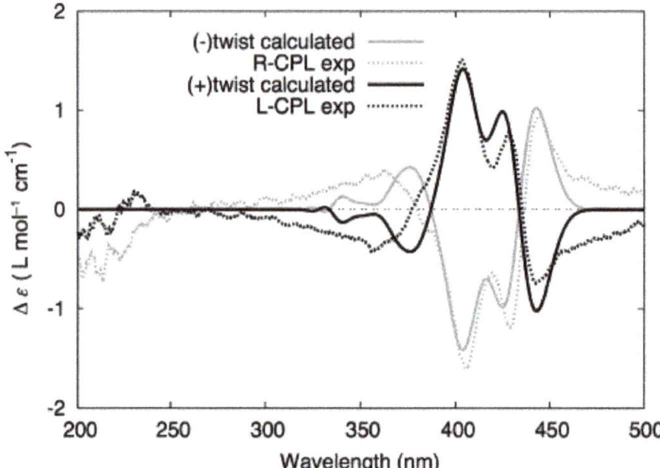

Figure 18. The ECD spectra for the calculated at the ZINDO level for the negative (−) and positive (+) twist basins of poly(fluoren-2,7-diyl) with the relating experimental spectra recorded after CPL irradiation for 6 min (reprinted with permission from Reference [82]. Copyright 2015 Wiley).

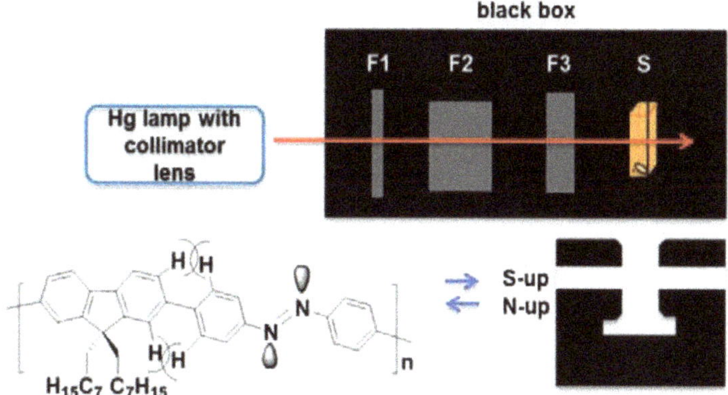

Figure 19. Schematic representation of setup for CPL triggered chirality induction and modulation in a solution of F8AZO particles by r-/l-CPL (reprinted with permission from Reference [83]. Copyright 2013 The Royal Society of Chemistry).

4. Conclusions

Homochirality is one of the universal geometric properties and has garnered remarkable interest in recent years. The circularly polarized light-triggered asymmetric polymerization and photo-modulation of chirality in polymers have gained considerable attention owing to the hypothesis that CPL could transfer single chirality signals to polymers. Moreover, the asymmetric chemical reaction based on CPL displayed several advantages including the purity of products without any chiral dopants or catalysts, and the facile adjustment of CPL parameters such as intensity, wavelength, polarization and interference. Outstanding examples of the asymmetric synthesis of homochirality in polymers, which are based on the effective CPL irradiation technique for control of the molecular asymmetry would help us acquire a better understanding of the mechanisms during single chirality formation, transfer, amplification and modulation. Although significant progress on single chirality induced by CPL has been made in recent years, the abundance of room for growth in this area also needs further research attention, including (1) the fabrication of monomers with a more efficient response to CPL with a large enantiomeric excess. (2) The utilization of CPL in the long-wavelength region to expand the number of chiral materials for potential applications, owing to the fact that CPL emitted from the universe is located in the infrared region. We hope this mini-review allows researchers to find new ways to fabricate chiral materials with more efficiency for applications of photolysis and photosynthesis as well as chiral recognition and to greater understand the probable origin of homochirality in living matter.

Author Contributions: Investigation, G.Y., S.Z.; writing—original draft preparation, G.Y., J.H.; writing—review and editing, G.Y., G.Z.; supervision, G.Z., M.F.; project administration, G.Y., G.Z.; funding acquisition, G.Y.

Funding: This research was funded by the National Natural Science Foundation of China (NSFC, No. 51803021), China Postdoctoral Science Foundation (2018M641790) and Fundamental Research Funds for the Central Universities (2572018BC12).

Conflicts of Interest: The authors declare no conflict of interest.

References

1. Qiu, M.; Zhang, L.; Tang, Z.X.; Jin, W.; Qiu, C.W.; Lei, D.Y. 3D Metaphotonic Nanostructures with Intrinsic Chirality. *Adv. Funct. Mater.* **2018**, *28*, 1803147–1803162. [CrossRef]
2. De los Santos, Z.A.; Lynch, C.C.; Wolf, C. Optical Chirality Sensing with an Auxiliary-Free Earth-Abundant Cobalt Probe. *Angew. Chem. Int. Ed.* **2019**, *58*, 1198–1202. [CrossRef]

3. Campbell, J.P.; Rajappan, S.C.; Jaynes, T.J.; Sharafi, M.; Ma, Y.T.; Li, J.N.; Schneebeli, S.T. Enantioselective Electrophilic Aromatic Nitration: AChiral Auxiliary Approach. *Angew. Chem. Int. Ed.* **2019**, *58*, 1035–1040. [CrossRef]
4. Liu, G.F.; Sheng, J.H.; Teo, W.L.; Yang, G.B.; Wu, H.W.; Li, Y.X.; Zhao, Y.L. Control on Dimensions and Supramolecular Chirality of Self-Assemblies through Light and Metal Ions. *J. Am. Chem. Soc.* **2018**, *140*, 16275–16283. [CrossRef] [PubMed]
5. Zor, E.; Bekar, N. Lab-in-a-syringe using gold nanoparticles for rapid colorimetric chiral discrimination of enantiomers. *Biosens. Bioelectron.* **2017**, *91*, 211–216. [CrossRef] [PubMed]
6. Seo, S.H.; Kim, S.; Han, M.S. Gold nanoparticle-based colorimetric chiral discrimination of histidine: Application to determining the enantiomeric excess of histidine. *Anal. Methods* **2014**, *6*, 73–76. [CrossRef]
7. Ngamdee, K.; Ngeonyae, W. Circular dichroism glucose biosensor based on chiral cadmium sulfide quantum dots. *Sens. Actuators B* **2018**, *274*, 402–411. [CrossRef]
8. Copura, F.; Bekarb, N.; Zorc, E.; Alpaydind, S.; Bingold, H. Nanopaper-based photoluminescent enantioselective sensing of L-Lysine by L-Cysteine modified carbon quantum dots. *Sens. Actuators B* **2019**, *279*, 305–312. [CrossRef]
9. Suzuki, N.; Wang, Y.C.; Elvati, P.; Qu, Z.B.; Kim, K.; Jiang, S.; Baumeister, E.; Lee, J.; Yeom, B.J.; Bahng, J.H.; et al. Chiral Graphene Quantum Dots. *ACS Nano* **2016**, *10*, 1744–1755. [CrossRef] [PubMed]
10. Sun, X.F.; Li, G.H.; Yin, Y.J.; Zhang, Y.Q.; Li, H.G. Carbon quantum dot-based fluorescent vesicles and chiral hydrogels with biosurfactant and biocompatible small molecule. *Soft Matter* **2018**, *14*, 6983–6993. [CrossRef]
11. Lu, Z.Y.; Lu, X.T.; Zhong, Y.H.; Hu, Y.F.; Li, G.K.; Zhang, R.K. Carbon dot-decorated porous organic cage as fluorescenct sensor for rapid discrimination of nitrophenol isomers and alcohols. *Anal. Chim. Acta* **2019**, *1050*, 146–153. [CrossRef] [PubMed]
12. Lee, T.D.; Yang, C.N. Question of parity conservation in weak interactions. *Phys. Rev.* **1956**, *104*, 254–258. [CrossRef]
13. Shukla, N.; Bartel, M.A.; Gellman, A.J. Enantioselective Separation on Chiral Au Nanoparticles. *J. Am. Chem. Soc.* **2010**, *132*, 8575–8580. [CrossRef]
14. Jose BA, S.; Matsushita, S.; Akagi, K. Lyotropic Chiral Nematic Liquid Crystalline Aliphatic Conjugated Polymers Based on Disubstituted Polyacetylene Derivatives That Exhibit High Dissymmetry Factors in Circularly Polarized Luminescence. *J. Am. Chem. Soc.* **2012**, *134*, 19795–19807. [CrossRef] [PubMed]
15. Spano, F.C.; Meskers SC, J.; Hennebicq, E.; Beljonne, D. Probing Excitation Delocalization in Supramolecular Chiral Stacks by Means of Circularly Polarized Light: Experiment and Modeling. *J. Am. Chem. Soc.* **2007**, *129*, 7044–7054. [CrossRef] [PubMed]
16. Oda, M.; Nothofer, H.G.; Scherf, U.; Šunjic', V.; Richter, D.; Regenstein, W.; Neher, D. Chiroptical Properties of Chiral Substituted Polyfluorenes. *Macromolecules* **2002**, *35*, 6792–6798. [CrossRef]
17. Gilot, B.J.; Abbel, R.; Lakhwani, G.; Meijer, E.W.; Schenning AP, H.J.; Meskers SC, J. Polymer Photovoltaic Cells Sensitive to the Circular Polarization of Light. *Adv. Mater.* **2010**, *22*, 131–134. [CrossRef]
18. Wu, D.T.; Yang, J.P.; Peng, Y.G.; Yu, Y.; Zhang, J.; Guo, L.L.; Kong, Y.A.; Jiang, J.L. Highly enantioselective recognition of various acids using polymerized chiral ionic liquid as electrode modifies. *Sens. Actuators B* **2019**, *282*, 164–170. [CrossRef]
19. Tian, Y.; Wang, G.X.; Ma, Z.Y.; Xu, L.; Wang, H. Homochiral Double Helicates Based on Cyclooctatetrathiophene: Chiral Self-Sorting with the Intramolecular S···N Interaction. *Chem. Eur. J.* **2018**, *24*, 15993–15997. [CrossRef]
20. Ding, J.W.; Zhang, M.; Dai, H.X.; Lin, C.M. Enantioseparation of chiral mandelic acid derivatives by supercritical fluid chromatography. *Chirality* **2018**, *30*, 1245–1256. [CrossRef]
21. Zhu, B.L.; Zhao, F.; Yu, J.; Wang, Z.K.; Song, Y.B.; Li, Q. Chiral separation and a molecular modeling study of eight azole antifungals on the cellulose tris(3,5-dichlorophenylcarbamate) chiral stationary phase. *New J. Chem.* **2018**, *42*, 13421–13429. [CrossRef]
22. Bruno, R.; Marino, N.; Bartella, L.; Donna, L.D.; Munno, G.D.; Pardo, E.; Armentano, D. Highly efficient temperature-dependent chiral separation with a nucleotide-based coordination polymer. *Chem. Commun.* **2018**, *54*, 6356–6359. [CrossRef]
23. Mao, B.; Mastral, M.F.; Feringa, B.L. Catalytic Asymmetric Synthesis of Butenolides and Butyrolactones. *Chem. Rev.* **2017**, *117*, 10502–105066. [CrossRef]

24. Chu, J.H.; Xu, X.H.; Kang, S.M.; Liu, N.; Wu, Z.Q. Fast Living Polymerization and Helix-Sense-Selective Polymerization of Diazoacetates Using Air-Stable Palladium (II) Catalysts. *J. Am. Chem. Soc.* **2018**, *140*, 17773–17781. [CrossRef]
25. Wang, Q.Y.; Jia, H.G.; Shi, Y.Q.; Ma, L.Q.; Yang, G.X.; Wang, Y.Z.; Xu, S.P.; Wang, J.J.; Zang, Y.; Aoki, T. [Rh(L-alaninate)(1,5-Cyclooctadiene)] Catalyzed Helix-Sense-Selective Polymerizations of Achiral Phenylacetylenes. *Polymers* **2018**, *10*, 1223. [CrossRef]
26. Suárez-Picado, E.; Quiñoá, E.; Riguera, R.; Freire, F. Poly(phenylacetylene) Amines: A General Route to Water-Soluble Helical Polyamines. *Chem. Mater.* **2018**, *30*, 6908–6914. [CrossRef]
27. Zola, R.S.; Bisoyi, H.K.; Wang, H.; Urbas, A.M.; Bunning, T.J.; Li, Q. Dynamic Control of Light Direction Enabled by Stimuli-Responsive Liquid Crystal Gratings. *Adv. Mater.* **2018**, *31*, 1806172. [CrossRef] [PubMed]
28. Kim, D.Y.; Yoon, W.J.; Choi, Y.J.; Lim, S.I.; Koob, J.; Jeong, K.U. Photoresponsive chiral molecular crystal for light-directing nanostructures. *J. Mater. Chem. C* **2018**, *6*, 12314–12320. [CrossRef]
29. Moran, M.J.; Magrini, M.; Walba, D.M.; Aprahamian, I. Driving a Liquid Crystal Phase Transition Using a Photochromic Hydrazone. *J. Am. Chem. Soc.* **2018**, *140*, 13623–13627. [CrossRef] [PubMed]
30. Venkatakrishnarao, D.; Mamonov, E.A.; Murzina, T.V.; Chandrasekar, R. Advanced Organic and Polymer Whispering-Gallery-Mode Microresonators for Enhanced Nonlinear Optical Light. *Adv. Opt. Mater.* **2018**, *6*, 1800343. [CrossRef]
31. Hadj Sadok, I.B.; Hajlaoui, F.; Ayed, H.B.; Ennaceur, N.; Nasri, M.; Audebrand, N.; Bataille, T.; Zouari, N. Crystal packing, high-temperature phase transition, second-order nonlinear optical and biological activities in a hybrid material: [(S)eC7H16N2][CuBr4]. *J. Mol. Struct.* **2018**, *1167*, 316–326. [CrossRef]
32. Yang, B.W.; Deng, J.P. Chiral PLLA particles with tunable morphology and lamellar structure for enantioselective crystallization. *Mater. Sci.* **2018**, *53*, 11932–11941. [CrossRef]
33. Kulikov, O.V.; Siriwardane, D.A.; Budhathoki-Uprety, J.; McCandless, G.T.; Mahmood, S.F.; Novak, B.M. The secondary structures of PEG-functionalized random copolymers derived from (R)- and (S)- families of alkyne polycarbodiimides. *Polym. Chem.* **2018**, *9*, 2759–2768. [CrossRef]
34. Aav, R.; Mishra, K.A. The Breaking of Symmetry Leads to Chirality in Cucurbituril-Type Hosts. *Symmetry* **2018**, *10*, 98. [CrossRef]
35. Sugahara, H.; Meinert, C.; Nahon, L.; Jones, N.C.; Hoffmann, S.V.; Hamase, K.; Takano, Y.; Meierhenrich, U.J. D-Amino acids in molecular evolution in space—Absolute asymmetric photolysis and synthesis of amino acids by circularly polarized light. *BBA Proteins Proteom.* **2018**, *1866*, 743–758. [CrossRef]
36. Kuhn, W.; Braun, E. Photochemische Erzeugung optisch aktiver Stoffe. *Nat. Chem.* **1929**, *17*, 227–228. [CrossRef]
37. Gopalaiah, K.; Kagan, H.B. Use of Nonfunctionalized Enamides and Enecarbamates in Asymmetric Synthesis. *Chem. Rev.* **2011**, *111*, 4599–4657. [CrossRef]
38. Noorduin, W.L.; Bode, A.A.C.; van der Meijden, M.; Meekes, H.; van Etteger, A.F.; van Enckevort, W.J.P.; Christianen, P.C.M.; Kaptein, B.; Kellogg, R.M.; Rasing, T.; et al. Complete chiral symmetry breaking of an amino acid derivative directed by circularly polarized light. *Nat. Chem.* **2009**, *1*, 729–732. [CrossRef]
39. Shibata, T.; Yamamoto, J.; Matsumoto, N.; Yonekubo, S.; Osanai, S.; Soai, K. Amplification of a Slight Enantiomeric Imbalance in Molecules Based on Asymmetric Autocatalysis: The First Correlation between High Enantiomeric Enrichment in a Chiral Molecule and Circularly Polarized Light. *J. Am. Chem. Soc.* **1998**, *120*, 12157–12158. [CrossRef]
40. Kawasaki, T.; Sato, M.; Ishiguro, S.; Saito, T.; Morishita, Y.; Sato, I.; Nishino, H.; Inoue, Y.; Soai, K. Enantioselective Synthesis of Near Enantiopure Compound by Asymmetric Autocatalysis Triggered by Asymmetric Photolysis with Circularly Polarized Light. *J. Am. Chem. Soc.* **2005**, *127*, 3274–3275. [CrossRef]
41. Nishino, H.; Kosaka, A.; Hembury, G.A.; Aoki, F.; Miyauchi, K.; Shitomi, H.; Onuki, H.; Inoue, Y. Absolute Asymmetric Photoreactions of Aliphatic Amino Acids by Circularly Polarized Synchrotron Radiation: Critically pH-Dependent Photobehavior. *J. Am. Chem. Soc.* **2002**, *124*, 11618–11627. [CrossRef]
42. Zou, G.; Jiang, H.; Kohn, H.; Manaka TIwamoto, M. Control and modulation of chirality for azobenzene-substituted polydiacetylene LB films with circularly polarized light. *Chem. Commun.* **2009**, *15*, 5627–5629. [CrossRef]
43. Meierhenrich, U.J.; Filippi, J.J.; Meinert, C.; Bredehöft, J.H.; Takahashi, J.I.; Nahon, L.; Jones, N.C.; Hoffmann, S.V. Circular Dichroism of Amino Acids in the Vacuum-Ultraviolet Region. *Angew. Chem. Int. Ed.* **2010**, *49*, 7799–7802. [CrossRef] [PubMed]

44. Bredehöft, J.H.; Jones, N.C.; Meinert, C.; Evans, A.C.; Hoffmann, S.V.; Meierhenrich, U.J. Understanding Photochirogenesis: Solvent Effects on Circular Dichroism and Anisotropy Spectroscopy. *Chirality* **2014**, *26*, 373–378. [CrossRef]
45. Meinert, C.; Hoffmann, S.V.; Chenaï, P.C.; Evans, A.C.; Giri, C.; Nahon, L.; Meierhenrich, U.J. Photonenergy-Controlled Symmetry Breaking with Circularly Polarized Light. *Angew. Chem. Int. Ed.* **2014**, *53*, 210–214. [CrossRef] [PubMed]
46. Yeom, J.; Yeom, B.; Chan, H.; Smith, K.; Medina, S.D.; Bahng, J.H.; Zhao, G.P.; Chang, S.W.; Chang, S.J.; Chuvilin, A.; et al. Chiral templating of self-assembling nanostructures by circularly polarized light. *Nat. Mater.* **2014**, *14*, 66–72. [CrossRef] [PubMed]
47. Kim, T.; Mori, T.; Aida, T.; Miyajima, D. Dynamic propeller conformation for the unprecedentedly high degree of chiral amplification of supramolecular helices. *Chem. Sci.* **2016**, *7*, 6689–6694. [CrossRef]
48. Wang, D.E.; Yan, J.H.; Jiang, J.J.; Liu, X.; Tian, C.; Xu, J.; Yuan, M.S.; Hana, X.; Wang, J.Y. Polydiacetylene liposomes with phenylboronic acid tags: A fluorescence turn-on sensor for sialic acid detection and cell-surface glycan imaging. *Nanoscale* **2018**, *10*, 4570–4578. [CrossRef] [PubMed]
49. Xu, Q.L.; Lee, S.; Cho, Y.; Kim, M.H.; Bouffard, J.; Yoon, J. Polydiacetylene-Based Colorimetric and Fluorescent Chemosensor for the Detection of Carbon Dioxide. *J. Am. Chem. Soc.* **2013**, *135*, 17751–17754. [CrossRef]
50. Xia, H.Y.; Li, J.G.; Zou, G.; Zhang, Q.J.; Jia, Q. A highly sensitive and reusable cyanide anion sensor based on spiropyran functionalized polydiacetylene vesicular receptors. *J. Mater. Chem. A* **2013**, *1*, 10713–10719. [CrossRef]
51. Wu, P.J.; Kuo, S.Y.; Huang, Y.C.; Chen, C.P.; Chan, Y.H. Polydiacetylene-Enclosed Near-Infrared Fluorescent Semiconducting Polymer Dots for Bioimaging and Sensing. *Anal. Chem.* **2014**, *86*, 4831–4839. [CrossRef]
52. Lee, J.; Pyo, M.; Lee, S.H.; Kim, J.; Ra, M.; Kim, W.Y.; Park, B.J.; Lee, C.W.; Kim, J.M. Hydrochromic conjugated polymers for human sweat pore mapping. *Nat. Commun.* **2014**, *5*, 3736–3745. [CrossRef]
53. Park, D.H.; Hong, J.; Park, I.S.; Lee, C.W.; Kim, J.M. A Colorimetric Hydrocarbon Sensor Employing a Swelling-Induced Mechanochromic Polydiacetylene. *Adv. Funct. Mater.* **2014**, *24*, 5186–5193. [CrossRef]
54. Wang, D.E.; Wang, Y.L.; Tian, C.; Zhang, L.L.; Han, X.; Tu, Q.; Yuan, M.S.; Chen, S.; Wang, J.Y. Polydiacetylene liposome-encapsulated alginate hydrogel beads for Pb2+ detection with enhanced sensitivity. *J. Mater. Chem. A* **2015**, *3*, 21690–21698. [CrossRef]
55. Wang, M.W.; Wang, F.; Wang, Y.; Zhang, W.; Chen, X.Q. Polydiacetylene-based sensor for highly sensitive and selective Pb^{2+} detection. *Dyes Pigments* **2015**, *120*, 307–313. [CrossRef]
56. Yang, G.; Hu, W.L.; Xia, H.Y.; Zou, G.; Zhang, Q.J. Highly selective and reproducible detection of picric acid in aqueous media, based on a polydiacetylene microtube optical waveguide. *J. Mater. Chem. A* **2014**, *2*, 15560–15565. [CrossRef]
57. Manaka, T.; Kon, H.; Ohshima, Y.; Zou, G.; Iwamoto, M. Preparation of Chiral Polydiacetylene Film from Achiral Monomers Using Circularly Polarized Light. *Chem. Lett.* **2006**, *35*, 1028–1029. [CrossRef]
58. Yang, G.; Han, L.; Jiang, H.; Zou, G.; Zhang, Q.J.; Zhang, D.G.; Wang, P.; Ming, H. Enantioselective synthesis of helical polydiacetylenes in the visible light region. *Chem. Commun.* **2014**, *50*, 2338–2340. [CrossRef] [PubMed]
59. Hao, C.L.; Xu, L.G.; Ma, W.; Wu, X.L.; Wang, L.B.; Kuang, H.; Xu, C.L. Unusual Circularly Polarized Photocatalytic Activity in Nanogapped Gold–Silver Chiroplasmonic Nanostructures. *Adv. Funct. Mater.* **2015**, *25*, 5816–5822. [CrossRef]
60. Kim, J.; Lee, J.; Kim, W.Y.; Kim, H.; Lee, S.; Lee, H.C.; Lee, Y.S.; Seo, M.; Kim, S.Y. Induction and control of supramolecular chirality by light in self-assembled helical nanostructures. *Nat. Commun.* **2015**, *6*, 6959–6966. [CrossRef] [PubMed]
61. Li, W.; Coppens, Z.J.; Besteuro, L.V.; Wang, W.Y.; Govorov, A.O.; Valentine, J. Circularly polarized light detection with hot electrons in chiral plasmonic metamaterials. *Nat. Commun.* **2015**, *6*, 8379–68385. [CrossRef] [PubMed]
62. Yang, G.; Zhu, L.F.; Hu, J.G.; Xia, H.Y.; Qiu, D.; Zhang, Q.J.; Zhang, D.G.; Zou, G. Near-Infrared Circularly Polarized Light Triggered Enantioselective Photopolymerization by Using Upconversion Nanophosphors. *Chem. Eur. J.* **2017**, *23*, 8032–8038. [CrossRef] [PubMed]
63. Xu, Y.Y.; Jiang, H.; Zhang, Q.J.; Wang, F.; Zou, G. Helical polydiacetylene prepared in the liquid crystal phase using circular polarized ultraviolet light. *Chem. Commun.* **2014**, *50*, 365–367. [CrossRef] [PubMed]

64. He, C.L.; Yang, G.; Kuai, Y.; Shan, S.Z.; Yang, L.; Hu, J.G.; Zhang, D.G.; Zhang, Q.J.; Zou, G. Dissymmetry enhancement in enantioselective synthesis of helical polydiacetylene by application of superchiral light. *Nat. Commun.* **2018**, *9*, 5117–5124. [CrossRef] [PubMed]
65. Wu, S.T.; Wu, Y.R.; Kang, Q.Q.; Zhang, H.; Long, L.S.; Zheng, Z.P.; Huang, R.B.; Zheng, L.S. Chiral Symmetry Breaking by Chemically Manipulating Statistical Fluctuation in Crystallization. *Angew. Chem. Int. Ed.* **2007**, *46*, 8475–8479. [CrossRef] [PubMed]
66. Wu, S.T.; Cai, Z.W.; Ye, Q.Y.; Weng, C.H.; Huang, X.H.; Hu, X.L.; Huang, C.C.; Zhuang, N.F. Enantioselective Synthesis of a Chiral Coordination Polymer with Circularly Polarized Visible Laser. *Angew. Chem. Int. Ed.* **2014**, *53*, 12860–12864. [CrossRef]
67. Zheng, Y.Q.; Kong, Z.P. A Novel 3D Framework Coordination Polymer based on Succinato bridged Helical Chains Connected by 4,4-Bipyridine: $[Cu(bpy)(H_2O)_2(C_4H_4O_4)] \cdot 2H_2O$. *Z. Anorg. Allg. Chem.* **2003**, *629*, 1469–1471. [CrossRef]
68. Yang, G.; Xu, Y.Y.; Zhang, Z.D.; Wang, L.H.; He, X.H.; Zhang, Q.J.; Hong, C.Y.; Zou, G. Circularly polarized light triggered enantioselective thiol-ene polymerization. *Chem. Commun.* **2017**, *53*, 1735–1738. [CrossRef]
69. Müller, M.; Zentel, R. Interplay of Chiral Side Chains and Helical Main Chains in Polyisocyanates reaction. *Macromolecules* **1996**, *29*, 1609–1617. [CrossRef]
70. Feringa, B.L.; Jager, W.F.; Lange, B.D. Chiroptical Molecular Switch. *J. Am. Chem. Soc.* **1991**, *13*, 5468–5469. [CrossRef]
71. Yamaguchi, T.; Uchida, K.; Irie, M. Asymmetric Photocyclization of Diarylethene Derivatives. *J. Am. Chem. Soc.* **1997**, *119*, 6066–6071. [CrossRef]
72. Eggers, L.; Buss, V. A Spiroindolinopyran with Switchable Optical Activity. *Angew. Chem. Int. Ed.* **1997**, *36*, 8. [CrossRef]
73. Nikolova, L.; Todorov, T.; Ivanov, M.; Andruzzi, F.; Hvilsted, S.; Ramanujam, P.S. Photoinduced circular anisotropy in side-chain azobenzene polyesters. *Opt. Mater.* **1997**, *8*, 255–258. [CrossRef]
74. Iftime, G.; Labarthet, F.L.; Natansohn, A.; Rochon, P. Control of Chirality of an Azobenzene Liquid Crystalline Polymer with Circularly Polarized Light. *J. Am. Chem. Soc.* **2000**, *122*, 12646–12650. [CrossRef]
75. Ivanov, M.; Naydenova, I.; Todorov, T.; Nikolova, L.; Petrova, T.; Tomova, N.; Dragostinova, V. Light-induced optical activity in optically ordered amorphous side-chain azobenzene containing polymer. *J. Mod. Opt.* **2000**, *47*, 861–867. [CrossRef]
76. Kim, M.J.; Shin, B.G.; Kim, J.J.; Kim, D.Y. Photoinduced Supramolecular Chirality in Amorphous Azobenzene Polymer Films. *J. Am. Chem. Soc.* **2002**, *124*, 3504–3505. [CrossRef]
77. Zheng, Z.; Xu, J.; Sun, Y.Y.; Zhou, J.L.; Chen, B.; Zhang, Q.J.; Wang, K.Y. Synthesis and Chiroptical Properties of Optically Active Polymer Liquid Crystals Containing Azobenzene Chromophores. *J. Polym. Sci. Part A Polym. Chem.* **2010**, *44*, 3210–3219. [CrossRef]
78. Zou, G.; Jiang, H.; Zhang, Q.J.; Kohn, H.; Manaka, T.; Iwamoto, M. Chiroptical switch based on azobenzene-substituted polydiacetylene LB films under thermal and photic stimuli. *J. Mater. Chem.* **2010**, *20*, 285–291. [CrossRef]
79. Burnham, K.S.; Schuster, G.B. Transfer of Chirality from Circularly Polarized Light to a Bulk Material Property: Propagation of Photoresolution by a Liquid Crystal Transition. *J. Am. Chem. Soc.* **1999**, *121*, 10245–10246. [CrossRef]
80. Li, J.; Schuster, G.B.; Cheon, K.S.; Green, M.M.; Selinger, J.V. Switching a Helical Polymer between Mirror Images Using Circularly Polarized Light. *J. Am. Chem. Soc.* **2000**, *122*, 2603–2612. [CrossRef]
81. Wang, Y.; Sakamoto, T.; Nakano, T. Molecular chirality induction to an achiral p-conjugated polymer by circularly polarized light. *Chem. Commun.* **2012**, *48*, 1871–1873. [CrossRef] [PubMed]
82. Pietropaolo, A.; Wang, Y.; Nakano, T. Predicting the Switchable Screw Sense in Fluorene-Based Polymers. *Angew. Chem. Int. Ed.* **2015**, *54*, 2688–2692. [CrossRef] [PubMed]
83. Fujiki, M.; Yoshida, K.; Suzuki, N.; Zhang, J.; Zhang, W.; Zhu, X.L. Mirror symmetry breaking and restoration within mmsized polymer particles in optofluidic media by pumping circularly polarised light. *RSC Adv.* **2013**, *3*, 5213–5219. [CrossRef]

Review

Overview of Low-Temperature Heat Capacity Data for $Zn_2(C_8H_4O_4)_2{\cdot}C_6H_{12}N_2$ and the Salam Hypothesis

Svetlana Kozlova [1,2,*], Maxim Ryzhikov [1,2], Denis Pishchur [1] and Irina Mirzaeva [1,2]

1 Nikolaev Institute of Inorganic Chemistry, Siberian Branch, Russian Academy of Sciences, Lavrentyev Av., 3, RU-630090 Novosibirsk, Russia; maxim.ryzhikov@gmail.com (M.R.); denispischur@ngs.ru (D.P.); dairdre@gmail.com (I.M.)

2 Novosibirsk State University, Pirogova Street, 2, RU-630090 Novosibirsk, Russia

* Correspondence: sgk@niic.nsc.ru

Received: 29 March 2019; Accepted: 9 May 2019; Published: 11 May 2019

Abstract: The review presents the progress in the analysis of low-temperature heat capacity of the metal-organic framework $Zn_2(C_8H_4O_4)_2{\cdot}C_6H_{12}N_2$ (Zn-DMOF). In Zn-DMOF, left-twisted D_3(S) and right-twisted D_3(R) DABCO molecules ($C_6H_{12}N_2$) can transform into each other by tunneling to form a racemate. Termination of tunneling leads to a phase transition in the subsystem of twisted molecules. It is suggested that Zn-DMOF may be considered a model system to study the mechanisms of phase transitions belonging to the same type as hypothetical Salam phase transitions.

Keywords: heat capacity; metal-organic framework; triethylenediamine (DABCO) molecules; racemate; Salam hypothesis

1. Introduction

According to the Salam hypothesis, a small parity-violating energy difference (PVED) between amino acid molecules along with the Bose-Einstein (BE) condensation, makes the less stable right enantiomers tunnel into the more stable left enantiomers, by changing their structural forms. This process was described as a second-order phase transition, which is an analog of the Bardeen-Cooper-Schrieffer (BCS) phase transition; therefore, physical properties such as heat capacity and magnetic susceptibility should change during this phase transition according to the BCS laws [1,2]. Even though no such phase transitions have been found in the crystals of known amino acids, the building material of living organisms, the systems demonstrating the BE condensation of chiral molecules are still of interest. A model of BE condensation was developed for a gas of non-interacting chiral molecules to determine the PVED contribution from low-temperature heat capacity data [3]. The present review summarizes low-temperature heat capacity data, which indicate that the BE condensation may work in a subsystem of triethylenediamine (DABCO) molecules ($C_6H_{12}N_2$) in the metal-organic framework $Zn_2(C_8H_4O_4)_2{\cdot}C_6H_{12}N_2$ (Zn-DMOF) and that the mechanism of Salam phase transitions remains possible. In Zn-DMOF, the enantiomers are represented by left- and right-twisted DABCO molecules, which transform into each other as a result of tunneling.

2. Structure of DABCO Molecule in $Zn_2(C_8H_4O_4)_2{\cdot}C_6H_{12}N_2$

Triethylenediamine (DABCO) appears in the form of two conformational isomers with D_{3h} and D_3 point group symmetries, depending on intermolecular interactions. Also, a quasi-D_{3h} form of DABCO is possible due to strong vibrations of the molecule around the C_3 axis. The molecules with the D_3 symmetry, which can be left-twisted D_3(S) or right-twisted D_3(R), are considered to be chiral isomers (enantiomers) [4].

Above 223 K, the crystal structure of the metal-organic framework Zn-DMOF is tetragonal, with space group P4/mmm [5]. The horizontal planes are formed by terephthalate anions $[C_8H_4O_4]^{2-}$

(BDC^{2-}) which are linked to {Zn_2} pairs by carboxylate anions. The vertical edges are formed by DABCO molecules (linkers), the point symmetry of which does not contain a 4-fold rotational symmetry axis (Figure 1). This is the reason why DABCO molecules are orientationally disordered; moreover, D_3(S) and D_3(R) forms can transform into each other (by activation or tunneling) [4,6]. Calorimetry, nuclear magnetic resonance, and X-ray structural analysis data provide evidence of the presence of phase transitions in Zn-DMOF at ~14, ~60, and ~130 K [7–12].

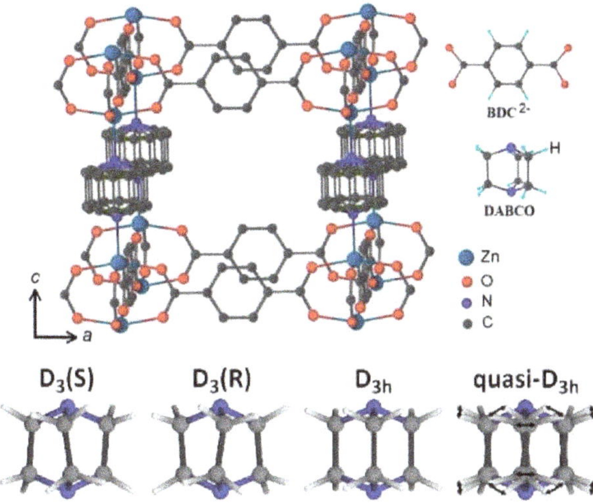

Figure 1. The structure of $Zn_2(C_8H_4O_4)_2{\cdot}C_6H_{12}N_2$ (Zn-DMOF), space group P4/mmm. Positions of carbon atoms in triethylenediamine (DABCO) molecules are disordered [5]. Hydrogen atoms are omitted for clarity. DABCO and BDC^{2-} structures are shown in the insets. (Compiled from Figure 1 in [10] and Figure 1 in [1]).

3. Mobility of DABCO Molecules in $Zn_2(C_8H_4O_4)_2{\cdot}C_6H_{12}N_2$

In Zn-DMOF, BDC^{2-} anions and DABCO molecules are involved in activation mobility. According to the nuclear magnetic resonance (NMR) studies of the activation mobility of BDC^{2-} anions, the [C_6H_4] groups of BDC^{2-} anions rotate about the C_2 axis through an angle of 180° (flipping) [13–15]. No effect of BDC^{2-} flipping on the mobility of DABCO in Zn-DMOF was discovered [13,16,17].

According to the detailed analysis of the temperature behavior of the spin-lattice relaxation times of hydrogen nuclei (^{1}H NMR T_1(T)), D_3(S) and D_3(R) forms of DABCO can make up a racemic mixture, and their mirror symmetry may be broken during the phase transition at ~60 K [6,11,12]. The time decay of nuclear magnetic moments (**M**) of hydrogen atoms in DABCO was analyzed to find the distribution of DABCO molecules over different states. Above ~165 K, the time decay of **M** is a single exponential function characterized by a single value T_1. In this case, DABCO molecules with D_3 and D_{3h} symmetries reorient similarly, their proton spins constitute a single spin system, the activation barrier is equal to ~4 kJ/mol. Between 165 and 60 K, the time decay of **M** is a biexponential function containing two values T_1, each corresponding to a certain fraction of nuclear spins in **M**. The ratio of these fractions is estimated to be ⅓:⅔. The ⅓·**M** fraction corresponds to ^{1}H spins of DABCO molecules of the D_{3h} symmetry, the mobility of which is characterized by a short value T_{1SH}. The ⅔·**M** fraction corresponds to ^{1}H spins of the sum of S- and R-forms of DABCO. In this case, these forms are indistinguishable due to tunneling transitions, so the above fraction (⅔·**M**) represents the racemic state of DABCO molecules and is characterized by a single value T_{1L} of a larger magnitude. During the phase transition at 60 K and down to 25 K, the behavior of T_{1L} is interpreted as the termination of tunneling between energy degenerate quantum states of R- and S- forms of DABCO, and their fractions in **M** remain equal to each other (⅓:⅓). Below 25 K, the decay **M** is nonexponential and can be conventionally characterized by three values T_1. So, the phase transition at ~14 K is associated with the redistribution of DABCO molecules over different energy states characterized by contributions ¼·**M**, ¼·**M**, and ½·**M** and the appearance of a chiral polarized state [11].

Note that the racemate state was also reported for 1,4-bis(carboxyethynyl)bicyclo[2.2.2]octane (BABCO) molecules, which are analogs of DABCO. It was shown that the ratio of left- and right-twisted forms of BABCO can be controlled by light [18,19]. The disorder of DABCO and its analogs, which causes phase transitions, is also observed in other systems [20–22].

Thus, the ^{1}H NMR $T_1(T)$ data testify that phase transitions are associated with the mobility of DABCO molecules. The analysis of the function **M** provides quantitative data on the distribution of DABCO molecules over different states at various temperatures. However, it is still unclear how these states are structurally realized in Zn-DMOF. Low-temperature heat capacity data for Zn-DMOF may be used to clarify this problem.

4. Low-Temperature Heat Capacity in $Zn_2(C_8H_4O_4)_2{\cdot}C_6H_{12}N_2$

All phase transitions in Zn-DMOF (at ~14, ~60, and ~130 K) are second-order phase transitions [10]. Table 1 shows the maximum values of the anomalous parts of heat capacity $\Delta C_p = C_p - C_p{}^L$, where C_p is the heat capacity of the substance and $C_p{}^L$ is the regular part of heat capacity "in the absence of phase transitions". The entropy of the phase transitions is shown in Table 2.

Table 1. ΔC_p (J/mol/K) values at the phase transitions in Zn-DMOF under various pressures of the heat-exchange gas ^{4}He (P·10^5,Pa).

P	~14 K	~60 K	~130 K
0.51	6.0 ± 0.4	8.0 ± 0.2	23.0 ± 0.3
1.52	5.0 ± 0.4	11.0 ± 0.2	23.0 ± 0.3

Table 2. Entropies ΔS/R of the phase transitions in the region of critical temperatures (T_c, K) under various pressures of the heat-exchange gas ^{4}He (P·10^5,Pa) in Zn-DMOF. R is the universal gas constant.

T_c	~14	~60	~130
P	ΔS/R	ΔS/R	ΔS/R
0.51	0.42 ± 0.05	0.14 ± 0.02	0.30 ± 0.04
1.52	0.28 ± 0.04	0.23 ± 0.02	0.30 ± 0.04

The obtained data indicate that the absorbed atoms of ^{4}He affect the states of D_3(S) and D_3(R) forms of DABCO (phase transitions ~14 and ~60 K) and do not affect the ordering and disordering of BDC^{2-} anions during the phase transition at 130 K. This result can be explained by the fact that the structure of DABCO is flexible [23,24] as compared to that of BDC^{2-} anions and can therefore be deformed in the presence of adjoining ^{4}He atoms, whereas the structure of BDC^{2-} anions remains unchanged.

The temperature dependence of heat capacity of Zn-DMOF is almost linear between the phase transitions at ~14, ~60, and ~130 K and above 130 K [8,10,11] to indicate the presence of a one-dimensional elastic continuum [8,12]. Figure 2 shows the comparison of experimental [8] and tabulated (Tarasov model) [25,26] heat capacity values using the fitting parameters obtained in [11].

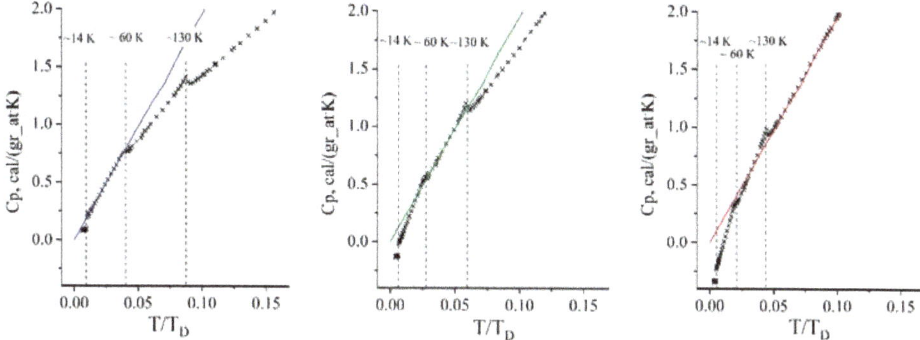

Figure 2. Tabulated (T/T_D) experimental (+) and calculated (solid lines) dependences of heat capacity (Cp) for Zn-DMOF and one-dimensional elastic continuums (Tarasov model [25]). Vertical dashed lines show the temperatures of phase transitions. Solid blue, green, and red lines corresponds to the Debye temperatures of 1490 K, 2230 K, and 2950 K, respectively (according to the data reported in [11]).

According to the XRD data, the crystal lattice of Zn-DMOF expands along the **a** and **b** axes (dL/LdT(**a,b**) = $-9.59 \cdot 10^{-6}$ K^{-1}) and shrinks along the **c** axis (dL/LdT(**c**) = $12.2 \cdot 10^{-6}$ K^{-1}) as the temperature decreases to ~130 K and below ~130 K |dL/LdT(**c**)| > |dL/LdT(**a,b**)|(dL/LdT is the coefficient of thermal expansion) [9]. Hence, the interactions in the -Zn-DABCO-Zn- chain directed along the **c** axis are assumingly stronger than BDC^{2-}-$[Zn_2]^{4+}$-BDC^{2-}- interactions in the **ab** plane, which determines a one-dimensional elastic continuum for the behavior of the heat capacity. The phase transition at ~130 K was interpreted as an order-disorder phase transition associated with a change in the relative spatial arrangement of BDC^{2-} anions, while the DABCO molecules preserve their activation mobility and remain disordered [9].

Linear regions of heat capacity in Zn-DMOF were analyzed using the Stockmayer-Hecht model for the heat capacity of chain crystalline polymers [26,27]. The model assumes that molecular groups in a chain vibrate as single units connected by strong intrachain bonds, while interchain interactions are neglected. The temperature dependence of the volumetric heat capacity C_v is expressed in terms of two relationships, $C_v/(3\mathbf{N}k)$ and T/T_m, where **N** is the number of repeated vibrating units, $T_m = h\nu_m/k$, h is the Planck constant, k is the Boltzmann constant, and ν_m is the maximum frequency of stretching vibrations in the chain. The repeated vibrating unit along the **c** axis in Zn-DMOF consists of two Zn atoms and one DABCO molecule ({Zn_2DABCO}) [5].

Experimental smoothed values C_p obtained as functions of temperature in [8] were represented on a log-log plot and fitted by best tabulated values $C_v/(3\mathbf{N}k)$ for each T/T_m assuming that $C_p - C_v$ is small [27] (Figure 3). As a result, it was found that the vibrating chain is formed by ~38–39 {Zn_2DABCO} units above 130 K, by ~30 units at 60–130 K, and by ~12 units at 14–60 K (Table 3). Below 14 K, the heat capacity obeys the ~T^3 law (Figure 3) to indicate that interchain interactions become stronger and the lattice vibrational modes become three-dimensional [8,25].

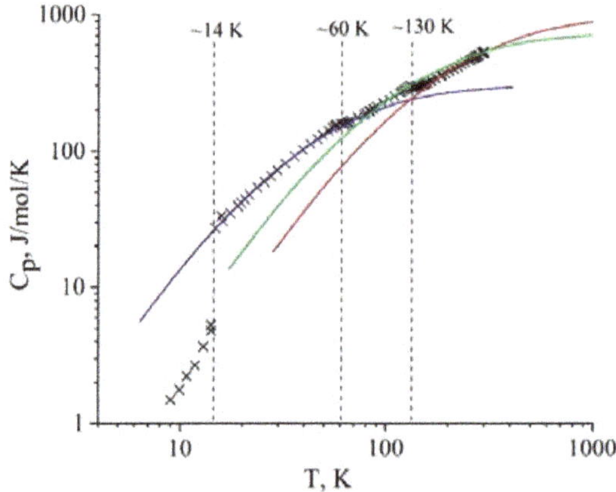

Figure 3. Log-log plot of the Zn-DMOF heat capacity versus temperature. Experimental (crosses) and calculated values of heat capacity at 14.7–57.4 K (blue lines), 130.1–72.6 K (green lines), and 299.6–141.6 K (red lines).

Table 3. Calculated parameters for Zn-DMOF. $\dot{\mathbf{M}}$ is the nuclear magnetic moment, $\dot{\mathbf{N}}$ is the number of {Zn_2DABCO} units normalized with respect to corresponding values above 130 K.

Region of Fit, K	299.6–141.6	130.1–72.6	57.4–14.7
ν_m, cm^{-1}	1250	765	285
N	~38.5	~28.9	~12.0
$\dot{\mathbf{N}}$	1	~0.75	~0.31
$\dot{\mathbf{M}}$	1	~0.67	~0.33

The values ν_m ~ 1250 cm^{-1} and ν_m ~ 765 cm^{-1} fall into the region of stretching vibrations of DABCO, and ν_m ~ 285 cm^{-1} fall into the region of Zn-N and Zn-Zn stretchings (Table 3) [28]. Thus, the obtained values ν_m correspond to the stretchings in the chains, in accordance with the model [27].

As can be seen, the values obtained from the analysis of C_p for **N** {Zn_2DABCO} units correlate with fractions (**M**) in different phases of Zn-DMOF, if **N** and **M** values above 130 K are taken as a unit (Table 3). The obtained quantitative agreement between NMR data and the analysis of heat capacity suggests the following conclusions. Above 130 K, the chains consisting of ~39 {Zn_2DABCO} units contain DABCO molecules with D_3(S), D_3(R), and D_{3h} symmetries. At 60-130 K, the longest chains (~29 {Zn_2DABCO} units) contain only D_3 forms in the racemic state. The vibrations of these chains make the largest contribution to the heat capacity, while the vibrations of the chains consisting of D_{3h} forms make no contribution practically, due to their shorter size. Finally, below ~60 K there are three types of chains (~12 {Zn_2DABCO} units) of the same length but containing three different DABCO forms (D_3(S), D_3(R), and D_{3h}). The size of the chains below 14 K cannot be estimated, since the heat capacity is no more linear at these temperatures.

5. Heat Capacity Behavior during the Phase Transition at 60 K and the Salam Hypothesis

The PVED values for the DABCO molecule and the $[Zn_2DABCO]^{4+}$ cation were obtained in [29]. The difference between the energies of mirror isomers is as small as ~$5\cdot10^{-16}$ kJ/mol (~$5.2\cdot10^{-18}$ eV) for DABCO and an order of magnitude higher (~$5\cdot10^{-15}$ kJ/mol or ~$5.2\cdot10^{-17}$ eV) for the $[Zn_2DABCO]^{4+}$ cation. Therefore, the contribution of PVED increases in the presence of Zn^{2+} cations and is determined mainly by the contribution of zinc cations. This contribution increases if Zn^{2+} cations are replaced by heavier cations Cd^{2+} and Hg^{2+} [30]. Hence, it can be assumed that the PVED breaking of mirror symmetry between $D_3(S)$ and $D_3(R)$ forms of DABCO may be caused by their external environment in the Zn-DMOF structure.

If the symmetry breaking during the phase transition at ~60 K takes place in the chains containing only $D_3(S)$ and $D_3(R)$ forms of DABCO, then, according to the Salam hypothesis, the behavior of heat capacity must correspond to the behavior of heat capacity during the superconducting phase transition [1,2].

In fact, it was discovered that the temperature behavior of heat capacity of Zn-DMOF is an exponential function below ~60 K (Figure 4) [31]. The behavior of heat capacity in the region of second-order phase transitions was studied using the values of the anomalous part of heat capacity $\Delta C_p = C_p - C_p{}^L$, and the behavior of $C_p{}^L$ was described using the Tarasov model [25]. Figure 3 shows the obtained ΔC_p values.

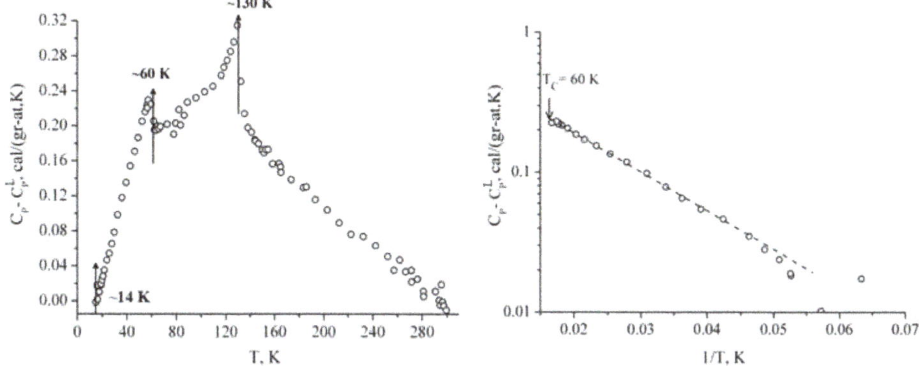

Figure 4. Temperature dependence of ΔC_p (in gram-atom units) for Zn-DMOF (left) and ΔC_p plotted as a function of 1/T below ~60 K (right). ΔC_p is shown on the logarithmic scale (according to the data from Figures 2 and 4 in [31]).

The region below ~60 K is of particular interest, since it is associated with the termination of tunneling between $D_3(S)$ and $D_3(R)$ forms of DABCO as the temperature decreases [11,12]. Based on the hypothesis suggested in [1,2], a study was carried out to verify the compliance of heat capacity ΔC_p to the exponential dependence ~ $\exp(-\Delta/T)$, где $\Delta = 1.76 \cdot T_c$ (Δ is the energy gap at 0 K). Figure 4 shows ΔC_p as a function of 1/T in the temperature region 15 K < T < 60 K. As can be seen, a good agreement with the exponential law is achieved for the parameter Δ equal to ~56 K (or ~$5 \cdot 10^{-3}$ eV) [31]. The obtained value Δ turned out to be almost twice as small as expected (~106 K for $T_c = 60$ K). There is probably some inaccuracy with the parameters determining function $C_p{}^L$, which may cause the error of determining the Δ value. However, the detected exponential behavior of ΔC_p below 60 K signifies the presence of a BE condensation. The amplitude of ΔC_p during the phase transition at is 60 K ≈ 10 J/mol/K (Table 1), which corresponds to the thermal energy jump ($\Delta C_p \cdot T_c$) ≈ 600 J/mol (or $6 \cdot 10^{-3}$ eV), which agrees well with Δ. The value of ($\Delta C_p \cdot T_c$) is 10^{15} times bigger than the PVED (~$5.2 \cdot 10^{-18}$ eV) of one DABCO molecule, but it can be explained by the phenomenon of BE condensation [1,2].

However, neither C_p data nor ^{1}H NMR T_1(T) data show any energy difference between $D_3(S)$ and $D_3(R)$ forms of DABCO below 60 K (according to the Salam hypothesis, the ratio between $D_3(S)$ and $D_3(R)$ forms of DABCO should change). Apparently, the energy difference between $D_3(S)$ and $D_3(R)$ forms remains negligible and can be observed only at lower temperatures, when the thermal energy of the crystal approaches zero [6]. Indeed, according to ^{1}H NMR T_1(T) data, the decay M as a function of time shows anomalous behavior below 25 K [12], but it is not manifested in the C_p behavior until the phase transition at ~14 K.

6. Conclusions

The Salam hypothesis is considered impossible during the phase transitions in amino acid crystals, since the barriers between L- and D-forms of alanine involve intramolecular bond breaking and are as high as ~200 kJ/mol [32]. In Zn-DMOF, the activation barrier between $D_3(S)$ and $D_3(R)$ forms of DABCO is estimated to be ~4 kJ/mol [7] and 5 kJ/mol [27] according to NMR data and quantum chemical calculations, respectively. Thus, this barrier is ~40 times smaller than the barrier between L- and D-forms of alanine [30]. The NMR data indicate the presence of tunneling between $D_3(S)$ and $D_3(R)$ forms of DABCO. The tunneling splitting for the DABCO molecule in the free state is estimated to be ~6 cm^{1-} (~8.6 K) [24], which is comparable to the temperature range of observed phase transitions in Zn-DMOF. The behavior of heat capacity below 60 K corresponds to the heat capacity during the BE condensation. According to the NMR data, still lower temperatures are associated with a redistribution of DABCO with different symmetries over energy states to form a chiral polarized state. In the model system $[Zn_2DABCO]^{4+}$, the R-form is most favorable due to the PVED [29,30], but it is currently unclear which symmetry of the chains built of $\{Zn_2DABCO\}$ units corresponds to the most energetically favorable state. The method of resonant X-ray diffraction with circularly polarized X-rays [33] or optical methods seem to be most preferable for use at low and extra-low temperatures.

We believe that metal-organic frameworks or related compounds containing enantiomers in the racemic state (not necessarily amino acid molecules) may be considered as model systems to study Salam phase transitions. Our studies were aimed at revealing the effects of chirality stabilization in isomeric molecules in solids at low temperatures with the goal of exploring the idea of the cold scenario of life origin on the Earth.

Author Contributions: S.K., analysis of heat capacity data, writing and editing of the manuscript, M.R., editing of the manuscript and discussion, D.P., discussion, I.M.: discussion.

Funding: This research received no external funding.

Conflicts of Interest: The authors declare no conflict of interest.

References

1. Salam, A. The role of chirality in the origin of life. *J. Mol. Evol.* **1992**, *33*, 105–113. [CrossRef]
2. Salam, A. Chirality, phase transitions and their induction in amino acids. *Phys. Lett. B* **1992**, *288*, 153–160. [CrossRef]
3. Bargueño, P.; Perez de Tudela, R.; Miret-Artes, S.; Gonzalo, I. An alternative route to detect parity violating energy differences through Bose–Einstein condensation of chiral molecules. *Phys. Chem. Chem. Phys.* **2011**, *13*, 806–810. [CrossRef] [PubMed]
4. Kozlova, S.G.; Mirzaeva, I.V.; Ryzhikov, M.R. DABCO molecule in the $M_2(C_8H_4O_4)_2{\cdot}C_6H_{12}N_2$ (M = Co, Ni, Cu, Zn) metal-organic frameworks. *Coord. Chem. Rev.* **2018**, *376*, 62–74. [CrossRef]
5. Dybtsev, D.N.; Chun, H.; Kim, K. Rigid and flexible: A highly porous metal–organic framework with unusual guest-dependent dynamic behavior. *Angew. Chem. Int. Ed.* **2004**, *43*, 5033–5036. [CrossRef]
6. Kozlova, S.G.; Gabuda, S.P. Thermal properties of $Zn_2(C_8H_4O_4)_2{\bullet}C_6H_{12}N_2$ metal-organic framework compound and mirror symmetry violation of dabco molecules. *Sci. Rep.* **2017**, *7*, 11505. [CrossRef]
7. Gabuda, S.P.; Kozlova, S.G.; Samsonenko, D.G.; Dybtsev, D.N.; Fedin, V.P. Quantum Rotations and Chiral Polarization of Qubit Prototype Molecules in a Highly Porous Metal–Organic Framework: ^1H NMR T_1 Study. *J. Phys. Chem. C* **2011**, *115*, 20460–20465. [CrossRef]
8. Paukov, I.E.; Samsonenko, D.G.; Pishchur, D.P.; Kozlova, S.G.; Gabuda, S.P. Phase transitions and unusual behavior of heat capacity in metal organic framework compound $Zn_2(C_8H_4O_4)_2\,N_2(CH_2)_6$. *J. Solid State Chem.* **2014**, *220*, 254–258. [CrossRef]
9. Kim, Y.; Haldar, R.; Kim, H.; Koo, J.; Kim, K. The guest-dependent thermal response of the flexible MOF $Zn_2(BDC)_2(DABCO)$. *Dalton Trans.* **2016**, *45*, 4187–4192. [CrossRef] [PubMed]
10. Pishchur, D.P.; Kompankov, N.B.; Lysova, A.A.; Kozlova, S.G. Order-disorder phase transitions in $Zn_2(C_8H_4O_4)_2{\cdot}C_6H_{12}N_2$ in atmospheres of noble gases. *J. Chem. Thermodyn.* **2019**, *130*, 147–153. [CrossRef]

11. Gabuda, S.P.; Kozlova, S.G. Chirality-related interactions and a mirror symmetry violation in handed nano structures. *J. Chem. Phys.* **2014**, *141*, 044701. [CrossRef]
12. Gabuda, S.P.; Kozlova, S.G. Abnormal difference between the mobilities of left- and right-twisted conformations of $C_6H_{12}N_2$ roto-symmetrical molecules at very low temperatures. *J. Chem. Phys.* **2015**, *142*, 234302. [CrossRef]
13. Sabylinskii, A.V.; Gabuda, S.P.; Kozlova, S.G.; Dybtsev, D.N.; Fedin, V.P. ^1H NMR refinement of the structure of the guest sublattice and molecular dynamics in the ultrathin channels of $[Zn_2(C_8H_4O_4)_2(C_6H_{12}N_2)]\cdot n(H_3C)_2NCHO$. *J. Struct. Chem.* **2009**, *50*, 421–428. [CrossRef]
14. Gallyamov, M.R.; Moroz, N.K.; Kozlova, S.G. NMR line shape for a rectangular configuration of nuclei. *Appl. Magn. Reson.* **2011**, *41*, 477–482. [CrossRef]
15. Khudozhitkov, A.E.; Kolokolov, D.I.; Stepanov, A.G.; Bolotov, V.A.; Dybtsev, D.N. Metal-cation-independent dynamics of phenylene ring in microporous MOFs: A ^2H solid-state NMR study. *J. Phys. Chem. C* **2015**, *119*, 28038–28045. [CrossRef]
16. Kozlova, S.G.; Pishchur, D.P.; Dybtsev, D.N. Phase transitions in a metal–organic coordination polymer: $[Zn_2(C_8H_4O_4)_2(C_6H_{12}N_2)]\cdot$with guest molecules. Thermal effects and molecular mobility. *Phase Trans.* **2017**, *90*, 628–636. [CrossRef]
17. Burtch, N.C.; Torres-Knoop, A.; Foo, G.S.; Leisen, J.; Sievers, C.; Ensing, B.; Dubbeldam, D.; Walton, K.S. Understanding DABCO nanorotor dynamics in isostructural metal–organic frameworks. *J. Phys. Chem. Lett.* **2015**, *6*, 812–816. [CrossRef] [PubMed]
18. Lemouchi, C.; Mézière, C.; Zorina, L.; Simonov, S.; Rodríguez-Fortea, A.; Canadell, E.; Wzietek, P.; Auban-Senzier, P.; Pasquier, C.; Giamarchi, T.; et al. Design and evaluation of a crystalline hybrid of molecular conductors and molecular rotors. *J. Am. Chem. Soc.* **2012**, *134*, 7880–7891. [CrossRef]
19. Lemouchi, C.; Iliopoulos, K.; Zorina, L.; Simonov, S.; Wzietek, P.; Cauchy, T.; Rodríguez-Fortea, A.; Canadell, E.; Kaleta, J.; Michl, J.; et al. Crystalline arrays of pairs of molecular rotors: Correlated motion, rotational barriers, and space-inversion symmetry breaking due to conformational mutations. *J. Am. Chem. Soc.* **2013**, *135*, 9366–9376. [CrossRef]
20. Shi, X.; Luo, J.; Sun, Z.; Li, S.; Ji, C.; Li, L.; Han, L.; Zhang, S.; Yuan, D.; Hong, M. Switchable dielectric phase transition induced by ordering of twisting motion in 1,4-diazabicyclo[2.2.2]octane chlorodifluoroacetate. *Cryst. Growth Des.* **2013**, *13*, 2081–2086. [CrossRef]
21. Yao, Z.-S.; Yamamoto, K.; Cai, H.-L.; Takahashi, K.; Sato, O. Above room temperature organic ferroelectrics: Diprotonated 1,4-diazabicyclo[2.2.2] octane shifts between two 2-chlorobenzoates. *J. Am. Chem. Soc.* **2016**, *138*, 12005–12008. [CrossRef]
22. Chen, L.-Z.; Huang, D.-D.; Ge, J.-Z.; Pan, Q.-J. Reversible ferroelastic phase transition of N-chloromethyl-1,4-diazabicyclo[2.2.2]octonium trichlorobromoaquo copper(II). *Inorg. Chem. Commun.* **2014**, *45*, 5–9. [CrossRef]
23. Nizovtsev, A.S.; Ryzhikov, M.R.; Kozlova, S.G. Structural flexibility of DABCO. Ab initio and DFT benchmark study. *Chem. Phys. Lett.* **2017**, *667*, 87–90. [CrossRef]
24. Mathivon, K.; Linguerri, R.; Hochlaf, M. Systematic theoretical studies of the interaction of 1,4-diazabicyclo [2.2.2]octane (DABCO) with rare gases. *J. Chem. Phys.* **2013**, *139*, 164306.
25. Tarasov, V.V. Heat Capacity of Anisotropic Solids. *Zhurnal Fiz. Khimii* **1950**, *24*, 111–128. (In Russian)
26. Wunderlich, B.; Baur, H. *Heat Capacities of Liner High Polymers*; Springer: Berlin, Germany, 1970.
27. Stockmayer, W.H.; Hecht, C.E. Heat capacity of chain polymeric crystals. *J. Chem. Phys.* **1953**, *21*, 1954–1958. [CrossRef]
28. Tan, K.; Nijem, N.; Canepa, P.; Gong, Q.; Li, J.; Thonhauser, T.; Chabal, Y.J. Stability and hydrolyzation of metal organic frameworks with paddle-wheel SBUs upon hydration. *Chem. Mater.* **2012**, *24*, 3153–3167. [CrossRef]
29. Mirzaeva, I.V.; Kozlova, S.G. Computational estimation of parity violation effects in a metal-organic framework containing DABCO. *Chem. Phys. Lett.* **2017**, *687*, 110–115. [CrossRef]
30. Mirzaeva, I.V.; Kozlova, S.G. Parity violating energy difference for mirror conformers of DABCO linker between two M^{2+} cations (M = Zn, Cd, Hg). *J. Chem. Phys.* **2018**, *149*, 214302. [CrossRef] [PubMed]
31. Kozlova, S.G. Behavior of the heat capacity at second-order phase transitions in the $[Zn_2(C_8H_4O_4)_2\cdot C_6H_{12}N_2]$ metal-organic framework compound. *JETP Lett.* **2016**, *104*, 253–256. [CrossRef]

32. Sullivan, R.; Pyda, M.; Pak, J.; Wunderlich, B.; Thompson, J.R.; Pagni, R.; Pan, H.J.; Barnes, C.; Schwerdtfeger, P.; Compton, R. Search for electroweak interactions in amino acid crystals. II. The Salam hypothesis. *J. Phys. Chem. A* **2003**, *107*, 6674–6680. [CrossRef]
33. Tanaka, Y.; Kojima, T.; Takata, Y.; Chainani, A.; Lovesey, S.W.; Knight, K.S.; Takeuchi, T.; Oura, M.; Senba, Y.; Ohashi, H.; et al. Determination of structural chirality of berlinite and quartz using resonant x-ray diffraction with circularly polarized x-rays. *Phys. Rev. B* **2010**, *81*, 144104. [CrossRef]

Article

Homochirality: A Perspective from Fundamental Physics

Anaís Dorta-Urra [1] and Pedro Bargueño [2,*]

[1] Departamento de Física, Facultad de Ciencias Básicas y Aplicadas, Universidad Militar Nueva Granada, Bogotá 110111, Colombia; anais.dorta@unimilitar.edu.co

[2] Departamento de Física, Universidad de los Andes, Apartado Aéreo 4976, Bogotá, Distrito Capital, Colombia

* Correspondence: p.bargueno@uniandes.edu.co

Received: 8 March 2019; Accepted: 16 April 2019; Published: 11 May 2019

Abstract: In this brief review, possible mechanisms which could lead to complete biological homochirality are discussed from the viewpoint of fundamental physics. In particular, the role played by electroweak parity violation, including neutrino-induced homochirality, and contributions from the gravitational interaction, will be emphasized.

Keywords: homochirality; parity violation; neutrinos; gravitation

---

1. Introduction

Life is not symmetric; i.e., left- and right-handed biological structures are not equivalent. In fact, there are almost only D-sugars and L-aminoacids in living systems. This remarkable fact is known as biological homochirality, this being one of the more intriguing fundamental problems of science for which an appropriate solution is still lacking [1]. Concerning possible routes which could led to complete homochirality, the idea of an extraterrestrial origin [2,3] for it has been reconsidered from the discovery of an enantiomeric excess of L-aminoacids in some meteorites [4]. Therefore, symmetry-breaking Earth-based mechanisms are actually not considered, these being superseded by universal mechanisms of chiral selection. Among these mechanisms, parity violation (PV) in (electro)weak interactions acquires special interest despite its tiny effects due to its ubiquity from particle physics to complex biological systems. We remark here that these effects have not been detected in molecular systems up till now, although several routes have been proposed in the past 40 years to succeed. Among the various proposals, here we remark on continuous efforts from several groups around the world, which include Quack [5], MacDermott [6,7], Chardonnet [8], Schwerdtfeger [9], Budker [10], DeMille [11,12], Hoekstra [13], Schnell [14] and Fujiki [15–17] groups and some proposals by Bargueño and coworkers [18–21] which were strongly influenced by the pioneering works of Harris [22]. In the context of blueautocatalysis and absolute asymmetric synthesis, the group of Soai has identified an interesting reaction [23] which was later interpreted by Lente [24] in the context of PV.

Concerning the evidence of the role played by PV effects in establishing biological homochirality, the works of MacDermott and coworkers have been decisive. They found [25] that the energy differences between two enantiomeric forms of all aminoacids found in the Murchison meteorite were negative due to PV (the so-called parity-violating energy differences (PVEDs)). Furthermore, they found intriguing correlations between the observed values for the enantiomeric excess (excess for the left enantiomer) and the calculated values for the PVEDs. Therefore, following these results, one could conclude that the PVED between enantiomers is, at least, consistent with the meteoritic enantiomeric excess [25]. At this point it is important to remark that an extremely small energy difference such as the PVED can only be interpreted statistically and it will not cause a deterministic excess of the favored enantiomer. Rather, it will cause a minor deviation from symmetry in the probability distribution, which has very important consequences as discussed, for example, in [26]. Among these consequences,

Lente concluded that the PVED is very unlikely to be relevant regarding the origin of homochirality, based on calculations at room temperature. However, if the temperature of the medium is very cold, as for instance in the interstellar medium, the PVED still persists as a valid candidate to produce complete enantioselection.

Interestingly, in a different context but also related to PV, neutrino-induced homochirality is being considered a plausible source for biological homochirality. From the early works of Cline [27,28], it has been suggested that neutrinos emitted in a supernova explosion could lead to certain amount of enantiomerism. Different suggestions, which explicitly depend on PV effects, involve the effects of cosmological neutrinos [29,30], neutrinos from supernovae [31,32], or even dark-matter candidates [30] on molecular electrons. In addition, there are some interesting works by Boyd and coworkers concerning a mechanism from creating aminoacid enantiomerism by taking into account the couplings of certain spins with the chirality of the molecules. In addition, for this mechanism to work, neutrinos and the magnetic field coming from the supernova progenitor should be considered [33–36].

Finally, we would like to remark that even though the electroweak force is the only one among the fundamental interactions that incorporates PV naturally, there are some interesting models that extend the usual gravitational theory (Newtonian or Einsteinian) by incorporating PV effects. Although their possible effects towards establishing complete enantioselection have not been considered until very recently [37], here we remark that some of the parity-violating extensions of general relativity proposed in [37,38] have been already tested [39], therefore paving the way for future experimental observations of gravity-induced homochirality.

The present work is intended to provide a brief review of the theoretical description, together with their experimental relevance, of the universal mechanisms described in this introduction which could be related to biological homochirality. Therefore, we will focus on electroweak- (including neutrino-) and gravitational PV.

2. Electroweak Parity Violation

One could think that both from the theoretical and from the experimental points of view, the main advances in basic questions (in physics) usually come hand in hand with high-energy physics. Although this is a generalized belief, here we will point out that this is not the general rule. However, fundamental importance should be given to very important and exciting achievements within the field of high-energy physics. The first symmetry violation was found by Wu [40] in the mid-1950s, after some pioneering theoretical works by Lee and Yang [41]. After that, PV was naturally incorporated into the Standard Model of Particle Physics (SMPP) by means of the electroweak unification developed by Glashow [42], Salam [43] and Weinberg [44], together with its corresponding renormalization by 't Hooft [45] and Veltman [46]. Coming back again to the experimental side of the history, the main ingredients of the SMPP were found by the discovery of the Z boson [47] and, finally, of the Higgs boson [48].

However, as first noticed in the 1970s at Novosibirsk, also table-top experiments could serve to ask big questions. Specifically, spontaneous optical activity of Bismuth atomic vapors was observed [49,50], extending the validity of the electroweak theory not only to the subatomic but to the atomic realm. After this important low-energy experiment, by improving low-temperature and high-resolution spectroscopic techniques, Wiemann and coworkers discovered the nuclear anapole moment of Cesium [51]. Here we remark that the anapole moment results from a parity-violating interaction between the nucleons and the electron. These and other low-energy experiments within PV are used presently in the main laboratories around the world to search for new physics beyond the SMMP [52–54]. Therefore, one can conclude that high-energy physics is not the only way of knowing Mother Nature. For a recent review, please see Ref. [55].

Therefore, we have arrived at a point where PV has been observed in several energy scales ranging from particles and nuclei to atoms. However, if we continue towards highly complex systems we find... molecules! Therefore, it is legitimate to ask: is there any role for PV in molecular systems?

Furthermore, could we gain valuable knowledge by studying it and by trying to observe it in the laboratory? In addition, finally, is the question: is there any connection between molecular PV and biological homochirality? Who knows?

2.1. Electron-Nucleon Interaction

What we already know is that with PV, there is a small enantiomeric energy difference between the corresponding molecular ground states, this being (mainly) due to the nuclear spin-independent interactions between nuclei and electrons [56]. Although these PVEDs are extremely small (of the order of 100 aeV for the two enantiomers of CHBrClF) [57,58], they are expected to be detected using different experimental techniques. Among them, we would like to point out rovibrational [8] and Mössbauer/NMR spectroscopies [9], dynamics in excited electronic states [5,59], spin–spin coupling [10], electronic spectroscopy [14] and a more recent technique that involves the use of cold molecules [13]. Finally, different proposals concerning measurements of the optical activity of a molecular sample with complete initial enantiomeric excess has been reported [18,19]. Despite all these efforts, no one has succeeded.

Up to this point we have mentioned the PVED several times. Now, it is time to define it. The PVED, ΔE^{ew}, between the L and R enantiomers is given by

$$\Delta E^{\mathrm{ew}} \equiv \langle L|V^{\mathrm{ew}}|L\rangle - \langle R|V^{\mathrm{ew}}|R\rangle = 2\langle L|V^{\mathrm{ew}}|L\rangle, \tag{1}$$

where V^{ew} is the electroweak parity-violating potential that uses a nonrelativistic approximation for the molecular electrons, reads [60,61]

$$V^{\mathrm{ew}} = \frac{G_F}{2\sqrt{2}m} \sum_{i=1}^{n} \sum_{A=1}^{N} Q_W(A) \{\mathbf{p}_i \cdot \mathbf{s}_i, \delta\,(\mathbf{r}_i - \mathbf{r}_A)\}. \tag{2}$$

Within this expression, G_F, Q_W and θ_W are Fermi's constant, the weak charge (corresponding to the considered nucleus), and Weinberg's angle, respectively. By m, $\mathbf{s}_i$ and $\mathbf{p}_i$ we denote the mass, spin, and momentum of the molecular electron. The delta function refers to the density of the nucleon, which has been considered to be point-like.

Please note that when only electromagnetic interactions are considered, as is usually done in molecular physics computations, the two enantiomers become degenerate and, thus, following simple energetic considerations, equally probable. However, this time, the molecular Hamiltonian contains a new term, given by Equation (2), which makes things very different. The most important point to remark here is the following:

The helicity operator, $h = \mathbf{s} \cdot \mathbf{p}$, is chiefly responsible for PV. This operator is P-odd, T-even and, therefore, PT-odd. Thus, following Barron's definition of what a truly chiral influence is [62–72], we see that h constitutes a universal truly chiral influence. Therefore, it lifts (as the PVED does, which in fact is based on the h operator) the degeneracy between enantiomers. Thus, if this small enantiomeric excess coming from P-odd effects could be amplified by some mechanisms such as, for example, the Kondepudi one [73] (for a review of amplification mechanisms with emphasis on stochastic models, see, for example, Ref. [74]) and references therein, at the levels seen in the Murchison meteorite, this would mean, at least, a big step towards establishing biological homochirality towards PV.

2.2. Electron-Neutrino Interaction

As previously mentioned in the introduction concerning the role of PV effects towards chiral selection, the electroweak-mediated interaction between both neutrino and dark-matter candidates (WIMPS) with molecular electrons have been reported in previous works [29–31]. In the first case, the interaction is also based on an interaction potential with depends on h but, in contrast with Equation (2), it crucially depends on the number-density difference between neutrinos and antineutrinos. In the

WIMP-mediated case, it depends on the number-density difference between left- and right-handed WIMPs [29–31].

As with the electron-nucleon interaction previously reviewed, one can obtain, for Dirac neutrinos and assuming nonrelativistic electrons, a P-odd potential which reads

$$V^{\nu-e} \sim \frac{G_F}{m_e}(n_\nu - n_{\bar{\nu}}) \sum_i \mathbf{p}_i \cdot \mathbf{s}_i. \tag{3}$$

As in the electron-nucleon case, this interaction causes a PVED between enantiomers because the helicity of a molecular electron has a different sign for each molecule depending of its chirality. Specifically, a surprisingly large energy split comparable with the thermal energy associated with the interstellar medium (10 K) was obtained when considering supernova neutrinos [31]. Therefore, although the model presented in Ref. [31] can be considerably improved, we think the large energy split between enantiomers due to supernova neutrinos is large enough to include it as a plausible mechanism for the origin of homochirality (we remark we are mainly reviewing the *origin* but not the *amplification* of homochirality). Concerning cosmological neutrinos and dark-matter candidates, the energy splits could reach, in the most favorable case, 10^{-21} eV [30].

3. Gravitational Parity Violation

The first ideas on gravitational PV appeared when Leitner and Okubo thought that if the weakness of the weak interaction had something to do with the violation of the parity symmetry, then, following the same reasoning, maybe there was some PV also present in the gravitational interaction [75]. After their proposal concerning a modified gravitational potential [75], Hari Dass extended it by writing a potential of the form ($c = 1$) [76]

$$V^{\mathrm{grav}}(r) = GM\left(\alpha_1 \frac{\mathbf{s}\cdot\mathbf{r}}{r^3} + \alpha_2 \frac{\mathbf{s}\cdot\mathbf{v}}{r^2} + \alpha_3 \frac{\mathbf{s}\times(\mathbf{r}\cdot\mathbf{v})}{r^3}\right). \tag{4}$$

In this equation, M stands for the mass of the gravitating object and $\mathbf{r}$ is its separation vector from a test particle whose spin and velocity are given by $\mathbf{s}$ and $\mathbf{v}$, respectively. It is interesting to note that under CPT conservation, only the α_2 term represents a true chiral interaction within this extension.

As pointed out in [37] and, as far as the author knows, the first (and only) application of PV within chiral molecules and the generalized gravitational potential of Equation (4) is Ref. [38]. The problem to compute the corresponding PVED between enantiomers, $\Delta E^{\mathrm{grav}} = 2\langle L|V^{\mathrm{grav}}|L\rangle$, is that α_2 is totally unknown. Despite this, what can be done is to put some bounds on the value of α_2 using non-conclusive experimental efforts towards establishing a clear signal of PV in chiral molecules ($\alpha_2 < 10^{17}$ [38]).

Although Leitner, Okubo, and Hari Dass's phenomenological ideas were appealing at that time, the quest for a complete quantum theory of gravitation has provided us with well-motivated physical mechanisms which naturally incorporate PV in the gravitational sector, as will be commented on in the next section.

3.1. Chern-Simons Modified General Relativity and Loop Quantum Gravity

Chern-Simons (CS) theory is a modified theory for gravity [77] that extends general relativity by including PV. This is done by considering not only the Einstein tensor (as usually done in general relativity) but also the C–tensor [78] and an extra pseudoscalar (as the h operator previously defined) field [77]. From the point of view of PV, one of the most important points to be remarked is that CS gives place to some kind of birefringence somehow analogous to its electromagnetic counterpart (left- and right-handed gravitational waves are selectively suppressed and, therefore, one could say that the CS theory has preference for a particular chirality) [79]. The interested reader can have a look at other signals of gravitational PV in Ref. [37] and references therein.

Regarding the experimental constraint for the CS energy scale (E_{cs}), which will be of interest when interpreted in terms of a possible enantioselection route, see Table 1. In view of these numbers, it is not surprising to say that CS effects remain elusive. However, we ask the interested reader to remain alert to the near future, in particular with relation to gravitational wave experiments.

Table 1. Experimental bounds for the CS energy scale. See text for details.

E_{cs} (eV)	Ref.	Method
10^{-14}	[80]	LAGEOS satellites
$5 \cdot 10^{-10}$	[81]	Double binary pulsar
10^{-14}	[82]	EMRIs

Other important candidates that incorporate P violation in the gravitational sector is Loop Quantum Gravity (LQG) [83–85], a theory which reconciles general relativity and quantum mechanics at the Planck scale. Without entering into mathematical details, here we note that there are some models within LQG [86] that give place to a nuclear spin-independent gravitational P-odd potential between electrons and nuclei of the form

$$V^{\mathrm{GPV}} = -\frac{9\pi\beta G_N}{2m} \sum_{i=1}^{n} \sum_{A=1}^{N} (Z+N)\{\mathbf{p}_i \cdot \mathbf{s}_i, \delta\left(\mathbf{r}_i - \mathbf{r}_A\right)\}. \tag{5}$$

Therefore, an *effective weak charge* appears when comparing Equations (2) and (5) [86] as

$$Q_\gamma = -9\pi\beta(Z+N)\frac{\sqrt{2}G_N}{G_F} \tag{6}$$

As the reader can see, the operator entered into Equation (5) is again h and, therefore, we have a short-ranged P-odd gravitational potential which constitutes a truly chiral influence.

3.2. *Gravitationally Selected Homochirality?*

As noted before, the comparison between the two charges, *weak* and *effective weak* of Equations (2) and (5) permits the opening of the way for treating PV in LQG as a possible candidate which could contribute to the selection of biological homochirality. However, extremely precise experimental constraints on β must be reported to finally see if the energy scale associated with it could reach the electroweak one (which is about 1 Hz $\simeq 10^{-14}$ eV). Concerning CS gravity, and as Table 1 shows, its corresponding energy scale could reach (or even supersede) the electroweak one. Therefore, CS gravity could also be also considered an interesting candidate towards establishing molecular homochirality.

4. Conclusions

In this work we have briefly reviewed possible ways to obtain complete biological homochirality for the point of view of fundamental physics. Emphasis has been given to electroweak, neutrino, and gravitational PV. Although the hypotheses here presented are well sustained from a theoretical point of view, specific calculations (quantum chemistry-like) would be desirable to test them. We remark that the work here presented refers to the origin but not to the amplification of molecular homochirality. In this sense, amplifications mechanisms adapted to the initial biases here described could be designed to see if the effects here presented remain realistic.

Author Contributions: Conceptualization, A.D.-U. and P.B.; investigation, A.D.-U. and P.B., writing—original draft preparation, A.D.-U. and P.B.

Funding: This research was funded by UNIVERSIDAD DE LOS ANDES grant number INV-2018-50-1378.

Acknowledgments: We thank Michiya Fujiki for his kind invitation to participate in this special issue on *Possible Scenarios for Homochirality on Earth*. Funding from Universidad de los Andes is acknowledged (P. B.). This work is dedicated to Lucía, Inés and Ana Bargueño-Dorta.

Conflicts of Interest: The authors declare no conflict of interest. The funders had no role in the design of the study; in the collection, analyses, or interpretation of data; in the writing of the manuscript, or in the decision to publish the results.

References

1. Guijarro, A.; Yus, M. *The Origin of Chirality in the Molecules of Life*; RSC Publishing: Cambridge, UK, 2009.
2. Engel, M.H.; Macko, S.A. Isotopic evidence for extraterrestrial non- racemic amino acids in the Murchison meteorite. *Nature* **1997**, *389*, 265–268. [CrossRef] [PubMed]
3. Pizzarello, S.; Huang, Y. The deuterium enrichment of individual amino acids in carbonaceous meteorites: A case for the presolar distribution of biomolecules precursors. *Geochim. Cosmochim. Acta* **2005**, *69*, 599–605. [CrossRef]
4. Cronin, J.R.; Pizzarello, S. Enantiomeric Excesses in Meteoritic Amino Acids. *Science* **1997**, *275*, 951–955. [CrossRef]
5. Quack, M. On the measurement of the parity violating energy difference between enantiomers. *Chem. Phys. Lett.* **1986**, *132*, 147–153. [CrossRef]
6. MacDermott, A.J.; Hegstrom, R.A. A proposed experiment to measure the parity-violating energy difference between enantiomers from the optical rotation of chiral ammonia-like "cat" molecules. *Chem. Phys.* **2004**, *305*, 55. [CrossRef]
7. MacDermott, A.J.; Hegstrom, R.A. Optical rotation of molecules in beams: The magic angle. *Chem. Phys.* **2004**, *305*, 47. [CrossRef]
8. Darquié, B.; Stoeffler, C.; Zrig, S.; Crassous, J.; Soulard, P.; Asselin, P.; Huet, T.R.; Guy, L.; Bast, R.; Saue, T.; et al. Progress toward a first observation of parity violation in chiral molecules by high-resolution laser spectroscopy. *Chirality* **2010**, *22*, 870–884. [CrossRef]
9. Lahamer, A.S.; Mahurin, S.M.; Compton, R.N.; House, D.; Laerdahl, J.K.; Lein ,M.; Schwerdtfeger, P. Search for a Parity-Violating Energy Difference between Enantiomers of a Chiral Iron Complex. *Phys. Rev. Lett.* **2000**, *85*, 4470. [CrossRef] [PubMed]
10. Ledbetter, M.P.; Crawford, C.W.; Pines, A.; Wemmer, D.E.; Knappe, S.; Kitching, J.; Budker, D. Optical detection of NMR J–spectra at zero magnetic field. *J. Magn. Reson.* **2009**, *199*, 25–29. [CrossRef]
11. DeMille, D.; Cahn, S.B.; Murphree, D.; Rahmlow, D.A.; Kozlov, M.G. Using Molecules to Measure Nuclear Spin-Dependent Parity Violation. *Phys. Rev. Lett.* **2008**, *100*, 023003. [CrossRef]
12. Altuntas, E.; Ammon, J.; Cahn, S.B.; DeMille, D. Demonstration of a Sensitive Method to Measure Nuclear-Spin-Dependent Parity Violation. *Phys. Rev. Lett.* **2018**, *120*, 142501. [CrossRef]
13. Quintero–Pérez, M.; Wall, T.E.; Hoekstra, S.; Bethlem, H.L. Preparation of an ultra–cold sample of ammonia molecules for precision measurements. *J. Mol. Spectrosc.* **2014**, *300*, 112–115. [CrossRef]
14. Schnell, M.; Meijer, G. Cold molecules: Preparation, applications, and challenges. *Angew. Chem. Int. Ed.* **2009**, *48*, 6010–6031. [CrossRef] [PubMed]
15. Fujiki, M. Experimental Tests of Parity Violation at Helical Polysilylene Level. *Macromol. Rapid Commun.* **2001**, *22*, 669. [CrossRef]
16. Fujiki, M. Mirror Symmetry Breaking in Helical Polysilanes: Preference between Left and Right of Chemical and Physical Origin. *Symmetry* **2010**, *2*, 1625–1652. [CrossRef]
17. Fujiki, M.; Koe, J.R.; Mori, T.; Kimura, Y. Questions of Mirror Symmetry at the Photoexcited and Ground States of Non-Rigid Luminophores Raised by Circularly Polarized Luminescence and Circular Dichroism Spectroscopy: Part 1. Oligofluorenes, Oligophenylenes, Binaphthyls and Fused Aromatics. *Molecules* **2018**, *23*, 2606. [CrossRef] [PubMed]
18. Bargueño, P.; Gonzalo, I.; de Tudela, R.P. Detection of parity violation in chiral molecules by external tuning of electroweak optical activity. *Phys. Rev. A* **2009**, *80*, 012110. [CrossRef]
19. Gonzalo, I.; Bargueño, P.; de Tudela, R.P.; Miret–Artés, S. Towards the detection of parity symmetry breaking in chiral molecules. *Chem. Phys. Lett.* **2010**, *489*, 127–129. [CrossRef]

20. Bargueño, P.; Pérez de Tudela, R.; Miret-Artés, S.; Gonzalo, I. An alternative route to detect parity violating energy differences through Bose-Einstein condensation of chiral molecules. *Phys. Chem. Chem. Phys.* **2011**, *13*, 806. [CrossRef]
21. Bargueño, P.; Sols, F. Macroscopic amplification of electroweak effects in molecular Bose-Einstein condensates. *Phys. Rev. A* **2012**, *85*, 021605(R). [CrossRef]
22. Harris, R.A.; Stodolsky, L. Quantum beats in optical activity and weak interactions. *Phys. Lett. B* **1978**, *78*, 313–317. [CrossRef]
23. Soai, K.; Sato, I.; Shibata, T.; Komiya, S.; Hayashi, M.; Matsueda, Y.; Imamura, H.; Hayase, T.; Morioka, H.; Tabira, H.; et al. Asymmetric synthesis of pyrimidyl alkanol without adding chiral substances by the addition of diisopropylzinc to pyrimidine-5-carbaldehyde in conjunction with asymmetric autocatalysis. *Tetrahedron Asymm.* **2003**, *14*, 185–188. [CrossRef]
24. Lente, G. Stochastic Interpretation of the Asymmetry of Enantiomeric Distribution Observed in the Absolute Asymmetric Soai Reaction. *Tetrahedron Asymm.* **2011**, *22*, 1595–1599. [CrossRef]
25. MacDermott, A.J.; Fu, T.; Nakatsuka, R.; Coleman, A.P.; Hyde, G.O. Parity–Violating Energy Shifts of Murchison L–Amino Acids are Consistent with an Electroweak Origin of Meteorite L–Enantiomeric Excesses. *Orig. Life Evol. Biosph.* **2009**, *39*, 459–478. [CrossRef] [PubMed]
26. Lente, G. Stochastic Analysis of the Parity-Violating Energy Differences between Enantiomers and Its Implications for the Origin of Biological Chirality. *J. Phys. Chem. A* **2006**, *110*, 12711–12713. [CrossRef] [PubMed]
27. Cline, D.B. (Ed.) *Proceedings of the 1st Symposium on the Physical Origins of Homochirality of Life, Santa Monica, CA, USA, February 1995*; AIP Press: Woodbury, NY, USA, 1996.
28. Cline, D.B. Supernova Antineutrino Interactions Cause Chiral Symmetry Breaking and Possibly Homochiral Biomaterials for Life. *Chirality* **2005**, *17*, S234. [CrossRef]
29. Bargueño, P.; Gonzalo, I. Effect of cosmological neutrinos on discrimination between the two enantiomers of a chiral molecule. *Orig. Life Evol. Biosph.* **2006**, *36*, 171–176. [CrossRef]
30. Bargueño, P.; Dobado, A.; Gonzalo, I. Could dark matter or neutrinos discriminate between the enantiomers of a chiral molecule? *EPL (Europhys. Lett.)* **2008**, *82*, 13002. [CrossRef]
31. Bargueño, P.; de Tudela, R.P. The role of supernova neutrinos on molecular homochirality. *Orig. Life Evol. Biosph.* **2007**, *37*, 253–257. [CrossRef]
32. Tsarev, V.A. Physical and Astrophysical Aspects of the Problem of Origin of Chiral Asymmetry of the Biosphere. *Phys. Part. Nucl.* **2009**, *40*, 998. [CrossRef]
33. Boyd, R.N.; Kajino, T.; Onaka, T. Supernovae and the Chirality of the Amino Acids. *Astrobiology* **2010**, *10*, 561. [CrossRef]
34. Boyd, R.N.; Kajino, T.; Onaka, T. Supernovae, Neutrinos and the Chirality of Amino Acids. *Int. J. Mol. Sci.* **2011**, *12*, 3432–3444. [CrossRef] [PubMed]
35. Famiano, M.; Boyd, R.; Kajino, T.; Onaka, T.; Koehler, K.; Hulbert, S. Determining Amino Acid Chirality in the Supernova Neutrino Processing Model. *Symmetry* **2014**, *6*, 909–925. [CrossRef]
36. Famiano, M.; Boyd, R.; Kajino, T.; Onaka, T. Selection of Amino Acid Chirality via Neutrino Interactions with ^{14}N in Crossed Electric and Magnetic Fields. *Astrobiology* **2018**, *18*, 1. [CrossRef] [PubMed]
37. Bargueño, P. Chirality and gravitational parity violation. *Chirality* **2015**, *27*, 375. [CrossRef]
38. Bargueño, P.; de Tudela, R.P. Constraining long–range parity violation in gravitation using high resolution spectroscopy of chiral molecules. *Phys. Rev. D* **2008**, *78*, 102004. [CrossRef]
39. Zhu, L.; Liu, Q.; Zhao, H.-H.; Gong, Q.-L.; Yang, S.-Q.; Luo, P.; Shao, C.-G.; Wang, Q.-L.; Tu, L.-C.; Luo, J. Test of the Equivalence Principle with Chiral Masses Using a Rotating Torsion Pendulum. *Phys. Rev. Lett.* **2018**, *121*, 261101. [CrossRef]
40. Lee, T.D.; Yang, C.N Question of parity violation in weak interactions. *Phys. Rev.* **1956**, *104*, 254–258. [CrossRef]
41. Wu, C.S.; Ambler, E.; Hayward, R.W.; Hoppes, D.D.; Hudson, R.P. An experimental test of parity conservation in beta decay. *Phys. Rev.* **1957**, *105*, 1413–1415. [CrossRef]
42. Glashow, S.L. Partial symmetries of weak interactions. *Nucl. Phys.* **1961**, *22*, 579–588. [CrossRef]
43. Weinberg, S. A model of leptons. *Phys. Rev. Lett.* **1967**, *19*, 1264–1266. [CrossRef]
44. Salam, A. Weak and electromagnetic interactions. In *Proceedings of the 8th Nobel Symposium, 15–19 May 1968*; Svartholom, N., Ed.; Almkvist und Wiksel: Stockholm, Sweden, 1968; pp. 367–377.

45. 't Hooft, G. iA onfrontation with infinity. *Rev. Mod. Phys.* **2000**, *72*, 333–339. [CrossRef]
46. Veltman, M.G.J. From weak interactions to gravitation. *Rev. Mod. Phys.* **2000**, *72*, 341–349. [CrossRef]
47. Groom, D.E.; Aguilar-Benitez, M.; Amsler, C.; Barnett, R.M.; Burchat, P.R.; Carone, C.D.; Caso, C.; Conforto, G.; Dahl, O.; Doser, M.; et al. Review of Particle Physics 2000. *Eur. Phys. J. C* **2000**, *15*, 1–878.
48. Statement from ATLAS. Available online: http://www.atlas.ch/news/2012/latest-results-from-higgs-search.html (accessed on 20 December 2012); Statement from CMS. Available online: http://cms.web.cern.ch/news/observation-new-particle-mass-125-gev (accessed on 12 December 2012); Aad, G.; et al. (ATLAS collaboration). Observation of a new particle in the search for the Standard Model Higgs Boson with the ATLAS detector at the LHC. *Phys. Lett. B* **2012**, *716*, 1–29; Chatrchyan, S.; et al. (CMS collaboration). Observation of a new Boson at a mass of 125 GeV with the CMS experiment at the LHC. *Phys. Lett. B* **2012**, *716*, 30–61.
49. Bouchiat, M.A.; Bouchiat, C. Parity violation induced by weak neutral currents in atomic physics. *J. Phys. (Fr.)* **1974**, *35*, 899–927. [CrossRef]
50. Khriplovich, I.B. *Parity Nonconservation in Atomic Phenomena*; Gordon and Breach: Philadelphia, PA, USA, 1991.
51. Wood, C.S.; Bennett, S.C.; Cho, D.; Masterson, B.P.; Roberts, J.L.; Tanner, C.E.; Wiemann, C.E. Measurement of Parity Nonconservation and an Anapole Moment in Cesium. *Science* **1997**, *275*, 1759–1763. [CrossRef]
52. Ginges, J.S.M.; Flambaum, V.V. Violations of fundamental symmetries in atoms and tests of unification theories of elementary particles. *Phys. Rep.* **2004**, *397*, 63–154. [CrossRef]
53. Langacker, P. The Physics of Heavy Z′ Gauge Bosons. *Rev. Mod. Phys.* **2009**, *81*, 1199–1228. [CrossRef]
54. DeMille, D.; Dyle, J.M.; Sushkov, A.O. Probing the frontiers of particle physics with tabletop-scale experiments. *Science* **2017**, *357*, 990. [CrossRef]
55. Safronova, M.S.; Budker, D.; DeMille, D.; Kimball, D.F.J.; Derevianko, A.; Clark, C.W. Search for new physics with atoms and molecules. *Rev. Mod. Phys.* **2018**, *90*, 025008. [CrossRef]
56. Bakasov, A.; Ha, T.K.; Quack, M. Ab initio calculation of molecular energies including parity violating interactions. *J. Chem. Phys.* **1999**, *109*, 7263–7285. [CrossRef]
57. Quack, M.; Stohner, J. Influence of parity violating weak nuclear potentials on vibrational and rotational frequencies in chiral molecules. *Phys. Rev. Lett.* **2000**, *84*, 3807–3810. [CrossRef]
58. Quack, M.; Stohner, J. Combined multidimensional anharmonic and parity violating effects in CDBrClF. *J. Chem. Phys.* **2003**, *119*, 11228–11240. [CrossRef]
59. Quack, M. Fundamental Symmetries and Symmetry Violations from High Resolution Spectroscopy. In *Handbook of High Resolution Spectroscopy*; Quack, M., Merkt, F., Eds.; Wiley: Chichester, UK; New York, NY, USA, 2011; Volume 1, Chapter 18, pp. 659–722.
60. Schwerdtfeger, P. *Computational Spectroscopy*; Grunenberg, J., Ed.; Wiley: Chichester, UK; New York, NY, USA, 2010; pp. 201–221.
61. Berger, R. Parity Violation Effects in Molecules. *Theor. Comput. Chem.* **2004**, *14*, 188–288.
62. Barron, L.D. Fundamental symmetry aspects of optical activity. *Chem. Phys. Lett.* **1981**, *79*, 392–394. [CrossRef]
63. Barron, L.D. Optical activity and time reversal. *Mol. Phys.* **1981**, *43*, 1395–1406. [CrossRef]
64. Barron, L.D. True and false chirality and absolute asymmetric synthesis. *J. Am. Chem. Soc.* **1986**, *108*, 5539–5542. [CrossRef]
65. Barron, L.D. Symmetry and molecular chirality. *Chem. Soc. Rev.* **1986**, *15*, 189–223. [CrossRef]
66. Barron, L.D. True and false chirality and parity violation. *Chem. Phys. Lett.* **1986**, *123*, 423–427. [CrossRef]
67. Barron, L.D. Reactions of chiral molecules in the presence of a time-non-invariant enantiomorphous influence: A new kinetic principle based on the breakdown of microscopic reversibility. *Chem. Phys. Lett.* **1987**, *135*, 1–8. [CrossRef]
68. Barron, L.D. Fundamental symmetry aspects of molecular chirality. In *New Developments in Molecular Chirality*; Mezey, P.G., Ed.; Kluwer Academic Publishers: Dordrecht, The Netherlands, 1991; pp. 1–5.
69. Avalos, M.; Babiano, R.; Cintas, P.; Jiménez, J.L.; Palacios, J.C.; Barron, L.D. Absolute asymmetric synthesis under physical fields: Facts and fictions. *Chem. Rev.* **1998**, *98*, 2391–2404. [CrossRef]
70. Barron, L.D. CP violation and molecular physics. *Chem. Phys. Lett.* **1994**, *221*, 311–316. [CrossRef]
71. Barron, L.D. Cosmic Chirality both True and False. *Chirality* **2012**, *24*, 957.

72. Barron, L.D. True and false chirality and absolute enantioselection. *Rend. Fis. Acc. Lincei* **2013**, *24*, 179–189. [CrossRef]
73. Kondepudi, D.K. Selection of molecular chirality by extremely weak chiral interactions under far from equilibrium conditions. *Biosystems* **1987**, *20*, 75. [CrossRef]
74. Lente, G. The Role of Stochastic Models in Interpreting the Origins of Biological Chiralit. *Symmetry* **2010**, *2*, 767–798. [CrossRef]
75. Leitner, J.; Okubo, S. Parity charge conjugation + time reversal in gravitational interaction. *Phys. Rev.* **1964**, *136*, B1542. [CrossRef]
76. Dass, N.D.H. Test for C, P, and T nonconservation in gravitation. *Phys. Rev. Lett.* **1976**, *36*, 393–395. [CrossRef]
77. Alexander, S.; Yunes, N. Chern–Simons modified general relativity. *Phys. Repts.* **2009**, *480*, 1–55. [CrossRef]
78. Jackiw, R.; Pi, S.Y. Chern-Ssimons modification of general relativity. *Phys. Rev. D* **2003**, *68*, 104012. [CrossRef]
79. Alexander, S.; Martin, J. Birefringent gravitational waves and the consistency check of inflation. *Phys. Rev. D* **2005**, *71*, 063526.
80. Smith, T.L.; Erickcek, A.L.; Caldwell, R.R.; Kamionkowski, M. The Effects of Chern–Simons gravity on bodies orbiting the Earth. *Phys. Rev. D* **2008**, *77*, 024015. [CrossRef]
81. Ali-Haimoud, Y. Revisiting the double–binary–pulsar probe of non–dynamical Chern–Simons gravity. *Phys. Rev. D* **2011**, *83*, 124050. [CrossRef]
82. Canizares, P.; Gair, J.R.; Sopuerta, C.F. Testing Chern–Simons Modified Gravity with Gravitational–Wave Detections of Extreme–Mass–Ratio Binaries. *Phys. Rev. D* **2012**, *86*, 044010. [CrossRef]
83. Rovelli, C. *Quantum Gravity*; Cambridge University Press: Cambridge, UK, 2004.
84. Ashtekar, A. Background Independent Quantum Gravity: A Status Report. *Class. Quantum Gravity* **2004**, *21*, R53. [CrossRef]
85. Thiemann, T. Lectures on loop quantum gravity. *Lect. Notes Phys.* **2003**, *631*, 41.
86. Freidel, L.; Minic, D.; Takeuchi, T. Quantum gravity, torsion, parity violation, and all that. *Phys. Rev. D* **2005**, *72*, 104002. [CrossRef]

Review

Electrochirogenesis: The Possible Role of Low-Energy Spin-Polarized Electrons in Creating Homochirality

Richard A Rosenberg

Advanced Photon Source, Argonne National Laboratory, Argonne, IL 60439, USA; rar@anl.gov

Received: 15 March 2019; Accepted: 8 April 2019; Published: 11 April 2019

Abstract: Electrochirogenesis deals with the induction of chirality by polarized electrons of which those with low energy (<15 eV) are seen to be the most effective. Possible sources of such electrons in the prebiotic universe are discussed and several examples where chiral induction by these electrons have been demonstrated are given. Finally, some possible scenarios where electrochirogenesis could have played a role in forming a chiral imbalance in a prebiotic setting have been speculated on and some possible future areas of research proposed.

Keywords: spin polarized electrons; homochirality; magnetism; prebiotic

---

1. Introduction

Pasteur first discovered the chiral nature of biological molecules in the 19th century. In the ensuing years numerous investigations have been devoted to trying to understand how chirality could have evolved in the prebiotic world. Despite all this effort, the editors of the 125th anniversary issue of *Science Magazine* in 2005 noted that among the 125 most important unanswered questions was the following:

> "What is the origin of chirality in nature? Most biomolecules can be synthesized in mirror-image shapes. Yet in organisms, amino acids are always left-handed, and sugars are always right-handed. The origins of this preference remain a mystery."

Fourteen years later, there is still no well accepted mechanism.

Numerous reviews of research in this area have been written including one in the present issue [1–8]. Mechanisms are generally classified as biotic or abiotic. Biotic theories presuppose that the evolution of living matter inevitably resulted in chiral selection and homogeneity, whereas abiotic theories presume that the origin of life requires the prior development of chirality. Bonner categorized abiotic mechanisms as, determinate, chance and amplification [2]. A determinate mechanism presupposes that interaction of relevant organic molecules with a chiral physical force [1] led to an enantiomeric excess (ee). Perhaps the most invoked determinate mechanism is photochirogenesis, where the interaction of prebiotic molecules with spin-polarized ultraviolet (UV) photons is conjectured to lead to an ee [9–12]. In the present paper we examine the possible role of spin-polarized electrons in prebiotic chemistry, electrochirogenesis.

The interaction of longitudinally spin-polarized electrons (SPEs) has long been considered as a possible determinate mechanism for creating a chiral imbalance through their interaction with a racemic mixture of chiral molecules. Although spin-polarized electrons are not chiral particles, per se, longitudinally spin-polarized electrons are chiral, since as they propagate they trace out a helical path. The spin direction determines the helicity. Initial research in this area was stimulated by Vester and Ulbricht who proposed that the high energy spin-polarized electrons produced by parity violation in nuclear β decay [13,14] could indirectly play a role (V–U hypothesis) [15,16]. They hypothesized that the irradiation of polarized electrons with matter produced circularly-polarized "Bremsstrahlen" photons, which subsequently induced stereoselective degradation or synthesis. Other researchers investigated the direct interaction of high energy SPEs with chiral molecules. Despite numerous investigations into both the V–U hypothesis and direct interaction of high-energy SPEs, no conclusive evidence was found to support either mechanism. Furthermore, Walker noted that the unpolarized secondary electrons produced by the high energy electrons would significantly dilute any chiral effects [17]. It is also worth noting that the high energy circularly-polarized "Bremsstrahlen" X-rays produced via the V–U hypothesis would have a very low absorption cross section for interacting with the low Z elements found in prebiotic molecules. Much of the research in this area was summarized in a 2011 review [18]. Although low energy electrons (<~15 eV) are very effective at inducing reactions in condensed molecular systems [19,20], the possible role of low energy SPEs at inducing chiral specific chemistry has been little discussed. In the present paper we will discuss recent progress in understanding the role of such low energy SPEs in creating an ee in an adsorbed chiral assembly.

2. Sources of Low-Energy Spin-Polarized Electrons

2.1. Valence-Level Photoionization by Ultraviolet (UV) Photons

In a ferromagnet the majority and minority valence electrons are unequally populated due to electron exchange interactions, which results in substantial spin polarization. Measurements on pure iron using 110 eV synchrotron radiation revealed polarizations as high as 45% [21]. Photoelectrons emitted from Ni single crystals show polarizations above 30% at threshold (5.2 eV) [22]. Even in the absence of no net magnetization, spin-orbit interactions can result in valence level polarized electrons. In the case of GaAs, spin polarizations of up to 40% have been observed following excitation with circularly polarized light in the range 1.5–3.6 eV [23]. As a result GaAs has found myriad uses as a source for SPEs in a variety of experiments. Polarizations as high as 15% have been found in gold using 6–9 eV CPL [24]. Since Au can also be used to self-assemble a variety of chiral molecules, it has been utilized as a substrate for spin-dependent transport measurements. Measurement using circular-, linear- and non- polarized light shown that there is an angular distribution in the emitted spin-polarized photoelectrons for a variety of non magnetic metals [25]. Using circularly polarized synchrotron radiation spin polarizations approaching 100% have been observed from adsorbed Kr and Xe excited by synchrotron radiation [26].

2.2. Secondary Electrons from a Magnetic Substrate

Any form of ionizing radiation (photons, electrons or ions) interacting with a solid will produce electrons. Primary electrons are emitted without any energy loss, but most electrons lose energy as a result of numerous collisions. The resulting secondary electrons (SEs) have very low energy, in the 0–10 eV range, as seen in a typical energy distribution curve shown in Figure 1b. In general, the electrons are unpolarized, but a high degree of polarization can be achieved if the material is ferromagnetic. The manner by which this is achieved is shown in Figure 1a. A ferromagnet has an imbalance between the number of majority and minority spin valence level electrons. The majority band in Figure 1a is totally filled while the minority band is partially filled. As the electrons scatter in the material, there is a higher density of minority band states available for occupation, so more electrons will scatter into them than into majority states. Hence the majority electrons have a longer mean free path than the minority ones, resulting in a net spin polarization of the former. Although subtleties exist in this model, it can be used to understand the basic process leading to SE polarization [27–31].

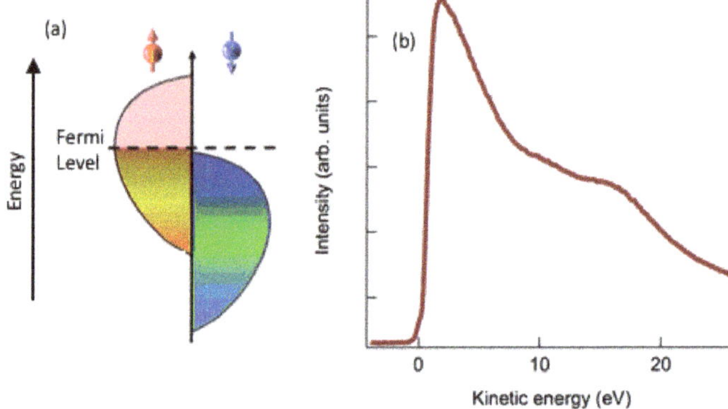

Figure 1. (**a**) Schematic illustration showing how spin polarization of secondary electrons in a ferromagnet is achieved. Preferential scattering of minority electrons (red) into empty states above the Fermi level leads to a decrease in those electrons and thus a net spin polarization of the majority electrons (blue). (**b**) Permalloy secondary electron yield curve following excitation by 1190 eV X-rays.

Numerous measurements of the SE polarization have been made on variety of substrates using ions, photons or electrons. Polarization can range between ~10 to 70%. For a tabulation of secondary electron yield studies the reader is referred to Table 1 in Ref. [18].

2.3. Chiral-Induced Spin Selectivity (CISS)

As mentioned earlier, gold is the substrate of choice for the self-assembly of organic molecules. This property has enabled Naaman and coworkers to study the production of spin-polarized electrons as a result of spin filtering of chiral molecules adsorbed on Au using both circularly and linearly polarized laser light [32–38]. As a result even initially unpolarized electrons can become polarized by filtering of an organized chiral assembly, with polarizations reaching as high as 70% [38]. The rationale for this effect may be understood in terms of an effective Lorentz force acting on the electron. For a chiral molecule, the moving electron experiences a centripetal force in the direction perpendicular to the electron's momentum. Furthermore, an effective magnetic field is produced along the axis of the chiral molecule, which introduces a spin-orbit coupling and thus a Zeeman splitting between the spin states of the moving electron. This results in a differential barrier height for the two spin states of the electron as it tunnels through the chiral bridge. Since the transmission probability depends exponentially on the barrier height, chiral molecules can act as very efficient spin filters. A number of recent articles have been devoted to detailing the mechanism by which this process occurs [39,40].

3. Chiral-Selective Chemistry via Low-Energy Spin-Polarized Electrons

Electrons have been shown to initiate chemical reactions in a variety of important biological molecules [41–43]. Dissociation primarily occurs by two main pathways. One possible dissociation process is a result of inelastic impact. A fast electron excites an electron in a molecule to a dissociative excited electronic state resulting in bond breaking. This is usually a minor process because the impact cross section is low, in the order of the absorption cross section, and there are relatively few electrons that have sufficiently high kinetic energy. The main route is called dissociative electron attachment (DEA). The incident electron becomes trapped in one of the lowest unoccupied molecular orbitals and forms a negative ion. Sometimes the incident electron also excites a second electron forming a one-hole, two-electron state or Feshbach resonance. These states are usually short lived and often these temporary negative ion states are dissociative, resulting in a negatively charged fragment and a neutral species. Dissociative electron attachment cross sections are about a factor of 100 or more higher than absorption cross sections in the hard and soft X-ray regimes and the cross section is maximum for *slow* electrons (<10 eV kinetic energy). As seen in Figure 1b there are a large number of secondary electrons in this energy range mainly because of inelastic scattering, which thermalizes the electron toward low energy, but also because such low energy electrons have a longer escape length so they are more likely to reach the interface region and react.

Since longitudinally SPEs are chiral particles, they should have chiral specific interactions with chiral molecules. The following sections will summarize recent experiments that demonstrate chiral specific reactions of adsorbed molecules initiated by SPEs.

3.1. Chiral-Specific Reactions Caused by Spin-Polarized Electrons from a Magnetic Substrate

Most experiments designed to see if a determinate mechanism can result in chiral specificity in a reaction start with a racemic mixture of molecules and then determine if the chiral physical force can impart an ee. The approach used in the work discussed here was to determine the reaction cross section, a more fundamental quantity, as a function of the chirality of the adsorbed molecule and the polarization of secondary electrons emitted from a magnetic substrate following X-ray irradiation. An easily magnetized permalloy ($Fe_{0.2}Ni_{0.8}$) substrate was chosen and a simple, model chiral molecule, (R)- or (S)-2-butanol ($CH_3CHOHC_2H_5$) was used as a reactant. Using X-ray photoelectron spectroscopy (XPS) changes in the intensity of the chemically shifted C–O peak were observed as a function of X-ray irradiation time. The chiral carbon on 2-butanol is bound to an –OH group. The intensity of the C–O peak decreases exponentially as a function of irradiation time as a result of cleavage of the C–O bond. The time constant (τ) (reciprocal of the reaction rate) was determined by using the simple kinetic relationship, $I = I_0 \exp(-t/\tau)$, where I_0 = initial C–O peak area; I = C–O peak area; t = time, as a function of the spin polarization (magnetization direction) of the substrate secondary electrons and the chirality of the adsorbed molecule.

Figure 2 shows a series of C 1s XPS spectra of ~2 monolayers of (S)-2-butanol adsorbed on an argon ion sputter cleaned Permalloy surface, cooled to 90 K, obtained sequentially during 1190-eV X-ray irradiation. The C–H and C–O peaks decrease in intensity, while there is an increase in the C-M (carbon-metal) intensity. Kinetic analysis was performed by tracking the intensity of the C–O peak, since it directly probes the state of the chiral carbon atom. The time dependence of the C–O component is shown in the inset in Figure 2. Data points are shown by the red dots and the solid line represents an exponential fit to the data. Such measurements were performed 6–8 times for each magnetization direction and chirality. Results of the analysis are summarized in Figure 3a. The average time constants (seconds), with error bars, for (R)-2-butanol are shown in red, while those for (S)-2-butanol are in blue and the magnetization direction is denoted by +/−. These results clearly demonstrate that a dramatic change in the reaction rate for a given chirality is obtained by a reversal of the secondary electron spin polarization. For (S)-2-butanol this difference is 8.9 ± 3.5%, while for (R)-2-butanol, it is 10.9 ± 3.8%. The results for each enantiomer are mirror images of the other which further validates the proposed mechanism.

The results in Figure 3a indicate an average difference between the time constants for the two enantiomers of ~10%. In Figure 3b results of the ee for an initial racemic mixture as a function of reaction time is shown. The time spent acquiring the data in Figure 2, was 6900 s which equals 4.3 time constants. After this much time there would be 1.4% of the initial concentration remaining and the ee would be 25%. The ee would be 12% with 14% of the initial population remaining after a time equivalent to 2 time constants. Such values for enantiomeric excesses are significantly higher than those reported for experiments involving irradiation by circularly-polarized light or high energy SPEs.

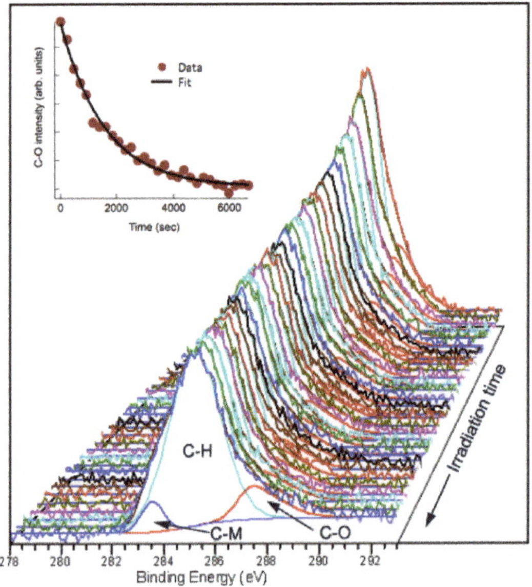

Figure 2. C 1s X-ray photoelectron spectroscopy (XPS) spectra of ~2 monolayers of (S)-2-butanol adsorbed on an argon ion sputter cleaned permalloy surface obtained sequentially during 1190-eV X-ray irradiation. The area of the C–O peak component as a function of irradiation time is shown in the inset. Data points are shown by the red dots and the solid line represents an exponential fit to the data.

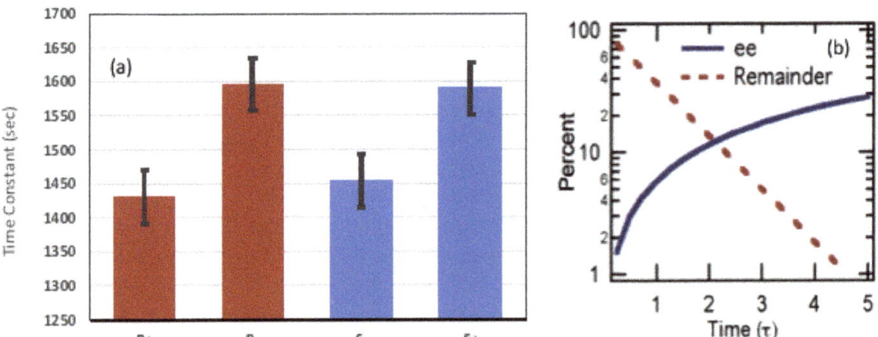

Figure 3. (**a**) Summary of the time constant results. The average time constants, with error bars, for (R)-2-butanol are shown in red, while those for (S)-2-butanol are in blue and the magnetization direction is denoted by +/–. (**b**) Calculation of the percentage enantiomeric excess (ee) (solid, blue line) and the percentage remaining concentration (dashed, red line) as a function of time, in terms of the dimensionless quantity (time constant (τ)/time (t)), based on a time constant difference of 10%. The right scale is a mirror of the left scale.

3.2. *Chiral-Specific Reactions Caused by Natural Selection of Spin-Polarized Electrons*

Section 2.3 discussed how the CISS mechanism can result in spin filtering of low energy (<1.5 eV) electrons by adsorbed chiral molecules resulting in spin polarizations as high as 70% [36,38]. Additional experimental and theoretical studies indicate that higher energy (<15 eV) electrons should be capable of filtering via the CISS effect as well [44]. When any substrate is subjected to ionizing radiation, secondary electrons will be produced. If chiral molecules are adsorbed on the surface then they could act as a spin filter for those initially unpolarized electrons. If additional chiral molecules are adsorbed on that layer then the SPEs should induce chiral specific reactions in those added molecules.

Figure 4 depicts as a schematic diagram of an experiment designed to test this idea. The additional chiral molecule was (R)- or (S)-epichlorohydrin (C_3H_5ClO, Epi), which was adsorbed on a self-assembled monolayer of 70 base-pair long double-stranded DNA (dsDNA) cooled to 90 K. The dsDNA layer acts as a spin filter for the initially unpolarized secondary electrons produced by X-ray irradiation of the gold substrate. These secondary electrons then react with adsorbed Epi and the reaction kinetics were determined by the following changes in the Cl 2p XPS spectra in a similar fashion to Section 3.1 and quantum yields (QYs) were determined. When the two enantiomers were adsorbed on bare Au, the QYs were the same, but when adsorbed on the dsDNA layer the QY for S-Epi was ~16% greater than for R-Epi [45].

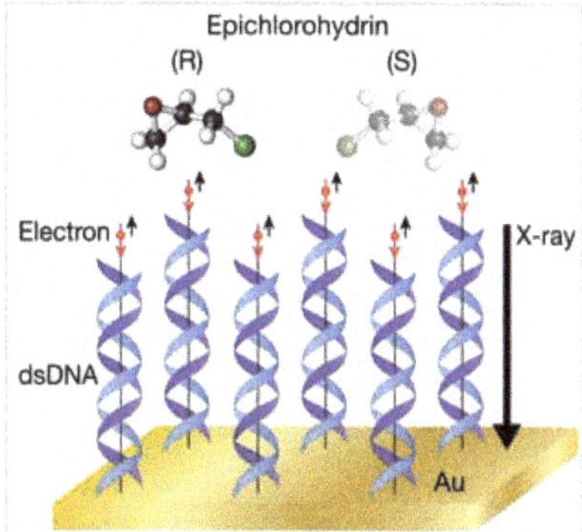

Figure 4. A schematic diagram showing how the Au secondary electrons produced by X-ray irradiation become spin polarized, with their spins aligned antiparallel to their velocity, and induce chiral selective chemistry in adsorbed (R)- or (S)- epichlorohdydrin. This figure was originally published in Ref. [45].

The above work shows that dsDNA can act as an effective spin filter for secondary electrons in the energy range 0–10 eV (Figure 1b). However, additional studies indicate that dsDNA does not effectively filter higher energy (>30 eV) electrons [44]. The cause for this can be understood by examining the theoretically predicted polarization for a chiral overlayer as a function of wave vector, k (E = 13.6 eV*k^2) shown in Figure 5. One can readily see that the relevant energy regime for chirality effects is k = 0 – 1.5 (0–30 eV) as shown in the inset and the effect decays very strongly at higher values [44].

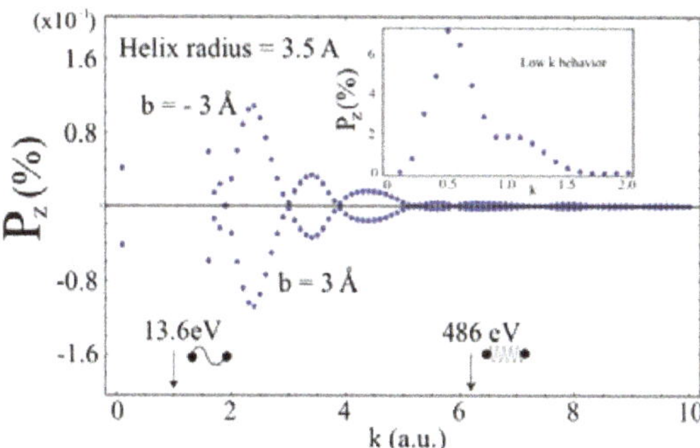

Figure 5. Predicted electron polarizations as a function of electron energy in terms of wave vector k (E = 13.6 eV*k^2) following scattering of electrons by a chiral overlayer. This figure was originally published in Ref. [44].

3.3. Chiral-Selective Adsorption

In a recent ground-breaking study, Banerjee-Ghosh, et al. demonstrated that the adsorption of a chiral molecule on a magnetic substrate is chiral specific [46]. They showed that if a substrate is magnetized in a particular direction one enantiomer adsorbs faster than the other, whereas if the magnetization is reversed, the opposite is true. This was determined by examining the initial adsorption rate of three different molecules: polyalanine, cysteine and DNA in solution, to a perpendicularly magnetized thin film covered with 5 nm of Au. In all cases the adsorption rate was dependent on the magnetization direction and the chirality.

In another recent paper Luque, et al. used XPS, ultraviolet photoelectron spectroscopy (UPS) and X-ray absorption spectroscopy (XAS) to monitor the bonding behavior of two different chiral molecules to a ferromagnetic thin film Co surface [47]. Using XAS they found that the ratio of the lowest unoccupied molecular orbital (LUMO) π^*/σ^* peaks for one enantiomer was significantly different than for the other, which indicates that the hybridization between the molecular orbitals and the surface is chiral-dependent. Previously it had been found that the density of LUMO orbitals in self-assembled DNA films can be related to the secondary electron-induced damage cross section [48]. Therefore, the differences in the density of LUMO levels as a function of chirality may indicate that there may be a chiral dependence of the reaction rate in this case. It is also interesting to note that Luque, et al. [47] using UPS, found a significantly higher secondary electron background for two different enantiomers adsorbed on the Co surface. This indicates that there is a higher interaction of the secondary electrons for one adsorbed enantiomer as opposed to the other, which should result in a higher damage cross section for the former [49].

4. Mechanisms

The explanation for the chiral selective adsorption on a magnetic substrate can be understood by referring to Figure 6. When a chiral molecule binds to a surface, charge transfer occurs resulting in electric dipole polarization and excess of electrons and holes on opposite ends of the molecule. This charge polarization is accompanied by spin polarization via a CISS type mechanism [39,50], with the spin orientation determined by the chirality of the molecule. The substrate–molecule interaction will be stabilized via an exchange interaction if the spin of the electron on the end of the molecule bound to the surface and that of the ferromagnet (FM) are antiparallel (low spin state, Figure 6a) and destabilized if they are parallel (high-spin state, Figure 6b). For further details see Refs. [40,46].

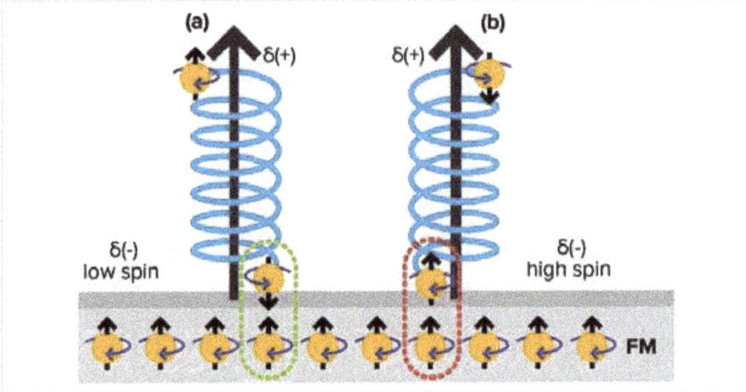

Figure 6. Schematic diagram depicting the spin-dependent interaction between a ferromagnet (FM) substrate and chiral molecules. In (**a**) is shown a low spin interface where the spins of the molecule's unpaired electron and that of the FM substrate are opposed whereas in (**b**) a high spin state is formed since the spins of the FM and the molecule are parallel.

In Sections 3.1 and 3.2 evidence was presented that demonstrated that low energy SPEs can produce chiral-specific reactions. By referring to Figure 6, symmetry arguments may be used to qualitatively understand the mechanism by which this may occur. The low energy secondary SPEs produced by irradiation of a magnetic substrate will have the same polarization as the majority spin electrons (spin up in Figure 6). The electron spin of the positively charged end of the chiral molecule is up in Figure 6a and down in Figure 6b. Thus, Pauli exclusion would favor the scattering of the free SPE into the same orbital as that of the electron in the positively charged end of the molecule in (b), which should stabilize the molecule. The opposite situation would exist for the chiral molecule in (a) so the SPE would tend to scatter into a higher lying orbital, producing a dissociative negative ion state via a Feshbach resonance [41]. Even in the absence of a strong adsorbate bond a chiral molecule will experience charge polarization and thus spin separation as a result of its dipole moment. Therefore, the SPEs produced by transmission through DNA (Section 3.2) will interact with an adsorbed chiral molecule in a similar fashion as discussed above, leading to chiral selective reactions.

5. Spin-Polarized Electrons in the Prebiotic Universe

Iron is one of the most common elements in the universe and numerous compounds based on it are magnetic. We have seen that irradiation of magnetic materials leads to low energy SPEs. Direct photoemission from a magnetic material by UV light will result in SPEs. If the UV light is circularly polarized, irradiation can result in SPEs from nonmagnetic material as well, although this effect is more dominant in higher Z materials.

A major component of the interstellar region is dust, which contains significant proportions of iron compounds, many of which are thought to be magnetic [51–54]. Analysis of meteorites has shown that there is a significant amount of iron- and nickel-based magnetic materials. Our nearest neighbor, Mars, is called the red planet due to the presence of various types of iron oxide in the soil, many of which have been determined to be magnetic [55]. Closer to home, numerous forms of magnetic materials are found on the earth's surface and magnetic iron sulfide is a major component of hydrothermal vents [56].

One can imagine numerous possible scenarios where an electrochirogenesis mechanism could play a role in the formation of chiral, prebiotic molecules. It has been shown that UV irradiation of basic condensed molecules, analogous to those that form on dust grains in interstellar molecular clouds, results in the formation of prebiotic molecules such as (racemic) amino acids [57,58]. Ionizing irradiation of magnetic components of the dust particles will yield low energy, secondary SPEs, which could selectively react with a particular enantiomer yielding an excess of the other. There are numerous sources of magnetic fields in the universe. In fact Greenberg and Bonner have posited that circularly polarized synchrotron light produced by relativistic electrons orbiting a neutron start could produce an ee by irradiation of molecules condensed on dust grains of a passing molecular cloud [59,60]. The electrons are circulating due to the enormous magnetic field produced by the neutron star. This field could also align the magnetic domains in the dust particles thereby producing secondary SPEs as a result of irradiation by the synchrotron light or electrons. Such low-energy secondary SPEs would be very effective at producing an ee from a racemic mixture of adsorbed chiral molecules. Furthermore, gas phase amino acids (glycine) and chiral molecules (propylene oxide) have been detected in the interstellar region [61,62], as well as in comets [63]. If a magnetized dust particle were to interact with a chiral molecule it would tend to accrete one enantiomer over another via a CISS type mechanism (Section 3.3) [46].

Analysis of the Murchison meteorite revealed the presence of a wide variety of prebiotic molecules and, most importantly, chiral amino acids [64–67]. Meteorites, including the Murchison, are known to contain significant proportions of iron and nickel in a variety of magnetic states [68–71]. It is quite possible that low-energy, secondary SPEs produced by irradiation of magnetic domains in the meteorite could have selectively reacted with a particular enantiomer in an initial racemic adsorbate mixture leading to an ee of the other.

In a recent article, researchers investigating the structure of Martian meteorites have found an inventory of organic carbon species and have hypothesized that interactions between spinel group minerals, sulfides and brine enabled the electrochemical reduction of aqueous CO_2 to organic molecules [72] on Mars. Previous studies have shown that electrochemical reactions can play a role in the chemistry of important prebiotic molecules, such as amino acids [73,74] and purines [75]. If the electrodes are magnetic then spin-dependent chemistry may occur [76]. The particular minerals thought to serve as the electrodes are titano-magnetite, magnetite, pyrite and pyrrhotite, which have magnetic properties. Therefore, somewhere during the reaction cycle, SPEs could have induced chiral selectivity in the reaction products.

Hydrothermal vent systems have been postulated to be potential sites for prebiotic chemistry on Earth and possibly Mars and Titan [77,78]. Wang has suggested that magnetic material such as Greigite (Fe_3S_4) present in the vents can act as a spin filter under the action of an external magnetic field to produce SPEs or that irradiation of ZnS in the vents by circularly polarized light could produce SPEs, which could induce the synthesis of chiral organic molecules [79]. Further down the evolutionary cycle, the electrons released from the coenzyme NAD(P)H of amino acid synthase could become spin-filtered and polarized when they pass through the chiral α-helix structure of the enzymes to the site of amino acid synthesis at the other end of the helix producing only "spin up" electrons. The SPEs induce the reductive reaction between ammonia and α-oxo acid. As a result of the Pauli exclusion principle only L-amino acids are produced [80].

6. Further Research

Electrochirogenesis using low energy SPEs is a relatively unexplored area of research. Although, numerous avenues for further studies exist, the following possibilities could be very illuminating:

- Determination of an ee following X-ray irradiation of a racemic mixture of chiral molecules adsorbed on a magnetic substrate. The X-rays will produce low-energy secondary SPEs which should then selectively react with one enantiomer, leaving an ee of the other. If the cross sections for the two enantiomers differ by ~10% then the plot in the inset of Figure 3 indicates that there should be an ee of ~12% after 2 time constants of irradiation with about 14% of the material remaining. Such differences are well within the sensitivity range of modern chromatography techniques.
- Perform a Miller–Urey type experiment with a magnetized electrode. The magnetized electrode will produce low energy SPEs in the discharge which should selectively react with one of the enantiomers formed in the process. This should yield an ee of the other enantiomer and reversing the magnetization should yield an opposite ee.
- Irradiation of molecules condensed on a magnetic substrate by UV light. Previous work has shown that racemic amino acids can be produced by UV irradiation of simple molecules condensed on a non-magnetic substrate [57,58]. If the same experiment were performed using a magnetic substrate, SPEs would be produced which could result in the direct formation of a chiral amino acid or the destruction of a particular enantiomer from an initially formed racemic mixture. Reversing the magnetization directions should result in an opposite ee.
- Theory: although there has been significant progress made in modelling the manner by which SPES are scattered by the gas phase [81] and adsorbed [82,83] chiral molecules, the mechanisms by which low energy SPEs cause chiral specific reactions are not well understood and could benefit from focused calculations using modern computational methods.

7. Conclusions

For more than 150 years scientists have been trying to determine what natural forces could have led to the prebiotic production of homochiral molecules which are essential for life as we know it. Electrons are the "glue" that hold molecules together, so it seems only natural that chiral electrons could have played a role. In this review we have discussed the possible role of electrochirogenesis, which deals with the induction of chirality by polarized electrons of which those with low energy (<15 eV) are seen to be the most effective. Possible sources of such electrons in the prebiotic universe have been discussed and several examples where chiral induction by such electrons have been demonstrated. Finally, some possible scenarios where electrochirogenesis could have played a role in forming a chiral imbalance in a prebiotic setting have been speculated about and some possible future areas of research proposed. This review is far from exhaustive and it is hoped that it will stimulate further activity in this field.

Funding: The work performed at the Advanced Photon Source was supported by the U.S. Department of Energy, Office of Science, Office of Basic Energy Sciences under contract No. DE-AC02-06CH11357.

Conflicts of Interest: The author declares no conflict of interest.

References

1. Avalos, M.; Babiano, R.; Cintas, P.; Jiménez, J.L.; Palacios, J.C.; Barron, L.D. Absolute asymmetric synthesis under physical fields: Facts and fictions. *Chem. Rev.* **1998**, *98*, 2391–2404. [CrossRef]
2. Bonner, W.A. The origin and amplification of biomolecular chirality. *Orig. Life Evol. Biosph.* **1991**, *21*, 59–111. [CrossRef] [PubMed]
3. Cintas, P. Chirality of living systems: A helping hand from crystals and oligopeptides. *Angew. Chem. Int. Ed.* **2002**, *41*, 1139–1145. [CrossRef]
4. Feringa, B.L.; Delden, R.A.v. Absolute asymmetric synthesis: The origin, control, and amplification of chirality. *Angew. Chem. Int. Ed.* **1999**, *38*, 3418–3438. [CrossRef]
5. Keszthelyi, L. Origin of the homochirality of biomolecules. *Q. Rev. Biophys.* **1995**, *28*, 473–507. [CrossRef] [PubMed]
6. Podlech, J. Origin of organic molecules and biomolecular homochirality. *Cell. Mol. Life Sci.* **2001**, *58*, 44–60. [CrossRef] [PubMed]
7. Tsarev, V. Physical and astrophysical aspects of the problem of origin of chiral asymmetry of the biosphere. *Phys. Part. Nucl.* **2009**, *40*, 998–1029. [CrossRef]
8. Davankov, A.V. Biological homochirality on the Earth, or in the universe? A selective review. *Symmetry* **2018**, *10*, 749. [CrossRef]
9. Cintas, P.; Viedma, C. On the physical basis of asymmetry and homochirality. *Chirality* **2012**, *24*, 894–908. [CrossRef] [PubMed]
10. Meierhenrich, U. *Amino Acids and the Asymmetry of Life*; Springer: Berlin/Heidelberg, Germany, 2008; p. 241.
11. Meierhenrich, U.J.; Thiemann, W.H.P. Photochemical concepts on the origin of biomolecular asymmetry. *Orig. Life Evol. Biosph.* **2004**, *34*, 111–121. [CrossRef]
12. Meinert, C.; de Marcellus, P.; d'Hendecourt, S.L.; Nahon, L.; Jones, N.C.; Hoffmann, S.V.; Bredehöft, J.H.; Meierhenrich, U.J. Photochirogenesis: Photochemical models on the absolute asymmetric formation of amino acids in interstellar space. *Phys. Life Rev.* **2011**, *8*, 307–330. [CrossRef] [PubMed]
13. Lee, T.D.; Yang, C.N. Question of parity conservation in weak interactions. *Phys. Rev.* **1956**, *104*, 254. [CrossRef]
14. Wu, C.S.; Ambler, E.; Hayward, R.W.; Hoppes, D.D.; Hudson, R.P. Experimental test of parity conservation in beta decay. *Phys. Rev.* **1957**, *105*, 1413. [CrossRef]
15. Ulbricht, T.L.V.; Vester, F. Attempts to induce optical activity with polarized β-radiation. *Tetrahedron* **1962**, *18*, 629. [CrossRef]
16. Vester, F.; Ulbricht, T.L.V.; Krauch, H. Optische aktivität und die paritätsverletzung im β-zerfall. *Naturwissenschaften* **1959**, *46*, 68. [CrossRef]
17. Walker, D.C. Leptons in chemistry. *Acc. Chem. Res.* **1985**, *18*, 167–173. [CrossRef]
18. Rosenberg, R.A. Spin-polarized electron induced asymmetric reactions in chiral molecules. In *Electronic and Mangetic Properties of Chiral Molecules and Supramolecular Architectures*; Naaman, R., Beratan, D.N., Waldeck, D.H., Eds.; Springer: Berlin/Heidelberg, Germany, 2011; Volume 298, pp. 279–306.
19. Arumainayagam, C.R.; Lee, H.-L.; Nelson, R.B.; Haines, D.R.; Gunawardane, R.P. Low-energy electron-induced reactions in condensed matter. *Surf. Sci. Rep.* **2010**, *65*, 1–44. [CrossRef]

20. Bass, A.D.; Sanche, L. Reactions induced by low energy electrons in cryogenic films. *Low Temp. Phys.* **2003**, *29*, 202–214. [CrossRef]
21. Xu, Y.B.; Greig, D.; Seddon, E.A.; Matthew, J.A.D. Spin-resolved photoemission of in situ sputtered iron and iron-yttrium alloys. *Phys. Rev. B* **1997**, *55*, 11442. [CrossRef]
22. Eib, W.; Alvarado, S.F. Spin-polarized photoelectrons from nickel single crystals. *Phys. Rev. Lett.* **1976**, *37*, 444–446. [CrossRef]
23. Pierce, D.T.; Meier, F. Photoemission of spin-polarized electrons from GaAs. *Phys. Rev. B* **1976**, *13*, 5484. [CrossRef]
24. Meier, F.; Pescia, D. Band-structure investigation of gold by spin-polarized photoemission. *Phys. Rev. Lett.* **1981**, *47*, 374. [CrossRef]
25. Heinzmann, U.; Dil, J.H. Spin–orbit-induced photoelectron spin polarization in angle-resolved photoemission from both atomic and condensed matter targets. *J. Phys. Condens. Matter* **2012**, *24*, 173001. [CrossRef] [PubMed]
26. Heinzmann, U. Angle-, energy- and spin-resolved photoelectron emission using circularly polarized synchrotron radiation. *Phys. Scr.* **1987**, *T17*, 77. [CrossRef]
27. Koike, K.; Kirschner, J. Primary energy dependence of secondary electron polarization. *J. Phys. D Appl. Phys.* **1992**, *25*, 1139. [CrossRef]
28. Penn, D.R.; Apell, S.P.; Girvin, S.M. Spin polarization of secondary electrons in transition metals: Theory. *Phys. Rev. B* **1985**, *32*, 7753. [CrossRef]
29. Penn, D.R.; Apell, S.P.; Girvin, S.M. Theory of spin-polarized secondary electrons in transition metals. *Phys. Rev. Lett.* **1985**, *55*, 518. [CrossRef]
30. Solleder, B.; Lemell, C.; Tőkési, K.; Hatcher, N.; Burgdörfer, J. Spin-dependent low-energy electron transport in metals. *Phys. Rev. B* **2007**, *76*, 075115. [CrossRef]
31. Tamura, K.; Yasuda, M.; Murata, K.; Koike, K.; Kotera, M. Analysis of the Spin polarization of secondary electrons emitted from permalloy polycrystals. *Jpn. J. Appl. Phys.* **1999**, *38*, 7173. [CrossRef]
32. Carmeli, I.; Leitus, G.; Naaman, R.; Reich, S.; Vager, Z. Magnetism induced by the organization of self-assembled monolayers. *J. Chem. Phys.* **2003**, *118*, 10372–10375. [CrossRef]
33. Naaman, R.; Vager, Z. Electron transmission through organized organic thin films. *Acc. Chem. Res.* **2003**, *36*, 291–299. [CrossRef]
34. Naaman, R.; Vager, Z. New electronic and magnetic properties emerging from adsorption of organized organic layers. *Phys. Chem. Chem. Phys.* **2006**, *8*, 2217–2224. [CrossRef] [PubMed]
35. Ray, K.; Ananthavel, S.P.; Waldeck, D.H.; Naaman, R. Asymmetric scattering of polarized electrons by organized organic films of chiral molecules. *Science* **1999**, *283*, 814–816. [CrossRef] [PubMed]
36. Ray, S.G.; Daube, S.S.; Leitus, G.; Vager, Z.; Naaman, R. Chirality-induced spin-selective properties of self-assembled monolayers of dna on gold. *Phys. Rev. Lett.* **2006**, *96*, 036101. [CrossRef] [PubMed]
37. Skourtis, S.S.; Beratan, D.N.; Naaman, R.; Nitzan, A.; Waldeck, D.H. Chiral control of electron transmission through molecules. *Phys. Rev. Lett.* **2008**, *101*, 238103-4. [CrossRef]
38. Göhler, B.; Hamelbeck, V.; Markus, T.Z.; Kettner, M.; Hanne, G.F.; Vager, Z.; Zacharias, H. Spin selectivity in electron transmission through self-assembled monolayers of double-stranded DNA. *Science* **2011**, *331*, 894–897. [CrossRef]
39. Michaeli, K.; Varade, V.; Naaman, R.; Waldeck, D.H. A new approach towards spintronics–spintronics with no magnets. *J. Phys. Condens. Matter* **2017**, *29*, 103002. [CrossRef]
40. Naaman, R.; Paltiel, Y.; Waldeck, D.H. Chirality and spin: A Different perspective on enantioselective interactions. *CHIMIA Int. J. Chem.* **2018**, *72*, 394–398. [CrossRef]
41. Alizadeh, E.; Orlando, T.M.; Sanche, L. Biomolecular damage induced by ionizing radiation: The direct and indirect effects of low-energy electrons on DNA. *Ann. Rev. Phys. Chem.* **2015**, *66*, 379–398. [CrossRef] [PubMed]
42. Boudaïffa, B.; Cloutier, P.; Hunting, D.; Huels, M.A.; Sanche, L. Resonant formation of DNA strand breaks by low-energy (3 to 20 eV) electrons. *Science* **2000**, *287*, 1658–1660.
43. Dugal, P.C.; Huels, M.A.; Sanche, L. Low-energy (5–25 eV) electron damage to homo-oligonucleotides. *Radiat. Res.* **1999**, *151*, 325–333. [CrossRef] [PubMed]
44. Rosenberg, R.A.; Haija, M.A.; Ryan, P.J. Chiral-selective chemistry induced by spin-polarized secondary electrons from a magnetic substrate. *Phys. Rev. Lett.* **2008**, *101*, 178301. [CrossRef] [PubMed]

45. Rosenberg, R.A.; Mishra, D.; Naaman, R. Chiral selective chemistry induced by natural selection of spin-polarized electrons. *Angew. Chem. Int. Ed.* **2015**, *54*, 7295–7298. [CrossRef] [PubMed]
46. Rosenberg, R.A.; Symonds, J.M.; Kalyanaraman, V.; Markus, T.; Orlando, T.M.; Naaman, R.; Mujica, V. Kinetic energy dependence of spin filtering of electrons transmitted through organized layers of DNA. *J. Phys. Chem. C* **2013**, *117*, 22307–22313. [CrossRef]
47. Banerjee-Ghosh, K.; Ben Dor, O.; Tassinari, F.; Capua, E.; Yochelis, S.; Capua, A.; Yang, S.H.; Parkin, S.S.P.; Sarkar, S.; Kronik, L.; et al. Separation of enantiomers by their enantiospecific interaction with achiral magnetic substrates. *Science* **2018**, *360*, 1331. [CrossRef]
48. Luque, F.J.; Nino, M.A.; Spilsbury, M.J.; Kowalik, I.A.; Arvanitis, D.; de Miguel, J.J. Enantiosensitive bonding of chiral molecules on a magnetic substrate investigated by means of electron spectroscopies. *CHIMIA Int. J. Chem.* **2018**, *72*, 418–423. [CrossRef]
49. Rosenberg, R.A.; Symonds, J.M.; Vijayalakshmi, K.; Mishra, D.; Orlando, T.M.; Naaman, R. The relationship between interfacial bonding and radiation damage in adsorbed DNA. *Phys. Chem. Chem. Phys.* **2014**, *16*, 15319–15325. [CrossRef] [PubMed]
50. Kumar, A.; Capua, E.; Kesharwani, M.K.; Martin, J.M.; Sitbon, E.; Waldeck, D.H.; Naaman, R. Chirality-induced spin polarization places symmetry constraints on biomolecular interactions. *Proc. Nat. Acad. Sci.* **2017**, *114*, 2474. [CrossRef]
51. Altobelli, N.; Postberg, F.; Fiege, K.; Trieloff, M.; Kimura, H.; Sterken, V.J.; Hsu, H.W.; Hillier, J.; Khawaja, N.; et al. Moragas-Klostermeyer, G.; et al. Flux and composition of interstellar dust at Saturn from Cassini's Cosmic Dust Analyzer. *Science* **2016**, *352*, 312. [CrossRef]
52. Brownlee, D.; Tsou, P.; Aléon, J.; Alexander, C.M.D.; Araki, T.; Bajt, S.; Borg, J. Comet 81P/wild 2 under a microscope. *Science* **2006**, *314*, 1711. [CrossRef]
53. Draine, B.T.; Hensley, B. Magnetic nanoparticles in the interstellar medium: Emission spectrum and polarization. *Astrophys. J.* **2013**, *765*, 159. [CrossRef]
54. Hoang, T.; Lazarian, A. Polarization of magnetic dipole emission and spinning dust emission from magnetic nanoparticles. *Astrophys. J.* **2016**, *821*, 91. [CrossRef]
55. Morris, R.V.; Golden, D.C.; Bell, J.F., III; Shelfer, T.D.; Scheinost, A.C.; Hinman, N.W.; Furniss, G.; Mertzman, S.A.; Bishop, J.L.; Ming, D.W.; et al. Mineralogy, composition, and alteration of Mars Pathfinder rocks and soils: Evidence from multispectral, elemental, and magnetic data on terrestrial analogue, SNC meteorite, and Pathfinder samples. *J. Geophys. Res. Planets* **2000**, *105*, 1757–1817. [CrossRef]
56. Tivey, M.A.; Dyment, J. The magnetic signature of hydrothermal systems in slow spreading environments. In *Slow Spreading Environments. In Diversity of Hydrothermal Systems on Slow Spreading Ocean Ridges*; Rona, P.A., Devey, C.W., Dyment, J., Murton, B.J., Eds.; American Geophysical Union: Washington, DC, USA, 2010; pp. 43–66.
57. Bernstein, M.P.; Dworkin, J.P.; Sandford, S.A.; Cooper, G.W.; Allamandola, L.J. Racemic amino acids from the ultraviolet photolysis of interstellar ice analogues. *Nature* **2002**, *416*, 401–403. [CrossRef] [PubMed]
58. Muñoz Caro, G.M.; Meierhenrich, U.J.; Schutte, WA.; Barbier, B.; Arcones Segovia, A.; Rosenbauer, H.; Thiemann, W.H.; Brack, A.; Greenberg, J.M. Amino acids from ultraviolet irradiation of interstellar ice analogues. *Nature* **2002**, *416*, 403–406. [CrossRef] [PubMed]
59. Bonner, W.A. Chirality and life. *Orig. Life Evol. Biosph.* **1995**, *25*, 175–190. [CrossRef]
60. Greenberg, J.M. Chirality in interstellar dust and in comets: Life from dead stars. *AIP Conf. Proc.* **1996**, *379*, 185–210.
61. McGuire, B.A.; Carroll, P.B.; Loomis, R.A.; Finneran, I.A.; Jewell, P.R.; Remijan, A.J.; Blake, G.A. Discovery of the interstellar chiral molecule propylene oxide (CH_3CHCH_2O). *Science* **2016**, *352*, 1449. [CrossRef]
62. Kuan, Y.J.; Charnley, S.B.; Huang, H.C.; Tseng, W.L.; Kisiel, Z. Interstellar glycine. *Astrophys. J.* **2003**, *593*, 848–867. [CrossRef]
63. Elsila, J.E.; Glavin, D.P.; Dworkin, J.P. Cometary glycine detected in samples returned by stardust. *Meteorit. Planet. Sci.* **2009**, *44*, 1323–1330. [CrossRef]
64. Cronin, J.R.; Pizzarello, S. Enantiomeric excesses in meteoritic amino acids. *Science* **1997**, *275*, 951–955. [CrossRef] [PubMed]
65. Pizzarello, S. The chemistry of life's origin: A carbonaceous meteorite perspective. *Acc. Chem. Res.* **2006**, *39*, 231–237. [CrossRef]

66. Pizzarello, S.; Huang, Y.; Alexandre, M.R. Molecular asymmetry in extraterrestrial chemistry: Insights from a pristine meteorite. *Proc. Natl. Acad. Sci. USA* **2008**, *105*, 3700–3704. [CrossRef] [PubMed]
67. Glavin, D.P.; Dworkin, J.P. Enrichment of the amino acid l-isovaline by aqueous alteration on CI and CM meteorite parent bodies. *Proc. Natl. Acad. Sci. USA* **2009**, *106*, 5487–5492. [CrossRef] [PubMed]
68. Bryson, J.F.; Nimmo, F.; Harrison, R.J. Magnetic meteorites and the early solar system. *Astron. Geophys.* **2015**, *56*, 4.36–4.42. [CrossRef]
69. Nagata, T. Meteorite magnetism and the early solar system magnetic field. *Phys. Earth Planet. Inter.* **1979**, *20*, 324–341. [CrossRef]
70. Pechersky, D.M.; Markov, G.P.; Tsel'movich, V.A. Pure iron and other magnetic minerals in meteorites. *Sol. Syst. Res.* **2015**, *49*, 61–71. [CrossRef]
71. BBryson, J.F.; Nichols, C.I.; Herrero-Albillos, J.; Kronast, F.; Kasama, T.; Alimadadi, H.; Harrison, R.J. Long-lived magnetism from solidification-driven convection on the pallasite parent body. *Nature* **2015**, *517*, 472. [CrossRef] [PubMed]
72. Steele, A.; Benning, L.G.; Wirth, R.; Siljeström, S.; Fries, M.D.; Hauri, E.; Needham, A. Organic synthesis on Mars by electrochemical reduction of CO_2. *Sci. Adv.* **2018**, *4*, eaat5118. [CrossRef] [PubMed]
73. Iwasaki, T.; Horikawa, H.; Matsumoto, K.; Miyoshi, M. An electrochemical synthesis of 2-acetoxy-2-amino acid and 3-acetoxy-3-amino acid derivatives. *J. Org. Chem.* **1977**, *42*, 2419–2423. [CrossRef] [PubMed]
74. Root, D.K.; Smith, W.H. Electrochemical behavior of selected imine derivatives, reductive carboxylation, α-amino acid synthesis. *J. Electrochem. Soc.* **1982**, *129*, 1231–1236. [CrossRef]
75. Smith, D.L.; Elving, P.J. Electrochemical reduction of purine, adenine and related compounds: Polarography and macroscale electrolysis. *J. Am. Chem. Soc.* **1962**, *84*, 1412–1420. [CrossRef]
76. Mondal, P.C.; Fontanesi, C.; Waldeck, D.H.; Naaman, R. Spin-dependent transport through chiral molecules studied by spin-dependent electrochemistry. *Acc. Chem. Res.* **2016**, *49*, 2560–2568. [CrossRef]
77. Farmer, J.D.; Des Marais, D.J. Exploring for a record of ancient Martian life. *J. Geophys. Res. Planets* **1999**, *104*, 26977–26995. [CrossRef]
78. Fortes, A.D. Exobiological implications of a possible ammonia–water ocean inside titan. *Icarus* **2000**, *146*, 444–452. [CrossRef]
79. Wang, W. Electron spin and the origin of Bio-homochirality II. Prebiotic inorganic-organic reaction model. *arXiv* **2014**, arXiv:1410.6555.
80. Wang, W. Electron spin and the origin of Bio-homochirality I. Extant enzymatic reaction model. *arXiv* **2013**, arXiv:1309.1229.
81. Musigmann, M.; Blum, K.; Thompson, D.G. Scattering of polarized electrons from anisotropic chiral ensembles. *J. Phys. B* **2001**, *34*, 2679. [CrossRef]
82. Medina, E.; López, F.; Ratner, M.A.; Mujica, V. Chiral molecular films as electron polarizers and polarization modulators. *EPL Europhys. Lett.* **2012**, *99*, 17006. [CrossRef]
83. Yeganeh, S.; Ratner, M.A.; Medina, E.; Mujica, V. Chiral electron transport: Scattering through helical potentials. *J. Chem. Phys.* **2009**, *131*, 014707. [CrossRef]

Article

Questions of Mirror Symmetry at the Photoexcited and Ground States of Non-Rigid Luminophores Raised by Circularly Polarized Luminescence and Circular Dichroism Spectroscopy. Part 2: Perylenes, BODIPYs, Molecular Scintillators, Coumarins, Rhodamine B, and DCM

Michiya Fujiki [1,*], Julian R. Koe [2,*] and Seiko Amazumi [1]

1 Division of Materials Science, Graduate School of Science and Technology, Nara Institute of Science and Technology (NAIST), 8916-5 Takayama, Ikoma, Nara 630-0192, Japan; amazumi@ms.naist.jp

2 Department of Natural Sciences, International Christian University (ICU), 3-10-2 Mitaka, Tokyo 181-8585, Japan

* Correspondence: fujikim@ms.naist.jp (M.F.); koe@icu.ac.jp (J.R.K.); Tel.: +81-743-72-6040 (M.F.); +81-422-33-3249 (J.R.K.)

Received: 2 February 2019; Accepted: 4 March 2019; Published: 11 March 2019

Abstract: We investigated whether semi-rigid and non-rigid π-conjugated fluorophores in the photoexcited (S_1) and ground (S_0) states exhibited mirror symmetry by circularly polarized luminescence (CPL) and circular dichroism (CD) spectroscopy using a range of compounds dissolved in achiral liquids. The fluorophores tested were six perylenes, six scintillators, 11 coumarins, two pyrromethene difluoroborates (BODIPYs), rhodamine B (RhB), and 4-(dicyanomethylene)-2-methyl-6-(4-dimethylaminostyryl)-4*H*-pyran (DCM). All the fluorophores showed negative-sign CPL signals in the ultraviolet (UV)–visible region, suggesting energetically non-equivalent and non-mirror image structures in the S_1 state. The dissymmetry ratio of the CPL (g_{lum}) increased discontinuously from approximately -0.2×10^{-3} to -2.0×10^{-3}, as the viscosity of the liquids increased. Among these liquids, C_2-symmetrical stilbene 420 showed $g_{\text{lum}} \approx -0.5 \times 10^{-3}$ at 408 nm in H_2O and D_2O, while, in a viscous alkanediol, the signal was amplified to $g_{\text{lum}} \approx -2.0 \times 10^{-3}$. Moreover, BODIPYs, RhB, and DCM in the S_0 states revealed weak (−)-sign CD signals with dissymmetry ratios (g_{abs}) $\approx -1.4 \times 10^{-5}$ at $\lambda_{\text{max}}/\lambda_{\text{ext}}$. The origin of the (−)-sign CPL and the (−)-sign CD signals may arise from an electroweak charge at the polyatomic level. Our CPL and CD spectral analysis could be a possible answer to the molecular parity violation hypothesis based on a weak neutral current of Z^0 boson origin that could connect to the origin of biomolecular handedness.

Keywords: circularly polarized luminescence; circular dichroism; symmetry breaking; parity violation; weak neutral current; tunneling; Z^0 boson; homochirality; precision measurement

---

1. Introduction

Since the mid-19th century, one of the greatest puzzles for scientists has been why life on Earth selected L-amino acids and D-sugars, because the corresponding L/D enantiomers were considered to be energetically identical [1–13]. Regarding other life forms that existed in the past or may exist now on exoplanets, solar planets, satellites, and comets, it is a matter of great curiosity as to whether these stereogenic centers and/or stereogenic bonds would be identical to those upon which our life is now based [14–16]. Living organisms can exist only in a far-from-equilibrium system allowing open flows of solar/thermal energy and low-entropy food [17]. This mystery is intimately connected to the origins of life [2–14] and the accelerating expansion of the universe [18,19].

In 1831, Faraday discovered that an oscillating magnetic (M) field induces an electric (E) current [20]. In 1861, Maxwell formulated this phenomenon as the theory of electromagnetism (EM) [20]. In 1845, Faraday observed a magneto-optical phenomenon, now called the "Faraday effect", in which a light–matter (LM) interaction causes linearly polarized (LP) light to be rotated clockwise (CW) and counterclockwise (CCW) (from the observer) by passing through achiral lead-containing glass and certain liquids under the influence of a static magnetic field. In 1860, Pasteur conjectured that molecular asymmetry is a consequence of dissymmetric forces of cosmological origin [1]. Possibly, the "Faraday effect" prompted Pasteur to attempt asymmetric crystallization under the influence of a magnetic field, but the quest failed [1]. In 1894, Curie considered that magnetic and electric fields need to align co-linearly or anti-colinearly to produce optically active substances [21].

Left (l)- or right (r)-handed circularly polarized (CP) light is a physical force carrying angular momentum ($\pm\hbar$) that is a pure chiral electromagnetic (EM) force, causing mirror-symmetric LM interactions [22]. Linearly polarized (LP) light is expressed as a superposition of r- and l-CP light. LeBel in 1874 and van't Hoff in 1894 postulated that CP light could catalyze asymmetric chemical reactions to produce chiral substances [21,23]. In 1896, Cotton, who discovered the anomalous phenomenon of circular dichroism (CD) and optical rotation dispersion (ORD) of potassium chromium (III) tartrate [24], attempted to degrade an alkaline aqueous solution of racemic copper tartrate using CP light but failed [21]. The first successful CP light-driven asymmetric synthesis, in the destruction mode, was reported by Kuhn and coworkers in 1929 and 1930, in which r- and l-CP light at 280 nm predominantly decomposed one enantiomer in racemic mixtures of 1-bromopropionic ethyl ester and azidopropionic acid dimethylamide, yielding corresponding optically active substances with small % enantioexcess (*ee*) values [21].

In recent years, modern photochemical reactions using unpolarized (UP) light under the influence of an intense static magnetic field afforded preferential degradation of one of two enantiomers, in which the product chirality is controllable according to the collinear or anti-collinear conditions [25]. Alternatively, CW and CCW swirling of molecular/supramolecular/polymer systems was found to result in mirror symmetry breaking (MSB) [26–28]; the product chirality was determined by the direction of the mechanical rotations, while no MSB happened under static conditions. The macroscopic mechano-physical rotation is assumed to impart a preferred twist direction to rotatable C–C single bonds. However, a recent experimental result suggested the occurrence of MSB even under static conditions via a thermal gradient at specific temperatures [29].

In 1927, Wigner formulated the principle of parity (P) symmetry (corresponding to mirror symmetry in chemistry), in which all interactions in nature are invariant with respect to space inversion [30]. This idea led to the categorization of seven symmetries, i.e., charge (C), P, time (T), CP, PT, CT, and CPT [31,32]. Until 1956, the seven symmetries were thought to be invariant and conserved. However, these ideas had to be partly revised because of two groundbreaking experiments: P-violated $\beta^{\pm}$-decay in 1957–1959 [33–38], and CP-violated decay from a neutral K^0 meson in 1964 [31,39] and neutral B^0 meson in 2001 [40,41]. Without doubt, P and CP symmetries were broken at subatomic levels, although CPT symmetry was conserved. More recently, an astonishing experiment conducted by the Tokai-to-Kamioka (T2K) particle physics team investigated the possibility of CP-symmetry breaking between neutrino and antineutrino due to generation mixing between an electron-like lepton (first generation) and a muon-like lepton (second generation) [42]. The idea of generation mixing is like the mixing between S_1 and triplet (T_1) states of luminophores [43,44]: the S_1 state involves a small fraction of the T_1 state and, conversely, the T_1 state is contaminated by a small fraction of the S_1 state, thereby permitting the occurrence of intersystem crossing [45]. Additionally, the mixing in degenerate coupling of three anthracene dimers in a double-well (DW) was detected as quantum coherence beats from a radiation process in the S_1 state at room temperature [46].

Currently, physicists concur that all events in the cosmos and material world are governed by the strong, EM, weak, and gravitational forces, known as the four fundamental physical forces [9,47]. Their relative strengths at 10^{-15} m are ~1:10^{-2}:10^{-13}:10^{-38}, respectively [47]. However, among these fundamental physical forces, the only cosmologically dissymmetric force, causing MSB, is the weak force, which is responsible for nuclear fusion and fission reactions, while the other three forces conserve symmetry. Hence, a circularly polarized light source is a P-conserved physical source [47]. In the 1980s, physicists at Conseil Europeen pour la Recherche Nucleaire or European Organization for Nuclear Research (CERN) succeeded in the detection of charged $W^{\pm}$ bosons (80.4 GeV) and the neutral Z^0 boson (91.2 GeV) [48–51]. This experiment proved that the P-conserved EM and P-violated weak forces are unified as an electroweak (EW) force with massive $W^{\pm}$ and Z^0 bosons and the massless photon (γ) according to the Weinberg–Salam theory [48]. $W^{\pm}$ and Z^0 bosons gain their masses from the Higgs boson (125 GeV) [48] while γ remains massless. Although $W^{\pm}$, Z^0, and γ are an equal family at high energies, Z^0/γ and $W^{\pm}$ bifurcate into neutral massive/massless and charged massive bosons, respectively.

Following 50 years of theoretical and experimental development, P- and CP-symmetry breaking in particle and atomic physics is now well established [51–60]. The EW force led to the further groundbreaking predictions theoretically [61–88] and experimentally [89–97] that paired L/D molecules are no longer enantiomers and should behave as diastereomers. To date, several theories invoke the P-violating weak neutral current (PV-WNC) via handed electron–nucleus interactions mediated by the Z^0 boson in the destabilization of one enantiomer by adding an extra energy bias (+E_{PV}) and, conversely, stabilizing the other by subtraction ($-E_{PV}$). This parity-odd energy bias is called a "parity violating energy difference" (PVED), ΔE_{PV}, called E_{PV}). The molecular parity violation (MPV) hypothesis definitively contradicts the accepted notion of enantiomers in the realm of modern stereochemistry and physical chemistry [98–100]. Molecular physicists have long argued whether P-symmetry of a molecular pair is *exactly* energetically equal and whether, if violated, E_{PV} is detectable [61–88].

However, the MPV theories teach us that E_{PV} between mirror-image molecules is very small: around 10^{-8}–10^{-14} kcal·mol^{-1} or 10^{-9}–10^{-15} % *ee* [61–88]. It is, thus, likely impossible that this radical hypothesis could be experimentally proven by ordinary UV–visible, infrared, microwave, or NMR spectrometers, or by enantioseparation column chromatography. If the potential barrier (E_b) between racemic molecules in a symmetrical DW is sufficiently small, the tunneling time between the two local minima is inversely proportional to the tunneling splitting energy, $\Delta E_{\pm}$ (hereafter called $E_{\pm}$), because of even- and odd-parity eigenstates [66,72,101–103].

In a previous paper [104] aiming to verify the MPV hypothesis experimentally, we used circularly polarized luminescence (CPL) and CD spectroscopy in an investigation of semi-rigid and non-rigid π-conjugated luminophores in symmetrical DW/multiple-well (MW) potentials with a smaller E_b in the lowest photoexcited (S_1) and ground (S_o) states. As we noted therein, a CPL spectropolarimeter may be regarded as a "low-energy spinning photon–molecule collider decelerator" to measure an inelastic scattering mode known as the Stokes' shift [104], allowing for the detection of the subtle difference between *l*- and *r*-handed light speeds and the radiative lifetimes of enantiomers in the S_1 state. We chose a series of luminophoric racemates, including oligofluorenes, linear and cyclic oligo-*p*-arylenes, binaphthyls, and fused aromatics carrying rotatable side groups. To control the E_b value in DW/MW of the luminophores in the S_1 and S_0 states, we used achiral solvents, including linear and branched alkanes, linear and branched alcohols, alkyl halides, linear cyclic ethers, and water (light and heavy) [104]. The solvent viscosity (η, in *cP*) was tunable, ranging from 0.21 to 71.0 at 20–25 °C. We observed that all the non-rigid luminophores showed negative-sign CPL signals in the UV–visible region, suggesting generation of non-mirror image structures in the S_1 state. The Kuhn's anisotropic parameter of CPL *vs.* photoluminescence (PL) signals (g_{lum}) of the non-rigid luminophores increased progressively but *discontinuously* in the range -0.2×10^{-3} to -2.0×10^{-3} as the solvent viscosity increased.

In this paper, to test the MPV hypothesis by further experiments, we investigated whether semi-rigid and non-rigid laser dyes, molecular scintillators, and other fluorophores in achiral liquids are mirror symmetrical by CPL and CD spectroscopy. Six perylenes, six scintillators, 11 coumarins, two 4,4-difluoro-4-bora-3a,4a-diaza-*s*-indacene derivatives (BODIPYs), rhodamine B (RhB), and 4-(dicyanomethylene)-2-methyl-6-(4-dimethylaminostyryl)-4*H*-pyran (DCM) were chosen. Negative-sign CPL signals were exhibited for all the fluorophores in the UV–visible region, suggesting the generation of non-mirror image structures in the S_1 state. Noticeably, BODIPY, RhB, and DCM in the S_0 states revealed clear (−)-sign CD signals with CD dissymmetry ratios (g_{abs}) of approximately -1.4×10^{-5} at λ_{ext}. The present comprehensive CPL/CD experimental datasets should support the long-argued MPV hypothesis regardless of PV-WNC scenarios [1].

2. Results

To validate the MPV hypothesis by CPL and CD spectroscopy, the crucial factors to choose are as follows: (i) semi-rigid and non-rigid racemic fluorophores carrying side chains allowing for rotatable freedom and/or flip-flop motional ability, (ii) fluorophores constituting only lighter atoms among the first three periods of the periodic table, and (iii) achiral liquids as solvents to continuously control the E_b value. Kasha's rule predicts that fluorescence occurs spontaneously from the lowest S_1 electronic-and-vibration coupling states (vibronic modes) associated with a significant structural reorganization at the photoexcited states via non-radiative, ro-vibrational, and translational pathways even if the fluorophores are excited at the S_2 and higher S_n states [43,44].

Previously, Quack et al. [72,80,84], MacDermott and Hegstrom [83], and Bargueño [86] argued three representative cases, (i) $E_{PV} >> E_{\pm}$, (ii) $E_{PV} << E_{\pm}$, and (iii) $E_{PV} \sim E_{\pm}$, for several rigid, semi-rigid, and non-rigid cases of molecular chirality. Rigid enantiomers consisting of tetrahedral stereocenters cannot interfere with the $E_{\pm}$ value due to the minuteness of the E_{PV} value. Quack et al. listed all the E_{PV} and $E_{\pm}$ values of nearly 20 non-rigid rotamers [80,84]; the sign and magnitude of E_{PV} in non-rigid XY–YX rotamers (X = H, D, T, Cl, and Y = O, S, Se, Te) definitively depend on the dihedral angles. The E_{PV} and $E_{\pm}$ values depend on the nature of the rotamers; the former changes by five orders of magnitude and the latter by considerably more (25 orders of magnitude). From this, an increase in the rotational barrier height of the rotamers may be inferred. Amongst the rotamers, only T_2S_2 can satisfy the $E_{PV} \sim E_{\pm}$ criterion (ideally $E_{PV} = E_{\pm}$), although radioactive T is not feasible in ordinary chemistry laboratories [82,84].

The PV-WNC model allows CPL-silent/CD-silent racemic molecular mixtures to become CPL-active/CD-active in the S_0 state. This model is also applicable to racemates in the S_1 state. The E_{PV} value of luminophores can be amplified by heavier atoms (e.g., Si, Ge, Sn, Pb, Se, Te, Cl, Br, I) in periods 3–7 of the Periodic Table obeying the V_{SO} ($\propto Z^2$) law [69,82,84]. However, luminophores containing such heavier atoms predominantly emit phosphorescence with a very low quantum yield (QY < 0.01). Herein, we focus on fluorophores without stereogenic centers, which are utilized as laser dyes and scintillators and have a high QY (typically, 0.1–0.9). Although these dyes and scintillators consist of only lighter C, N, O, F, and S atoms, their spin–orbit interactions (ζ) are non-zero and notably large with ζ = 0.1, 0.2, 0.4, 0.7, and 1.0 kcal·mol^{-1}, respectively [44]. If a huge number of fluorophoric molecules (>10^{10}–10^{16}) in a cuvette are photoexcited simultaneously by focusing on them an incident laser beam, the faint 10^{-9}–10^{-15} % *ee* or PVED of 10^{-8}–10^{-14} kcal·mol^{-1} is expected to be resonantly boosted to a level that is detectable using an ordinary CPL spectrometer during spontaneous radiation in a synchronized fashion [104].

When non-rigid enantiomers drop into one well in preference to the other in a dissymmetrical DW at the S_1/S_0 states by ceasing to oscillate, we postulate that the diastereomeric characteristics may accord with one of the three following scenarios [104]: (i) the fluorophore does not reveal any CPL or CD signals; (ii) the fluorophore at the S_1 state does not reveal CPL signals but, at the S_0 state, shows CD signals; and (iii) the fluorophore reveals CPL signals at the S_1 state and CD signals at the S_0 state.

Detectable signals, as (+)- or (−)-sign CPL or CD, are considered to arise due to handed rotational and/or flip-flop motions.

2.1. D_{2h}/D_2 Symmetrical Perylene and C_2/C_1 Symmetrical Derivatives

Firstly, to ascertain the achirality of unsubstituted rigid flat π-conjugated aromatics with D_{2h} symmetry such as the fluorophores naphthalene, anthracene, tetracene, and pyrene (see Chart 1), in the S_1 and S_0 states, we measured their CPL and CD spectra in the low-viscosity solvent methanol ($[\eta]$ = 0.55 *cP*) and several other solvents, since it was pointed out that artefact-free precision measurements are serious concerns if CPL and CD spectrometers are operated using a single 50-kHz photoelastic modulator (PEM) [105,106]. We confirmed that no obvious CPL or CD signals in the corresponding PL and UV–visible spectral regions are detected by our CPL-200 [JASCO (Hachioji, Tokyo, Japan) model CPL-200]] and CD (JASCO model J-820) spectrometers [104].

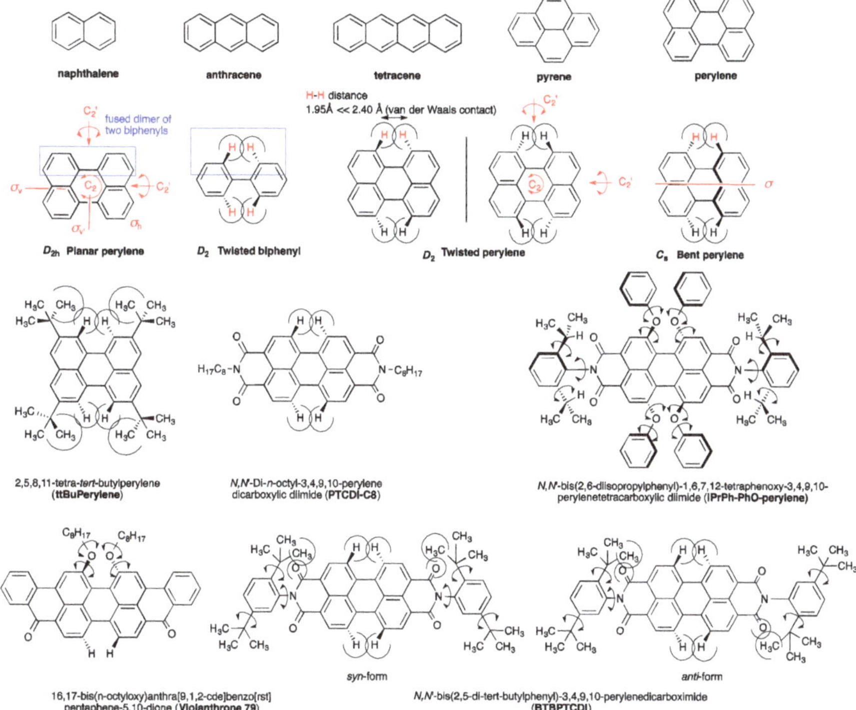

Chart 1. Chemical structures of five unsubstituted fused aromatics (naphthalene, anthracene, tetracene, pyrene, and perylene) and five perylene derivatives with substituents, 2,5,8,11-tetra-*tert*-butylperylene (ttBuperylene), *N,N*′-bis(2,6-diisopropylphenyl)-1,6,7,12-tetra-phenoxy-3,4,9,10-perylenetetracarboxylic diimide (iPrPh-PhO-perylene), *N,N*′-bis(2,5-di-*tert*-butyl-phenyl)-3,4,9,10-perylenedicarboximide (BTBPTCDI), 16,17-bis(*n*-octyloxy)-anthrax[9,1,2-cde]-benzo[*rst*]-pentaphene-5,10-dione (Violanthrone 79), and *N,N*′-di-*n*-octyl-3,4,9,10-perylenetetra- carboxylic diimide (PTCDI-C8).

Perylene has long been believed to adopt a D_{2h} symmetrical planar and achiral framework and is postulated as one of the polyaromatic hydrocarbons (PAHs) existing in molecular clouds of the interstellar universe [107]. In actuality, the interstellar PAHs emit infrared (IR) radiation in bright HII regions, and planetary and reflection nebulae. The interstellar IR spectral radiation upon excitation of vacuum–UV and UV–visible spectral lines (for example, Lyman and Balmer series) from ionized atomic hydrogen dominate most radiation sources of the galaxy and extragalaxies [108,109]. Our fundamental question, however, pertains to whether perylene truly remains achiral in the S_1 and S_0 states when all the hydrogen atoms attached to the framework are considered. To address this apparently naive query, we measured the CPL and CD spectra of unsubstituted perylene and five related derivatives carrying rotatable side groups, 5,8,11-tetra-*tert*-butylperylene (ttBuperylene), N'-bis(2,6-diisopropylphenyl)-1,6,7,12-tetraphenoxy-3,4,9,10-perylenetetracarboxylic diimide (iPrPhPhOperylene), N,N'-bis(2,5-di-*tert*-butylphenyl)-3,4,9,10-perylenedicarboximide (BTBPT-CDI), 16,17-bis(n-octyloxy)-anthrax[9,1,2-cde]-benzo[rst]-pentaphene-5,10-dione (Violanthrone 79), and N,N'-dioctyl-3,4,9,10-perylenedicarboxylic diimide (PTCDI-C8) (Chart 1).

It is possible to regard unsubstituted perylene as a fused dimer of l- and r-twistable biphenyl substructures; however, if there is a twist, then perylene should no longer exhibit D_{2h} symmetry but should exist as a mixture of D_2-symmetrical l- and r-twists and/or a C_S-symmetrical achiral folded framework. The structural hypothesis at the S_0 state is obvious for the five cases of (a) four bulky substituents in the 1-, 6-, 7- and 12-positions of iPrPh-PhO-perylene, (b) two bulky alkoxy substituents in the 16- and 17-positions in Violanthrone 79, (c) two bulky aromatic groups in the N,N'-positions of BTBPTCDI, (d) four bulky alkyl substituents in the 2-, 5-, 8- and 11-positions of ttBuperylene, and (e) two less-bulky alkyl groups in the N,N'-positions of PTCDI-C8. This query at the S_1 state is still unanswered.

Figure 1a–j display comparisons of the CPL/PL (photoluminescence) spectra of perylene and five derivatives in alcoholic solvents and chloroform at room temperature.

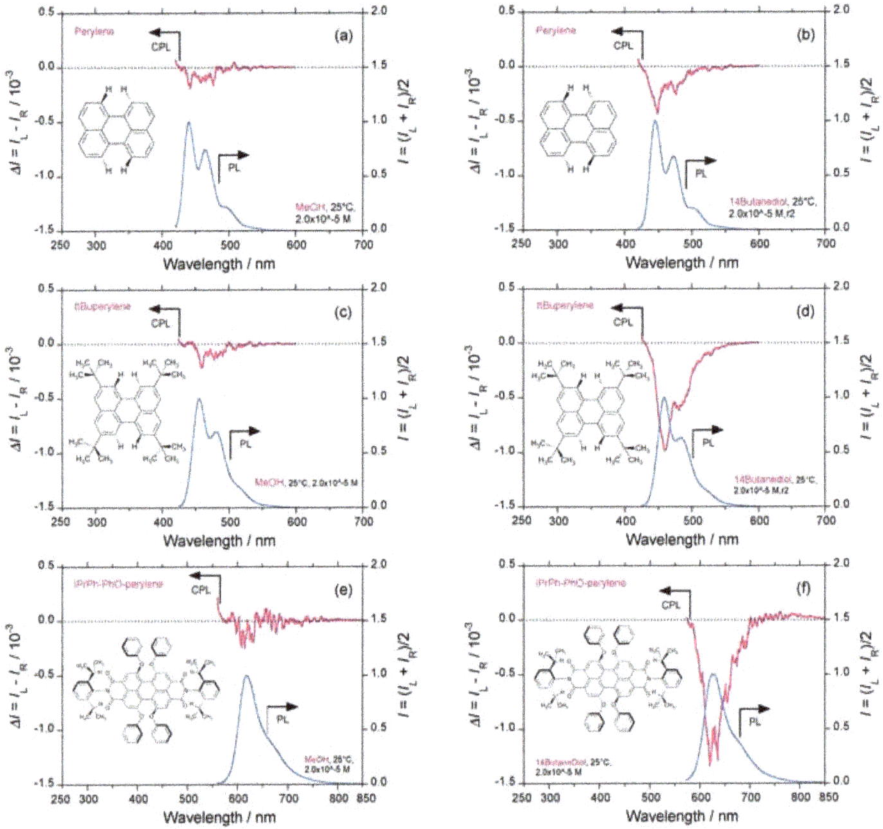

Figure 1. *Cont.*

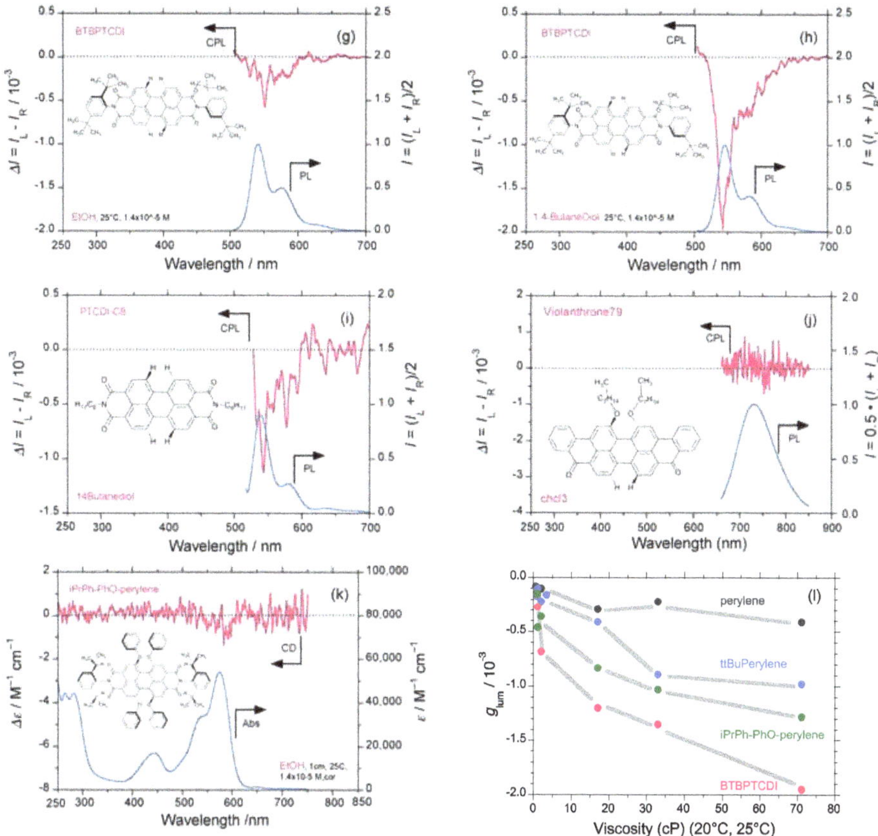

Figure 1. Comparison of circularly polarized luminescence/photoluminescence (CPL/PL) and circular dichroism (CD)/ultraviolet (UV)–visible spectra of perylene and five derivatives in alcoholic solvents at room temperature (path length: 10 mm, cylindrical cuvette, concentration 1–2 × 10^{-5} M. CPL/PL spectra of perylene excited at 390 nm in (**a**) methanol, and (**b**) 1,4-butanediol. CPL/PL spectra of ttBuperylene excited at 395 nm in (**c**) methanol, and (**d**) 1,4-butanediol. CPL/PL spectra of iPrPh-PhO-perylene excited at 470 nm in (**e**) methanol, and (**f**) 1,4-butanediol. CPL/PL spectra of BTBPTCDI excited at 525 nm in (**g**) methanol, and (**h**) 1,4-butanediol. (**i**) CPL/PL spectra of PTCDI-C8 excited at 490 nm in 1,4-butanediol. (**j**) CPL/PL spectra of Violanthrone 79 excited at 625 nm in chloroform. (**k**) CD/UV-visible spectra of iPrPh-PhO-perylene in methanol. (**l**) The g_{lum} value of perylene, ttBuperylene, iPrPh-PhO-perylene, and BTBPTCDI as a function of solvent viscosity.

From Figure 1a,b, unsubstituted perylene in low-viscosity methanol (η, 0.55 cP) reveals a weak vibronic CPL signal at the corresponding PL emission at approximately 400–500 nm. The vibronic CPL band becomes more obvious, and g_{lum} reaches -0.45×10^{-3} at the first vibronic band when methanol is replaced with the more viscous 1,4-butanediol (η, 71.0 cP). The magnitude of the vibronic CPL band at the 0–0 and 0–1 peaks increases as solvent viscosity increases (Figure 1l and Figure S1a–f, Supplementary Materials). However, CD signals at the corresponding UV–visible bands cannot be distinguished due to π–π* transitions (Figure S1u, Supplementary Materials). These results imply that perylene in the S_1 state temporarily adopts a chiral twisted geometry. However, the observed chirality of perylene disappears in the S_0 state. Possibly, perylene in the S_0 state exists as a mixture of l- and r-twisted geometries, thereby resulting in a CD-silent pair of opposite chirality twisted conformers. Photoexcited perylene may be optically active.

Research on biaryl-sensitized terbium (III) complexes showed that fluorescence lifetime enhancement in these systems is due to solvent polarity and oxygen sensitivity rather than viscosity effects [110]. Nevertheless, oxygen solubilization does not significantly influence the fluorescence quantum yield of most organic luminophores including laser dyes, but does significantly suppress phosphorescence. Solvent viscosity is, therefore, considered to be a critical factor in the fluorescence lifetimes in our systems.

In ttBuperylene, when the four hydrogen atoms in the 2-, 5-, 8- and 11-positions of the perylene framework are replaced by four bulky three-fold symmetric *tert*-butyl groups as rotors, (−)-sign CPL signals for the 0–0 and 0–1 peaks at approximately 450–500 nm in methanol, 1,4-butanediol, and other solvents are more clearly evident (Figure 1c–d and Figure S1g–k, Supplementary Materials). Similarly, in *i*PrPh-PhO-perylene, when the four isopropyl groups in the 2-, 5-, 8-, and 11-positions of the perylene framework are replaced by four two-fold symmetric sterically hindered phenoxy groups as rotors, the (−)-sign CPL signals redshift to 530–570 nm and are without doubt more obvious in methanol, 1,4-butanediol, and other solvents (Figure 1e–f and Figure S1p–t, Supplementary Materials).

Similarly, BTBPTCDI, with two sterically hindered phenyl groups in the *N,N*′-positions of the perylene diimide framework unambiguously exhibits (−)-sign vibronic CPL signals at approximately 550–590 nm in methanol, 1,4-butanediol, and other solvents (Figure 1g–h and Figure S1l–n, Supplementary Materials). PTCDI-C8, bearing two less bulky *n*-octyl groups in the *N,N*′-positions of the perylene diimide framework, reveals obvious (−)-sign vibronic CPL signals at approximately 540–580 nm in 1,4-butanediol and *n*-dodecane (Figure 1i and Figure S1o, Supplementary Materials). CPL signals for Violanthrone 79 in chloroform were not detected (Figure 1j). A faint (−)-sign CD signal for iPrPh-PhO-perylene in methanol may be seen on the order of $g_{abs} \approx 10^{-6}$, but it is not obvious (Figure 1k). No detectable CD signals for perylene and ttBuperylene were observed (Figure S1f,g, Supplementary Materials).

Figure 1l summarizes the g_{lum} values of perylene, ttBuperylene, iPrPh-PhO-perylene, and BTBPTCDI as a function of solvent viscosity for the solvents methanol, ethanol (1.1 *cP*), *n*-propanol (2.0 *cP*), *n*-undecanol (17.0 *cP*), 1,3-propandiol (33.0 *cP*), and 1,4-butanediol (71.0 *cP*). We can thus conclude that the perylene and perylene diimide frameworks in the S1 state adopt a twisted geometry due to steric repulsion in the 2-, 5-, 8-, and 11-positions of the perylene framework. CPL signals of (−)-sign are apparent, and the g_{lum} value reaches a maximum of -2.0×10^{-3}. Perylene, thus, does not adopt an achiral framework in the S_1 state, and the same is possibly true for the S_0 state.

To see the effect of twisted perylene in snapshot mode, we simulated the CD/UV–visible spectra with 0.20 eV full width at half maximum (fwhm) and electron density mapping at (c) the first first lowest unoccupied molecular orbital (1st LUMO) and (d) the first highest occupied molecular orbital (1st HOMO) for a hypothetical model of perylene twisted by 30° (Figure 2a–d). The twisted perylene clearly shows negative CD spectra at 443 nm and bisignate CD bands at approximately 300 nm. The value of g_{abs} at 443 nm is found to be 4×10^{-4}. A closed-chiral-loop current (reddish zone) (Figure 2c) is obvious for the LUMO, while the same is not true for the HOMO (Figure 2d). The closed-loop current may interfere with the one-handed chiral WNC postulated.

2.2. Rigid Luminophores Bearing Multiple Three-Fold Symmetrical Alkyl Substituents

C_{2v}-symmetrical pyrromethene-difluoroborate (BODIPY) and its derivatives are established as excellent emitters with a high QY and a very small Stokes' shift (350–500 cm^{-1}) [111,112]. Incorporation of alkyl groups into BODIPYs improves their solubility in common organic solvents. It is conceivable that certain BODIPYs carrying three-fold symmetrical alkyl groups may reveal optical activity due to handed gear motions between these alkyl groups. We, therefore, designed several C_{2v} BODIPYs as candidates to address the question of whether the handed gear motions occur with exactly equal energies and opposite senses [113]. If the gear-like motions in CW and CCW directions at the S_0 and S_1 states are equally operational, CD and CPL signals will not be detectable due to mutual cancellation of the opposite chiroptical signed bands, but they will have the same absolute magnitudes. However, if the gear-like motions in the S_0 and/or S_1 states are occurring unidirectionally due to unequal intramolecular gear energies, this should be detectable as CD and CPL signals. Unidirectional gear-like motions drive unidirectional molecular motors.

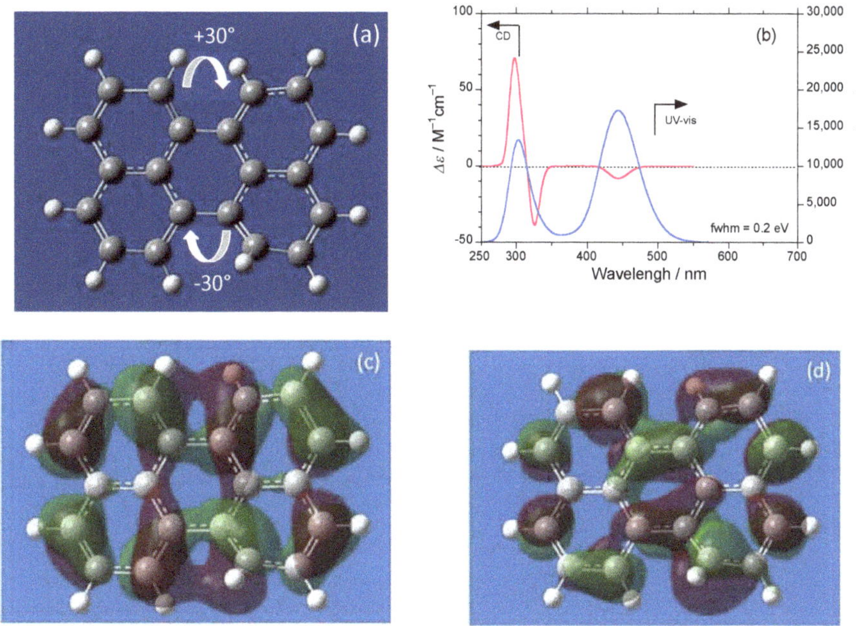

Figure 2. (**a**) Hypothetical model of perylene twisted by 30°; (**b**) simulated CD/UV–visible spectra with 0.2 eV full width at half maximum (fwhm) and electron density mapping at (**c**) first lowest unoccupied molecular orbital (1st LUMO) and (**d**) first highest occupied molecular orbital (1st HOMO). For this time dependent density functional theory (TD-DFT) and Becke, three-parameter, Lee-Yang-Parr exchange-correlation functional (B3LYP) with 6-31G(d) basis set, Gaussian09 rev D.01 (GaussView5 package)-calculated structures, a closed-chiral-loop current (red) is obvious for the LUMO, although this is not the case for the HOMO.

Firstly, we tested the gear-motion behaviors of pyrromethene 597 (BODIPY 597) and pyrromethene 546 (BODIPY 546) (Chart 2), which both have five three-fold symmetry methyl groups in the 1-, 3-, 5-, 7- and 8-positions of the BODIPY frameworks. Pyrromethene 597 has two additional three-fold symmetrical *tert*-butyl groups in the 4,4′-positions. The *tert*-butyl group itself consists of three methyl groups with three-fold symmetry. Both methyl and *tert*-butyl groups are assumed to act as gears.

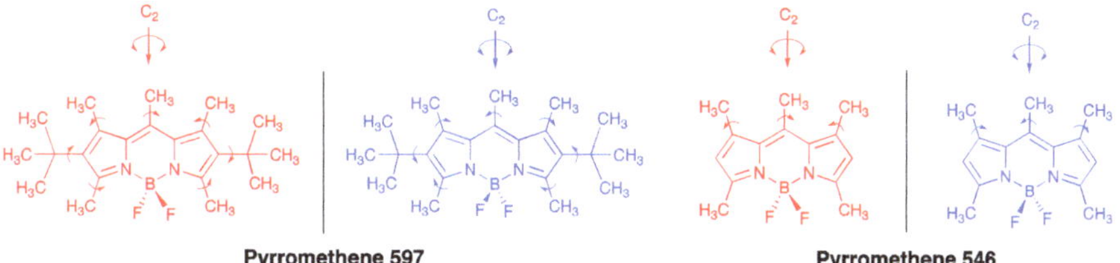

Chart 2. Chemical structures of [[(4-*tert*-butyl-3,5-dimethyl-1*H*-pyrrol-2-yl)(4-*tert*-butyl-3,5-dimethyl -2*H*-pyrrol-2-ylidene)methyl]methane](difluoroborane) (pyrromethene 597) and [[(3,5-dimethyl-1*H*-pyrrol-2-yl)3,5-dimethyl-2*H*-pyrrol-2-ylidene]methyl]methane](difluoroborane) (pyrromethene 546).

Pyrromethene 597 (Chart 2) in methanol showed weakly vibronic CPL bands at 550–600 nm; however, in 1,4-butanediol, these were amplified significantly to $g_{\mathrm{lum}} = -1.0 \times 10^{-3}$ at 562 nm (Figure 3a,d). Similarly, pyrromethene 546 (Chart 2) in methanol showed weakly vibronic CPL bands; however, in 1,4-butanediol, these bands were magnified to $g_{\mathrm{lum}} = -0.4 \times 10^{-3}$ at 515 nm (Figure 3g,h). Pyrromethene 597 in *n*-undecanol, ethylene glycol, and other solvents showed similar vibronic CPL bands at 550–600 nm (Figure 3b,c). These (−)-sign CPL characteristics of pyrromethenes 597 and 546 depend on the nature of solvents (Figure S2a–h, Supplementary Materials).

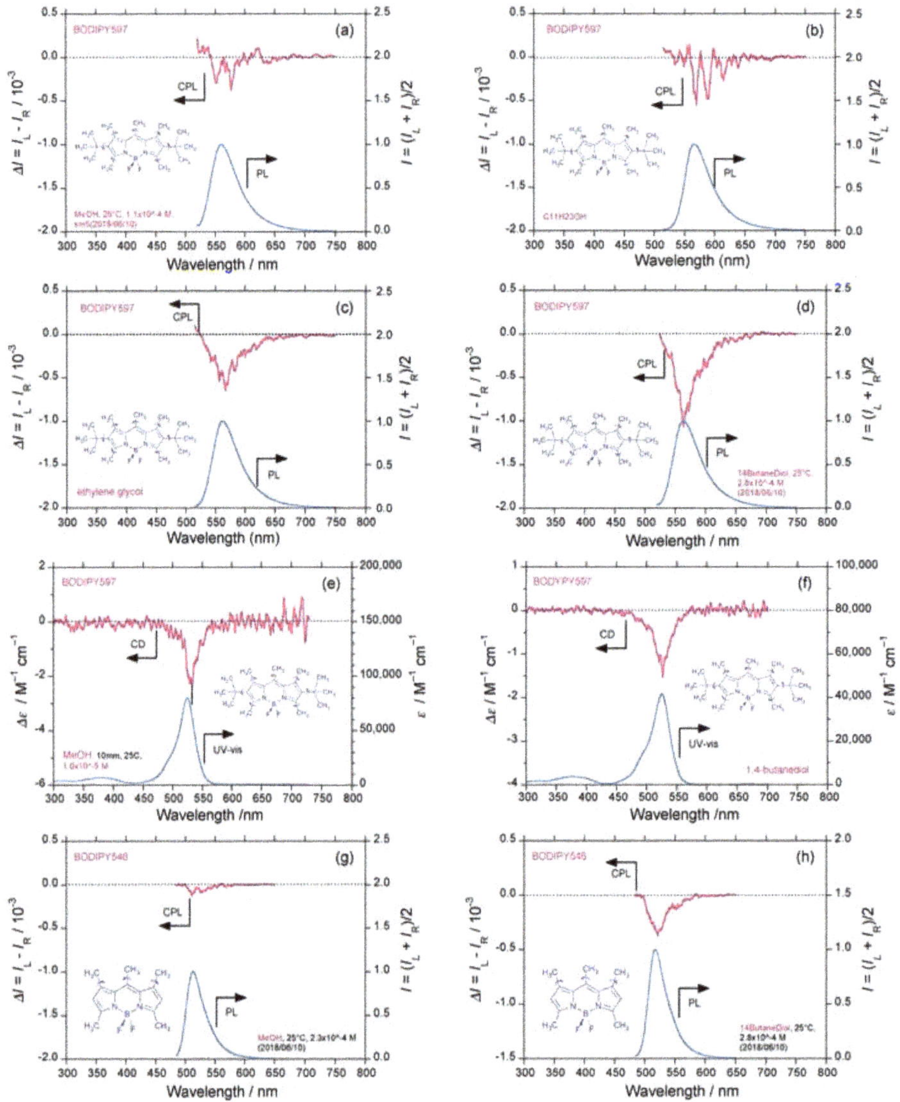

Figure 3. *Cont.*

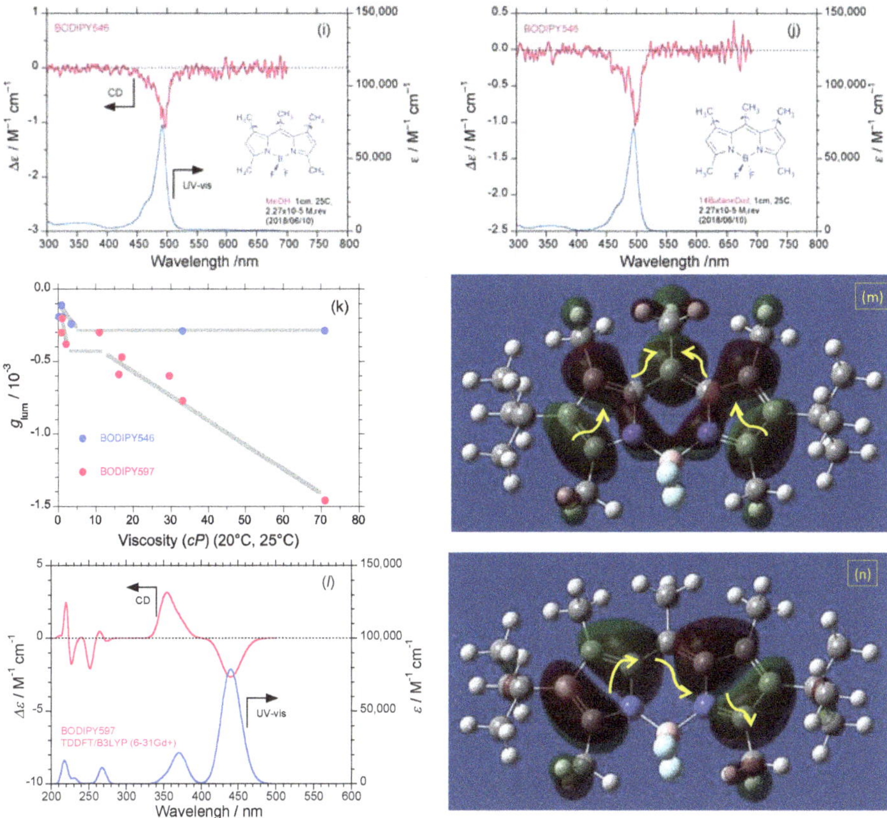

Figure 3. Comparison of CPL/PL and CD/UV–visible spectra of two pyrromethenes bearing rotatable alkyl groups (pyrromethene 597 and pyrromethene 546) in solution at room temperature (path length: 10 mm, cylindrical cuvette, concentration 1–3 × 10^{-5} M, and path length 0.1 cm and 1–3 × 10^{-4} M. CPL/PL spectra of pyrromethene 597 excited at 490 nm in (**a**) methanol, (**b**) *n*-undecanol, (**c**) ethylene glycol, and (**d**) 1,4-butanediol. CD/UV–visible spectra of pyrromethene 597 in (**e**) methanol, and (**f**) 1,4-butanediol. CPL/PL spectra of pyrromethene 546 excited at 450 nm in (**g**) methanol, and (**h**) 1,4-butanediol. CD/UV–visible spectra of pyrromethene 546 in (**i**) methanol, and (**j**) 1,4-butanediol. (**k**) g_{lum} value of pyrromethene 597 and pyrromethene 546 as a function of solvent viscosity. (**l**) Simulated CD/UV–visible spectra (fwhm = 0.20 eV) of hypothetical model of pyrromethene 597 twisted weakly by 2°, electron density mapping at (**m**) first LUMO and (**n**) first HOMO, obtained with TD-DFT, B3LYP functional with 6-31+G(d,p) basis set using Gaussian09 rev D.01 and GaussView5 package. A semi-closed-chiral-loop current (green) is seen for the HOMO, although this is not obvious for the LUMO.

More surprisingly, pyrromethene 597 in methanol revealed a clear (−)-sign CD band ($\Delta\varepsilon$ = 0.8 $M^{-1}\cdot cm^{-1}$ at 530 nm, $g_{abs} = -1.5 \times 10^{-5}$), whilst the λ_{max} of the visible band was 525 nm (Figure 3e). The Cotton CD band in viscous 1,4-butanediol showed a more intense (−)-sign CD band ($\Delta\varepsilon$ = 1.5 $M^{-1}\cdot cm^{-1}$, $g_{abs} = -3.8 \times 10^{-5}$ at 527 nm), with λ_{max} of the visible band at 525 nm (Figure 3f). Similarly, pyrromethene 546 in methanol revealed a clear (−)-sign CD band ($\Delta\varepsilon$ = 1.0 $M^{-1}\cdot cm^{-1}$ at 495 nm, $g_{abs} = -1.5 \times 10^{-5}$) with λ_{max} is 492 nm (Figure 3i). The Cotton CD band in viscous 1,4-butanediol revealed a (−)-sign CD band ($\Delta\varepsilon$ = 1.0 $M^{-1}\cdot cm^{-1}$, $g_{abs} = -1.4 \times 10^{-5}$ at 496 nm) (Figure 3j). These (−)-CD characteristics are unchanged and independent of the solvent (Figure S3a–h,

Supplementary Materials). These (−)-sign g_{abs} values for the S_0 state are smaller by two orders of magnitude compared to those of the corresponding (−)-sign g_{lum} values for the S_1 state.

Pyrromethenes 597 and 546 (Chart 2) preferentially exhibit (−)-sign CD and CPL signals, indicating that they preferentially absorb and emit *l*-CP light over *r*-CP light. These unexpected chiroptical results imply the occurrence of handed gear-like motions between multiple alkyl rotors in the S_0 and S_1 states, causing a subtle distortion of the framework of the BODIPY ring. Gaussian09 simulations usng time dependent density functional theory (TD-DFT) and Becke, three-parameter, Lee-Yang-Parr exchange-correlation functional (B3LYP) with 6-31+G(d,p) basis set indicate that one of the subtly distorted pyrromethenes 597 has a negative Cotton CD band at 450 nm (Figure 3l). Its LUMO and HOMO orbitals correspond to symmetrical and anti-symmetrical electron density with respect to the C_2 molecular axis (Figure 3m,n). Green- and red-colored electron density maps merely indicate the phase of the electron wavefunctions. If one assumes that a handed closed loop WNC flows only in regions of the same phase (green regions) indicated by yellow arrows, pyrromethene 597 becomes a handed chiral π-electron system at the C_2 axis in the S_0 state. Although a handed current flow by the yellow arrows is not obvious for the S_1 state, the framework of photoexcited pyrromethene 597 may be more twisted and, thus, associated with more rapid gear motions of the seven three-fold symmetrical alkyl (methyl and *tert*-butyl) rotors, suggesting a potential application of UP-driven one-way alkyl rotors without chiral chemical entities detectable by CPL and CD spectroscopy.

2.3. Organic Scintillators

Spontaneous radiation produced by free electrons at the valence bands of molecules is responsible for the production of scintillation light in π-conjugated organic molecules in crystalline forms and molecularly disperse solutions [114]. Scintillation light is fluorescence from the S_1 state. This scenario should obey the Jablonski diagram [43,44]. Highly emissive fluorophores with a high QY are candidates for molecular scintillators.

In fact, the Kamioka Liquid Scintillator Antineutrino Detector (KamLAND) used a molecular scintillator to detect anti-neutrinos generated geologically from the β^--decay of ^{238}U and ^{232}Th in the Earth's crust, but it cannot detect anti-neutrinos from ^{40}K due to the low energy [115,116]. Three radioactive nuclei (^{40}K, ^{238}U, and ^{232}Th) are responsible for geothermal power. These radioactive atoms are considered probes of supernova explosions that followed the nucleosynthesis of heavy elements and the birth of the Earth [117].

In KamLAND, 1000 tons of liquid scintillator was composed of *n*-dodecane (80 vol.%), 1,2,4-trimethylbenzene (20 vol.%), and 150 kg of diphenyloxazole (PPO, Chart 3) [115,116], although it is unclear to us why *n*-dodecane was considered the best solvent for the liquid scintillator. Anyway, fortuitous or otherwise, based on the g_{lum}–η relationship shown above and discussed in our previous paper [104], we assume that a viscous fluid medium such as *n*-dodecane is crucial to magnify the efficiency of scintillation light.

Non-rigid scintillators, 1,4-bis(2-methylstyryl)benzene (bis-MSB), 1,4-bis(5-phenyloxazol-2-yl)-benzene (POPOP), 1,4-bis(4-methyl-5-phenyloxazol-2-yl)benzene (DMPOPOP), 2,5-bis(5-*t*-butyl-2-benzoxalyl)thiophene (BBOT), 2,5-bis(4-biphenylyl)thiophene (BBT), are soluble in organic solvents, and 2,2″-([1,1′-biphenyl]-4.4′-diyldi-2,1-ethenediyl)bis-benzenesulfonic acid disodium salt (stilbene 420) is soluble in water and alcoholic solvents [118–121]. These molecular scintillators can adopt a polar C_2-symmetrical conformation in the S_1 and S_0 states due to their rotatable main axes. However, due to multiple C–C bonds with low rotational barrier (~1.5 kcal·mol^{-1}) between the aromatic and *trans*-vinylene moieties, and due to the proximity effect of these C–H/H–C repulsions [122], those molecules cannot adopt planar structures. Thus, these molecules should exist as a mixture of many rotamers with an equal population of P- and M-twisted molecules in solution at ambient temperature.

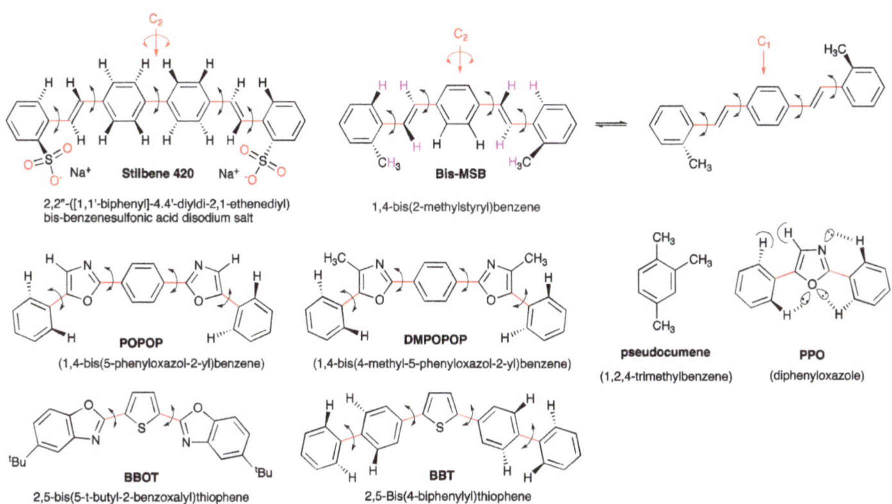

Chart 3. Chemical structures of non-rigid scintillators, 2,2″-([1,1′-biphenyl]-4.4′-diyldi-2,1-ethenediyl) bis-benzenesulfonic acid disodium salt (stilbene 420), 1,4-bis(2-methylstyryl)benzene (bis-MSB), 1,4-bis(5-phenyloxazol-2-yl)benzene (POPOP), 1,4-bis(4-methyl-5-phenyloxazol-2-yl)benzene (DMPOPOP), 2,5-bis(5-*t*-butyl-2-benzoxalyl)thiophene (BBOT), 2,5-bis(4-biphenylyl)thiophene (BBT), diphenyloxazole (PPO), and 1,2,4-trimethylbenzene (pseudocumene).

Stilbene 420, as a *trans-p*-biphenylenevinylene-type oligomer, in H_2O (0.96 *cP*) and D_2O (0.96 *cP*) emitted (−)-sign CPL with $g_{lum} = -0.5 \times 10^{-3}$ at 430 nm (Figure 4a,b), possibly at the second vibronic 0–1′ band, indicating no marked isotope effect between H and D. The g_{lum} value increased to -2.0×10^{-3} at 410 nm at the first vibronic 0–0′ band when 1,4-butanediol was employed as a solvent. The g_{lum}–η relationships showed several transitions when η = 0.96–2.5 *cP*, 2.5-6 *cP*, and >22 *cP* (Figure 4e,f). This feature arises from 512° of rotational freedom of stilbene 420 with the five rotatable C–C bonds. From the (−)-sign in a vacuum in g_{lum} value extrapolated at η = 0.0 *cP* and the (−)-sign g_{lum} value in water and heavy water, water-soluble non-rigid PAHs in the interstellar universe could spontaneously favor a handed chiral and/or helical geometry that is radiating (−)-sign CP light.

CPL/PL spectra and g_{lum}–η characteristics of bis-MSB as a *trans-p*-phenylenevinylene-type oligomer are similar to those of stilbene 420 (Figure 5a–f). Moreover, bis-MSB in low-viscosity solvents (*n*-pentane, diethyl ether, and methanol) revealed several weak but clearly detectable vibronic CPL signals with (−)-sign at 400–450 nm (Figure 5a,c and Figure S5a–p, Supplementary Materials). The CPL signals were further amplified to -0.72×10^{-3} and -0.90×10^{-3} at 400 nm when the more viscous *n*-undecanol and squalane were employed, respectively (Figure 5b,d). The g_{lum} value progressively and discontinuously increased when carbon numbers increased in two series of *n*-alkanes and *n*-alkanols including ethanol (Figure 5e). The g_{lum}–η characteristics showed step-like transitions with at least three plateaus between η = 5–10 *cP*, 10–17 *cP*, and >30 *cP*, while the g_{lum} value changed linearly in response to the η value when η = 0.22–5 *cP* (Figure 5f). The g_{lum} value extrapolated to η = 0 *cP* is -0.5×10^{-3}.

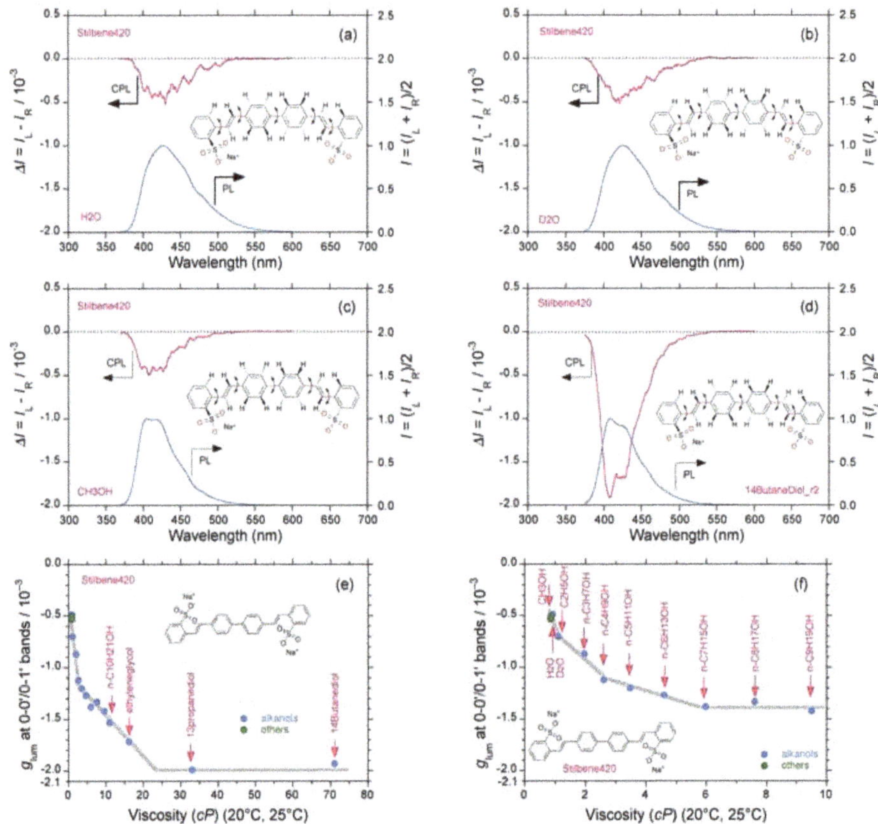

Figure 4. CPL/PL spectra of stilbene 420 (water-soluble *trans*-*p*-biphenylenevinylene-type scintillator) in (**a**) H_2O, (**b**) D_2O, (**c**) methanol, and (**d**) 1,4-butanediol (path length: 10 mm, cylindrical cuvette, concentration 2.5–10 × 10^{-5} M); (**e**) g_{lum} value of stilbene 420 as a function of solvent viscosity (η = 0–71 *cP*); (**f**) g_{lum} value of stilbene 420 as a function of solvent viscosity (η = 0–10 *cP*).

The scintillators bis-MSB, POPOP, DMPOP, BBOT, BBT, and stilbene 420 have rotational freedom along five, four, four, four, two, and five C–C bonds, respectively, producing huge numbers of rotamers. In actuality, these π-conjugated organic scintillators do not reveal noticeable CD bands at the corresponding π–π* transitions in UV–visible regions exemplified in Figures 6e,f and 7c, and Figures S4k,l and S5q,r (Supplementary Materials). However, without exception, bis-MSB, POPOP, DMPOP, BBOT, BBT, and stilbene 420 revealed intense (−)-sign CPL signals at the corresponding PL bands in various solvents (Figures 4–7 and Figures S4a–p and S5a–p, Supplementary Materials).

The (−)-sign CPL signals and PL spectra of POPOP in chloroform and *n*-hexadecane are very similar to those of DMPOPOP in chloroform and *n*-hexadecane (Figure 6a–d). The only difference is the luminescence wavelength; the λ_{lum} values of the CPL/PL bands at the 0-0′ band are 399 nm for POPOP and 407 nm for DMPOPOP (Figure 6a,b). Similarly, thiophene ring-containing scintillators BBOT and BBT in *n*-hexadecane showed (−)-sign CPL signals with g_{lum} = -0.6×10^{-3} at 428 nm (0–1′ band) and -1.0×10^{-3} at 409 nm (0–0′ band) (Figure 7a,b,d). No detectable CD signal of BBT in chloroform was confirmed (Figure 7c). This difference in g_{lum} value between BBOT and BBT should arise from the number of rotatable C–C bonds between the aromatic rings; BBOT has two, while BBT has four.

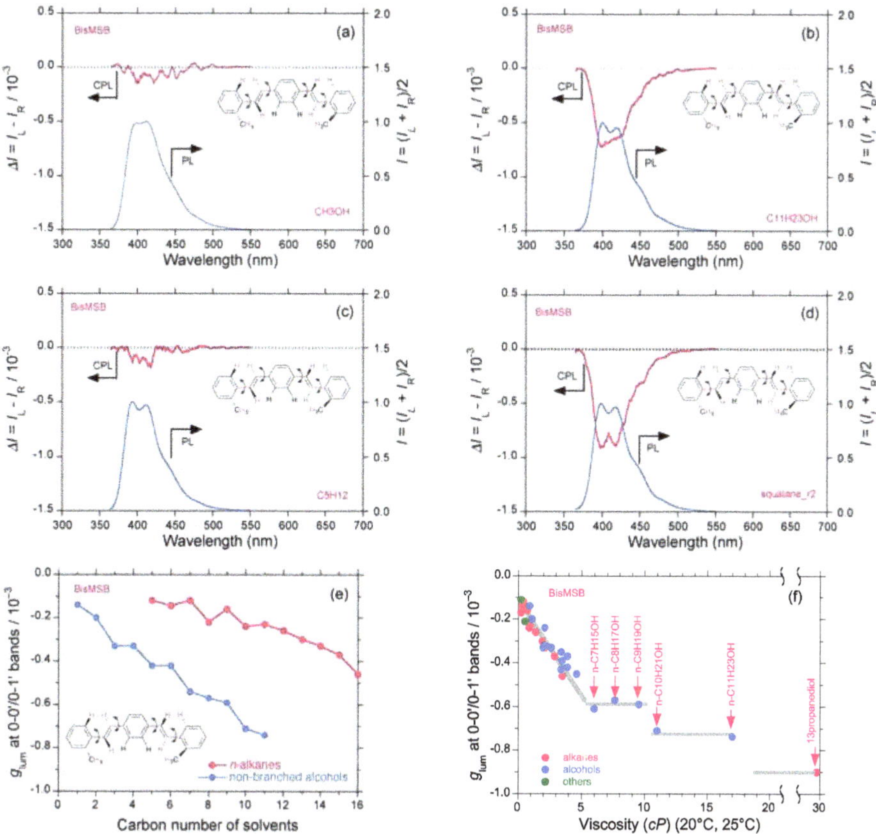

Figure 5. CPL/PL spectra of bis-MSB (*trans-p*-phenylenevinylene-type scintillator) in (**a**) methanol, (**b**) *n*-undecanol, (**c**) *n*-pentane, and (**d**) squalane (path length: 10 mm, cylindrical cuvette, concentration 2.5–10 × 10^{-5} M); (**e**) g_{lum} value of bis-MSB as a function of carbon number in two series of *n*-alkanes and *n*-alkanols (including methanol and ethanol); (**f**) g_{lum} value of bis-MSB as a function of solvent viscosity (η = 0–30 *cP*).

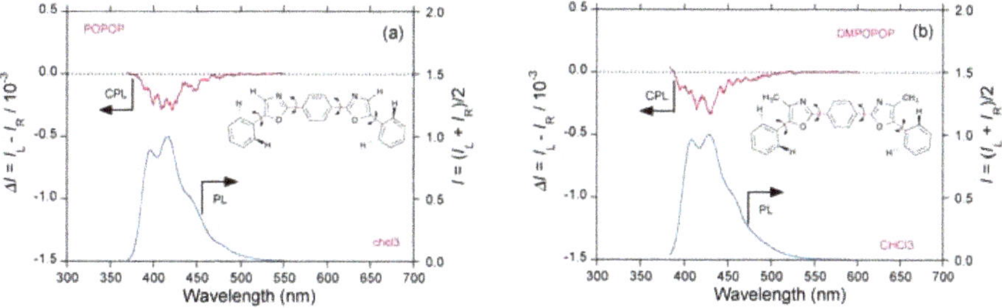

Figure 6. *Cont.*

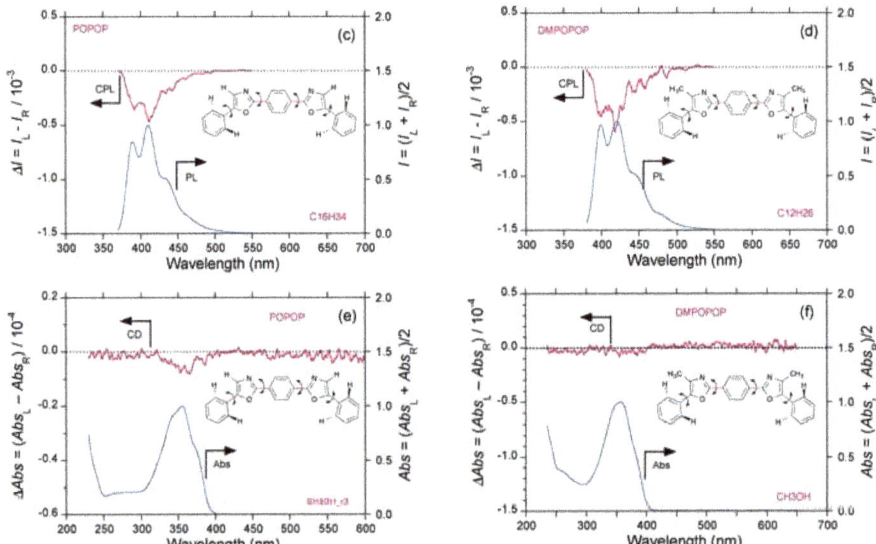

Figure 6. CPL/PL spectra of POPOP, a 1,3-oxazole ring-containing scintillator, in (**a**) chloroform and (**c**) *n*-hexadecane, and CD/UV–visible spectra in (**e**) methanol (path length: 10 mm, cylindrical cuvette, concentration 2.5–10 × 10^{-5} M). For comparison, CPL/PL spectra of DMPOPOP in (**b**) chloroform, and (**d**) *n*-dodecane, and normalized CD/UV–visible spectra in (**f**) methanol (path length: 10 mm, cylindrical cuvette, concentration 2.5–10 × 10^{-5} M).

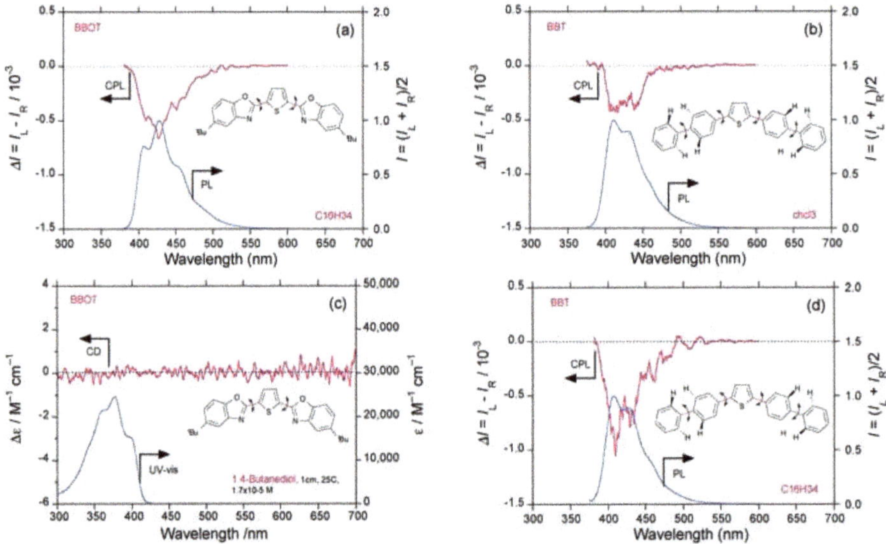

Figure 7. CPL/PL spectra of BBOT, a thiophene ring-containing scintillator, in (**a**) *n*-hexadecane and CD/UV–visible spectra in (**c**) 1,4-butanediol at room temperature. CPL/PL spectra of BBT in (**b**) chloroform and (**d**) *n*-hexadecane. Path length: 10 mm, cylindrical cuvette, concentration 2.5–10 × 10^{-5} M.

2.4. Luminophores Carrying Dialkylamino Group with Flip-Flop and/or Rotatable Motions

MacDermott and Hegstrom proposed that ammonia-type molecules ($R_1R_2R_3N$ with a lone pair) able to undergo flip-flop motion are well suited to test the MPV hypothesis [83] experimentally. Coumarin derivatives, DCM, and RhB [118–121], which are C_1-symmetrical π-conjugated luminophores, are candidates because the frameworks of coumarin, rhodamine, 4-(dicyanomethylene)-6-styryl-4*H*-pyran possess dialkylamino groups, which are susceptible to flip-flop and/or rotatable motions in the S_1 and S_0 states. The temporal generation at the S_1 state and/or persistent generation at the S_0 state are detectable as CPL and/or CD signals if certain chiral geometries are indeed generated. Most researchers do not think that coumarins, DCM, and rhodamine B are optically inactive because of the lack of chiral stereocenters. Chemical structures of 11 coumarin derivatives, DCM, and rhodamine B, which all carry dialkylamino group(s) as side chains, are shown in Chart 4.

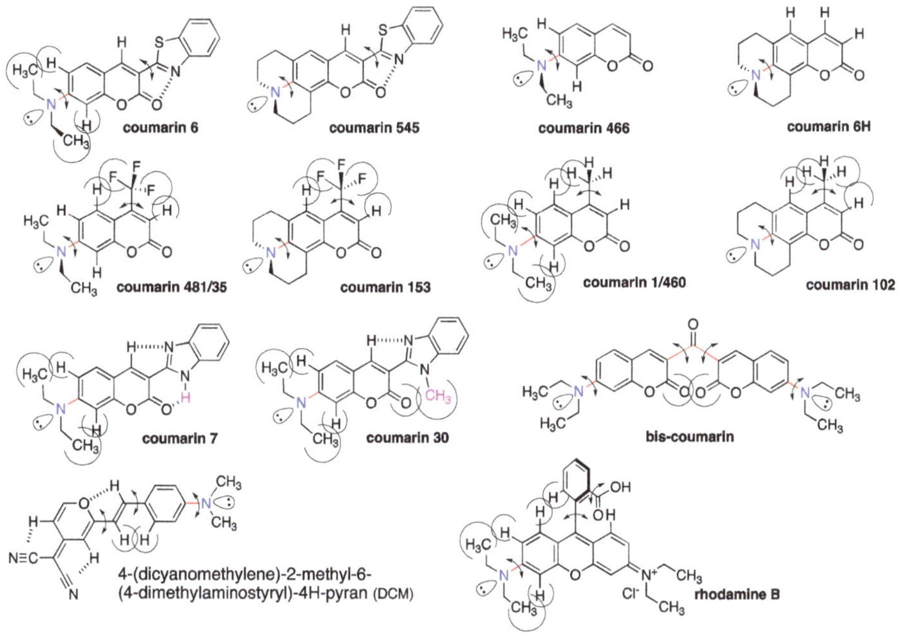

coumarin 6 · coumarin 545 · coumarin 466 · coumarin 6H · coumarin 481/35 · coumarin 153 · coumarin 1/460 · coumarin 102 · coumarin 7 · coumarin 30 · bis-coumarin · 4-(dicyanomethylene)-2-methyl-6-(4-dimethylaminostyryl)-4H-pyran (DCM) · rhodamine B

Chart 4. Chemical structures of coumarin 6, coumarin 545, coumarin 466, coumarin 6H, coumarin 481/35, coumarin 153, coumarin 1/460, coumarin 102, coumarin 7, coumarin 30, 3,3′-carbonyl-bis(7-diethylaminocoumarin) (bis-coumarin), 4-(dicyanomethylene)-2-methyl-6-(4-dimethylaminostyryl)-4*H*-pyran (DCM), and rhodamine B (RhB).

Firstly, we measured the CPL/PL spectra of coumarin 6 and coumarin 545 in several solvents (Figure 8a–f and Figure S6a–j, Supplementary Materials), and, for comparison, we measured the CD/UV–visible spectra in methanol (Figure 8g,h). Coumarin 6 and coumarin 545 showed very weak green-colored (−)-sign CPL signals on the order of $g_{lum} = -0.1 \times 10^{-3}$ at 505 nm and 520 nm, respectively (Figure 8a,b). When 1,4-butanediol was employed as the solvent, the weak (−)-sign CPL signals increased substantially to $g_{lum} = -1.2 \times 10^{-3}$ at 509 nm and $g_{lum} = -1.3 \times 10^{-3}$ at 526 nm, respectively (Figure 8e,f). The magnitude of the (−)-sign CPL signals in *n*-hexadecane (η = 3.47 *cP*) was between those in methanol and in 1,4-butanediol (Figure 8c,d). Coumarin 6 and coumarin 545 also showed very weak (−)-sign CD signals on the order of $g_{abs} = -1.3 \times 10^{-5}$ at 464 nm and $g_{abs} = -1.3 \times 10^{-5}$ at 479 nm, respectively.

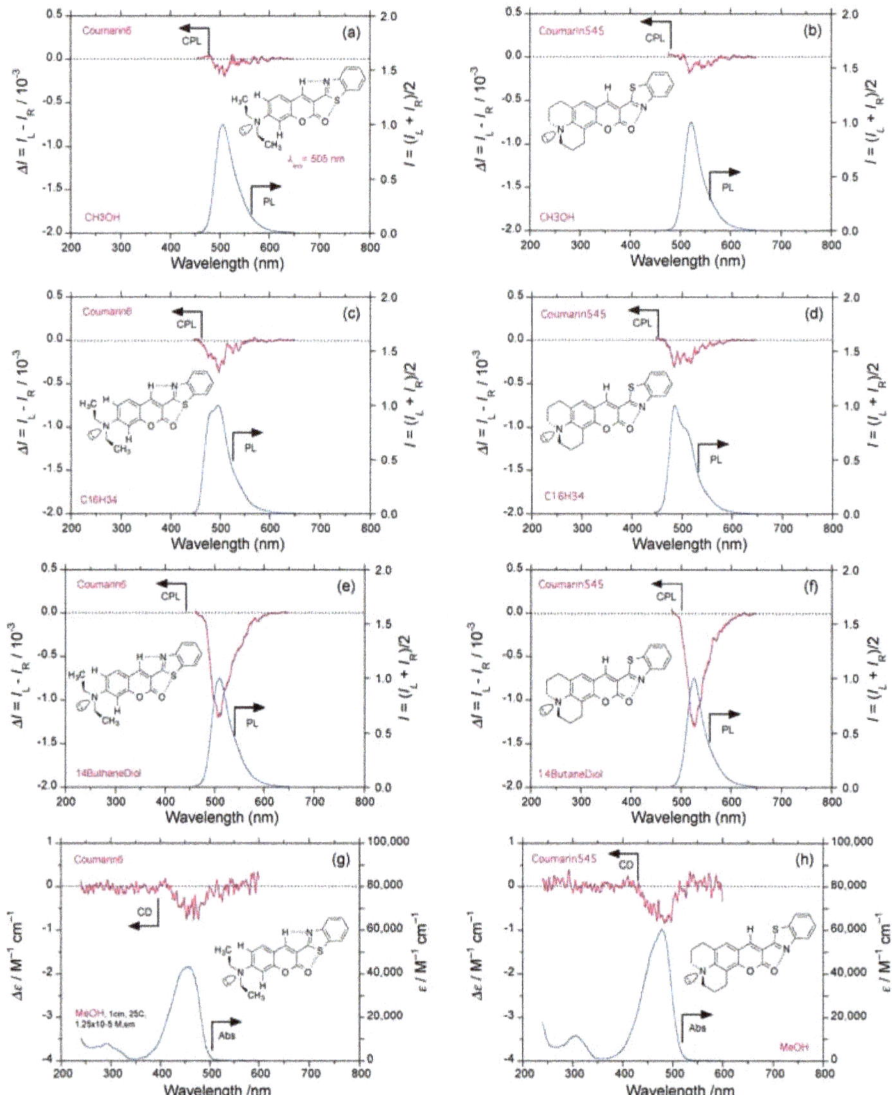

Figure 8. Comparison of CPL/PL and UV–visible spectra of coumarin 6 and coumarin 545. CPL/PL spectra of coumarin 6 excited at 420 nm in (**a**) methanol, (**c**) *n*-hexadecane, and (**e**) 1,4-butanediol at room temperature. CPL/PL spectra of coumarin 545 excited at 420 nm in (**b**) methanol, (**d**) *n*-hexadecane, and (**f**) 1,4-butanediol at room temperature (path length: 10 mm, cylindrical cuvette, concentration 2.5–10 $\times$ 10^{-5} M. (**g**) CD/UV–visible spectra of coumarin 6 in methanol. (**h**) CD/UV–visible spectra of coumarin 545 in methanol.

These unexpected CD signals can be seen in ethanol and 1,4-butanediol. The major difference between coumarin 6 and coumarin 545 is that the former allows for free-rotation and flip-flop motions of the dialkylamino group, while, in the latter, free rotation is restricted, although the flip-flop motion is still permitted. We assume that certain flip-flop twists of the dialkylamino group commonly induce optically active conformations at the S_1/S_0 states of coumarin 6 and coumarin 545, giving rise to an optically active intramolecular charge transfer (ICT) state arising due to electron donation by the dialkylamino group to the electron-accepting benzothiazole ring.

Next, to clarify the effect of the benzothiazole ring, we measured the CPL/PL spectra of coumarin 466 and coumarin 6H in several solvents (Figure 9a–d and Figure S7a–f, Supplementary Materials), and, for comparison, the CD/UV–visible spectra in methanol (Figure 9e,f). Similarly, coumarin 466 and coumarin 6H showed very weak (−)-sign CPL signals on the order of $g_{\text{lum}} = -0.1 \times 10^{-3}$ at 457 nm and 481 nm (Figure 9a,b). In 1,4-butanediol, the (−)-sign CPL signals were enhanced to $g_{\text{lum}} = -0.67 \times 10^{-3}$ at 459 nm and $g_{\text{lum}} = -0.53 \times 10^{-3}$ at 482 nm, respectively (Figure 9c,d). These (−)- sign CPL magnitudes are half those of coumarin 6 and coumarin 545. Introduction of the benzothiazole ring appears, thus, to result in CPL signal amplification by a factor of two. Coumarin 466 and coumarin 6H also showed very weak bisignate-like CD signals although they are not obvious (Figure 9e,f). Similarly, the twisted flip-flop motion of the dialkylamino group may be crucial in inducing optically active conformations at the S_1 states of coumarin 466 and coumarin 6H.

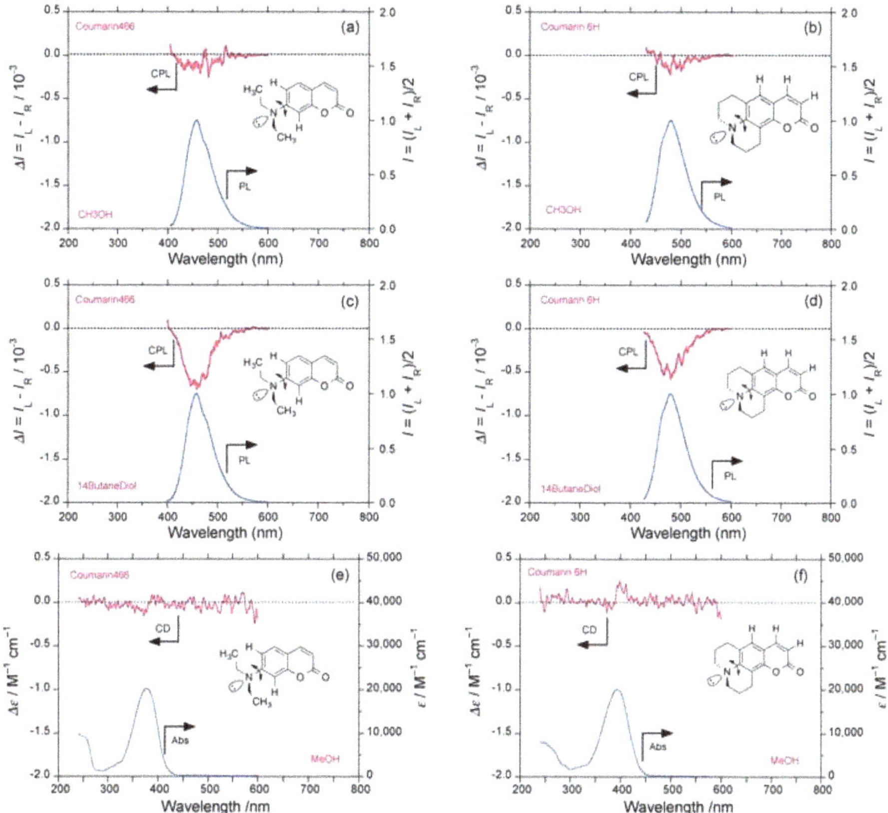

Figure 9. Comparison of CPL/PL spectra of coumarin 466 and coumarin 6H at room temperature (path length: 10 mm, cylindrical cuvette, concentration 2.5–10 × 10^{-5} M. CPL/PL spectra of coumarin 466 excited at 370 nm in (**a**) methanol, and (**c**) 1,4-butanediol at room temperature. CPL/PL spectra of coumarin 6H excited at 400 nm in (**b**) methanol, and (**d**) *n*-hexadecane. CD/UV–visible spectra of (**e**) coumarin 466 and (**f**) coumarin 6H in methanol.

To clarify the effect of the three-fold symmetrical but electron-accepting CF_3 group, we measured the CPL/PL spectra of coumarin 481/35 and coumarin 153 in several solvents (Figure 10a–f and Figure S8a–j, Supplementary Materials), and, for comparison, the CD/UV–visible spectra in methanol

(Figure 10g,h). Coumarin 481/35 and coumarin 153 in methanol showed weak green-colored (−)-CPL signals at 507 nm and 527 nm, respectively (Figure 10a,b). Interestingly, these coumarins in squalane emitted blue-colored (−)-CPL signals with increased $g_{lum} = -0.28 \times 10^{-3}$ at 437 nm and $g_{lum} = -0.39 \times 10^{-3}$ at 455 nm, respectively (Figure 10c,d). In 1,4-butanediol, these (−)-CPL signals became enhanced to $g_{lum} = -1.17 \times 10^{-3}$ at 509 nm and $g_{lum} = -0.59 \times 10^{-3}$ at 527 nm, respectively (Figure 10e,f).

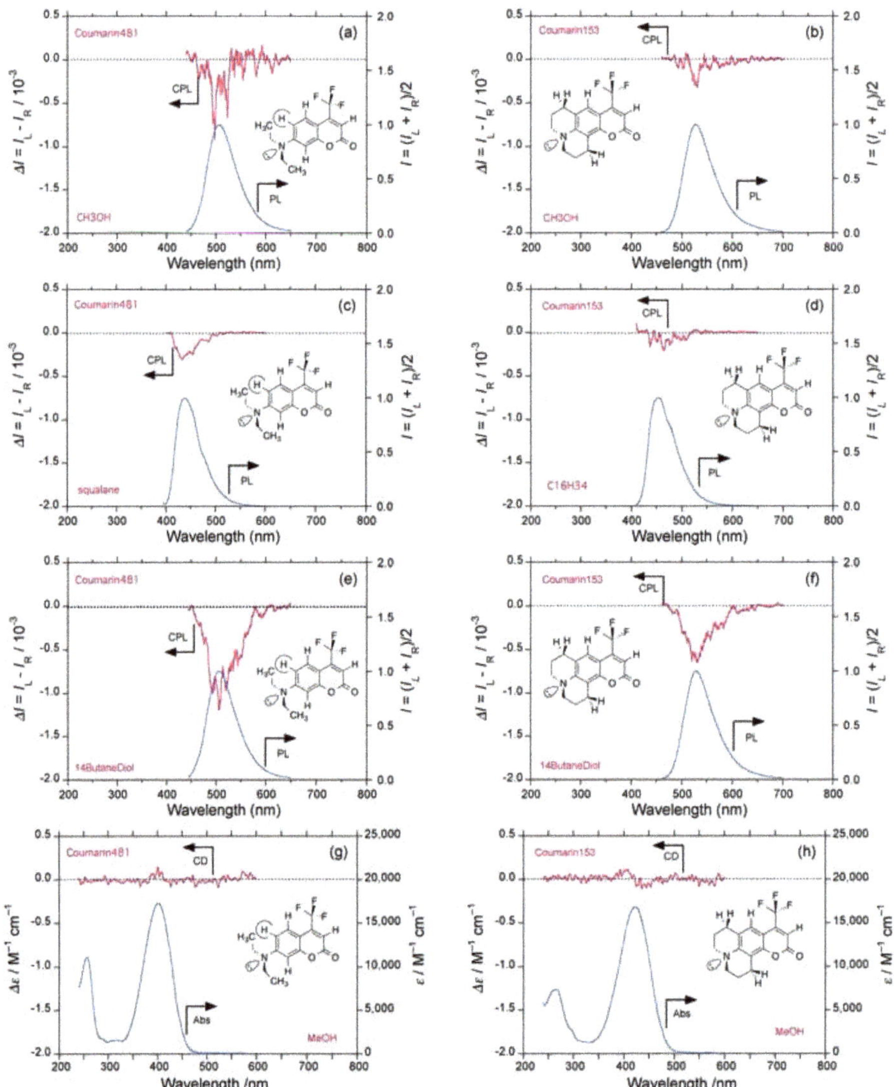

Figure 10. Comparisons of CPL/PL and UV–visible spectra of coumarin 481/35 and coumarin 153 at room temperature (path length: 10 mm, cylindrical cuvette, concentration 2.5–10 × 10^{-5} M. CPL/PL spectra of coumarin 481/35 in (**a**) methanol (exited at 410 nm), (**c**) squalane (excited at 370 nm), and (**e**) 1,4-butanediol (excited at 415 nm). CPL/PL spectra of coumarin 153 in (**b**) methanol (excited at 435 nm), (**d**) squalane (excited at 385 nm), and (**f**) 1,4-butanediol (excited at 435 nm). CD/UV–visible spectra of (**g**) coumarin 481/35, and (**h**) coumarin 153 in methanol.

We tested the effect of the CH_3 group at the peripheral position of the coumarin framework in place of the CF_3 group. The CPL/PL spectra of coumarin 1/460 and coumarin 102 were recorded in several solvents (Figure 11a–d and Figure S9a–d, Supplementary Materials), and, for comparison, the CD/UV–visible spectra were recorded in methanol (Figure 11e,f). Coumarin 1/460 and coumarin 102 in methanol showed weak blue (−)-CPL signals at 450 nm and 480 nm (Figure 11a,b). In 1,4-butanediol, these increased to $g_{lum} = -0.72 \times 10^{-3}$ at 449 nm and -0.65×10^{-3} 465 nm, respectively. The CD signals of coumarin 1/460 and coumarin 102 in methanol are not obvious (Figure 11e,f).

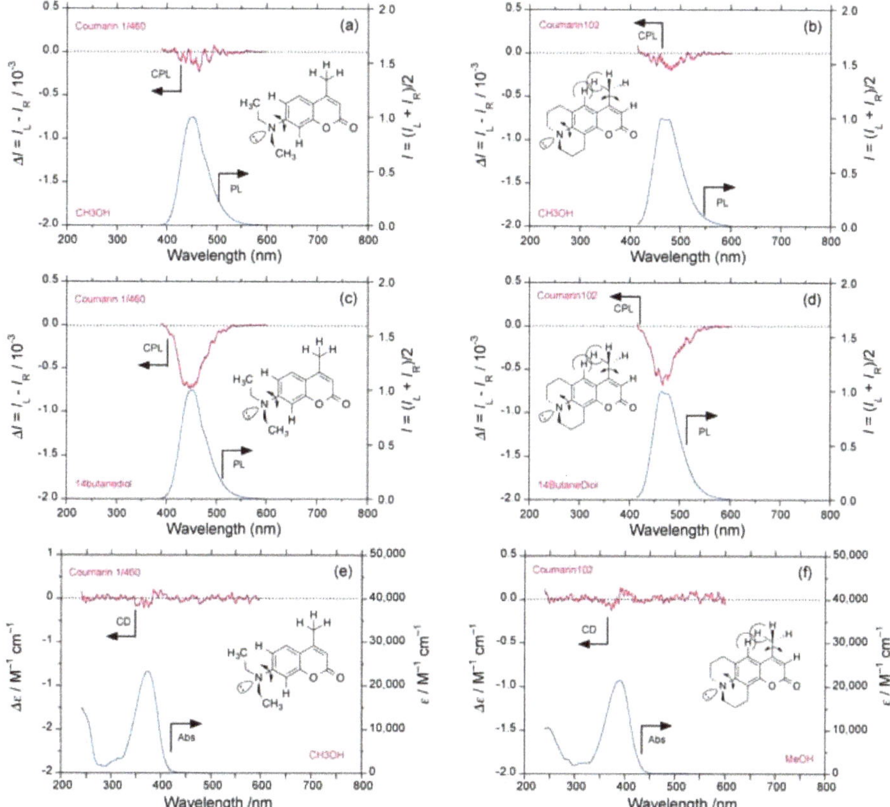

Figure 11. Comparisons of CPL/PL and UV–visible spectra of coumarin 1/460 and coumarin 102 at room temperature (path length: 10 mm, cylindrical cuvette, concentration 2.5–10 $\times$ 10^{-5} M. CPL/PL spectra of coumarin 1/460 in (**a**) methanol (excited at 365 nm), and (**c**) 1,4-butanediol (excited at 365 nm). CPL/PL spectra of coumarin 102 in (**b**) methanol (excited at 380 nm), and (**d**) 1,4-butanediol (excited at 380 nm). UV–visible spectra of: (**e**) coumarin 1/460, and (**f**) coumarin 102 in methanol.

To view the effect of the *N*-methyl group at benzimidazole, the CPL/PL spectra of coumarin 7 and coumarin 30 were recorded in several solvents (Figure 12a–d and Figure S10a–p, Supplementary Materials). Coumarin 7 and coumarin 30 in methanol showed weak blue-green vibronic (−)-CPL signals associated with $g_{lum} = -0.16 \times 10^{-3}$ at 465 nm and -0.22×10^{-3} at 482 nm, respectively (Figure 12a,b). In 1,4-butanediol, these (−)-CPL signals increased to $g_{lum} = -1.21 \times 10^{-3}$ at 497 nm and -1.20×10^{-3} at 452 nm, respectively (Figure 12c,d). Coumarin 7 and coumarin 30 in methanol may show similar bisignate features in their CD signals, but the spectral profile of coumarin 7 is the opposite of coumarin 30 (Figure 12e,f). Although the effect of the methyl group is minimal, the presence of the benzimidazole ring markedly affects the g_{lum} values in 1,4-butanediol, when compared to coumarins 466, 6H, 481/35, 153, 1/460, and 102. The presence of benzimidazole and benzothiazole groups in coumarins with dialkylamino groups may possibly be another crucial factor in photoinduced CPL signals.

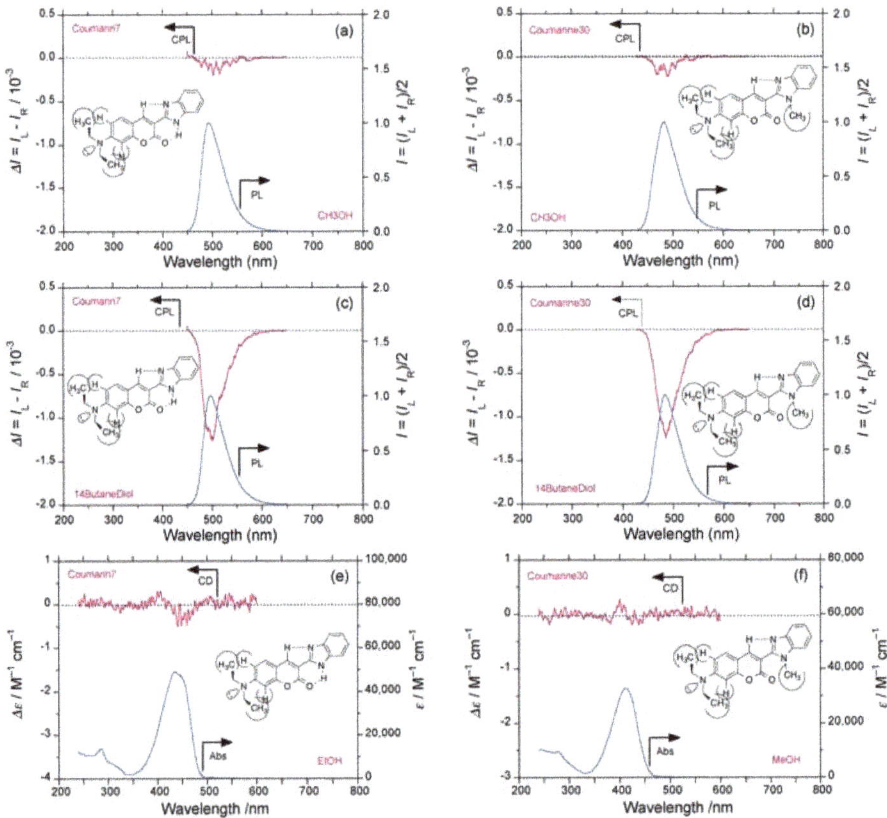

Figure 12. Comparisons of CPL/PL and UV–visible spectra of coumarin 7 and coumarin 30 at room temperature (path length: 10 mm, cylindrical cuvette, concentration 2.5–10 × 10^{-5} M. CPL/PL spectra of coumarin 7 in (**a**) methanol (excited at 420 nm), and (**c**) 1,4-butanediol (excited at 420 nm). CPL/PL spectra of coumarin 30 in (**b**) methanol (excited at 430 nm), and (**d**) 1,4-butanediol (excited at 430 nm). UV–visible spectra of: (**e**) coumarin 7, and (**f**) coumarin 30 in methanol.

As candidates of photoinduced red-light CPL emitters without stereocenters, we investigated whether 4-(dicyanomethylene)-2-methyl-6-(4-dimethylaminostyryl)-4*H*-pyran (DCM) and rhodamine 6 in several solvents reveal CPL signals at the corresponding PL bands. DCM and rhodamine 6 are representative red-light emitters, and both bear flip-flop dialkylamino groups.

Surprisingly, DCM showed weak (−)-sign CD signals on the order of $g_{\text{abs}} = -1.3 \times 10^{-5}$ at 474 nm (Figure 13f). This weak CD signal was reproducible when measured on several different occasions and unchanged in ethanol, *n*-propanol, and 1,4-butanediol. Although DCM showed weak (−)-CPL signals in methanol on the order of $g_{\text{lum}} = -0.18 \times 10^{-3}$ at 615 nm (Figure 13a), the CPL signal increased to $g_{\text{lum}} = -0.95 \times 10^{-3}$ at 615 nm in ethylene glycol (Figure 13c), -1.44×10^{-3} at 617 nm in 1,4-butanediol (Figure 13d), and, more surprisingly, -1.17×10^{-3} at 549 nm in the low-viscosity solvent, 1,4-dioxane (η = 1.10 *cP*) (Figure 13b). The CPL/PL wavelengths of DCM in 1,4-dioxane greatly blue-shifted by ca. 70 nm and showed spectral narrowing compared to the alcoholic solvents. For reasons which are unclear, ethylene glycol and 1,4-dioxane are recommended for DCM when it is used as a laser dye.

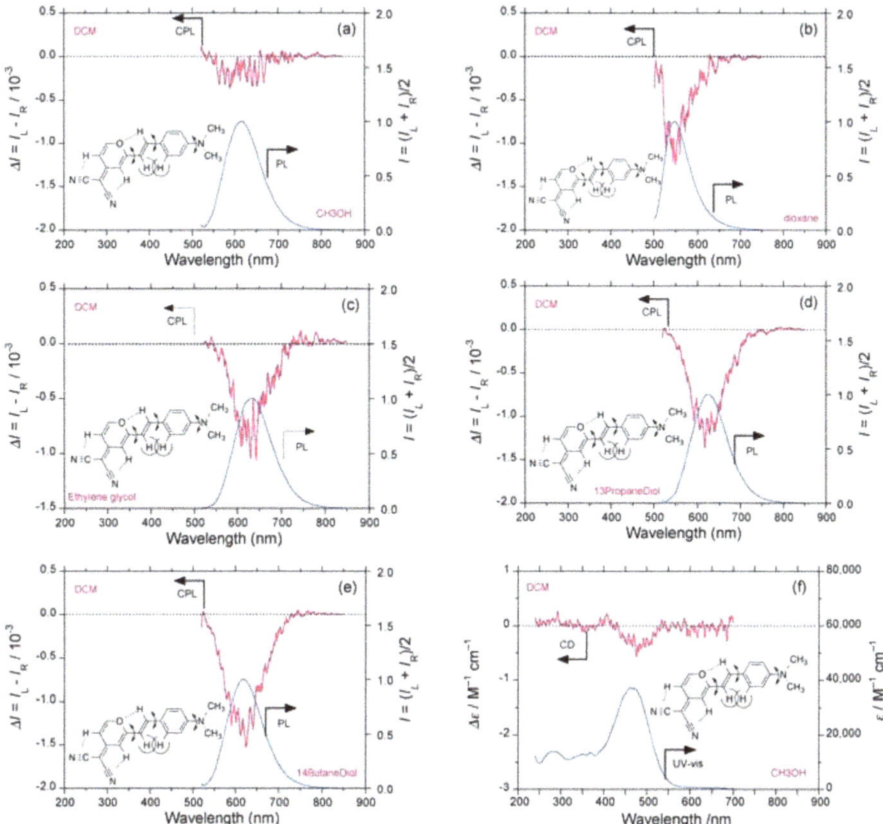

Figure 13. Comparisons of CPL/PL spectra of 4-(dicyanomethylene)-2-methyl-6-(4-dimethyl-aminostyryl)-4*H*-pyran (DCM) in (**a**) methanol (**b**) 1,4-dioxane (**c**) ethylene glycol, (**d**) 1,3-propanediol, and (**e**) 1,4-butanediol. (**f**) CD/UV–visible spectra of DCM in methanol at room temperature (path length: 10 mm, cylindrical cuvette, concentration 2.5–10 $\times$ 10^{-5} M.

More surprisingly, RhB had a clear (−)-sign CD signal on the order of $g_{abs} = -2.0 \times 10^{-5}$ at 550 nm (Figure 14f). This CD signal was reproducible when measured on several different occasions and was confirmed to be unchanged in ethanol, *n*-propanol, and 1,4-butanediol. Although RhB in methanol showed weak vibronic CPL signals at 572 nm (Figure 14a), it magnified abruptly to g_{lum} = -0.74×10^{-3} at 574 nm in *n*-$C_{11}H_{23}OH$ (Figure 14c), $g_{lum} = -0.72 \times 10^{-3}$ at 581 nm in ethylene glycol (Figure 14d), $g_{lum} = -0.83 \times 10^{-3}$ at 596 nm in 1,4-dioxane (Figure 14b), and $g_{lum} = -1.01 \times 10^{-3}$ at 576 nm in 1,4-butanediol (Figure 14e). Similarly, ethylene glycol and 1,4-dioxane are recommended for RhB when it is used as a laser dye. Ethylene glycol and 1,4-dioxane are not the only solvents for RhB and are the key to magnified (−)-sign CPL signals in fluidic media with a higher viscosity.

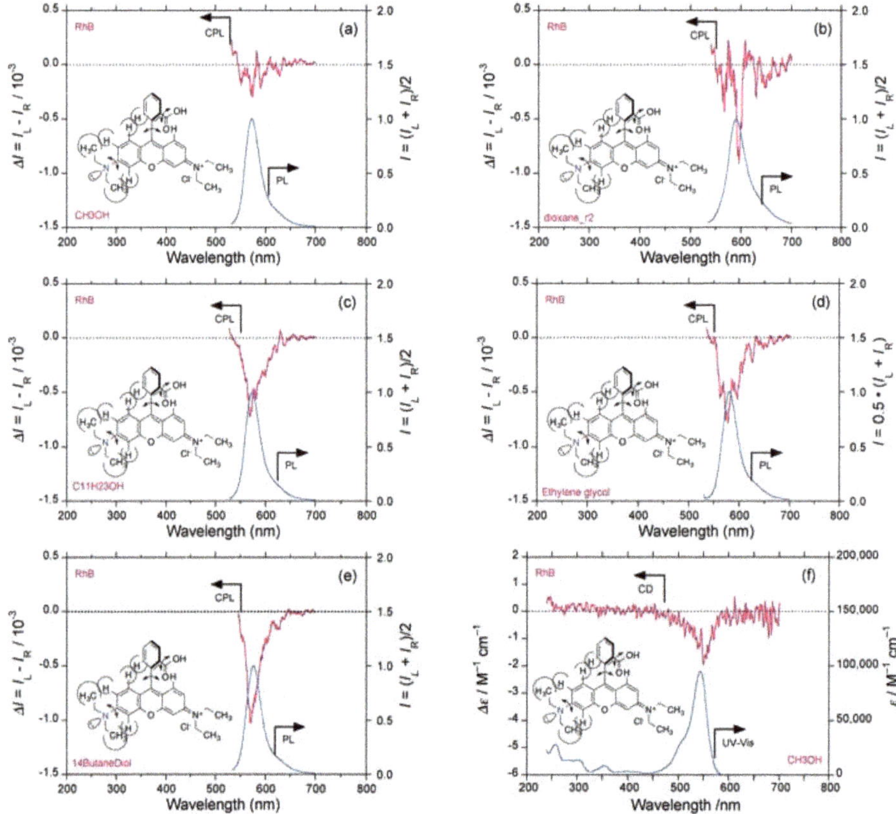

Figure 14. Comparisons of CPL/PL spectra of rhodamine B (RhB) in (**a**) methanol, (**b**) 1,4-dioxane, (**c**) *n*-undecanol, (**d**) ethylene glycol, and (**e**) 1,4-butanediol. (**f**) CD/UV–visible spectra in methanol at room temperature (path length: 10 mm, cylindrical cuvette, concentration 2.5–10 × 10^{-5} M.

Finally, we checked the CPL/PL and CD/UV–visible spectra of 3,3′-carbonylbis(7-diethylamino-coumarin) (bis-coumarin). As a result, bis-coumarin in 1,4-butanediol showed clear but broader (−)-sign CPL signals with $g_{\text{lum}} = -0.65 \times 10^{-3}$ at 531 nm (Figure 15a) and a clearly associated (−)-sign CD signal with $g_{\text{abs}} = -1.0 \times 10^{-5}$ at 463 nm (Figure 15b).

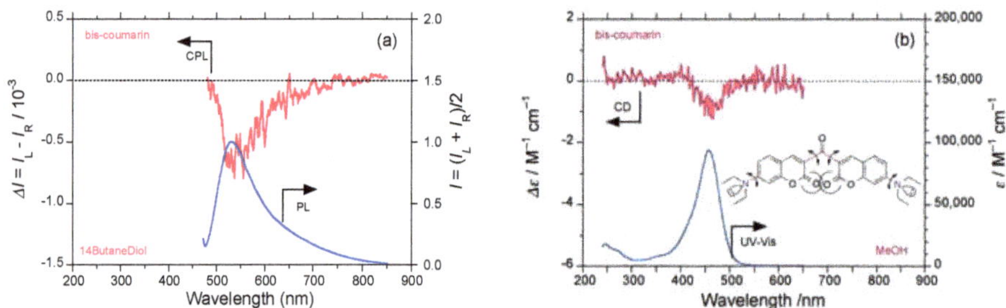

Figure 15. (**a**) CPL/PL spectra of bis-coumarin excited at 430 nm in 1,4-butandiol, and (**b**) CD/UV–visible spectra in methanol at room temperature (path length: 10 mm, cylindrical cuvette, concentration 2.5 and 10 × 10^{-5} M.

As exemplified in the cases of the fused aromatic rings with substituents, BINOL derivatives [104], BODIPY, and organic scintillators shown in the sections above, the magnitudes of the (−)-sign CPL signals in a series of coumarin dyes, DCM, and RhB are greatly amplified in response to the viscosity of the solvents. The g_{lum}–η relationships for ten sets of coumarin dyes, DCM, and RhB are summarized in Figures 16a–j and 17a–b. The data show that, in most cases, the absolute g_{lum} values leveled off at specific values when $\eta > 30\ cP$. The leveled-off g_{lum} values are highly dependent on the nature of the substituents such as, for example, the presence or absence of benzimidazole or benzothiazole as electron-accepting groups and the position of the alkyl substituents. Moreover, in all cases, the g_{lum} values extrapolated to $\eta = 0.0\ cP$ are non-zero values, -0.2×10^{-3}, suggesting that these luminophores should emit (−)-CPL signals under solvent-free conditions, such as in a collision-free vacuum.

The non-zero g_{lum} values with (−)-sign extrapolated at $\eta = 0.0\ cP$ suggest that coumarins bearing dialkylamino group(s) with flip-flop capability adopt a handed chiral geometry preferentially by radiating (−)-sign CP light even in solvent-free, collision-free conditions. We conjecture that twisted flip-flop motions of the dialkylamino group in these luminophores may play a key role in the emergent photoinduced (−)-sign CPL signals at the S_1 state with inherent handedness dictated by the PV-WNC mediated by Z^0 boson.

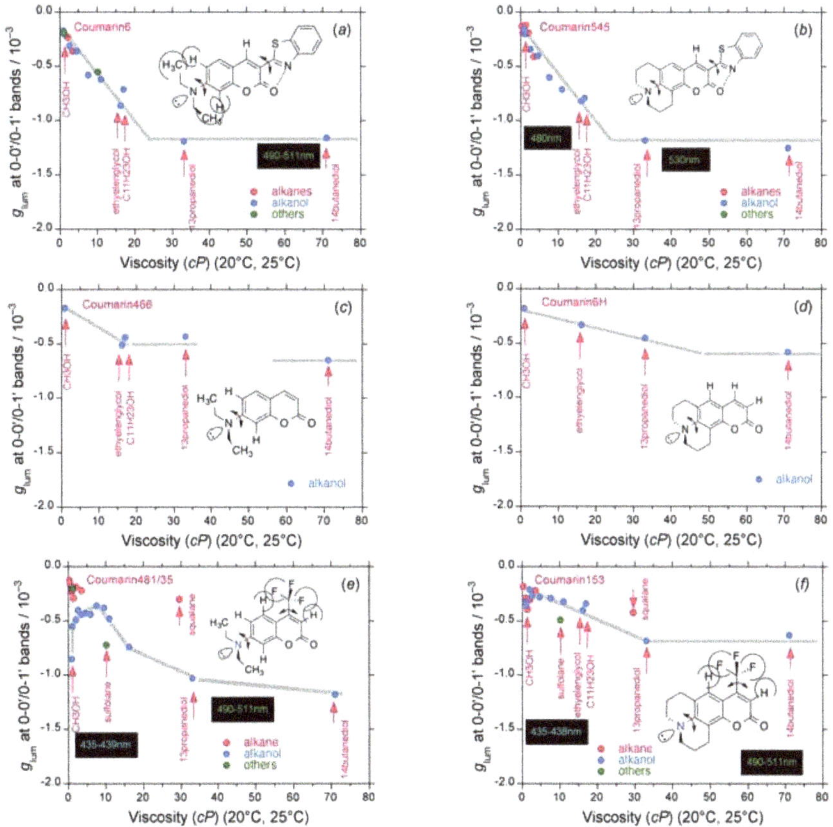

Figure 16. *Cont.*

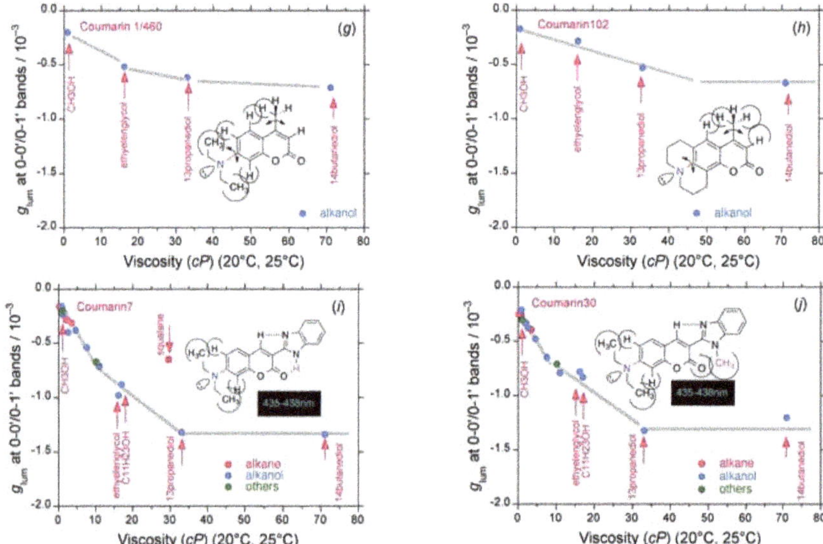

Figure 16. Comparison of g_{lum} values of (**a**) coumarin 6, (**b**) coumarin 545, (**c**) coumarin 466, (**d**) coumarin 6H, (**e**) coumarin 48/135, (**f**) coumarin 153, (**g**) coumarin 1/460, (**h**) coumarin 102, (**i**) coumarin 7, and (**j**) coumarin 30 as a function of solvent viscosity at room temperature (path length: 10 mm, cylindrical cuvette, concentration 2.5–10 × 10^{-5} M.

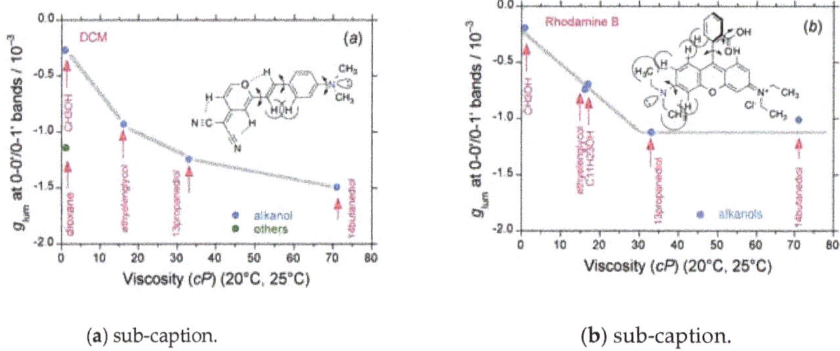

(**a**) sub-caption. (**b**) sub-caption.

Figure 17. Comparison of g_{lum} values of (**a**) DCM and (**b**) RhB as a function of solvent viscosity at room temperature (path length: 10 mm, cylindrical cuvette, concentration (2.5–10) × 10^{-5} M.

In Figure 18, we show the CD/UV–visible spectra and HOMO–LUMO electron density of coumarin 545 optimized by Gaussian09 (DFT, B3LYP/6-31G(d) level), followed by 20 singlet states by TD-DFT calculation at the B3LYP/6-31G(d) level. Optimized coumarin 545 adopts a chiral conformation such that the dihedral angle between the benzothiazole and coumarin rings is 27.7°, and the two dihedral angles between the nitrogen atom and the two nearest carbons are 4.1° and 7.8° (Figure 18a). In fact, the chiral coumarin 545 reveals CD signals at the corresponding UV–visible bands (fwhm = 0.20 eV). The g_{abs} value at the first Cotton band (397 nm) is calculated to be -8×10^{-5} $M^{-1}\cdot cm^{-1}$ (Figure 18b). From the estimated g_{abs} value, the experimental g_{abs} (= -1.3×10^{-5} at 479 nm) implies an enantiomeric excess (*ee*) of 16% in methanol. However, no CD signals are detected at the S_2 state (~300 nm). The weak (−)-CD with $g_{\text{abs}} = -1.3 \times 10^{-5}$ at the S_1 state (~480 nm) may be interpreted as the postulated PV-WNC under a zero magnetic field.

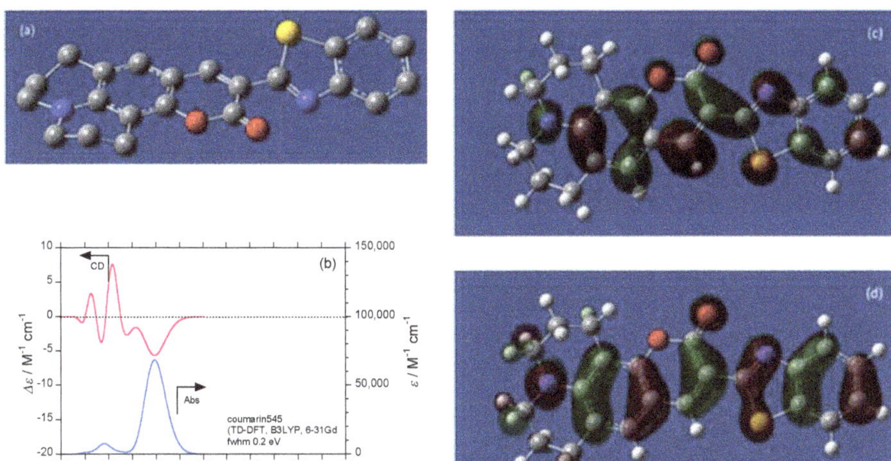

Figure 18. (**a**) Chemical structure of coumarin 545 optimized by Gaussian09 (DFT, B3LYP/6-31G(d) level) (**b**) Simulated CD/UV spectra with fwhm 0.20 eV, (**c**) 1st LUMO, and (**d**) 1st HOMO of coumarin 545 obtained with Gaussian09 (DFT, B3LYP/6-31G(d) level), followed by 20 singlet states by TD-DFT calculation at the B3LYP/6-31G(d) level.

The mapping of electron densities in the first LUMO and first HOMO of the optimized coumarin 545 are displayed in Figure 18c,d, respectively. The nitrogen atom in the dialkylamino group retains high electron density at both HOMO and LUMO levels. However, the phases (from green to red or vice versa) differ between HOMO and LUMO. Coumarin 545 at the S_1 state may possibly adopt a more distorted geometry, simultaneously allowing a twisted flip-flop motion at the nitrogen atom.

3. Discussion

Weak interactions occur between all the six quarks and six leptons at the first., second, and third generations in the framework of elemental particle physics [47,81,123,124]. The weak interactions generate a "weak charge", leading to a charged weak current and the weak neural current (WNC). Only left-handed particles and right-handed anti-particles carry the weak charge. The weak charge is analogous to Coulomb charges (+ and −) and Mulliken charges (δ+ and $\delta-$) established in chemistry arising from the parity-conserving EM force. The weak charge is unique and is responsible for the basis of handedness at subatomic and atomic levels and, presumably, even at artificial molecular, oligomer, and polymer levels. The hierarchy in the handedness could be connected to the origin of biomolecular and biopolymer handedness, and beyond.

Charged left-particles and right-anti-particles can take part in the parity violating β^- decay process of neutrons in the $n \rightarrow p + e^-$ + anti-ν_e reaction. This event arises from a left-handed spinning electron (or right-handed electron from the observer) and right-handed spinning anti-neutrino (or left-handed anti-neutrino from the observer's perspective).

According to Fermi's theory, β^- decay is the result of current–current interactions, leading to the vector (V)–axial vector (A) components of the charged weak currents. The term V is a polar vector and A is carrying angular momentum. Although the charged currents may be considered analogously to the cationic or anionic charges in molecules, no suggestion of dynamically flowing current is applied to cationic or anionic static charges. The V–A terms can generate different electric charges between $p \rightarrow n$ and $e^- \rightarrow \nu_e$ reactions, leading to the charged weak current.

Weinberg, Salam, and Glashow [47,123,124] formulated a unified theory of parity-conserving EM and parity-violating weak forces among the four fundamental physical forces based on gauge symmetry group SU(2) × U(1), while SU stands for special unitary (triplet states) and U unitary (singlet state) groups. The unified theory is popularly called the "Standard Model", and formulates an electroweak (EW) force.

A spontaneous symmetry-breaking process with handedness is a result of three massive bosons (W^+, W^-, and Z_0) and the massless photon (γ) [47]. The *WNC* is coupled with the massive neutral Z^0 boson, and the EM neutral current is coupled with γ. The unified theory connects the electric charge e to the effective weak coupling g_W, given by $g_W = e/2\sqrt{2}\ \sin\theta_W$, where θ_W is the Winberg's weak mixing angle. Experimental datasets are m_W (for $W^\pm$ bosons) = 81.0 GeV, m_Z (for Z^0 boson) = 92.4 GeV, and $\sin^2\theta_W$ = 0. 223, because $\cos\theta_W = m_W/m_Z$, while γ is kept massless.

Certainly, exchange of γ and Z^0 occurs between electrons and the nucleus. The PV interaction in elemental particles can cause atomic parity violation (APV) effects as observable values in photon-induced absorption and radiation modes. Several APV theories invoked the idea that negatively charged electrons bonded to a positively charged nucleus by Coulomb force (γ) and the weak force (Z^0), as illustrated in Figure 1.3 (Reference [47]) can involve three γ–γ–γ, Z^0–γ–γ, and γ–γ–Z^0 processes. PV potential is expressed as

$$V_{PV} = \frac{G_F}{4\sqrt{2}}\frac{Q_w}{m_e c}\left[\sigma \cdot \overleftarrow{p}\delta^3(r) + \delta^3(r)\sigma \cdot \overrightarrow{p}\right], \tag{1}$$

where G_F is Fermi's weak coupling constant (1.16637×10^{-11} eV or 2.68971×10^{-10} kcal·mol^{-1}), $\boldsymbol{r}$ is the position of the electron, $\boldsymbol{\sigma}$ is the spin operator, m_e is the electron mass, c is the light speed, $\boldsymbol{p} = -i\hbar\nabla$ momentum operator, and δ is Dirac's delta function. The admixing factor between the $s_{1/2}$ and $p_{1/2}$ states of the atom is on the order of $10^{-17}\ Z^3K_r$, whereby K_r is a relativistic correction factor (K_r ~ 3 for Cs and ~ 10 for Bi) [47]. The weak charge Q_W is expressed as

$$Q_w = 2\sum_{p,n}(C_{VP} + C_{Vn}) = Z(1 - 4\sin^2\theta_w) - N \approx -N, \tag{2}$$

where C_{Vp} and C_{Vn} are the coupling constants of vectoral V of proton and neutron, and N and Z are the numbers of neutrons and protons. Because $1 - 4\sin^2\theta_W = 0.116 < 1$, $Q_W \approx -N$. In actuality, chiroptical rotation in visible and near infrared (NIR) regions due to APV effects for vapors of heavy atoms (Bi, Pb, Tl, and Cs) is always of (−)-sign, regardless of atomic mass, indicating the dominance of the $-N$ term of Equation (2) [47].

A proton is constituted by two up quarks and one down quark, while a neutron is constituted by one up quark and two down quarks. All atoms are, thus, made of multiple up and down quarks and electrons. We radically postulate that the (−)-sign Q_W value of all molecules is linearly [61] and nonlinearly amplified by huge numbers of neutrons because molecules are polyatomic and polyneutron systems constituting parity-violating atoms: unsubstituted perylene contains 20 × 6 = 120 neutrons, while pyrromethene 597 and stilbene 420 have 169 neutrons and 248 neutrons, respectively. We conjecture that enantiomeric pairs are no longer equivalent energetically, and their characteristics behave diastereomerically, owing to hundreds of neutrons within the nuclei that leads to non-mirror-symmetric LM interactions as a consequence of the (−)-sign electroweak charge.

The consistent observation of (−)-sign CPL and (−)-sign CD signals may arise from the inherent (−)-sign of Q_W. If the negative-sign Q_W in Equation (2) can be applied to spontaneous radiation and non-radiative processes from photoexcited non-rigid fluorophores, the CPL (and CD) signals are postulated to be commonly the same (−)–sign regardless of their chemical structures. Because the CPL signals from the observer are defined as PL(left) − PL(right), the non-rigid fluorophores in the present results and previous study [104] are primarily radiating PL(right) over PL(left) during the non-radiating reorganization process at the S_1 state. Although the predicted signs in CPL and CD signals are inverted by dihedral angles of multiple C–C bonds in the non-rigid fluorophores [80], the weak (−)-charge Q_W may be efficiently coupled with dipolar (δ^+ and δ^-) molecular structures that can adopt a significantly polar V-shape and *syn*-form (pseudo-C_2 symmetric rotamer), and a polar rod-like shape (C_1-symmetric rotamer) in the S_1 and S_0 states, as schematically illustrated in Charts 1–4. The negative value Q_W is additive to δ^- (then, $\delta^- - |Q_W|$) but is subtractive to δ^+ (then, $\delta^+ + |Q_W|$), then $|\delta^- - |Q_W|| \neq |\delta^+ + |Q_W||$ for an enantiomeric pair of rotamers.

From the viewpoint of molecular dynamics, the PV-WNC force causes parity-odd rotational and/or flip-flop motions. The motions enforce (*R*)- and (*S*)-forms in the same direction (CW or CCW) that facilitates radiation with only (−)-sign CPL. However, the EM force, a parity-even, parity-conserved force, allows plural C–C bonds in the *(R)*-form to rotate and/or undergo flip-flop motions with *CW* motion and, conversely, those with the *(S)*-form to rotate and/or flip-flop *CCW* or vice versa. These motional dynamics should be mirror-symmetric. The handedness of motional dynamics by handed elemental particles can be recognized as chiral crystallization of achiral molecules: longitude polarized electrons and positrons that mirror image leptons oppositely affect an L/R preference in the crystallization of sodium chlorate and bromates in water solution [81].

In our previous paper [104], we grouped the apparent CPL and CD spectral characteristics with their signs, magnitudes, and wavelengths and associated barrier heights (E_b) in double-well and multi-well potentials into four categories as follows:

(i) *Case 1.* The value of E_b between rigid enantiomers is relatively high >30 kcal·mol^{-1} in the S_0 and S_1 states. Mirror-image CD and CPL spectra are evident for the enantiomers. The parity-conserved EM force is a determining factor. Racemization rate obeys the Arrhenius equation with activation energy (E_a).

(ii) *Case 2.* When $10 < E_b < 30$ kcal·mol^{-1} at the S_0 and S_1 states, non-mirror-image CPL and CD spectra are often observed [15,27,90,95,120,122,125–134]. Although (+)- and (−)-signs in CPL and CD are primarily determined by atrope and point chirality, the absolute magnitudes and wavelengths at the CPL and CD bands differ considerably from each other.

(iii) *Case 3.* When $1 < E_b < 10$ kcal·mol^{-1} at the S_1 and S_0 states, only (−)-sign CPL and (−)-sign CD spectra should be observed. The parity-violating weak force might be a determining factor in the S_1 and S_0 states [104].

(iv) *Case 4.* When $0 < E_b < 1$ kcal·mol^{-1} in the S_1 and S_0 states, no detectable CD bands are observed, although (−)-sign CPL signals are obvious. Resonance quantum tunneling without E_a is responsible for dynamic racemization, oscillating chirality, and quantum beat [66,72,101]. The parity-violating weak force is a determining factor in the S_1 state, while the parity-conserved EM force is a determining factor in the S_0 state.

Moreover, regarding hidden molecular chirality, in 1970s, Mislow argued the cryptochirality of mirror-image molecules in which optical activity is non-detectable [135]. In 2006 and 2009, approaches to chemically decipher cryptochiral molecules and polymers were reported [136,137]. Additionally, with the help of CPL and CD spectroscopy, a photophysical deciphering approach was applied to the EM-originating cryptochirality of several CD-silent molecules [138–142].

The previous paper did not report Case 3, though we reported examples of Cases 1, 2, and 4 [104]. The present paper reports Case 3 for the first time, i.e., that pyrromethene 546, pyrromethene 597, DCM, RhB, and bis-coumarin all reveal (−)-sign CPL and (−)-sign CD spectra, even in low-viscosity solvents. EW-perturbed quantum chemistry [143], EW-perturbed photophysics and EW-perturbed photochemical reactions should be considered when open questions of unexpected L/R preference and their detectable L/R differences in non-rigid and semi-rigid artificial molecules, and supramolecules and biomolecules in the S_0 and S_1, S_n ... states are raised.

Other plausible scenarios for the L/R preference are possible. The handedness of non-rigid molecules in an ultra-tiny % *ee* can be increased to ~100% *ee* upon photoexcitation of parity-conserved (PC) EM force-driven circularly polarized (CP) light carrying a single angular momentum ($\pm n\hbar$, $n = 1$) in the broad range of γ-ray, X-ray, vacuum–UV, UV–visible–IR, far-IR far-THz, and cosmic microwave radiation according to certain nonlinear amplification scenarios [97–109], known as autocatalytic self-replication [144], sergeant-and-soldier and majority rules [145], and polymerization [146].

Parity-conserving EM force-originating macroscopic MSB was comprehensively reviewed for a large number of molecules, polymers, supramolecules, colloids, gels, and crystals [147–150]. These alternative modern scenarios provide other possible answers to the greatest mystery on the origin of homochirality on Earth.

Recently, lightning was found to be a natural particle accelerator, ubiquitously generating γ-rays. Lightning causes atmospheric photonuclear reactions. The γ-ray energy is captured by N_2 molecules, followed by producing, possibly, weak force-origin handed neutrons, right-handed positrons, and left-handed neutrinos [151]. Additionally, cosmological-origin right-handed anti-ν interacting with ^{14}N in molecular clouds in star-forming regions of supernovae and neutron stars [87,152], gravitational origin parity violation [153,154], and hydrodynamic vortex flows with the opposite handedness in the north and south hemispheres on Earth [26,27,155] are of specific interest. In recent years, vortex light (alternatively called optical spanner, spiral light, twisted light, and helical light) [156–169] was recognized as a new sort of chiral light carrying multiple orbital angular momenta (OAM) with $\pm$ $l\hbar$, l = 1, 2, 3, 4, 5, 10 ... up to 200. Vortex light with l = 0 is achiral. Vortex light can generate a torque enabling the rotation of molecular droplets, polymeric solids, and metallic particles in CW or CCW directions [158,161,164]. The helical wavefront for the Laguerre–Gaussian mode of vortex light allows for sculpturing spiral relief and motifs and for rotating small objects in CCW or CW directions [156–170]. Like CP light–matter interactions [171,172], vortex light can discriminate between enantiomers [164], and it is possible to predominantly generate handed chiral motifs from achiral polymers [162,163,169]. Astrophysical origin vortex light [156], CP light, and right-handed solar neutrinos may, thus, be connected to the L/R preference of biomolecular substances.

Recently, astonishing findings seeking source materials connecting with the handedness of biomolecules on Earth were reported. In 2009, a National Aeronautics and Space Administration (NASA) team characterized extraterrestrial-origin glycine-embedded samples returned from comet 81P/Wild2 using liquid chromatography and spectrometry [173]. In 2016, other researchers determined glycine, phosphorus and several organic substances involving O, S, and F in specimens collected from the coma of 67P/Churyumov–Gerasimenko using a double-focusing mass spectrometer [174]. Moreover, in 2016, radio astronomers found the first astronomical-origin chiral propylene oxide and achiral *n*-propanol in the Sagittarius B2 star-forming region of the Milky Way galaxy, although the existence of any L/R preference remains to be elucidated [175]. Comets and interstellar materials could deliver biomolecules or their precursors and water to Earth. Although it is possible to synthesize mirror-image DNA and proteins in laboratories [176,177], it is challenging to directly detect the L/R preference, possibly associated with (−)-sign circularly polarized radiation from the observation of interstellar PAHs at the S_1/S_0 states in the UV–visible region [175], and rigid and non-rigid non-π-conjugated organics at the near-IR/mid-IR/far-IR/microwave regions [108,109]. It remains a great challenge and a great curiosity to provide more realistic scenarios for biomolecular handedness.

4. Materials and Methods

Instrumentation details, lists of solvents and fluorophores and their vendors, preparation of sample solutions, and chiroptical analytical data [99,104] are described below.

4.1. Instrumentation

Using a JASCO (Tokyo, Japan) J-820 spectropolarimeter, UV–visible and CD spectra were simultaneously recorded at ambient temperature using a cylindrical quartz cuvette with a path length of 10 mm. The cylindrical cuvette assured a precise CD measurement compared to the rectangular cuvettes that are often used in routine experiments. Precise CD/UV–visible spectra were obtained by using a bandwidth of 2 nm, with one or two accumulations at scanning rates of 50 or 100 nm·min^{-1} with a response time of 2 s. The CD signals of the two BODIPYs and RhB were triply confirmed under the following conditions: bandwidth = 2 nm, response time = 8 s, scanning rate = 20 nm·min^{-1} with four accumulations. To minimize drifts in the light source and power supply, the instrument was aged for at least 2 h prior to measurements. CPL and PL spectra were likewise collected using a JASCO CPL-200 spectrofluoropolarimeter (Hachioji, Tokyo, Japan) employing cylindrical quartz cuvettes with path lengths of 10 mm at ambient temperature. The best experiment parameters were as follows: bandwidth = 10 nm for excitation and detection; response time of PMT = 8–16 s during measurements; two to eight accumulations with scanning rate = 20–50 nm·min^{-1}.

4.2. Materials

4.2.1. Luminophores (vendor)

Section 1: Perylene (Tokyo Chemical Company (TCI), Tokyo, Japan), 5,8,11-tetra-*tert*- butylperylene (TCI), *N,N′*-bis(2,6-diisopropylphenyl)-1,6,7,12-tetraphenoxy-3,4,9,10-perylene-tetra-carboxylic diimide (TCI), *N,N′*-bis(2,5-di-*tert*-butylphenyl)-3,4,9,10-perylenedicarboximide (Sigma-Aldrich, St. Louis, MO, USA), 16,17-bis(*n*-octyloxy)-anthrax[9,1,2-*cde*]-benzo[*rst*]-pentaphene-5,10-dione (TCI), and *N,N′*-di-*n*-octyl-3,4,9,10-perylenetetracarboxylic diimide (Sigma-Aldrich) were obtained as indicated.

Section 2: Pyrromethene 546 (TCI) and pyrromethene 597 (TCI) were obtained as indicated.

Section 3: Stilbene 420 (Exciton, Tokyo Instruments Inc. (Tokyo, Japan)), bis-MSB (Exciton), POPOP (TCI), DMPOPOP (Dotite, Kumamoto, Japan), BBOT (Dotite), and BBT (TCI) were obtained as indicated.

Section 4: Coumarin 6 (TCI), coumarin 545 (TCI), coumarin 466 (TCI), coumarin 6H (Sigma-Aldrich), coumarin 481/35 (TCI), coumarin 153 (Sigma-Aldrich), coumarin 1/460 (Sigma-Aldrich), coumarin 102 (Sigma-Aldrich), coumarin 7 (Sigma-Aldrich), coumarin 30 (Sigma-Aldrich), biscoumarin) (TCI), DCM (Sigma-Aldrich), and RhB (TCI) were obtained as indicated.

4.2.2. Solvents

Vendor, viscosity in cP, and temperature in °C are provided in brackets [178–183]; in each series, entries are given in order of increasing viscosity as follows:

(1) *n*-Alkanes: *n*-pentane (FUJIFILM Wako, 0.21 (25)), *n*-hexane (FUJIFILM Wako, 0.30 (25)), *n*-heptane (Sigma-Aldrich, 0.39 (25)), *n*-octane (Sigma-Aldrich, 0.51 (25), *n*-nonane (Sigma-Aldrich, 0.71 (20)), *n*-decane (Sigma-Aldrich, 0.85 (25)), *n*-undecane (Sigma-Aldrich, 0.93 (20)), *n*-dodecane (Sigma-Aldrich, 1.36 (25), *n*-tridecane (Sigma-Aldrich, 1.88 (20)), *n*-tetradecane (Fluka, 2.08 (25), *n*-pentadecane (Sigma-Aldrich, 2.86 (20)), and *n*-hexadecane (Sigma-Aldrich, 3.71 (20)).

(2) Branched and cyclic alkanes: isooctane (Dotite, 0.50 (25)), cyclohexane (Dotite, 0.93 (22)), and squalane (2,6,10,15,19,23-hexamethyltetracosane) (Sigma-Aldrich, 29.50 (25)).

(3) Non-branched and *n*-alcohols: methanol (FUJIFILM Wako, 0.55 (25)), ethanol (FUJIFILM Wako, 1.09 (25)), *n*-propanol (Sigma-Aldrich, 1.96 (25)), *n*-butanol (FUJIFILM Wako, 2.59 (25)), *n*-pentanol (Sigma-Aldrich, 3.47 (25)), *n*-hexanol (Sigma-Aldrich, 4.59 (25)), *n*-heptanol (Wako, 5.97 (25)), *n*-octanol (Wako, 7.59 (25)), *n*-nonanol (Sigma-Aldrich, 9.51 (25)), *n*-decanol (Sigma-Aldrich, 11.50 (25)), ethylene glycol (FUJIFILM Wako, 16.1 (25), *n*-undecanol (Sigma-Aldrich, 16.95 (25)), 1,3-propandiol (FUJIFILM Wako, 33.0 (25)), and 1,4-butanediol (FUJIFILM Wako, 71.0 (25)).

(4) Branched alcohols: isopropanol (Dotite, 2.07 (25), isobutanol (Sigma-Aldrich, 3.38 (25)), and isopentanol (Sigma-Aldrich, 3.86 (25)).

(5) Chlorinated hydrocarbons: dichloromethane (Dotite, 0.41 (25)) and chloroform (Dotite, 0.55 (25)).

(6) Other solvents: diethyl ether (FUJIFILM Wako, 0.22 (25)), acetone (FUJIFILM Wako, 0.31 (25), acetonitrile (FUJIFILM Wako, 0.34 (25)), tetrahydrofuran (Dotite, 0.46 (25)), benzene (FUJIFILM Wako, 0.60 (25)), water (Wako, 1.00 (20)), 1,4-dioxane (Dotite, 1.10 (25)), anisole (TCI, 1.09 (25)), heavy water (Wako, 1.25 (20)), and sulfolane (TCI, 10.10 (25)).

4.3. Preparation of Sample Solutions

Firstly, a representative stock solution (10^{-3} M) of luminophore dissolved in spectroscopic grade $CHCl_3$ (Dotite, Kumamoto, Japan) was prepared. For RhB and stilbene 420, ethanol was used as the stock solution solvent. A small quantity of the stock solution was added to the desired liquid (1.9–2.1 mL) in the cylindrical quartz cuvette using a microsyringe. The CD/UV–visible and CPL/PL spectra were then recorded. Oxygen was not purged from the solvents or solutions in CPL and CD measurements since it does not significantly influence the fluorescence quantum yield of most organic luminophores.

4.4. Chiroptical Analysis

The dissymmetry factor of the circular polarization at the S_0 state (g_{abs}) was evaluated as $g_{abs} = (\varepsilon_L - \varepsilon_R)/(1/2(\varepsilon_L+\varepsilon_R))$, where ε_L and ε_R are the extinction coefficients for *l*- and *r*-CP light, respectively [99]. The dissymmetry factor of the circular polarization at the S_1 state (g_{lum}) was evaluated as $g_{lum} = (I_L - I_R)/(1/2(I_L + I_R))$, where I_L and I_R are the intensities of the signals for *l*- and *r*-CP light respectively, under the incident UP light [99]. The parameter g_{abs} was experimentally determined using the expression $\Delta\varepsilon/\varepsilon$ = (ellipticity (in mdeg)/32980)/absorbance at the CD extremum, similar to the parameter g_{lum}, calculated as $\Delta I/I$ = (ellipticity (in mdeg)/(32980/ln10))/total PL intensity (in volts) at the CPL extremum.

5. Conclusions

We tested whether or not semi-rigid and non-rigid π-conjugated fluorophores in the S_1 and S_0 states in a series of achiral liquids with η ranging from 0.22 *cP* to 71.0 *cP* are optically inactive and have mirror symmetry as measured by CPL and CD spectroscopy. The fluorophores included six perylenes with and without substituents, two BODIPYs, six scintillators, RhB, DCM, and 11 coumarins. Perylenes were models of interstellar small and large PAHs radiating IR spectra of bright HII regions, planetary nebulae, and reflection nebulae. Without exception, all the non-rigid fluorophoric enantiomers, and possibly also the highly twisted perylene derivatives, showed (−)-sign CPL signals radiating from the vibronic photoexcited state in support of the molecular parity-violating hypothesis based on the Z^0 boson origin PV-weak neutral current mechanism. The fluorophore emission intensities increased *progressively* and *discontinuously* to approximately –0.2 × 10^{-3} and −2.0) × 10^{-3} as a function of the solvent viscosity. Of specific interest was the detection of weak but clear CD signals with g_{abs} values of -1.4×10^{-5} at $\lambda_{max}/\lambda_{ext}$ for two pyrromethene derivatives, RhB, DCM, and bis-coumarin at the S_0 states. The results of the present CPL and CD spectral characteristics should provide a possible answer to the parity violation hypothesis at the molecular level based on a handed weak neutral current mediated by the Z^0 boson. The present comprehensive and previous experimental datasets [104] led us to address the "Ozma problem" posed by Gardner [1]. The query was how we can correctly communicate the left-and-right issue to intellectually advanced alien lifeforms. Our answer is that, when an unpolarized UV light source is applied to excite semi-rigid and non-rigid π-conjugated luminophores, we define (−)-sign CPL signals from the observer as "right" without exception.

Supplementary Materials: The following are available online at http://www.mdpi.com/2073-8994/11/3/363/s1.

Author Contributions: M.F. and J.R.K. have long discussed searching for experimental verification of the MPV hypothesis over 20 years. M.F. designed the application of CPL/CD spectroscopy to test the MPV hypothesis of the π-conjugated luminophores. M.F. and S.A. measured and analyzed CPL and CD datasets. M.F. and J.R.K. cowrote the manuscript. Requests for all original and processed CPL/CD datasets (saved in several file formats JASCO ##.jws, their converted ##.txt, followed by processed data using KaleidaGraph (mac, ver 4.53) ##.qpc, and ##.qda) should be sent to M.F. (fujikim@ms.naist.jp).

Funding: This work was supported by Grants-in-Aid for Scientific Research (16H04155 (FY2016-2018), 23651092 (FY2014-2016), 22350052 (FY2010-2013), 16655046 (FY2003-2005)), the Sekisui Chemical Foundation (FY2009), and the NAIST Foundation (FY2009).

Acknowledgments: Firstly, we owe a debt of gratitude to Reiko Kuroda (Tokyo Science University) for giving us the opportunity to disclose our hitherto unpublished CPL/CD spectroscopic datasets accumulated at NTT Basic Research Laboratory (Tokai, Musashino and Atsugi) and NAIST over 25 years to test the MPV hypothesis. Great inspiration for this work came from her book written in Japanese and titled "Seimei-Sekai-No-Hi-Taishosei" (Broken Symmetry in the Biological World—Why Nature Loves Imbalance) (Chuko-Shinsho, Tokyo, Japan, 1992). This intuitive book prompted M.F. to officially start an in-house research project titled "Study on Artificial Helix—Quest for the Possibility of Parity Nonconservation (PNC) of Helical Polymers and Organic Substances" at NTT Basic Research Laboratory (proposed on 19 May 1994 and 20 September 1995). We are grateful to the Kento Okoshi (Chitose Institute of Science and Technology), Eiji Yashima (Nagoya University), Yoshio Okamoto (Nagoya University), Yosef Scolnik (Weizmann Institute of Science), the late Meir Shinitzky (Weizmann Institute of Science), Victor Borovkov (South-Central University for Nationalities), Katsuya Inoue (Hiroshima University), Ullrich Scherf (Bergische Universität Wuppertal), Kazuo Yamaguchi (Kanagawa University), Daisuke Uemura (Kanagawa University), Masakatsu Matsumoto (Kanagawa University), Nobuhiro Kihara (Kanagawa University), Takayoshi Kawasaki (Tokuyama Corporation), Anubhav Saxena (Pidilite Industries, India), Tohru Asahi (Waseda University), Kenso Soai (Tokyo Science of University), Tamaki Nakano (Hokkaido University), Kenji Monde (Hokkaido University), Yoshitane Imai (Kindai University), and Mikiharu Kamachi (Osaka University). M.F. is also grateful to his colleagues at NTT Basic Research Laboratory, Kyozaburo Takeda (Waseda University), Hiroyuki Teramae (Jousai University), Kazuaki Furukawa (Meisei University), Keisuke Ebata, Tsutomu Horiuchi (Institute of Technology), Hiroaki Takayanagi (Tokyo Science University), Kei-ichi Torimitsu (Tohoku University), Nobuo Matsumoto (Shonan Institute of Technology), Masao Morita (Tohoku University), Osamu Niwa (Saitama Institute of Technology), Noriyuki Hatakenaka (Hiroshima University), Seiji Toyota (NTT group), Hisao Tabei, Hiromi Takigawa-Tamoto, Masaie Fujino (National Institute of Technology, Gunma College), and Hiroaki Isaka (NTT group). Also, M.F., as a principal investigator of a JST-CREST program (FY 1998–2003), is grateful to his intuitive mentors, Hideki Sakurai (Tohoku University), Toyoki Kunitake (Kyushu University), Masaki Hasegawa (the University of Tokyo), Shinji Murai (Osaka University), Hiizu Iawamura (University of Tokyo), Shigeyuki Kimura (National institute for Research in Inorganic Materials), the late Kenji Koga (University of Tokyo), Tadashi Imaki (Mitsubishi Chemical Corporation), the late Katsuhiko Kuroda (Mitsubishi Chemical Corporation), the late Akio Teramoto (Osaka University), Takahiro Sato (Osaka University), Masashi Kunitake (Kumamoto University), Junji Watanabe (Tokyo Institute of Technology), Hiroshi Nakashima (NTT Basic Research Laboratory), Hong-Zhi Tang (University of Michigan), Masao Motonaga, Zhong-Biao Zhang (Tianjin Normal University), and his colleagues. Also, M.F. is grateful to his students at NAIST (since FY 2002), Masaaki Ishikawa, Fumiko Ichiyanagi, Yoshihiro Kimura, Yoshifumi Kawagoe, Yoko Nakano, Takashi Mori, Woojung Chung, Ayako Nakao, Kana Yoshida, Makoto Taguchi, Yuri Donguri, Nozomu Suzuki, Keisuku Yoshida, Yuka Kato, Shosei Yoshimoto, Duong Thi Sang, Sibo Guo, Yota Katsurada, Hiroki Kamite, Ai Yokokura, Toshiki Nagai, Nor Azura Abdul Rahim, Jalilah Binti Abd Jalil, Shun Okazaki, Nanami Ogata, and Asuka Okubo for critical comments and constructive discussion on the MPV hypothesis and our radical ideas. M.F. learned about the unique nature of vortex light from Masahiro Katoh (Institute for Molecular Science (IMS)), Masaki Fujimoto (IMS), Daisuke Tadokoro (Kyoto University), and Takashiro Akitsu (Tokyo University of Science). Most of all, students at NAIST and several visiting researchers and students recognized the inherent imbalance in chiroptical properties (sign, magnitude, and wavelength) of the non-rigid luminophores using our CPL-200 spectrophotometer even after precision maintenance by a JASCO engineer, including replacement and tuning of the high-pressure Xe light source, power supply, concave/flat mirror sets, and two focused lenses. We are thankful to Takashi Takakuwa (JASCO), Yoshirou Kondo (JASCO), Hiroshi Kiyonaga (JASCO), and Koushi Nagamori (JASCO) for their technical advice for many years, and particular thanks are due to Nobuyuki Sakayanagi (JASCO), who designed the original CPL-200 spectrophotometer and released the first commercial model to M.F.'s lab in March 1999. Also, thanks are given to Yasuo Nakanishi (JASCO Engineering Co.), who continuously maintained the instrument in top condition with a high S/N ratio and wavelength calibration. Without such care and maintenance of the spectrophotometers, we could not test the MPV hypothesis.

Conflicts of Interest: The authors have no competing interests or other interests that might be perceived to influence the results and/or discussion reported in this article.

References

1. Gardner, M. *The New Ambidextrous Universe—Symmetry and Asymmetry from Mirror Reflections to Superstrings*; 3rd ed.; Freeman: New York, NY, USA, 1990; ISBN 9780486442440.
2. Miller, S.L. A Production of Amino Acids under Possible Primitive Earth Conditions. *Science* **1953**, *117*, 528–529. [CrossRef] [PubMed]
3. Hanafusa, H.; Akabori, S. Polymerization of Aminoacetonitrile. *Bull. Chem. Soc. Jpn.* **1959**, *32*, 626–630. [CrossRef]
4. Harada, K.; Fox, S.W. Thermal Synthesis of Natural Amino-Acids from a Postulated Primitive Terrestrial Atmosphere. *Nature* **1964**, *201*, 335–336. [CrossRef] [PubMed]
5. Seckbach, J.; Chela-Flores, J.; Owen, T.; Raulin, F. (Eds.) *Life in the Universe: From the Miller Experiment to the Search for Life on other Worlds*; Kluwer: Dordrecht, Germany, 2004; ISBN 1-4020-2371-5.

6. Breslow, R. A Likely Possible Origin of Homochirality in Amino Acids and Sugars on Prebiotic Earth. *Tetrahedron Lett.* **2011**, *52*, 2028–2032. [CrossRef]
7. Mason, S.F. *Chemical Evolution: Origin of the Elements, Molecules, and Living Systems*; Oxford University Press: New York, NY, USA, 1991; ISBN 0-19-855272-6.
8. Wagnière, G.H. *On Chirality and the Universal Asymmetry: Reflections on Image and Mirror Image*; Wiley-VCH: Zülich, Switzerland, 2007; ISBN 978-3-90639-038-3.
9. Guijarro, A.; Yus, M. *Origin of Chirality in the Molecules of Life: A Revision from Awareness to the Current Theories and Perspectives of this Unsolved Problem*; RSC Publishing: Cambridge, UK, 2008; ISBN 978-0-85404-156-5.
10. Rauchfuss, H. *Chemical Evolution and the Origin of Life*; Mitchell, T.N., Translator; Springer: Berlin, Germany, 2008; ISBN 978-3-540-78822-5.
11. Soai, K. (Ed.) *Amplification of Chirality*; Springer: Berlin, Germany, 2008; ISBN 978-3-540-77868-4.
12. Meierhenrich, U. *Amino Acids and the Asymmetry of Life*; Springer: Berlin, Germany, 2010; ISBN 978-3-540-76885-2.
13. Boyd, R. *Stardust, Supernovae and the Molecules of Life: Might We All Be Aliens?* Springer: New York, NY, USA, 2012; ISBN 978-1-4614-1331-8.
14. MacDermott, A.J. The Ascent of Parity-Violation: Exochirality in the Solar System and Beyond. *Enantiomer* **2000**, *5*, 153–168. [PubMed]
15. Fujiki, M.; Yoshida, K.; Suzuki, N.; Rahim, N.A.A.; Jalil, J.A. Tempo-Spatial Chirogenesis. Limonene-Induced Mirror Symmetry Breaking of Si–Si Bond Polymers During Aggregation in Chiral Fluidic Media. *J. Photochem. Photobiol. A Chem.* **2016**, *331*, 120–129. [CrossRef]
16. Schwieterman, E.W.; Kiang, N.Y.; Parenteau, M.N.; Harman, C.E.; DasSarma, S.; Fisher, T.M.; Arney, G.N.; Hartnett, H.E.; Reinhard, C.T.; Olson, S.L.; et al. Exoplanet Biosignatures: A Review of Remotely Detectable Signs of Life. *Astrobiology* **2018**, *18*, 663–708. [CrossRef] [PubMed]
17. Schrodinger, E. *What Is Life? With Mind and Matter and Autobiographical Sketches*; Reprint Version; Cambridge University Press: Cambridge, UK, 2012; ISBN 1107683653.
18. Freedman, R.; Geller, R.; Kaufmann, W.J. *Universe*, 10th ed.; Freeman, W.H., Ed.; W.H. Freeman and Company, Now an Imprint of Macmillan Higher Education, a Division of Macmillan Publishers: London, UK, 2015; ISBN 1319042384.
19. Accelerating Expansion of the Universe. Available online: https://en.wikipedia.org/wiki/Accelerating_expansion_of_the_universe (accessed on 11 November 2018).
20. Forbes, N.; Mahon, B. *Faraday, Maxwell, and the Electromagnetic Field: How Two Men Revolutionized Physics*; Prometheus Books: Amherst, NY, USA, 2014; ISBN 9781616149420.
21. Lennartson, A. Appendix. Absolute Asymmetric Synthesis 1874–2009. In *Absolute Asymmetric Synthesis*; University of Gothenburg: Gothenburg, Sweden, 2011; pp. 59–74. ISBN 978-91-628-7836-8.
22. Beth, R.A. Mechanical Detection and Measurement of the Angular Momentum of Light. *Phys. Rev.* **1936**, *50*, 115–125. [CrossRef]
23. Inoue, Y. Asymmetric Photochemical Reactions in Solution. *Chem. Rev.* **1992**, *92*, 741–770. [CrossRef]
24. Laur, P. The First Decades After the Discovery of CD and ORD by Aimé Cotton in 1895. In *Comprehensive Chiroptical Spectroscopy: Applications in Stereochemical Analysis of Synthetic Compounds, Natural Products, and Biomolecules*; Berova, N., Polavarapu, P.L., Nakanishi, K., Woody, R.W., Eds.; Wiley: Hoboken, NJ, USA, 2000; Volume 2, Chapter 1, pp. 1–35. [CrossRef]
25. Rikken, G.L.J.A.; Raupach, E. Enantioselective Magnetochiral Photochemistry. *Nature* **2000**, *405*, 932–935. [CrossRef] [PubMed]
26. Ribó, J.M.; Blanco, C.; Crusats, J.; El-Hachemi, Z.; Hochberg, D.; Moyano, A. Absolute Asymmetric Synthesis in Enantioselective Autocatalytic Reaction Networks: Theoretical Games, Speculations on Chemical Evolution and Perhaps A Synthetic Option. *Chem. Eur. J.* **2014**, *20*, 17250–17271. [CrossRef]
27. Okano, K.; Taguchi, M.; Fujiki, M.; Yamashita, T. Circularly Polarized Luminescence of Rhodamine B in a Supramolecular Chiral Medium Formed by a Vortex Flow. *Angew. Chem. Int. Ed.* **2011**, *50*, 12474–12477. [CrossRef] [PubMed]
28. Sun, J.; Li, Y.; Yan, F.; Liu, C.; Sang, Y.; Tian, F.; Feng, Q.; Duan, P.; Zhang, L.; Shi, X.; et al. Control Over the Emerging Chirality in Supramolecular Gels and Solutions by Chiral Microvortices in Milliseconds. *Nat. Commun.* **2018**, *9*, 2599. [CrossRef]
29. Mineo, P.; Villari, V.; Scamporrino, E.; Micali, N. New Evidence about the Spontaneous Symmetry Breaking: Action of an Asymmetric Weak Heat Source. *J. Phys. Chem. B* **2015**, *119*, 12345–12353. [CrossRef] [PubMed]

30. Eugene Wigner. Available online: https://en.wikipedia.org/wiki/Eugene_Wigner (accessed on 17 November 2018).
31. Wigner, E.P. Violation of Symmetry in Physics. *Sci. Am.* **1965**, *213*, 28–36. [CrossRef]
32. Gross, D.J. Symmetry in Physics: Wigner's Legacy. *Phys. Today* **1995**, *48*, 46–50. [CrossRef]
33. Lee, T.D.; Yang, C.N. Question of Parity Conservation in Weak Interactions. *Phys. Rev.* **1956**, *104*, 254–258. [CrossRef]
34. Wu, C.S.; Ambler, E.; Hayword, R.W.; Hoppes, D.D.; Hudson, R.P. Experimental Test of Parity Conservation on Beta Decay. *Phys. Rev.* **1957**, *105*, 1413–1415. [CrossRef]
35. Schopper, H. Circular Polarization of γ-Rays: Further Proof for Parity Failure in β Decay. *Philos. Mag.* **1957**, *2*, 710–713. [CrossRef]
36. Goldhaber, M.; Grodzins, L.; Sunyar, A. Helicity of Neutrinos. *Phys. Rev.* **1958**, *109*, 1015–1017. [CrossRef]
37. Fagg, L.W.; Hanna, S.S. Polarization Measurements on Nuclear Gamma Rays. *Rev. Mod. Phys.* **1959**, *31*, 711–758. [CrossRef]
38. Wu, C.S. Parity Experiments in Beta Decay. *Rev. Mod. Phys.* **1959**, *31*, 783–790. [CrossRef]
39. Christenson, J.H.; Cronin, J.W.; Fitch, V.L.; Turlay, R. Evidence for the 2π Decay of the ${K_2}^0$ Meson. *Phys. Rev. Lett.* **1964**, *13*, 138–140. [CrossRef]
40. Aubert, B.; et al. [BABAR Collaboration]. Observation of CP Violation in the B_0 Meson System. *Phys. Rev. Lett.* **2001**, *87*, 091801. [CrossRef] [PubMed]
41. Abe, K.; et al. [Belle Collaboration]. Observation of Large CP Violation in the Neutral *B* Meson System. *Phys. Rev. Lett.* **2001**, *87*, 091802. [CrossRef] [PubMed]
42. Abe, K.; et al. [T2K Collaboration]. Combined Analysis of Neutrino and Antineutrino Oscillations at T2K. *Phys. Rev. Lett.* **2017**, *118*, 151801. [CrossRef] [PubMed]
43. Calvert, J.G.; Pitts, J.N., Jr. *Photochemistry*, 2nd ed.; John Wiley & Sons, Inc.: New York, NY, USA, 1966; Chapter 4-2D Selection rule; pp. 258–260. ISBN 0471130907.
44. Turro, N.J. *Modern Molecular Photochemistry*; University Science Books: Sausalito, CA, USA, 1991; Chapter 5.6 State Mixing: Breakdown of the Single Orbital Configuration and Pure Multiplicity Approximations; pp. 96–103. ISBN 0935702717.
45. Görling, C.; Jalviste, E.; Ohta, N.; Ottinger, C. Lifetime Measurements of the Collision-Free Slow Fluorescence from Glyoxal S1/T1 Gateway Levels in a Beam. *J. Phys. Chem. A* **1998**, *102*, 10620–10629. [CrossRef]
46. Yamazaki, I.; Aratani, N.; Akimoto, S.; Yamazaki, T.; Osuka, A. Observation of Quantum Coherence for Recurrence Motion of Exciton in Anthracene Dimers in Solution. *J. Am. Chem. Soc.* **2003**, *125*, 7192–7193. [CrossRef]
47. Latal, H. Parity Violation in Atomic Physics. In *Chirality—From Weak Bosons to the α-Helix*; Janoschek, R., Ed.; Springer: Berlin, Germany, 1991; pp. 1–17. ISBN 978-3-642-76569.
48. Van der Meer, S. Stochastic Cooling and the Accumulation of Antiprotons. *Rev. Mod. Phys.* **1985**, *57*, 689–698. [CrossRef]
49. Rubbia, C. Experimental Observation of the Intermediate Vector Bosons W^+, W^- and Z^0. *Rev. Mod. Phys.* **1985**, *57*, 699–722. [CrossRef]
50. Walgate, R. What Will Come After the Z^0? *Nature* **1983**, *303*, 473. [CrossRef]
51. Bouchiat, M.A.; Bouchiat, C.C. Weak Neutral Currents in Atomic Physics. *Phys. Lett. B* **1974**, *48*, 111–114. [CrossRef]
52. Baied, P.E.G.; Brimcombe, M.W.S.M.; Roberts, G.J.; Sandars, P.G.H.; Soreide, D.C.; Fortson, E.N.; Lewis, L.L.; Lindahl, E.G.; Soreide, D.C. Search for Parity Non-Conserving Optical Rotation in Atomic Bismuth. *Nature* **1976**, *264*, 528–529. [CrossRef]
53. Forte, M.; Heckel, B.R.; Ramsey, N.F.; Green, K.; Greene, G.L.; Byrne, J.; Pendlebury, J.M. First measurement of parity-nonconserving neutron-spin rotation: The tin isotopes. *Phys. Rev. Lett.* **1980**, *45*, 2088–2091. [CrossRef]
54. Bucksbaum, P.H.; Commins, E.D.; Hunter, L.R. Observations of Parity Non-Conservation in Atomic Thallium. *Phys. Rev. D* **1981**, *24*, 1134–1148. [CrossRef]
55. Emmons, T.P.; Reeves, J.M.; Fortson, E.N. Parity-Non-Conserving Optical Rotation in Atomic Lead. *Phys. Rev. Lett.* **1983**, *51*, 2089–2091. [CrossRef]
56. Bouchiat, M.-A.; Pottier, L. Optical Experiments and Weak Interactions. *Science* **1986**, *234*, 1203–1210. [CrossRef]

57. Bouchiat, M.A.; Bouchiat, C.C. Parity Violation in Atoms. *Rep. Prog. Phys.* **1997**, *60*, 1351–1396. [CrossRef]
58. Wood, C.S.; Bennett, S.C.; Cho, D.; Masterson, B.P.; Roberts, J.L.; Tanner, C.E.; Wieman, C.E. Measurement of Parity Nonconservation and an Anapole Moment in Cesium. *Science* **1997**, *275*, 1759–1763. [CrossRef]
59. Mitchell, G.E.; Bowman, J.D.; Penttilä, S.I.; Sharapov, E.I. Parity Violation in Compound Nuclei: Experimental Methods and Recent Results. *Phys. Rep.* **2001**, *354*, 157–241. [CrossRef]
60. Guéna, J.; Lintz, M.; Bouchiat, M.-A. Atomic Parity Violation: Principles, Recent Results, Present Motivations. *Mod. Phys. Lett. A* **2005**, *20*, 375–390. [CrossRef]
61. Yamagata, Y. A Hypothesis for the Asymmetric Appearance of Biomolecules on Earth. *J. Theor. Biol.* **1966**, *11*, 495–498. [CrossRef]
62. Rein, D.W. Some Remarks on Parity Violating Effects of Intramolecular Interactions. *J. Mol. Evol.* **1974**, *4*, 15–22. [CrossRef]
63. Letokhov, V.S. On Difference of Energy Levels of Left and Right Molecules Due to Weak Interactions. *Phys. Lett. A* **1975**, *53*, 275–276. [CrossRef]
64. Zel'Dovich, Ya. B.; Saakyan, D.B.; Sobel'Man, I.I. Energy Difference between Right-Hand and Left-Hand Molecules due to Parity Nonconservation in Weak Interactions of Electrons with Nuclei. *JETP Lett.* **1977**, *25*, 94–97.
65. Keszthelyi, L. Origin of the Asymmetry of Biomolecules and Weak Interaction. *Orig. Life* **1977**, *8*, 299–340. [CrossRef]
66. Harris, R.A.; Stodolsky, L. Quantum Beats in Optical Activity and Weak Interactions. *Phys. Lett. B* **1978**, *78*, 313–317. [CrossRef]
67. Hegstrom, R.A.; Rein, D.W.; Sandars, P.G.H. Calculation of the Parity Nonconserving Energy Difference between Mirror-Image Molecules. *J. Chem. Phys.* **1980**, *73*, 2329–2341. [CrossRef]
68. Mason, S.F.; Tranter, G.E. Energy Inequivalence of Peptide Enantiomers from Parity Non-Conservation. *Chem. Commun.* **1983**, 117–119. [CrossRef]
69. Mason, S.F.; Tranter, G.E. The Parity-Violating Energy Difference between Enantiomeric Molecules. *Mol. Phys.* **1984**, *53*, 1091–1111. [CrossRef]
70. Barron, L.D. Symmetry and molecular chirality. *Chem. Soc. Rev.* **1986**, *15*, 189–223. [CrossRef]
71. Kondepudi, D. Parity Violations and the Origin of Bimolecular Handedness. In *Entropy, Information and Evolution: New Perspective on Physical and Biological Evolution*; Weber, B.H., Depew, D.J., Smith, J.D., Eds.; MIT Press: Cambridge, MA, USA, 1988; ISBN 0262731681.
72. Quack, M. Structure and Dynamics of Chiral Molecules. *Angew. Chem. Int. Ed.* **1989**, *28*, 571–586. [CrossRef]
73. Hegstrom, R.A.; Kondepudi, D.K. The Handedness of the Universe. *Sci. Am.* **1990**, *262*, 108–115. [CrossRef]
74. Salam, A. The Role of Chirality in the Origin of Life. *J. Mol. Evol.* **1991**, *33*, 105–113. [CrossRef]
75. Macdermott, A.J. Electroweak Enantioselection and the Origin of Life. *Orig. Life Evol. Biosph.* **1995**, *25*, 191–199. [CrossRef]
76. Kikuchi, O.; Kiyonaga, H. Parity-Energy Shift of Helical *n*-Alkanes. *J. Mol. Struct. (Theochem.)* **1994**, *312*, 271–274. [CrossRef]
77. Avetisov, V.; Goldanskii, V. Mirror Symmetry-Breaking at the Molecular Level. *Proc. Natl. Acad. Sci. USA* **1996**, *93*, 11435–11442. [CrossRef]
78. Bonner, W.A. Enantioselective Autocatalysis. IV. Implications for Parity Violation Effects. *Orig. Life Evol. Biosph.* **1996**, *26*, 27–45. [CrossRef]
79. Szabó-Nagy, A.; Keszthelyi, L. Demonstration of the Parity-Violating Energy Difference between Enantiomers. *Proc. Natl. Acad. Sci. USA* **1999**, *96*, 4252–4255. [CrossRef]
80. Gottselig, M.; Luckhaus, D.; Quack, M.; Stohner, J.; Willeke, M. Mode Selective Stereomutation and Parity Violation in Disulfane Isotopomers H_2S_2, D_2S_2, T_2S_2. *Helv. Chim. Acta* **2001**, *84*, 1846–1861. [CrossRef]
81. Compton, R.N.; Pagni, R.M. The Chirality of Biomolecules. *Adv. At. Mol. Opt. Phys.* **2002**, *48*, 219–261. [CrossRef]
82. Schwerdtfeger, P.; Gierlich, J.; Bollwein, T. Large Parity-Violation Effects in Heavy-Metal-Containing Chiral Compounds. *Angew. Chem. Int. Ed.* **2003**, *42*, 1293–1296. [CrossRef]
83. MacDermott, A.J.; Hegstrom, R.A. A Proposed Experiment to Measure the Parity-Violating Energy Difference between Enantiomers from the Optical Rotation of Chiral Ammonia-Like "Cat" Molecules. *Chem. Phys.* **2004**, *305*, 55–68. [CrossRef]

84. Quack, M.; Stohner, J.; Willeke, M. High-Resolution Spectroscopic Studies and Theory of Parity Violation in Chiral Molecules. *Annu. Rev. Phys. Chem.* **2008**, *59*, 741–769. [CrossRef]
85. Bargueño, P.; Gonzalo, I.; de Tudela, R.P. Detection of Parity Violation in Chiral Molecules by External Tuning of Electroweak Optical Activity. *Phys. Rev. A* **2009**, *80*, 012110. [CrossRef]
86. Dorta-Urra, A.; Peñate-Rodríguez, H.C.; Bargueño, P.; Rojas-Lorenzo, G.; Miret-Artés, S. Dissipative Geometric Phase and Decoherence in Parity-Violating Chiral Molecules. *J. Chem. Phys.* **2012**, *136*, 174505. [CrossRef]
87. Famiano, M.A.; Boyd, R.N.; Kajino, T.; Onaka, T.; Mo, Y. Amino Acid Chiral Selection via Weak Interactions in Stellar Environments: Implications for the Origin of Life. *Sci. Rep.* **2018**, *8*, 8833. [CrossRef]
88. Daussy, Ch.; Marrel, T.; Amy-Klein, A.; Nguyen, C.T.; Bordé, C.J.; Chardonnet, C. Limit on the Parity Nonconserving Energy Difference between the Enantiomers of a Chiral Molecule by Laser Spectroscopy. *Phys. Rev. Lett.* **1999**, *83*, 1554–1557. [CrossRef]
89. Wang, W.; Yi, F.; Ni, Y.; Zhao, Z.; Jin, X.; Tang, Y. Parity Violation of Electroweak Force in Phase Transitions of Single Crystals of *D*- and *L*-Alanine and Valine. *J. Biol. Phys.* **2000**, *26*, 51–65. [CrossRef]
90. Fujiki, M. Experimental Tests of Parity Violation at Helical Polysilylene Level. *Macromol. Rapid Commun.* **2001**, *22*, 669–674. [CrossRef]
91. Pagni, R.M.; Compton, R.N. Asymmetric Synthesis of Optically Active Sodium Chlorate and Bromate Crystals. *Cryst. Growth Des.* **2002**, *2*, 249–253. [CrossRef]
92. Scolnik, T.; Portnaya, I.; Cogan, U.; Tal, S.; Haimovitz, R.; Fridkin, M.; Elitzur, A.C.; Deamer, D.W.; Shinitzky, M. Subtle Differences in Structural Transitions between Poly-L- and Poly-D-Amino Acids of Equal Length in Water. *Phys. Chem. Chem. Phys.* **2006**, *8*, 333–339. [CrossRef]
93. Kodona, E.K.; Alexopoulos, C.; Panou-Pomonis, E.; Pomonis, P.J. Chirality and Helix Stability of Polyglutamic Acid Enantiomers. *J. Colloid Interface Sci.* **2008**, *319*, 72–80. [CrossRef]
94. Darquié, B.; Stoeffler, C.; Shelkovnikov, A.; Daussy, C.; Amy-Klein, A.; Chardonnet, C.; Zrig, S.; Guy, L.; Crassous, J.; Soulard, P. Progress Toward the First Observation of Parity Violation in Chiral Molecules by High-Resolution Laser Spectroscopy. *Chirality* **2010**, *22*, 870–884. [CrossRef]
95. Fujiki, M. Mirror Symmetry Breaking in Helical Polysilanes: Preference between Left and Right of Chemical and Physical Origin. *Symmetry* **2010**, *2*, 1625–1652. [CrossRef]
96. Albert, S.; Arn, F.; Bolotova, I.; Chen, Z.; Fábri, C.; Grassi, G.; Lerch, P.; Quack, M.; Seyfang, G.; Wokaun, A.; et al. Synchrotron-Based Highest Resolution Terahertz Spectroscopy of the ν_{24} Band System of 1,2-Dithiine ($C_4H_4S_2$): A Candidate for Measuring the Parity Violating Energy Difference between Enantiomers of Chiral Molecules. *J. Phys. Chem. Lett.* **2016**, *7*, 3847–3853. [CrossRef]
97. Kozlova, S.G.; Gabuda, S.P. Thermal Properties of $Zn_2(C_8H_4O_4)_2{\cdot}C_6H_{12}N_2$ Metal-Organic Framework Compound and Mirror Symmetry Violation of Dabco Molecules. *Sci. Rep.* **2017**, *7*, 11505. [CrossRef]
98. Lightner, D.A.; Gurst, J.E. *Organic Conformational Analysis and Stereochemistry from Circular Dichroism Spectroscopy*; Wiley-VCH: Weinheim, Germany, 2000; ISBN 978-0-471-35405-5.
99. Eliel, E.L.; Wilen, S.H. *Stereochemistry of Organic Compounds*, 1st ed.; Wiley-Interscience: Hoboken, NJ, USA, 1994; ISBN 9780471016700.
100. Anslyn, E.V.; Dougherty, D.A. *Modern Physical Organic Chemistry*; University Science: Mill Valley, CA, USA, 2005; ISBN 9781891389313.
101. Hund, F. Symmetriecharaktere von Termen bei Systemen mit Gleichen Partikeln in der Quantenmechanik. *Z. Phys.* **1927**, *43*, 788–803. [CrossRef]
102. Bell, R.P. *The Tunnel Effect in Chemistry*; Chapman and Hall: London, UK, 1980; ISBN 0-412-21340-0.
103. Laane, J. Vibrational Potential Energy Surfaces in Electronic Excited States. In *Frontiers of Molecular Spectroscopy*; Laane, J., Ed.; Elsevier: New York, NY, USA, 2009; Chapter 4, pp. 63–132, ISBN 978-0-444-53175-9.
104. Fujiki, M.; Koe, J.R.; Mori, T.; Kimura, Y. Questions of Mirror Symmetry at the Photoexcited and Ground States of Non-Rigid Luminophores Raised by Circularly Polarized Luminescence and Circular Dichroism Spectroscopy: Part 1. Oligofluorenes, Oligophenylenes, Binaphthyls and Fused Aromatics. *Molecules* **2018**, *23*, 2606. [CrossRef]
105. Shindo, Y.; Nakagawa, M. On the Artifacts in Circularly Polarized Emission Spectroscopy. *Appl. Spectrosc.* **1985**, *39*, 32–38. [CrossRef]
106. Blok, P.M.L.; Dekkers, H.P.J.M. Measurement of the Circular Polarization of the Luminescence of Photoselected Samples under Artifact-free Conditions. *Appl. Spectrosc.* **1990**, *44*, 305–309. [CrossRef]

107. Kaur, S. A Review on Electronic Spectroscopy of Perylene. Master's Thesis, San Jose State University, San Jose, CA, USA, 1999. Available online: http://scholarworks.sjsu.edu/etd_theses/1879 (accessed on 1 February 2019).
108. Allamandola, L.J.; Tielens, A.G.G.M.; Barker, J.R. Interstellar Polycyclic Aromatic Hydrocarbons—The Infrared Emission Bands, the Excitation/Emission Mechanism, and the Astrophysical Implications. *Astrophys. J. Suppl. Ser.* **1989**, *71*, 733–775. [CrossRef]
109. Tielens, A.G.G.M. Interstellar Polycyclic Aromatic Hydrocarbon Molecules. *Annu. Rev. Astron. Astrophys.* **2008**, *46*, 289–337. [CrossRef]
110. Walter, E.R.H.; Williams, J.A.G.; Parker, D. Solvent polarity and oxygen sensitivity, rather than viscosity, determine lifetimes of biaryl-sensitised terbium luminescence. *Chem. Commun.* **2017**, *53*, 13344–133347. [CrossRef]
111. Arbeloa, F.L.; Bañuelos, J.; Martínez, V.; Arbeloa, T.; López Arbeloa, I. Structural, Photophysical and Lasing Properties of Pyrromethene Dyes. *Int. Rev. Phys. Chem.* **2005**, *24*, 339–374. [CrossRef]
112. Cerdán, L.; García-Moreno, S.; Costela, A.; García-Moreno, I.; de la Moya, A.S. Circularly Polarized Laser Emission Induced in Isotropic and Achiral Dye Systems. *Sci. Rep.* **2016**, *6*, 28740. [CrossRef]
113. Ebata, K.; Inada, T.; Kabuto, C.; Sakurai, H. Hexakis(fluorodimethylsilyl)benzene, Hexakis(methoxy-dimethylsilyl)benzene, and Related Compounds. Novel Neutral Pentacoordinate Structures for Silicon and Merry-Go-Round Degenerate Fluorine Migration. *J. Am. Chem. Soc.* **1994**, *116*, 3595–3596. [CrossRef]
114. Organic Scintillators Energy. Available online: https://en.wikipedia.org/wiki/Scintillator#Organic_scintillators (accessed on 13 October 2018).
115. Araki, T.; Enomoto, S.; Furuno, K.; Gando, Y.; Ichimura, K.; Ikeda, H.; Inoue, K.; Kishimoto, Y.; Koga, M.; Koseki, Y.; et al. Experimental Investigation of Geologically Produced Antineutrinos with KamLAND. *Nature* **2005**, *436*, 499–503. [CrossRef]
116. KamLAND. Available online: https://en.wikipedia.org/wiki/Kamioka_Liquid_Scintillator_Antineutrino_Detector (accessed on 7 November 2018).
117. Nucleosynthesis Reactions. Available online: https://en.wikipedia.org/wiki/Nucleosynthesis (accessed on 7 November 2018).
118. Jones, G., II; Jackson, W.R.; Choi, C.Y.; Bergmark, W.R. Solvent Effects on Emission Yield and Lifetime for Coumarin Laser Dyes. Requirements for a Rotatory Decay Mechanism. *J. Phys. Chem.* **1985**, *89*, 294–300. [CrossRef]
119. Weber, M.J. *Handbook of Laser Wavelengths*; CRC Press: Boca Raton, FL, USA, 1998; ISBN 0849335086.
120. Birks, J.B. *The Theory and Practice of Scintillation Counting*; Elsevier: Amsterdam, The Netherlands, 1964; ISBN 978-0-08-010472-0.
121. Duarte, F.J. (Ed.) *Tunable Lasers Handbook (Optics and Photonics)*; Academic Press: Cambridge, MA, USA, 1995; ISBN 012222695X.
122. Fujiki, M.; Jalilah, A.J.; Suzuki, N.; Taguchi, M.; Zhang, W.; Abdellatif, M.M.; Nomura, K. Chiral Optofluidics: Gigantic Circularly Polarized Light Enhancement of All-*trans*-poly(9,9-di-*n*-octylfluorene-2,7-vinylene) during Mirror-symmetry-breaking Aggregation by Optically Tuning Fluidic Media. *RSC Adv.* **2012**, *2*, 6663–6671. [CrossRef]
123. Weinberg Angle. Available online: https://en.wikipedia.org/wiki/Weinberg_angle (accessed on 2 October 2018).
124. Electroweak Interaction. Available online: https://en.wikipedia.org/wiki/Electroweak_interaction (accessed on 30 October 2018).
125. Nakashima, H.; Koe, J.R.; Torimitsu, K.; Fujiki, M. Transfer and Amplification of Chiral Molecular Information to Polysilylene Aggregates. *J. Am. Chem. Soc.* **2001**, *123*, 4847–4848. [CrossRef] [PubMed]
126. Fujiki, M. Mirror Symmetry Breaking of Silicon Polymers—From Weak Bosons to Artificial Helix. *Chem. Rec.* **2009**, *9*, 271–298. [CrossRef] [PubMed]
127. Nakano, Y.; Fujiki, M. Circularly Polarized Light Enhancement by Helical Polysilane Aggregates Suspension in Organic Optofluids. *Macromolecules* **2011**, *44*, 7511–7519. [CrossRef]
128. Fujiki, M.; Yoshida, K.; Suzuki, N.; Zhang, J.; Zhang, W.; Zhu, X. Mirror Symmetry Breaking and Restoration within μm-sized Polymer Particles in Optofluidic Media by Pumping Circularly Polarised Light. *RSC Adv.* **2013**, *3*, 5213–5219. [CrossRef]

129. Fujiki, M.; Kawagoe, Y.; Nakano, Y.; Nakao, A. Mirror-Symmetry-Breaking in Poly[(9,9-di-*n*-octylfluorenyl-2,7-diyl)-*alt*-biphenyl] (PF8P2) is Susceptible to Terpene Chirality, Achiral Solvents and Mechanical Stirring. *Molecules* **2013**, *18*, 7035–7057. [CrossRef] [PubMed]
130. Wang, L.; Suzuki, N.; Liu, J.; Matsuda, T.; Rahim, N.A.A.; Zhang, W.; Fujiki, M.; Zhang, Z.; Zhou, N.; Zhu, X. Limonene Induced Chiroptical Generation and Inversion during Aggregation of Achiral Polyfluorene Analogs: Structure-Dependence and Mechanism. *Polym. Chem.* **2014**, *5*, 5920–5927. [CrossRef]
131. Fujiki, M.; Donguri, Y.; Zhao, Y.; Nakao, A.; Suzuki, N.; Yoshida, K.; Zhang, W. Photon Magic: Chiroptical Polarisation, Depolarisation, Inversion, Retention and Switching of Non-Photochromic Light-Emitting Polymers in Optofluidic Medium. *Polym. Chem.* **2015**, *6*, 1627–1638. [CrossRef]
132. Nakano, Y.; Ichiyanagi, F.; Naito, M.; Yang, Y.; Fujiki, M. Chiroptical Generation and Inversion During the Mirror-Symmetry-Breaking Aggregation of Dialkylpolysilanes due to Limonene Chirality. *Chem. Commun.* **2012**, *48*, 6636–6638. [CrossRef]
133. Duong, T.S.; Fujiki, M. The Origin of Bisignate Circularly Polarized Luminescence (CPL) Spectra from Chiral Polymer Aggregates and Molecular Camphor: Anti-kasha's Rule Revealed by CPL Excitation (CPLE) Spectra. *Polym. Chem.* **2017**, *8*, 4673–4679. [CrossRef]
134. Jalilah, A.J.; Asanoma, F.; Fujiki, M. Unveiling Controlled Breaking of the Mirror Symmetry of Eu(fod)$_3$ with α-/β-Pinene and BINAP by Circularly Polarised Luminescence (CPL), CPL Excitation, and ^{19}F-/^{31}P{1H}-NMR Spectra and Mulliken Charges. *Inorg. Chem. Front.* **2018**, *5*, 2718–2733. [CrossRef]
135. Mislow, K. Absolute Asymmetric Synthesis: A Commentary. *Collect. Czech. Chem. Commun.* **2003**, *68*, 849–864. [CrossRef]
136. Kawasaki, T.; Tanaka, H.; Tsutsumi, T.; Kasahara, T.; Sato, I.; Soai, K.J. Chiral discrimination of cryptochiral saturated quaternary and tertiary hydrocarbons by asymmetric autocatalysis. *Am. Chem. Soc.* **2006**, *128*, 6032–6033. [CrossRef]
137. Kawasaki, T.; Hohberger, C.; Araki, Y.; Hatase, K.; Beckerle, K.; Okuda, J.; Soai, K. Discrimination of Cryptochirality in Chiral Isotactic Polystyrene by Asymmetric Autocatalysis. *Chem. Commun.* **2009**, 5621–5623. [CrossRef]
138. Amako, T.; Nakabayashi, K.; Suzuki, N.; Guo, S.; Rahim, N.A.A.; Harada, T.; Fujiki, M.; Imai, Y. Pyrene Magic: Chiroptical Enciphering and Deciphering 1,3-Dioxolane Bearing Two Wirepullings to Drive Two Remote Pyrenes. *Chem. Commun.* **2015**, *51*, 8237–8240. [CrossRef]
139. Nakanishi, S.; Nakabayashi, K.; Mizusawa, T.; Suzuki, N.; Guo, S.; Fujiki, M.; Imai, Y. Cryptochiral Binaphthyl–Bipyrene Luminophores Linked with Alkylene Esters: Intense Circularly Polarised Luminescence, But Ultraweak Circular Dichroism. *RSC Adv.* **2016**, *6*, 99172–99176. [CrossRef]
140. Nakabayashi, K.; Kitamura, S.; Suzuki, N.; Guo, S.; Fujiki, M.; Imai, Y. Non-Classically Controlled Signs in a Circularly Polarised Luminescent Molecular Puppet: The Importance of the Wire Structure Connecting Binaphthyl and Two Pyrenes. *Eur. J. Org. Chem.* **2016**, 64–69. [CrossRef]
141. Hara, N.; Yanai, M.; Kaji, D.; Shizuma, M.; Tajima, N.; Fujiki, M.; Imai, Y. A Pivotal Biaryl Rotamer Bearing Two Floppy Pyrenes that Exhibits Cryptochiral Characteristics in the Ground State. *ChemistrySelect* **2018**, *3*, 9970–9973. [CrossRef]
142. Maeda, K.; Hirose, D.; Okoshi, N.; Shimomura, K.; Wada, Y.; Ikai, T.; Kanoh, S.; Yashima, E. Direct Detection of Hardly Detectable Hidden Chirality of Hydrocarbons and Deuterated Isotopomers by a Helical Polyacetylene through Chiral Amplification and Memory. *J. Am. Chem. Soc.* **2018**, *140*, 3270–3276. [CrossRef]
143. Bakasov, A.; Ha, T.-K.; Quack, M. Ab Initio Calculation of Molecular Energies including Parity Violating Interactions. *J. Chem. Phys.* **1998**, *109*, 7263–7285. [CrossRef]
144. Soai, K.; Kawasaki, T. Asymmetric Autocatalysis with Amplification of Chirality. *Top. Curr. Chem.* **2008**, *284*, 1–33. [CrossRef]
145. Green, M.M.; Jain, V. Homochirality in Life: Two Equal Runners, One Tripped. *Orig. Life Evol. Biosph.* **2010**, *40*, 111–118. [CrossRef] [PubMed]
146. Sandars, P.G.H. A Toy Model for the Generation of Homochirality During Polymerization. *Orig. Life Evol. Biosph.* **2003**, *33*, 575–587. [CrossRef] [PubMed]
147. Viedma, C. Selective Chiral Symmetry Breaking during Crystallization: Parity Violation or Cryptochiral Environment in Control? *Cryst. Growth Des.* **2007**, *7*, 553–556. [CrossRef]

148. McLaughlin, D.T.; Nguyen, T.P.T.; Mengnjo, L.; Bian, C.; Leung, Y.H.; Goodfellow, E.; Ramrup, P.; Woo, S.; Cuccia, L.A. Viedma Ripening of Conglomerate Crystals of Achiral Molecules Monitored Using Solid-State Circular Dichroism. *Cryst. Growth Des.* **2014**, *14*, 1067–1076. [CrossRef]
149. Liu, M.; Zhang, L.; Wang, T. Supramolecular Chirality in Self-Assembled Systems. *Chem. Rev.* **2015**, *115*, 7304–7397. [CrossRef]
150. Yashima, E.; Ousaka, N.; Taura, D.; Shimomura, K.; Ikai, T.; Maeda, K. Supramolecular Helical Systems: Helical Assemblies of Small Molecules, Foldamers, and Polymers with Chiral Amplification and Their Functions. *Chem. Rev.* **2016**, *22*, 13752–13990. [CrossRef]
151. Enoto, T.; Wada, Y.; Furuta, Y.; Nakazawa, K.; Yuasa, T.; Okuda, K.; Makishima, K.; Sato, M.; Sato, Y.; Nakano, T.; et al. Photonuclear Reactions Triggered by Lightning Discharge. *Nature* **2017**, *551*, 481–484. [CrossRef]
152. Bargueño, P.; de Tudela, R.P. The Role of Supernova Neutrinos on Molecular Homochirality. *Orig. Life Evol. Biosph.* **2007**, *37*, 253–257. [CrossRef]
153. Alexander, S.; Marcianò, A.; Smolin, L. Gravitational Origin of the Weak Interaction's Chirality. *Phys. Rev. D* **2014**, *89*, 065017. [CrossRef]
154. Bargueño, P. Gravitational Origin Parity Violation. *Chirality* **2015**, *27*, 375–381. [CrossRef]
155. Ribó, J.M.; Crusats, J.; Sagúes, F.; Claret, J.; Rubires, R. Chiral Sign Induction by Vortices During the Formation of Mesophases in Stirred Solutions. *Science* **2001**, *292*, 2063–2066. [CrossRef]
156. Tamburini, F.; Thidé, B.; Molina-Terriza, G.; Gabriele Anzolin, G. Twisting of Light Around Rotating Black Holes. *Nat. Phys.* **2011**, *7*, 195–197. [CrossRef]
157. Higurashi, E.; Ohguchi, O.; Tamamura, T.; Ukita, H.; Sawada, R. Optically Induced Rotation of Dissymmetrically Shaped Fluorinated Polyimide Micro-Objects in Optical Traps. *J. Appl. Phys.* **1997**, *82*, 2773–2779. [CrossRef]
158. Simpson, N.B.; Dholakia, K.; Allen, L.; Padgett, M.J. Mechanical Equivalence of Spin and Orbital Angular Momentum of Light: An Optical Spanner. *Opt. Lett.* **1997**, *22*, 52–54. [CrossRef]
159. Friese, M.E.J.; Rubinsztein-Dunlop, H.; Gold, J.; Hagberg, P.; Hanstorp, D. Optically Driven Micromachine Elements. *Appl. Phys. Lett.* **2001**, *78*, 547–549. [CrossRef]
160. Curtis, J.E.; Koss, B.A.; Grier, D.G. Dynamic Holographic Optical Tweezers. *Opt. Commun.* **2002**, *207*, 169–175. [CrossRef]
161. Brasselet, E.; Murazawa, N.; Misawa, H.; Juodkazis, S. Optical Vortices from Liquid Crystal Droplets. *Phys. Rev. Lett.* **2009**, *103*, 103903. [CrossRef]
162. Ambrosio, A.; Marrucci, L.; Borbone, F.; Roviello, A.; Maddalena, P. Light-Induced Spiral Mass Transport in Azo-Polymer Films Under Vortex-Beam Illumination. *Nat. Commun.* **2012**, *3*, 989. [CrossRef]
163. Watabe, M.; Juman, G.; Miyamoto, K.; Omatsu, T. Light Induced Conch-Shaped Relief in An Azo-Polymer Film. *Sci. Rep.* **2014**, *4*, 4281. [CrossRef]
164. Brullot, W.; Vanbel, M.K.; Swusten, T.; Verbiest, T. Resolving Enantiomers Using the Optical Angular Momentum of Twisted Light. *Sci. Adv.* **2016**, *2*, e1501349. [CrossRef]
165. Shen, Z.; Su, L.; Yuan, X.-C.; Shen, Y.-C. Trapping and Rotating of a Metallic Particle Trimer with Optical Vortex. *Appl. Phys. Lett.* **2016**, *109*, 241901. [CrossRef]
166. Katoh, M.; Fujimoto, M.; Mirian, N.S.; Konomi, T.; Taira, Y.; Kaneyasu, T.; Kuroda, K.; Miyamoto, A.; Miyamoto, K.; Sasaki, S. Helical Phase Structure of Radiation from an Electron in Circular Motion. *Sci Rep.* **2017**, *7*, 6130. [CrossRef]
167. Taira, Y.; Masahiro Katoh, M. Gamma-Ray Vortices Emitted from Nonlinear Inverse Thomson Scattering of a Two-Wavelength Laser Beam. *Phys. Rev. A* **2018**, *98*, 052130. [CrossRef]
168. Chen, Y.; Gao, J.; Jiao, X.-Q.; Sun, K.; Shen, W.-G.; Qiao, L.-F.; Tang, H.; Lin, X.-F.; Jin, X.-M. Mapping Twisted Light into and Out of a Photonic Chip. *Phys. Rev. Lett.* **2018**, *121*, 233602. [CrossRef]
169. Samlan, C.T.; Suna, R.R.; Naik, D.N.; Viswanathan, N.K. Spin-orbit Beams for Optical Chirality Measurement. *Appl. Phys. Lett.* **2018**, *112*, 031101. [CrossRef]
170. Torrres, J.P.; Torner, L. (Eds.) *Twisted photons: Applications of Light with Orbital Angular Momentum*; Wiley-VCH: Weinheim, Germany, 2011; ISBN 3527635378.
171. Inoue, Y.; Ramamurthy, V. (Eds.) *Chiral Photochemistry: Molecular and Supramolecular Photochemistry*; CRC Press: Tokyo, Japan, 2004; ISBN 9780824757106.

172. Fujiki, M. Creation and Controlling Asymmetric Small Molecules, Polymers, Colloids, and Small Objects Endowed with Polarized Light and Spin Polarized Particles. *Kobunshi Ronbunshu* **2017**, *74*, 114–133. [CrossRef]
173. Elsila, J.E.; Glavin, D.P.; Dworkin, J.P. Cometary Glycine Detected in Samples Returned by Stardust. *Meteorit. Planet. Sci.* **2009**, *44*, 1323–1330. [CrossRef]
174. Altwegg, K.; Balsiger, H.; Bar-Nun, A.; Berthelier, Je.; Bieler, A.; Bochsler, P.; Briois, C.; Calmonte, U.; Combi, M.R.; Cottin, H.; et al. Prebiotic Chemicals—Amino Acid and Phosphorus—In the Coma of Comet 67P/Churyumov-Gerasimenko. *Sci. Adv.* **2016**, *2*, e1600285. [CrossRef]
175. McGuire, B.A.; Carroll, B.P.; Loomis, R.A.; Finneran, I.A.; Jewell, R.P.; Remijan, A.J.; Blake, G.A. Discovery of the Interstellar Chiral Molecule Propylene Oxide (CH_3CHCH_2O). *Science* **2016**, *352*, 1449–1452. [CrossRef]
176. Urata, H.; Shinohara, K.; Ogura, E.; Uweda, Y.; Akagi, M. Mirror-image DNA. *J. Am. Chem. Soc.* **1991**, *113*, 8174–8175. [CrossRef]
177. Zawadzke, L.E.; Berg, J.M. A Racemic Protein. *J. Am. Chem. Soc.* **1992**, *114*, 4002–4003. [CrossRef]
178. Riddick, J.A.; Bunger, W.B.; Sakano, T.K. *Organic Solvents: Physical Properties and Methods of Purification*, 4th ed.; John Wiley & Sons: New York, NY, USA, 1986; ISBN 0-471-08467-0.
179. Lide, D.R. *Handbook of Organic Solvents*; CRC Press: Boca Raton, FL, USA, 1994; ISBN 0849389305.
180. Viswanath, D.S.; Ghosh, T.; Prasad, D.H.L.; Dutt, N.V.K.; Rani, K.Y. *Viscosity of Liquids; Theory, Estimation, Experiment and Data*; Springer: Berlin, Germany, 2007; ISBN 9048173787.
181. Properties of Organic Solvents. Available online: http://murov.info/orgsolvents.htm (accessed on 12 June 2018).
182. Hardy, R.C.; Cottington, R.L. Viscosity of Deuterium Oxide and Water in the Range 5°C to 125 °C. *J. Res. Natl. Bureau Stand.* **1949**, *42*, 573–578. [CrossRef]
183. Cho, C.H.; Urquidi, J.; Singh, S.; Robinson, G.W. Thermal Offset Viscosities of Liquid H_2O, D_2O, and T_2O. *J. Phys. Chem. B* **1999**, *103*, 1991–1994. [CrossRef]

Review

Role of Asymmetric Autocatalysis in the Elucidation of Origins of Homochirality of Organic Compounds

Kenso Soai [1,*], Tsuneomi Kawasaki [1] and Arimasa Matsumoto [2]

1 Department of Applied Chemistry, Tokyo University of Science, Kagurazaka, Shinjuku-ku, Tokyo 162-8601, Japan; tkawa@rs.tus.ac.jp

2 Department of Chemistry, Biology and Environmental Science, Nara Women's University, Kita-Uoya Nishi-machi, Nara 630-8506, Japan; a-matsumoto@cc.nara-wu.ac.jp

* Correspondence: soai@rs.kagu.tus.ac.jp; Tel.: +81-(0)3-5228-8261

Received: 29 April 2019; Accepted: 16 May 2019; Published: 20 May 2019

Abstract: Pyrimidyl alkanol and related compounds were found to be asymmetric autocatalysts in the enantioselective addition of diisopropylzinc to pyrimidine-5-carbaldehyde and related aldehydes. In the asymmetric autocatalysis with amplification of enantiomeric excess (ee), the very low ee (ca. 0.00005%) of 2-alkynyl-5-pyrimidyl alkanol was significantly amplified to >99.5% ee with an increase in the amount. By using asymmetric autocatalysis with amplification of ee, several origins of homochirality have been examined. Circularly polarized light, chiral quartz, and chiral crystals formed from achiral organic compounds such as glycine and carbon ($^{13}C/^{12}C$), nitrogen ($^{15}N/^{14}N$), oxygen ($^{18}O/^{16}O$), and hydrogen (D/H) chiral isotopomers were found to act as the origin of chirality in asymmetric autocatalysis. And the spontaneous absolute asymmetric synthesis was also realized without the intervention of any chiral factor.

Keywords: asymmetric autocatalysis; homochirality; chirality; asymmetric synthesis; Soai reaction

1. Introduction

The origins of biological homochirality of L-amino acids and D-sugars have attracted considerable attention ever since Pasteur discovered molecular dissymmetry in 1848 [1]. Although several theories of the origins of homochirality of organic compounds have been proposed [2–10], the enantiomeric excesses induced by these have usually been very low. For organic compounds to achieve homochirality, an amplification process from low enantiomeric excess (ee) to very high ee is required [11–23]. Therefore, asymmetric autocatalysis with amplification of chirality has been envisaged as the efficient process. We describe the discovery of asymmetric autocatalysis with amplification of ee. We also describe the study on the elucidation of the origin of homochirality of organic compounds by using asymmetric autocatalysis [24–36].

Asymmetric autocatalysis involves a process where a chiral product serves as the catalyst for its own production (Scheme 1). The reaction is a catalytic self-replication, i.e., automultiplication of a chiral compound. The superiority of asymmetric autocatalysis over the conventional non-autocatalytic asymmetric catalysis is as follows: (1) Because of the process of self-replication, the efficiency is high. (2) During the reaction, the amount of catalyst increases as the product increases. The catalytic activity and amount of catalyst does not decrease. (3) Because the structure of the product and the catalyst is the same, the separation of product from catalyst is not necessary.

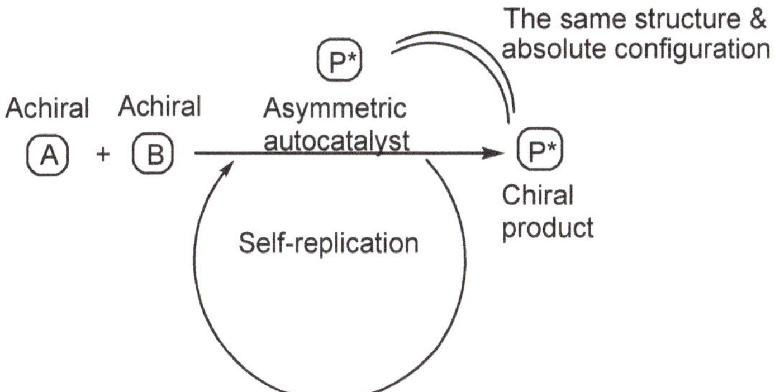

Scheme 1. Principle of asymmetric autocatalysis.

Frank proposed a mechanism, i.e., a mathematical equation, of asymmetric autocatalysis without showing any chemical structure in 1953 [21]. However, no real asymmetric autocatalysis had been reported until we first reported on the asymmetric autocatalysis of 3-pyridyl alkanol in 1990 [37].

2. Discovery of Asymmetric Autocatalysis with Amplification of Enantiomeric Excess

After the examination of the chiral diol system [38], we found in 1995 an efficient asymmetric autocatalysis of 5-pyrimidyl alkanol **1** with amplification of ee from 2% ee to 88% ee in the reaction between diisopropylzinc (*i*-Pr_2Zn) and pyrimidine-5-carbaldehyde **2a** (Scheme 2) [39,40]. In that reaction, pyrimidyl alkanol **1a** with 2% ee serves as an asymmetric autocatalyst to produce more of itself with an amplified ee. The consecutive asymmetric autocatalysis enables the amplification from 2 to 88% ee [39]. 2-Alkynylpyrimidyl alkanol **1c** with >99.5% ee was found to be an efficient asymmetric autocatalyst affording itself, **1c**, with >99.5% ee and with >99% yield [41]. It was also found that the asymmetric autocatalysis of pyrimidyl alkanol **1c** exhibit significant amplification of ee (Scheme 3). Indeed, starting from a very low (ca. 0.00005%) ee of (*S*)-pyrimidyl alkanol **1c** as an asymmetric autocatalyst, three cycles of asymmetric autocatalysis enabled the amplification of ee of alkanol **1c** to >99.5%. During the reaction, the amount of (*S*)-**1c** increased by a factor of ca. 630,000 times [42]. 2-Alkenylpyrimidyl alkanol **1e** [43], 3-quinolyl alkanol **4** [44–46], and 5-carbamoylpyridyl alkanol **5** [47,48] are also highly enantioselective asymmetric autocatalysts with amplification of ee (Scheme 2). The unique aspect of amplification of ee by asymmetric autocatalysis is that it is accomplished without the intervention of any other chiral factor. The only chiral factor is the initial enantiomeric imbalance of alkanol **1** itself as an asymmetric autocatalyst. In addition, asymmetric autocatalytic self-multiplication of multi-functionalized pyrimidyl alkanol **3** [49] and ultra-remote intramolecular asymmetric autocatalysis [50] were reported.

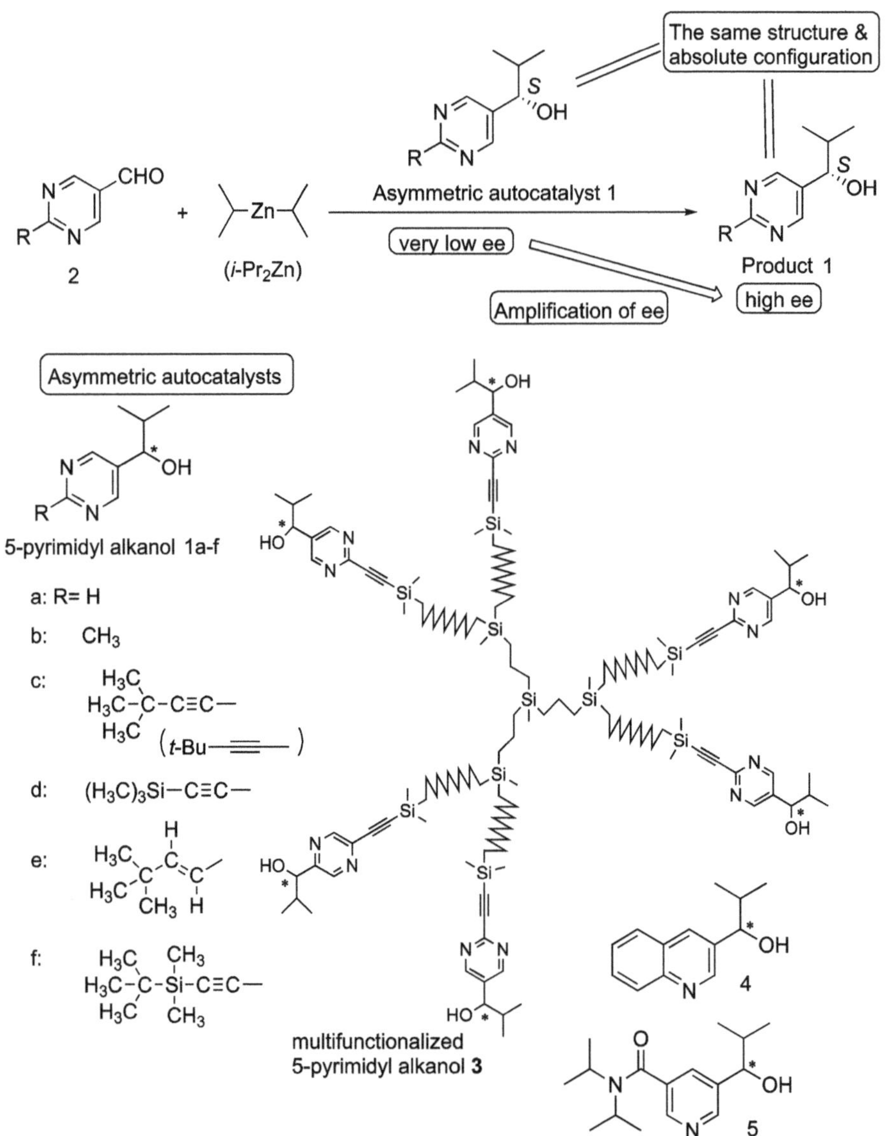

Scheme 2. Asymmetric autocatalysis. Structures of the autocatalysts of pyrimidyl alkanols **1a–f**, multi-functionalized pyrimidyl alkanol, **3**; 3-quinolyl alkanol, **4**; and 5-carbamoyl-3-pyridyl alkanol, **5**.

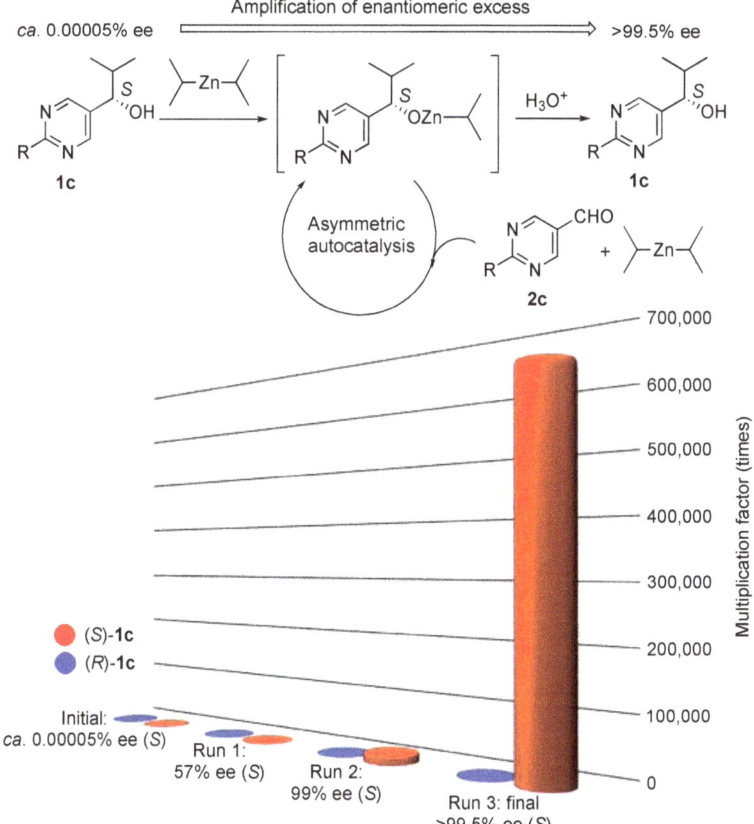

Scheme 3. Asymmetric autocatalysis of 5-pyrimidyl alkanol, **1c**, with amplification of enantiomeric excess from ca. 0.00005% to >99.5% ee.

Thus, it was proved that a chemical reaction exists in which very low enantioenrichment is amplified to almost enantiopure (>99.5% ee).

3. Study on the Mechanism of Asymmetric Autocatalysis

As described in the preceding section, asymmetric autocatalysis exhibits enormous amplification of ee during the self-replication. Thus, mechanistic insights into the asymmetric autocatalysis have attracted great attention. For the non-autocatalytic, non-linear effect in asymmetric catalysis, the dimer mechanism by Noyori [51] and MLn mechanism by Kagan [52] have been proposed.

We revealed the relationship between the reaction time and yield in the asymmetric autocatalysis using pyrimidyl alkanol **1c** with >99.5% ee [53]. A sigmoidal curve of product formation was observed. We also reported the relationship between the time, yield, and ee of the product by using chiral HPLC [54], which suggested dimeric or higher order aggregated catalytic species.

Several groups also investigated the mechanism of asymmetric autocatalysis. Heat flow measurement by microcalorimeter revealed the relationship between a reaction rate and the progress of the reaction. This suggested the dimeric catalyst model [55]. The dimeric and tetrameric species were proposed by the NMR measurement of the reaction solution [56,57]. The structure of catalyst aggregates has been proposed by density functional theory (DFT) calculation [58–61]. Reaction models have also been presented based on spontaneous mirror-symmetry breakage. These works proposed the mechanistic frameworks of asymmetric autocatalysis of pyrimidyl alkanol [62–69]. We clarified the crystal structures of asymmetric autocatalyst **1c** based on X-ray diffraction [70,71]. It was revealed that the structures are either tetrameric or oligomeric. The tetrameric crystal structure is formed in the presence of an excess molar amount of i-Pr_2Zn, while the higher order aggregate is formed in the presence of an equimolar or slightly excess amount of i-Pr_2Zn. Recently, reaction modeling was reported which suggests that the tetramer or higher order aggregates work for the asymmetric autocatalysis [72]. The clarification of the entire reaction pathway of asymmetric autocatalysis awaits further investigation.

4. Elucidation of the Origins of Homochirality by Using Asymmetric Autocatalysis

As described in the preceding section, asymmetric autocatalysis amplified ee from very low to very high. We then examined the origins of homochirality by using asymmetric autocatalysis. We envisaged that the low ee induced by the origin of chirality could be amplified by asymmetric autocatalysis. The origins of chirality so far proposed have usually induced only very low ees. To explain the very high ees observed in nature, the amplification of very low ee of organic compounds is necessary. We employed asymmetric autocatalysis of amplification of ee to examine the several proposed mechanisms of the origin of chirality.

4.1. Circularly Polarized Light

One of the representative chiral physical forces is circularly polarized light (CPL). Left (*l*) and right (*r*)-CPL have long been considered as the origin of chirality. In some of the star-forming regions, the occurrence of relatively strong CPL has been observed [73]. It is known that only ca. 2% ee is induced by irradiation of CPL to racemic organic compounds such as leucine. Asymmetric photosynthesis of hexa-helicen by CPL irradiation has been reported [5]. The induced low ee in leucine was correlated, for the first time, to the very high ee of organic compounds by using asymmetric autocatalysis [74].

The direct irradiation of *l*-CPL to racemic (*rac*) pyrimidyl alkanol **1c**, and the subsequent asymmetric autocatalysis, gave (*S*)-alkanol **1c** with >99.5% ee (Scheme 4) as a result of the amplification of ee [75]. On the other hand, *r*-CPL irradiation affords (*R*)-**1c** with >99.5% ee. The relationship between the handedness of *l*- and *r*-CPL and (*S*)-**1c** and (*R*)-**1c** is explained by the following consideration: The cotton effects of the circular dichroism (CD) spectra of the solid state of (*R*)-**1** and (*S*)-**1c** are plus (+) and minus (-) at 313 nm, respectively. Thus, when *l*-CPL is irradiated on *rac*-**1c**, the asymmetric photodecomposition of (*R*)-**1c** is induced because *l*-CPL is absorbed preferentially. Then, the less reactive (*S*)-**1c** becomes the predominant enantiomer over (*R*)-**1c**. The asymmetric autocatalysis of the remaining alkanol increases the ee of (*S*)-**1c** to >99.5% ee. Thus, the direct correlation is accomplished between the handedness of CPL and that of highly enantioenriched organic compound.

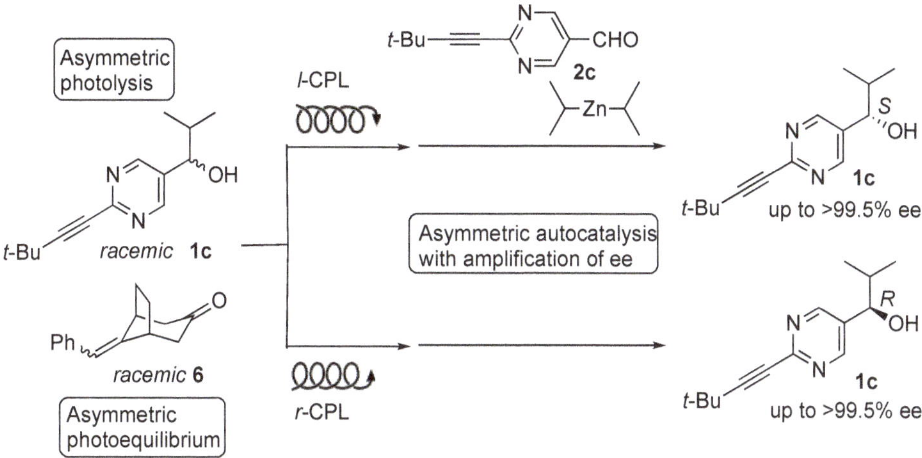

Scheme 4. Circularly polarized light (CPL) triggers asymmetric autocatalysis.

The asymmetric photoequilibrium of *rac*-olefin **6** using CPL, and the subsequent asymmetric autocatalytic reaction, gave pyrimidyl alkanol **1c** of the correlated absolute configuration to CPL [76]. Recently, under CPL irradiation, a Viedma-type racemization-crystallization of an amino acid derivative was reported [77].

4.2. Chiral Inorganic Crystals of Quartz, Sodium Chlorate, Cinnabar, and Retgersite, and the Enantiotopic Face of the Achiral Crystal of Gypsum

A chiral single crystal of silicon dioxide is known as quartz, and it exhibits enantiomorphism. Chiral minerals including quartz have been proposed as the origin of homochirality [6]. There are many reports attempting to induce chirality in organic compounds by using quartz [78]. However, no significant asymmetric induction has yet been reported by using quartz.

We thought that the asymmetric autocatalysis amplifies significantly the very low ee of the product initially induced by chiral *d*- and *l*-quartz [79]. Indeed, in the presence of *d*-quartz, asymmetric autocatalysis using pyrimidine-5-carbaldehyde **2c** and *i*-Pr_2Zn afforded (*S*)-**1c** with 97% ee in a yield of 95% (Scheme 5). On the other hand, *l*-quartz afforded (*R*)-**1c** with 97% ee. It was clearly shown by these results that *d*- and *l*-quartz act as chiral initiators of asymmetric autocatalysis. The initially formed slightly enriched (S)-(zinc alkoxide) of pyrimidyl alkanol **1c** serves as an asymmetric autocatalyst and automultiplies with amplification of ee. Thus, the chirality of *d*- and *l*-quartz is correlated to the chirality of a near enantiopure organic compound.

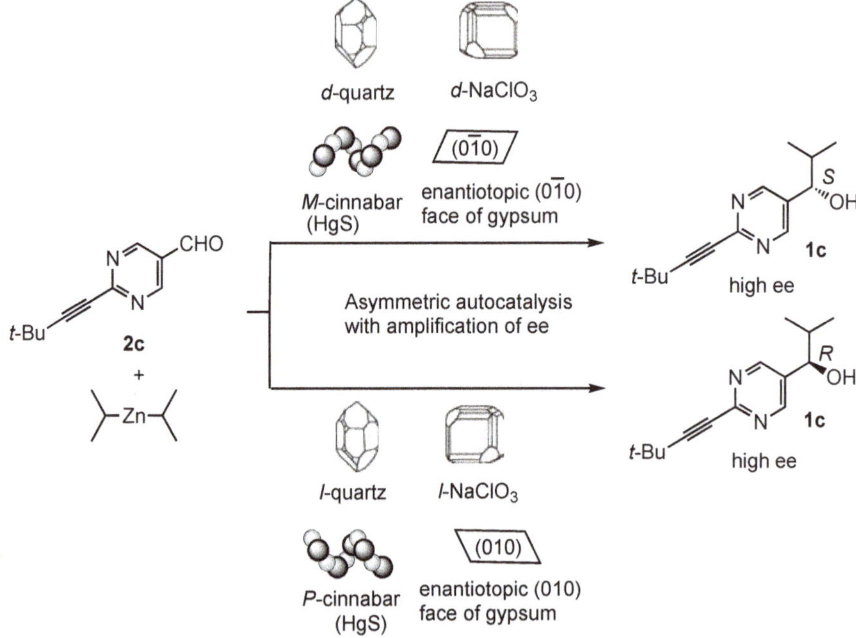

Scheme 5. Asymmetric autocatalysis triggered by chiral quartz, sodium chlorate, cinnabar and enantiotopic face of achiral crystal of gypsum.

Sodium chlorate ($NaClO_3$) and sodium bromate ($NaBrO_3$) are chiral inorganic ionic crystals [14,80,81]. It was also found that *d*-$NaClO_3$ triggers asymmetric autocatalysis to give (*S*)-**1c**, while *l*-$NaClO_3$ gives (*R*)-**1c** [82]. On the other hand, *d*-$NaBrO_3$ and *l*-$NaBrO_3$ trigger the formation of (*R*)- and (*S*)-**1c**, respectively [83]. Note that *d*-$NaClO_3$ and *l*-$NaBrO_3$ with the opposite signs of optical activity have the same type of enantiomorph. Enantiomorphic *P*- and *M*-crystals of cinnabar, mercury(II) sulfide (HgS), are composed of –Hg–S–Hg–S helical chains. We found that *P*-cinnabar acts as a chiral trigger of asymmetric autocatalysis to give (*R*)-**1c**. In contrast, *M*-HgS triggers the formation of (*S*)-**1c** [84]. Retgersite ($NiSO_4$ $6H_2O$) of [CD(+)390_{Nujol}] triggers asymmetric autocatalysis to afford (*S*)-**1c**. In contrast, retgersite of [CD(−)390_{Nujol}] affords (*R*)-**1c** [85].

Gypsum (calcium sulfate dihydrate) is a common mineral which has been widely used. The crystal structure is not chiral. However, gypsum exhibits two-dimensional enantiotopic cleavage (010) and (0−10) face. Pyrimidine-5-carbaldehyde **2c** was put on the enantiotopic (010) face. Then, the reaction of aldehyde **2c** on gypsum with the vapor of *i*-Pr_2Zn gave (*R*)-pyrimidyl alkanol **1c** [86]. In contrast, the reaction by exposing on the opposite (0−10) face gave (*S*)-alkanol **1c**. Thus, it was shown that the enantiotopic face of achiral gypsum works as an origin of chirality.

In combination with asymmetric autocatalysis, chiral inorganic crystals serve as the origin of chirality to give enantioenriched organic compounds of the correlated absolute configurations.

4.3. Chiral Crystals Formed from Achiral Organic Compounds

Achiral organic compounds often form achiral crystals. However, it is known that some of the achiral organic compounds form chiral crystals [87]. In some stereospecific reactions, these chiral organic crystals have been used as reactants [10]. However, in enantioselective synthesis, chiral crystals composed of achiral organic compounds have seldom been used as inducers. We used chiral crystals formed from achiral organic compounds as chiral inducers of asymmetric autocatalysis (Schemes 6 and 7).

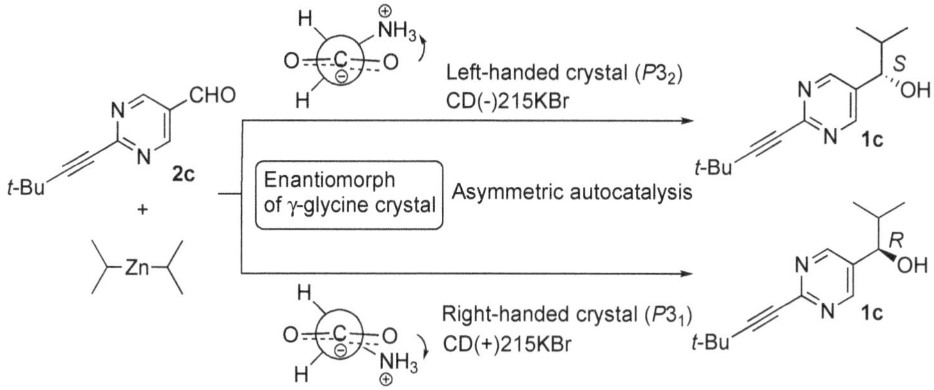

Scheme 6. Asymmetric autocatalysis triggered by chiral γ-polymorph of achiral glycine.

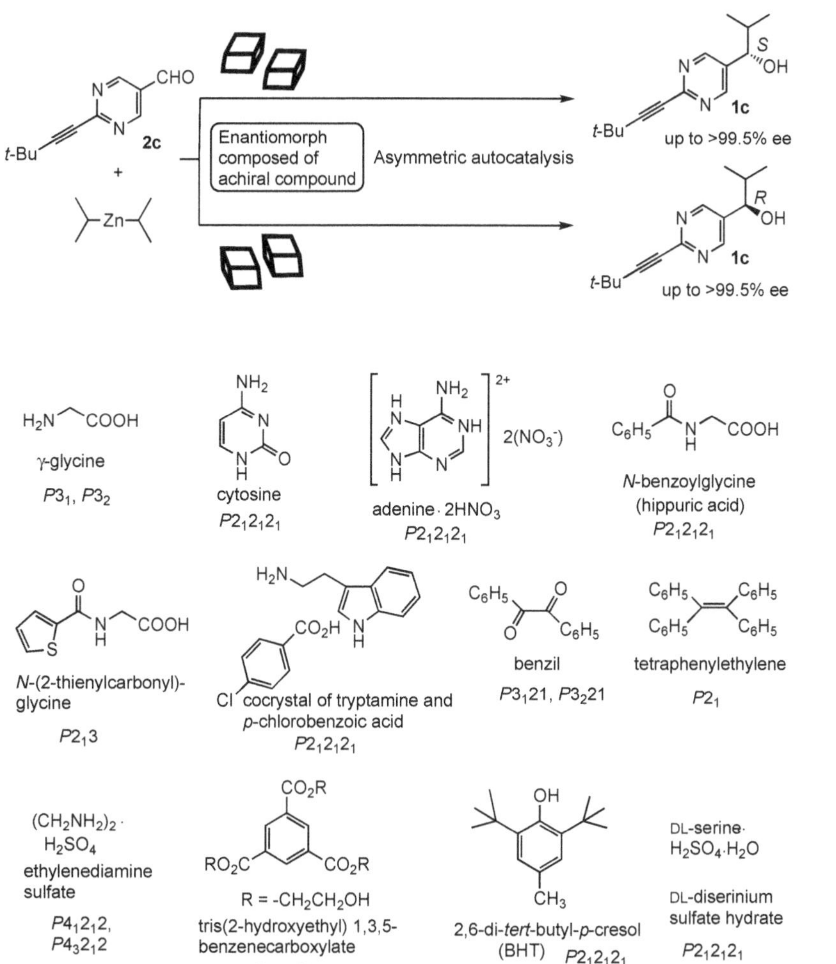

Scheme 7. Asymmetric autocatalysis initiated by chiral crystals composed of achiral organic compounds.

Natural proteinogenic amino acids, except glycine, exhibit L-form. Glycine stands as the only achiral amino acid that possesses no asymmetric carbon atoms. Although it is known that the stable crystal structure of the γ-glycine polymorph is chiral, it took years to determine the absolute crystal structure of the γ-glycine polymorph. Recently, the absolute crystal structure of the γ-glycine polymorph was correlated with optical rotatory dispersion (ORD) [88]. Guillemin reported CD spectra of γ-glycine [89].

We have correlated the absolute crystal structure of γ-glycine and have used the γ-glycine crystal as a chiral trigger of asymmetric autocatalysis [90]. It was found that the $P3_2$ crystal (left-handed) of γ-glycine triggers the formation of (*S*)-pyrimidyl alkanol **1c** with up to >99.5% ee (Scheme 6). In contrast, the $P3_1$ crystal afforded (*R*)-alkanol **1c** with up to >99.5% ee.

Thus, in conjunction with asymmetric autocatalysis, achiral glycine as its chiral γ-polymorph acts as the origin of homochirality.

Cytosine is a nucleobase and achiral. It may be formed under plausible prebiotic conditions [91]. When cytosine is crystallized from methanol, chiral crystals form. Chiral crystals of cytosine trigger asymmetric autocatalysis. When cytosine crystals of [CD(+)310_{Nujol}] were used as chiral initiators of the reaction of aldehyde **2c** with *i*-Pr_2Zn, (*R*)-alkanol **1c** was formed in combination with asymmetric autocatalysis (Scheme 7). In contrast, a [CD(−)310_{Nujol}]-cytosine crystal afforded (*S*)-**1c** [92]. Thus, the chiral cytosine crystal serves as the origin of chirality.

Cytosine forms achiral crystals of cytosine monohydrate when it is crystallized from water. When it is heated from one of the enantiotopic faces, the crystal water is eliminated by heating and chiral dehydrated cytosine is formed [93]. Interestingly, the chirality of the dehydrated crystal is determined by the enantiotopic face of the crystal from which the heating is applied. It is worth noting that the dehydration of the crystal water of cytosine monohydrate under reduced pressure conditions [94] also gives the chiral cytosine crystal with the opposite chirality to that dehydrated by heating. Thus, by removal of crystal water from an achiral crystal of cytosine monohydrate either by heating or under reduced pressure, the formation of chiral crystals with controlled absolute chirality was achieved.

Adenine is another achiral nucleobase. Chiral crystals of adenine dinitrate act as chiral initiators of asymmetric autocatalysis (Scheme 7) [95]. Thus, achiral nucleobases, i.e., cytosine and adenine, can serve as the origin of homochirality in conjunction with asymmetric autocatalysis.

Enantiomorphous crystals formed from achiral *N*-benzoylglycine (hippuric acid) [96], 2-thenoylglycine [97], certain chiral cocrystals consisting of two achiral compounds [98], benzil [99], tetraphenylethylene [100], ethylenediammonium sulfate [101], aromatic triester [102], and 2,6-di-*tert*-butyl-*p*-cresol (BHT) [103] serve as chiral initiators of asymmetric autocatalysis (Scheme 7). It should be added that a chiral crystal composed of a racemic serine initiates asymmetric autocatalysis. Asymmetric autocatalysis using the *M*-crystals of DL-diserinium sulfate hydrate as the chiral initiator afford (*R*)-pyrimidyl alkanol **1c**, while *P*-crystals afford (*S*)-alkanol **1c** [104].

4.4. Enantiotopic Face of Achiral Organic Crystal Composed of Achiral Organic Compound

Some of the crystal faces of achiral organic crystals formed from achiral compounds become enantiotopic. Achiral 2-(*tert*-butyldimethylsilylethynyl) pyrimidine-5-carbaldehyde **2f** forms an achiral crystal (*P*-1) that has enantiotopic faces. When the *Re*-face of the crystal was exposed to *i*-Pr_2Zn, (*R*)-pyrimidyl alkanol, **1f** was formed (Scheme 8) [105]. In contrast, exposure of *i*-Pr_2Zn on the *Si*-face gave (*S*)-alkanol **1f**. The ees of alkanol **1f** were amplified to >99.5% ee by asymmetric autocatalysis. Thus, it was shown that the enantiotopic faces of achiral crystals act as the origin of homochirality in conjunction with asymmetric autocatalysis.

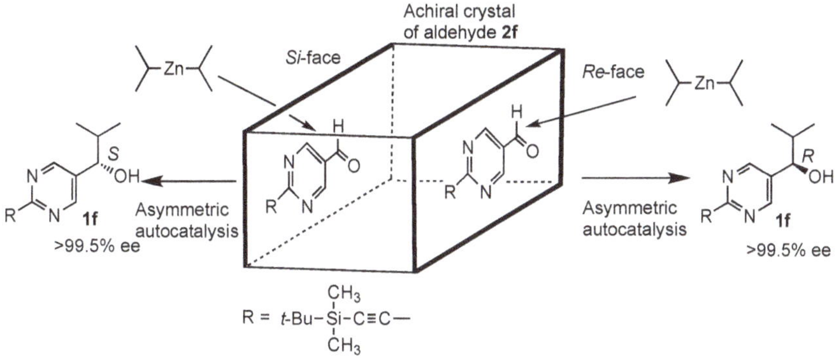

Scheme 8. Asymmetric autocatalysis initiated on the enantiotopic face of an achiral 2-(*tert*-butyldimethylsilylethynyl) pyrimidine-5-carbaldehyde **2f**.

4.5. Spontaneous Absolute Asymmetric Synthesis by Asymmetric Autocatalysis

As described in the preceding section, asymmetric autocatalysis of pyrimidyl alkanol enhances extremely low ca. 0.00005% ee to near enantiopure >99.5% ee [42]. We reasoned that if *i*-Pr_2Zn is reacted with pyrimidine-5-carbaldehyde **2** without using any chiral factor, the product with low ee based on the statistical fluctuation would form. The subsequent asymmetric autocatalysis may enhance the initial low ee to the detectable high ee (Scheme 9).

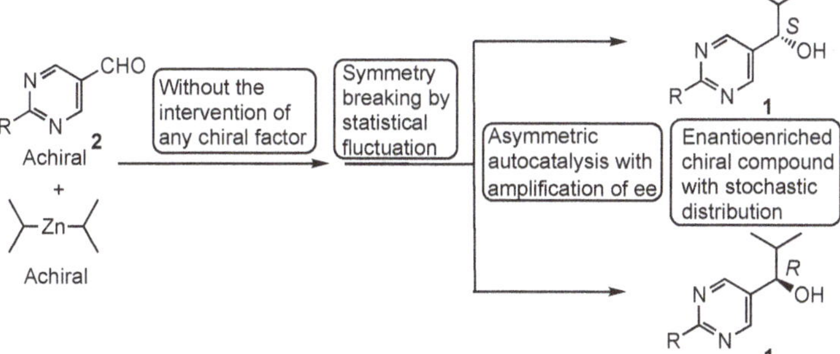

Scheme 9. Spontaneous absolute asymmetric synthesis by asymmetric autocatalysis without the intervention of any chiral factor.

Although the term "absolute asymmetric synthesis" had been used for the asymmetric synthesis "without the use of any chiral chemical substance," Mislow newly defined absolute asymmetric synthesis as "the formation of an enantioenriched compound from achiral compounds without the intervention of any chiral factor [3]." The spontaneous absolute asymmetric synthesis, based on the statistical fluctuation, has been thought of as one of the origins of chirality. However, it is known that the reaction between achiral reagents without any chiral factor always gives so-called racemic product. However, there are statistical fluctuations in the numbers of enantiomers [3]. Let us consider the situation of flipping a coin one hundred times: there is an 8% probability of 50 heads and 50 tails. The remaining 92% are results with either heads or tails being in excess: 49 to 51, 53 to 47, etc. Pályi et al. described the distribution of ee by statistical fluctuations of various amounts of so-called racemic molecules [106–108].

We found spontaneous absolute asymmetric synthesis in the reaction between pyrimidine-5-carbaldehyde **2** and *i*-Pr_2Zn without the addition of any chiral substance. In 1996, we applied patent for this absolute asymmetric synthesis [109,110]. The reaction afforded enantioenriched (*S*)-pyrimidyl alkanol **1** or (*R*)-alkanol **1** [109]. When aldehyde **2c** and *i*-Pr_2Zn were reacted in a mixed solvent of ether-toluene, enantioenriched product was formed in situ by statistical fluctuation. The subsequent asymmetric autocatalysis gave (*S*) or (*R*)-**1** with detectable enantioenrichments. The formation of (*S*)-alkanol **1c** occurred 19 times and (*R*)-**1c** occurred 18 times in a total of 37 reactions (Figure 1a) [110]. The absolute configurations of **1c** formed exhibits a stochastic distribution of *S* and *R* enantiomers. Moreover, by using achiral amorphous silica gel (Figure 1b) [111] and achiral amines (Figure 1c) [112], enantioenriched **1c** was obtained and the distribution of (*S*)- and (*R*)-handedness was stochastic. The absolute asymmetric synthesis has also been reported between pyrimidine-5-carbaldehyde **2b** and *i*-Pr_2Zn (*S*)-**1b** or (*R*)-**1b** in a stochastic distribution [113]. As described, the results fulfill the conditions necessary for spontaneous absolute asymmetric synthesis [62,65,114–117].

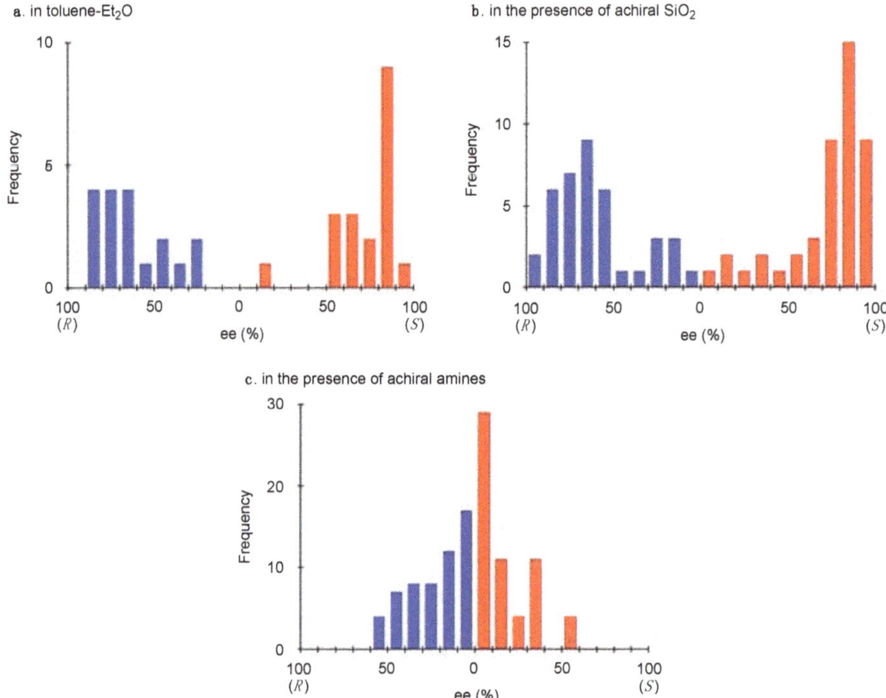

Figure 1. Spontaneous absolute asymmetric synthesis of pyrimidyl alkanol **1**. Histograms of the absolute configuration and ee of products.

Very recently, absolute asymmetric synthesis under heterogeneous solid-vapor phase conditions has been reported by us (Scheme 10) [118]. The powder of pyrimidine-5-carbaldehyde **2c** in test tubes was exposed to the vapor of *i*-Pr_2Zn and toluene in a desiccator. In 129 reactions, (*R*)-pyrimidyl alkanol **1c** was formed 61 times. On the other hand, (*S*)-alkanol **1c** was formed 58 times (10 times the formation of **1c** of <0.5% ee was assigned as below the detection level). Thus, the results show that the distribution of (*S*) and (*R*)-alkanol **1c** is stochastic. Although the ee values of alkanol **1c** varied, these ee could be enhanced to >99.5% ee during the subsequent asymmetric autocatalysis. The present heterogeneous absolute asymmetric synthesis under solid vapor phase conditions could be possible in a more spacious platform.

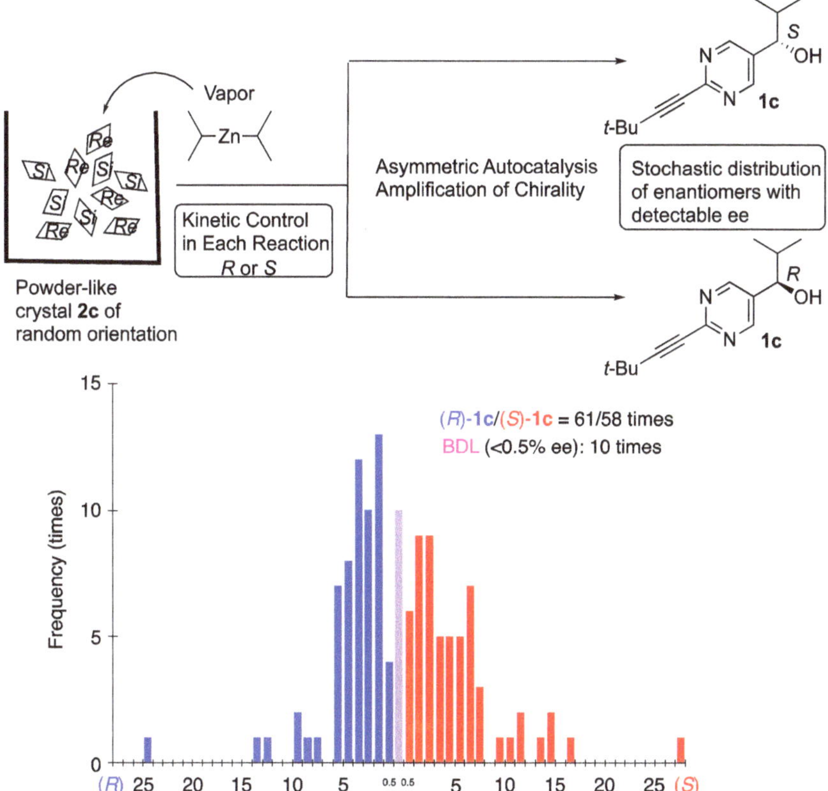

Scheme 10. Absolute asymmetric synthesis of pyrimidyl alkanol **1c** under solid-vapor phase conditions.

4.6. Asymmetric Autocatalysis Triggered by Hydrogen, Carbon, Oxygen, and Nitrogen Chiral Isotopomers

Many apparent achiral organic compounds become chiral by substitution of carbon (^{12}C), nitrogen (^{14}N), and oxygen (^{16}O) for their isotopes of ^{13}C, ^{15}N, and ^{18}O, respectively. For example, dimethylphenylmethanol **8** is an achiral compound because it has the same two methyl groups. However, when one of the carbon atoms of the methyl group is labelled with ^{13}C, the alkanol becomes a chiral (*R*)-alkanol **8**(^{13}C) or (*S*)-alkanol **8**(^{13}C) (Scheme 11). Because the difference of carbon (^{13}C/^{12}C) isotopomers between enantiomers is so small, no report has appeared before on the asymmetric induction by using chiral carbon (^{13}C/^{12}C) isotopomers.

We found that in the presence of chiral carbon (^{13}C/^{12}C) isotopomer and (*R*) or (*S*)-**8**(^{13}C), as a chiral trigger, pyrimidine-5-carbaldehyde **2c** reacts with *i*-Pr_2Zn to give pyrimidyl alkanol **1c** with a very high ee of the absolute configuration correlated to that of the carbon isotopomer (Scheme 11). (*R*)-Carbon isotopomer **8**(^{13}C) triggered the formation of (*S*)-pyrimidyl alkanol **1c** with high ee. In contrast, (*S*)-carbon isotopomer **8**(^{13}C) gave (*R*)-pyrimidyl alkanol [119]. Other carbon (^{13}C/^{12}C) isotopomers also serve as chiral triggers on asymmetric autocatalysis. Chiral nitrogen (^{15}N/^{14}N) isotopomer, [^{15}N](*S*) and [^{15}N](*R*)-diamine **9**(^{15}N) were also found to work as chiral triggers of asymmetric autocatalysis [120]. In addition, oxygen (^{18}O/^{16}O) isotopomer, [^{18}O](*R*), and [^{18}O](*S*)-diol **10**(^{18}O), trigger asymmetric autocatalysis to give pyrimidyl alkanol **1c** of high ee with the correlated absolute configuration to that of oxygen isotopomer [121,122]. As described, carbon, nitrogen, and oxygen isotopomers were found to act as the origin of homochirality in conjunction with asymmetric autocatalysis.

As to chiral hydrogen (D/H) isotopomers, there are a few examples of low asymmetric induction by hydrogen isotopomers [123,124]. It was found that chiral hydrogen isotopomers act as chiral initiators of asymmetric autocatalysis [125,126]. It should be noted that achiral glycine **7** becomes chiral by substituting one of the hydrogen atoms of the methylene group for deuterium (D). In the presence of chiral (*S*)-glycine-α-*d* **7**(D), (*S*)-pyrimidyl alkanol **1c** of high ee was formed with the correlated absolute configuration to that of chiral glycine-α-*d* [127].

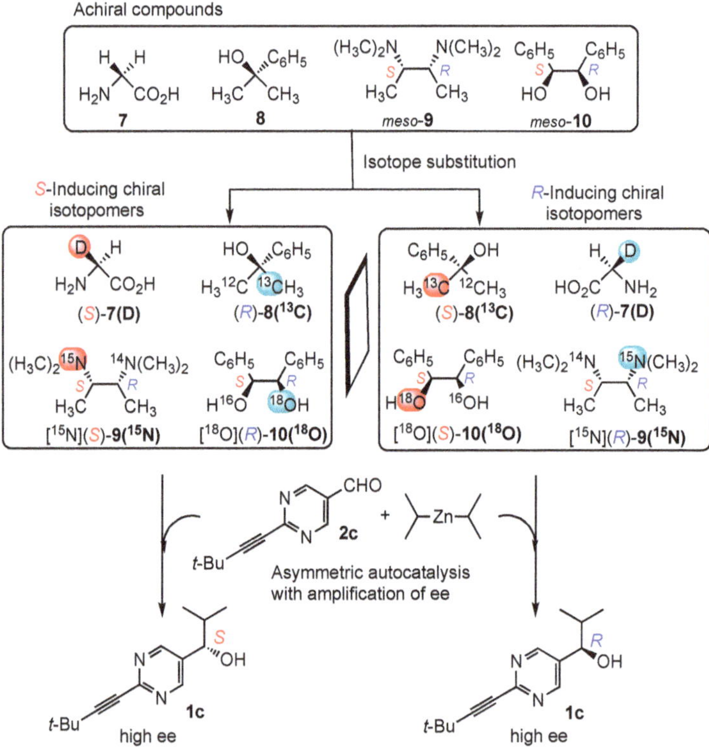

Scheme 11. Asymmetric autocatalysis triggered by carbon ($^{13}C/^{12}C$), nitrogen ($^{15}N/^{14}N$), oxygen ($^{18}O/^{16}O$), and hydrogen (D/H) isotope chirality.

5. Various Chiral Compounds as Triggers of Asymmetric Autocatalysis

Various chiral compounds work as chiral initiators of asymmetric autocatalysis. Amino acids even with low ee [128], such as hexa-helicene [129], tetrathia-hepta-helicene [130], and 2-aza-hexa-helicene [131], initiate asymmetric autocatalysis to give alkanol **1c** of the correlated absolute configuration to those of the chiral initiators. It is known that the value of optical rotation of a chiral saturated quaternary hydrocarbon, 5-ethyl-5-propylundecane, is below detection level because the differences in the structures of the four substituents are so small. The compound is called cryptochiral. It was found that 5-ethyl-5-propylundecane triggers asymmetric autocatalysis [132]. Cryptochiral isotactic polystyrene also works as a chiral trigger [133]. Artificially designed helical [134] silica and mesoporous helical silica [135] are also chiral triggers.

6. Conclusions

Asymmetric autocatalysis of the enantioselective addition of *i*-Pr_2Zn to pyrimidine-5-carbaldehyde was discovered by us. In this reaction, the very low ca. 0.00005% ee of (*S*)-2-alkynylpyrimidyl alkanol

1 was enhanced to >99.5% ee by consecutive asymmetric autocatalyses. Mislow first mentioned this reaction as the Soai reaction [3]. The asymmetric autocatalysis with amplification of ee is unique because no chiral substance other than the asymmetric autocatalyst itself is required.

To elucidate the origins of homochirality, asymmetric autocatalysis with amplification of ee was applied. By using asymmetric autocatalysis, the initially induced low ee by the proposed origin of chirality was enhanced significantly by the asymmetric autocatalysis. The racemic pyrimidyl alkanol was irradiated with *l* or *r*-circularly polarized light. The subsequent asymmetric autocatalysis correlated the chirality of CPL with that of the formed alkanol **1**. Thus, for the first time, the correlation was made possible between the chirality of CPL and that of a chiral organic compound of very high ee. Chiral minerals such as quartz and cinnabar were found to act as chiral triggers of asymmetric autocatalysis. Thus, chirality of quartz was correlated to that of a highly enantioenriched organic compound. It was also found that chiral organic crystals composed of achiral compounds, i.e., glycine, cytosine, and adenine, serve as chiral triggers of asymmetric autocatalysis. Spontaneous absolute asymmetric synthesis without the intervention of any chiral factors was realized using the asymmetric autocatalysis of pyrimidyl alkanol with amplification of ee. Asymmetric autocatalysis was initiated by chiral compounds resulting from carbon ($^{13}C/^{12}C$), nitrogen ($^{15}N/^{14}N$), and oxygen ($^{16}O/^{18}O$) isotopomers. X-ray crystallographic analysis revealed the structure of asymmetric autocatalysts. It should be mentioned that bio-reactions should be studied looking for asymmetric autocatalysis.

Author Contributions: Conceptualization, K.S.; Writing-Original Draft Preparation, K.S. and T.K.; Writing—Review and Editing, K.S., T.K. and A.M.; Supervision, K.S.

Funding: KAKENHI: 19K05482 from Japan Society for the Promotion of Science.

Acknowledgments: The authors gratefully acknowledge their collaborators whose names appear in the literature cited.

Conflicts of Interest: The authors declare no conflict of interest.

References

1. Pasteur, L. Recherches sur les relations qui peuvent exister entre la forme crystalline, la composition chimique et le sens de la polarisation rotatoire. *Ann. Chim. Phys.* **1848**, *24*, 442–459.
2. Guijarro, A.; Yus, M. *The Origin of Chirality in the Molecules of Life*; The Royal Society of Chemistry: Cambridge, UK, 2009.
3. Mislow, K. Absolute asymmetric synthesis: A commentary. *Collect. Czech. Chem. Commun.* **2003**, *68*, 849–864. [CrossRef]
4. Feringa, B.L.; van Delden, R.A. Absolute asymmetric synthesis: The origin, control, and amplification of chirality. *Angew. Chem. Int. Ed.* **1999**, *38*, 3418–3438.
5. Inoue, Y. Asymmetric photochemical reactions in solution. *Chem. Rev.* **1992**, *92*, 741–770. [CrossRef]
6. Hazen, R.M.; Sholl, D.S. Chiral selection on inorganic crystalline surfaces. *Nat. Mater.* **2003**, *2*, 367–374. [PubMed]
7. Bolli, M.; Micura, R.; Eschenmoser, A. Pyranosyl-RNA: Chiroselective self-assembly of base sequences by ligative oligomerization of tetranucleotide-2′, 3′-cyclophosphates (with a commentary concerning the origin of biomolecular homochirality). *Chem. Biol.* **1997**, *4*, 309–320. [CrossRef]
8. Ribó, J.M.; Crusats, J.; Sagués, F.; Claret, J.; Rubires, R. Chiral sign induction by vortices during the formation of mesophases in stirred solutions. *Science* **2001**, *292*, 2063–2066.
9. Ernst, K.-H. Molecular chirality at surfaces. *Phys. Status Solidi* **2012**, *249*, 2057–2088. [CrossRef]
10. Weissbuch, I.; Lahav, M. Crystalline architectures as templates of relevance to the origins of homochirality. *Chem. Rev.* **2011**, *111*, 3236–3267. [CrossRef] [PubMed]
11. Kitamura, M.; Okada, S.; Suga, S.; Noyori, R. Enantioselective addition of dialkylzincs to aldehydes promoted by chiral amino alcohols. Mechanism and nonlinear effect. *J. Am. Chem. Soc.* **1989**, *111*, 4028–4036.
12. Satyanarayana, T.; Abraham, S.; Kagan, H.B. Nonlinear effects in asymmetric catalysis. *Angew. Chem. Int. Ed.* **2009**, *48*, 456–494. [CrossRef]

13. Kondepudi, D.K.; Asakura, K. Chiral autocatalysis, spontaneous symmetry breaking, and stochastic behavior. *Acc. Chem. Res.* **2001**, *34*, 946–954. [CrossRef] [PubMed]
14. Viedma, C. Chiral symmetry breaking during crystallization: Complete chiral purity induced by nonlinear autocatalysis and recycling. *Phys. Rev. Lett.* **2005**, *94*, 065504. [CrossRef] [PubMed]
15. Soloshonok, V.A.; Ueki, H.; Yasumoto, M.; Mekala, S.; Hirschi, J.S.; Singleton, D.A. Phenomenon of optical self-purification of chiral non-racemic compounds. *J. Am. Chem. Soc.* **2007**, *129*, 12112–12113. [CrossRef] [PubMed]
16. Hayashi, Y.; Matsuzawa, M.; Yamaguchi, J.; Yonehara, S.; Matsumoto, Y.; Shoji, M.; Hashizume, D.; Koshino, H. Large nonlinear effect observed in the enantiomeric excess of proline in solution and that in the solid state. *Angew. Chem. Int. Ed.* **2006**, *45*, 4593–4597. [CrossRef] [PubMed]
17. Córdova, A.; Engqvist, M.; Ibrahem, I.; Casas, J.; Sundén, H. Plausible origins of homochirality in the amino acid catalyzed neogenesis of carbohydrates. *Chem. Commun.* **2005**, 2047–2049. [CrossRef]
18. Green, M.M.; Park, J.-W.; Sato, T.; Teramoto, A.; Lifson, S.; Selinger, R.L.B.; Selinger, J.V. The macromolecular route to chiral amplification. *Angew. Chem. Int. Ed.* **1999**, *38*, 3138–3154. [CrossRef]
19. Noorduin, W.L.; Vlieg, E.; Kellogg, R.M.; Kaptein, B. From Ostwald ripening to single chirality. *Angew. Chem. Int. Ed.* **2009**, *48*, 9600–9606. [CrossRef]
20. Saito, Y.; Hyuga, H. Colloquium: Homochirality: Symmetry breaking in systems driven far from equilibrium. *Rev. Mod. Phys.* **2013**, *85*, 603–621. [CrossRef]
21. Frank, F.C. On spontaneous asymmetric synthesis. *Biochim. Biophys. Acta* **1953**, *11*, 459–463. [CrossRef]
22. Han, J.; Kitagawa, O.; Wzorek, A.; Klia, K.D.; Soloshonok, V.A. The self-disproportionation of enantiomers (SDE): A menace or an opportunity? *Chem. Sci.* **2018**, *9*, 1718–1739. [CrossRef]
23. Moberg, C. Recycling in asymmetric catalysis. *Acc. Chem. Res.* **2016**, *49*, 2736–2745. [CrossRef] [PubMed]
24. Soai, K.; Shibata, T.; Sato, I. Enantioselective automultiplication of chiral molecules by asymmetric autocatalysis. *Acc. Chem. Res.* **2000**, *33*, 382–390. [CrossRef] [PubMed]
25. Soai, K.; Kawasaki, T. Discovery of asymmetric autocatalysis with amplification of chirality and its implication in chiral homogeneity of biomolecules. *Chirality* **2006**, *18*, 469–478. [CrossRef] [PubMed]
26. Soai, K.; Kawasaki, T. Asymmetric autocatalysis with amplification of chirality. *Top. Curr. Chem.* **2008**, *284*, 1–31.
27. Kawasaki, T.; Soai, K. Amplification of chirality as a pathway to biological homochirality. *J. Fluor. Chem.* **2010**, *131*, 525–534. [CrossRef]
28. Kawasaki, T.; Soai, K. Asymmetric induction arising from enantiomerically enriched carbon-13 isotopomers and highly sensitive chiral discrimination by asymmetric autocatalysis. *Bull. Chem. Soc. Jpn.* **2011**, *84*, 879–892. [CrossRef]
29. Kawasaki, T.; Soai, K. Asymmetric autocatalysis triggered by chiral crystals formed from achiral compounds and chiral isotopomers. *Isr. J. Chem.* **2012**, *52*, 582–590. [CrossRef]
30. Soai, K.; Kawasaki, T. Asymmetric Autocatalysis—Discovery and State of The Art. In *The Soai Reaction and Related Topic*; Palyi, G., Zicchi, C., Caglioti, C., Eds.; Academia Nationale di Scienze Lettere e Arti Modena, Edizioni Artestampa: Modena, Italy, 2012; pp. 9–34.
31. Soai, K.; Kawasaki, T.; Matsumoto, A. The origins of homochirality examined by using asymmetric autocatalysis. *Chem. Rec.* **2014**, *14*, 70–83. [CrossRef]
32. Soai, K.; Kawasaki, T.; Matsumoto, A. Asymmetric autocatalysis of pyrimidyl alkanol and its application to the study on the origin of homochirality. *Acc. Chem. Res.* **2014**, *47*, 3643–3654. [CrossRef]
33. Soai, K.; Kawasaki, T.; Matsumoto, A. Asymmetric autocatalysis of pyrimidyl alkanol and related compounds. Self-replication, amplification of chirality and implication for the origin of biological enantioenriched chirality. *Tetrahedron* **2018**, *74*, 1973–1990. [CrossRef]
34. Soai, K. Asymmetric autocatalysis. Chiral symmetry breaking and the origins of homochirality of organic molecules. *Proc. Jpn. Acad. Ser. B* **2019**, *95*, 89–110. [CrossRef]
35. Podlech, J.; Gehring, T. New aspects of Soai's asymmetric autocatalysis. *Angew. Chem. Int. Ed.* **2005**, *44*, 5776–5777. [CrossRef]
36. Gehring, T.; Busch, M.; Schlageter, M.; Weingand, D. A concise summary of experimental facts about the Soai reaction. *Chirality* **2010**, *22*, E173–E182. [CrossRef]
37. Soai, K.; Niwa, S.; Hori, H. Asymmetric self-catalytic reaction. Self-production of chiral 1-(3-pyridyl) alkanols as chiral self-catalysts in the enantioselective addition of dialkylzinc reagents to pyridine-3-carbaldehyde. *J. Chem. Soc. Chem. Commun.* **1990**, 982–983. [CrossRef]

38. Soai, K.; Hayase, T.; Shimada, C.; Isobe, K. Catalytic asymmetric synthesis of chiral diol, bis[2-(l-hydroxyalkyl)phenylether, an asymmetric autocatalytic reaction. *Tetrahedron Asymm.* **1994**, *5*, 789–792. [CrossRef]
39. Soai, K.; Shibata, T.; Morioka, H.; Choji, K. Asymmetric autocatalysis and amplification of enantiomeric excess of a chiral molecule. *Nature* **1995**, *378*, 767–768. [CrossRef]
40. Shibata, T.; Morioka, H.; Hayase, T.; Choji, K.; Soai, K. Highly enantioselective catalytic asymmetric automultiplication of chiral pyrimidylalcohol. *J. Am. Chem. Soc.* **1996**, *118*, 471–472. [CrossRef]
41. Shibata, T.; Yonekubo, S.; Soai, K. Practically perfect asymmetric autocatalysis using 2-alkynyl-5-pyrimidylalkanol. *Angew. Chem. Int. Ed.* **1999**, *38*, 659–661. [CrossRef]
42. Sato, I.; Urabe, H.; Ishiguro, S.; Shibata, T.; Soai, K. Amplification of chirality from extremely low to greater than 99.5% ee by asymmetric autocatalysis. *Angew. Chem. Int. Ed.* **2003**, *42*, 315–317. [CrossRef]
43. Sato, I.; Yanagi, T.; Soai, K. Highly enantioselective asymmetric autocatalysis of 2-alkenyl- and 2-vinyl-5-pyrimidyl alkanols with significant amplification of anantiomeric axcess. *Chirality* **2002**, *14*, 166–168. [CrossRef]
44. Shibata, T.; Choji, K.; Morioka, H.; Hayase, T.; Soai, K. Highly enantioselective synthesis of a chiral 3-quinolylalkanol by an asymmetric autocatalytic reaction. *Chem. Commun.* **1996**, 751–752. [CrossRef]
45. Shibata, T.; Choji, K.; Hayase, T.; Aizu, Y.; Soai, K. Asymmetric autocatalytic reaction of 3-quinolylalkanol with amplification of enantiomeric excess. *Chem. Commun.* **1996**, 1235–1236. [CrossRef]
46. Sato, I.; Nakao, T.; Sugie, R.; Kawasaki, T.; Soai, K. Enantioselective synthesis of substituted 3-quinolyl alkanols and their application to asymmetric autocatalysis. *Synthesis* **2004**, 1419–1428. [CrossRef]
47. Shibata, T.; Morioka, H.; Tanji, S.; Hayase, T.; Kodaka, Y.; Soai, K. Enantioselective synthesis of chiral 5-carbamoyl-3-pyridyl alcohols by asymmetric autocatalytic reaction. *Tetrahedron Lett.* **1996**, *37*, 8783–8786. [CrossRef]
48. Tanji, S.; Kodaka, Y.; Ohno, A.; Shibata, T.; Sato, I.; Soai, K. Asymmetric autocatalysis of 5-carbamoyl-3-pyridyl alkanols with amplification of enantiomeric excess. *Tetrahedron Asymm.* **2000**, *11*, 4249–4253. [CrossRef]
49. Kawasaki, T.; Nakaoda, M.; Takahashi, Y.; Kanto, Y.; Kuruhara, N.; Hosoi, K.; Sato, I.; Matsumoto, A.; Soai, K. Self-replication and amplification of enantiomeric excess of chiral multi-functionalized large molecule by asymmetric autocatalysis. *Angew. Chem. Int. Ed.* **2014**, *53*, 11199–11202. [CrossRef]
50. Kawasaki, T.; Ishikawa, Y.; Minato, Y.; Otsuka, T.; Yonekubo, S.; Sato, I.; Shibata, T.; Matsumoto, A.; Soai, K. Point-to-point ultra-remote asymmetric control with flexible linker. *Chem. A Eur. J.* **2017**, *23*, 282–285. [CrossRef]
51. Kitamura, M.; Suga, S.; Oka, H.; Noyori, R. Quantitative analysis of the chiral amplification in the amino alcohol-promoted asymmetric alkylation of aldehydes with dialkylzincs. *J. Am. Chem. Soc.* **1998**, *120*, 9800–9809. [CrossRef]
52. Guillaneux, D.; Zhao, S.-H.; Samuel, O.; Rainford, D.; Henri, B.; Kagan, H.B. Nonlinear effects in asymmetric catalysis. *J. Am. Chem. Soc.* **1994**, *116*, 9430–9439. [CrossRef]
53. Sato, I.; Omiya, D.; Tsukiyama, K.; Ogi, Y.; Soai, K. Evidence of asymmetric autocatalysis in the enantioselective addition of diisopropylzinc to pyrimidine-5-carbaldehyde using chiral pyrimidyl alkanol. *Tetrahedron Asymm.* **2001**, *12*, 1965–1969. [CrossRef]
54. Sato, I.; Omiya, D.; Igarashi, H.; Kato, K.; Ogi, Y.; Tsukiyama, K.; Soai, K. Relationship between the time, yield, and enantiomeric excess of asymmetric autocatalysis of chiral 2-alkynyl-5-pyrimidyl alkanol with amplification of enantiomeric excess. *Tetrahedron Asymm.* **2003**, *14*, 975–979. [CrossRef]
55. Blackmond, D.G.; McMillan, C.R.; Ramdeehul, S.; Schorm, A.; Brown, J.M. Origins of asymmetric amplification in autocatalytic alkylzinc additions. *J. Am. Chem. Soc.* **2001**, *123*, 10103–10104. [CrossRef]
56. Quaranta, M.; Gehring, T.; Odell, B.; Brown, J.M.; Blackmond, D.G. Unusual inverse temperature dependence on reaction rate in the asymmetric autocatalytic alkylation of pyrimidyl aldehydes. *J. Am. Chem. Soc.* **2010**, *132*, 15104–15107. [CrossRef]
57. Gehring, T.; Quaranta, M.; Odell, B.; Blackmond, D.G.; Brown, J.M. Observation of a transient intermediate in Soai's asymmetric autocatalysis: Insights from ^{1}H NMR turnover in real time. *Angew. Chem. Int. Ed.* **2012**, *51*, 9539–9542. [CrossRef]
58. Schiaffino, L.; Ercolani, G. Unraveling the mechanism of the Soai asymmetric autocatalytic reaction by first-principles calculations: Induction and amplification of chirality by self-assembly of hexamolecular complexes. *Angew. Chem. Int. Ed.* **2008**, *47*, 6832–6835. [CrossRef]

59. Ercolani, G.; Schiaffino, L. Putting the mechanism of the Soai reaction to the test: DFT study of the role of aldehyde and dialkylzinc structure. *J. Org. Chem.* **2011**, *76*, 2619–2626. [CrossRef]
60. Gridnev, I.D.; Vorobiev, A.K. Quantification of sophisticated equilibria in the reaction pool and amplifying catalytic cycle of the Soai reaction. *ACS Catal.* **2012**, *2*, 2137–2149. [CrossRef]
61. Gridnev, I.D.; Vorobiev, A.K. On the origin and structure of the recently observed acetal in the Soai reaction. *Bull. Chem. Soc. Jpn.* **2015**, *88*, 333–340. [CrossRef]
62. Barabás, B.; Caglioti, L.; Micskei, K.; Pályi, G. Data-based stochastic approach to absolute asymmetric synthesis by autocatalysis. *Bull. Chem. Soc. Jpn.* **2009**, *82*, 1372–1376. [CrossRef]
63. Micheau, J.C.; Cruz, J.M.; Coudret, C.; Buhse, T. An autocatalytic cycle model of asymmetric amplification and mirror-symmetry breaking in the Soai reaction. *ChemPhysChem* **2010**, *11*, 3417–3419. [CrossRef]
64. Micheau, J.C.; Coudret, C.; Cruz, J.M.; Buhse, T. Amplification of enantiomeric excess, mirror-image symmetry breaking and kinetic proofreading in Soai reaction models with different oligomeric orders. *Phys. Chem. Chem. Phys.* **2012**, *14*, 13239–13248. [CrossRef]
65. Micskei, K.; Rábai, G.; Gál, E.; Caglioti, L.; Pályi, G. Oscillatory symmetry breaking in the Soai reaction. *J. Phys. Chem. B* **2008**, *112*, 9196–9200. [CrossRef]
66. Maioli, M.; Micskei, K.; Caglioti, L.; Zucchi, C.; Pályi, G. Evolution of chirality in consecutive asymmetric autocatalytic reaction cycles. *J. Math. Chem.* **2008**, *43*, 1505–1515. [CrossRef]
67. Crusats, J.; Hochberg, D.; Moyano, A.; Ribó, J.M. Frank model and spontaneous emergence of chirality in closed systems. *Chem. Phys. Chem.* **2009**, *10*, 2123–2131. [CrossRef]
68. Dóka, É.; Lente, G. Mechanism-based chemical understanding of chiral symmetry breaking in the Soai reaction. A combined probabilistic and deterministic description of chemical reactions. *J. Am. Chem. Soc.* **2011**, *133*, 17878–17881. [CrossRef]
69. Lavabre, D.; Micheau, J.-C.; Islas, J.R.; Buhse, T. Enantioselectivity Reversal by achiral additives in the Soai reaction: A kinetic understanding. *J. Phys. Chem. A* **2007**, *111*, 281–286. [CrossRef]
70. Matsumoto, A.; Abe, T.; Hara, A.; Tobita, T.; Sasagawa, T.; Kawasaki, T.; Soai, K. Crystal structure of isopropylzinc alkoxide of pyrimidyl alkanol: Mechanistic insights for asymmetric autocatalysis with amplification of enantiomeric excess. *Angew. Chem. Int. Ed.* **2015**, *54*, 15218–15221. [CrossRef]
71. Matsumoto, A.; Fujiwara, S.; Abe, T.; Hara, A.; Tobita, T.; Sasagawa, T.; Kawasaki, T.; Soai, K. Elucidation of the structures of asymmetric autocatalyst based on X-ray crystallography. *Bull. Chem. Soc. Jpn.* **2016**, *89*, 1170–1177. [CrossRef]
72. Noble-Teran, M.E.; Cruz, J.-M.; Micheau, J.-C.; Buhse, T.W. A quantification of the Soai reaction. *ChemCatChem* **2018**, *10*, 642–648. [CrossRef]
73. Bailey, J.; Chrysostomou, A.; Hough, J.H.; Gledhill, T.M.; McCall, A.; Clark, S.; Ménard, F.; Tamura, M. Circular polarization in star-formation regions: Implications for biomolecular homochirality. *Science* **1998**, *281*, 672–674. [CrossRef] [PubMed]
74. Shibata, T.; Yamamoto, J.; Matsumoto, N.; Yonekubo, S.; Osanai, S.; Soai, K. Amplification of a slight enantiomeric imbalance in molecules based on asymmetric autocatalysis—The first correlation between high enantiomeric enrichment in a chiral molecule and circularly polarized light. *J. Am. Chem. Soc.* **1998**, *120*, 12157–12158. [CrossRef]
75. Kawasaki, T.; Sato, M.; Ishiguro, S.; Saito, T.; Morishita, Y.; Sato, I.; Nishino, H.; Inoue, Y.; Soai, K. Enantioselective synthesis of near enantiopure compound by asymmetric autocatalysis triggered by asymmetric photolysis with circularly polarized light. *J. Am. Chem. Soc.* **2005**, *127*, 3274–3275. [CrossRef] [PubMed]
76. Sato, I.; Sugie, R.; Matsueda, Y.; Furumura, Y.; Soai, K. Asymmetric synthesis utilizing circularly polarized light mediated by the photoequilibrium of chiral olefins in conjunction with asymmetric autocatalysis. *Angew. Chem. Int. Ed.* **2004**, *43*, 4490–4492. [CrossRef] [PubMed]
77. Noorduin, W.L.; Bode, A.C.; van der Meijden, M.; Meekes, H.; van Etteger, A.F.; van Enckevort, W.J.P.; Christianen, P.C.M.; Kaptein, B.; Kellogg, R.M.; Rasing, T.; Vlieg, E. Complete chiral symmetry breaking of an amino acid derivative directed by circularly polarized light. *Nat. Chem.* **2009**, *1*, 729–732. [CrossRef]
78. Bonner, W.A.; Kavasmaneck, P.R.; Martin, F.S.; Flores, J.J. Asymmetric adsorption of alanine by quartz. *Science* **1974**, *186*, 143–144. [CrossRef]

79. Soai, K.; Osanai, S.; Kadowaki, K.; Yonekubo, S.; Shibata, T.; Sato, I. *d*- and *l*-Quartz-promoted highly enantioselective synthesis of a chiral organic compound. *J. Am. Chem. Soc.* **1999**, *121*, 11235–11236. [CrossRef]
80. Kondepudi, D.K.; Kaufman, R.J.; Singh, N. Chiral symmetry breaking in sodium chlorate crystallizaton. *Science* **1990**, *250*, 975–976. [CrossRef]
81. McBride, J.M.; Carter, R.L. Spontaneous resolution by stirred crystallization. *Angew. Chem. Int. Ed.* **1991**, *30*, 293–295. [CrossRef]
82. Sato, I.; Kadowaki, K.; Soai, K. Asymmetric synthesis of an organic compound with high enantiomeric excess induced by inorganic ionic sodium chlorate. *Angew. Chem. Int. Ed.* **2000**, *39*, 1510–1512. [CrossRef]
83. Sato, I.; Kadowaki, K.; Ohgo, Y.; Soai, K. Highly enantioselective asymmetric autocatalysis induced by chiral ionic crystals of sodium chlorate and sodium bromate. *J. Mol. Cat. A Chem.* **2004**, *216*, 209–214. [CrossRef]
84. Shindo, H.; Shirota, Y.; Niki, K.; Kawasaki, T.; Suzuki, K.; Araki, Y.; Matsumoto, A.; Soai, K. Asymmetric autocatalysis induced by cinnabar: Observation of the enantioselective adsorption of a 5-pyrimidyl alkanol on the crystal surface. *Angew. Chem. Int. Ed.* **2013**, *52*, 9135–9138. [CrossRef]
85. Matsumoto, A.; Ozawa, H.; Inumaru, A.; Soai, K. Asymmetric induction by retgersite, nickel sulfate hexahydrate, in conjunction with asymmetric autocatalysis. *New. J. Chem.* **2015**, *39*, 6742–6745. [CrossRef]
86. Matsumoto, A.; Kaimori, Y.; Uchida, M.; Omori, H.; Kawasaki, T.; Soai, K. Achiral inorganic gypsum acts as an origin of chirality through its enaniotopic surface in conjunction with asymmetric autocatalysis. *Angew. Chem. Int. Ed.* **2017**, *56*, 545–548. [CrossRef]
87. Matsuura, T.; Koshima, H. Introduction to chiral crystallization of achiral organic compounds. Spontaneous generation of chirality. *J. Photochem. Photobiol. C Photochem. Rev.* **2005**, *6*, 7–24. [CrossRef]
88. Ishikawa, K.; Tanaka, M.; Suzuki, T.; Sekine, A.; Kawasaki, T.; Soai, K.; Shiro, M.; Lahav, M.; Asahi, T. Absolute chirality of the gamma-polymorph of glycine: Correlation of the absolute structure with the optical rotation. *Chem. Commun.* **2012**, *48*, 6031–6033. [CrossRef] [PubMed]
89. Tarasevych, A.V.; Sorochinsky, A.E.; Kukhar, V.P.; Toupet, L.; Crassous, J.; Guillemin, J.-C. Attrition-induced spontaneous chiral amplification of the γ polymorphic modification of glycine. *Cryst. Eng. Comm.* **2015**, *17*, 1513–1517. [CrossRef]
90. Matsumoto, A.; Ozaki, H.; Tsuchiya, S.; Asahi, T.; Lahav, M.; Kawasaki, T.; Soai, K. Achiral amino acid glycine acts as an origin of homochirality in asymmetric autocatalysis. *Org. Biomol. Chem.* **2019**, *17*, 4200–4203. [CrossRef]
91. Robertson, M.P.; Miller, S.L. An efficient prebiotic synthesis of cytosine and uracil. *Nature* **1995**, *375*, 772–774. [CrossRef]
92. Kawasaki, T.; Suzuki, K.; Hakoda, Y.; Soai, K. Achiral nucleobase cytosine acts as an origin of homochirality of biomolecules in conjunction with asymmetric autocatalysis. *Angew. Chem. Int. Ed.* **2008**, *47*, 496–499. [CrossRef]
93. Kawasaki, T.; Hakoda, Y.; Mineki, H.; Suzuki, K.; Soai, K. Generation of absolute controlled crystal chirality by the removal of crystal water from achiral crystal of nucleobase cytosine. *J. Am. Chem. Soc.* **2010**, *132*, 2874–2875. [CrossRef]
94. Mineki, H.; Kaimori, Y.; Kawasaki, T.; Matsumoto, A.; Soai, K. Enantiodivergent formation of a chiral cytosine crystal by removal of crystal water from an achiral monohydrate crystal under reduced pressure. *Tetrahedron Asymm.* **2013**, *24*, 1365–1367. [CrossRef]
95. Mineki, H.; Hanasaki, T.; Matsumoto, A.; Kawasaki, T.; Soai, K. Asymmetric autocatalysis initiated by achiral nucleic acid base adenine: Implications on the origin of homochirality of biomolecul. *Chem. Commun.* **2012**, *48*, 10538–10540. [CrossRef]
96. Kawasaki, T.; Suzuki, K.; Hatase, K.; Otsuka, M.; Koshima, H.; Soai, K. Enantioselective synthesis mediated by chiral crystal of achiral hippuric acid in conjunction with asymmetric autocatalysis. *Chem. Commun.* **2006**, 1869–1871. [CrossRef]
97. Carter, D.J.; Rohl, A.L.; Shtukenberg, A.; Bian, S.D.; Hu, C.-H.; Baylon, L.; Kahr, B.; Mineki, H.; Abe, K.; Kawasaki, T. Prediction of Soai reaction enantioselectivity induced by crystals of *N*-(2-thienylcarbonyl) glycine. *Cryst. Growth Des.* **2012**, *12*, 2138–2145. [CrossRef]
98. Kawasaki, T.; Jo, K.; Igarashi, H.; Sato, I.; Nagano, M.; Koshima, H.; Soai, K. Asymmetric amplification using chiral co-crystal f formed from achiral organic molecules by asymmetric autocatalysis. *Angew. Chem. Int. Ed.* **2005**, *44*, 2774–2777. [CrossRef]

99. Kawasaki, T.; Harada, Y.; Suzuki, K.; Tobita, T.; Florini, N.; Palyi, G.; Soai, K. Enantioselective synthesis utilizing enantiomorphous organic crystal of achiral benzils as a source of chirality in asymmetric autocatalysis. *Org. Lett.* **2008**, *10*, 4085–4088. [CrossRef]
100. Kawasaki, T.; Nakaoda, M.; Kaito, N.; Sasagawa, T.; Soai, K. Asymmetric autocatalysis induced by chiral crystals of achiral tetraphenylethylenes. *Orig. Life Evol. Biosph.* **2010**, *40*, 65–78. [CrossRef]
101. Matsumoto, A.; Ide, T.; Kaimori, Y.; Fujiwara, S.; Soai, K. Asymmetric autocatalysis triggered by chiral crystal of achiral ethylenediamine sulfate. *Chem. Lett.* **2015**, *44*, 688–690. [CrossRef]
102. Kawasaki, T.; Uchida, M.; Kaimori, Y.; Sasagawa, T.; Matsumoto, A.; Soai, K. Enantioselective synthesis induced by the helical molecular arrangement in the chiral crystal of achiral tris(2-hydroxyethyl)-1,3,5-benzenetricarboxylate in conjunction with asymmetric autocatalysis. *Chem. Lett.* **2013**, *42*, 711–713. [CrossRef]
103. Matsumoto, A.; Takeda, S.; Harada, S.; Soai, K. Determination of the absolute structure of the chiral crystal consisting of achiral dibutylhydroxytoluene and asymmetric autocatalysis triggered by this chiral crystal. *Tetrahedron Asymm.* **2016**, *27*, 943–946. [CrossRef]
104. Kawasaki, T.; Sasagawa, T.; Shiozawa, K.; Uchida, M.; Suzuki, K.; Soai, K. Enantioselective synthesis induced by chiral crystal composed of DL-serine in conjunction with asymmetric autocatalysis. *Org. Lett.* **2011**, *13*, 2361–2363. [CrossRef]
105. Kawasaki, T.; Kamimura, S.; Amihara, A.; Suzuki, K.; Soai, K. Enantioselective C-C bond formation as a result of the oriented prochirality of an achiral aldehyde at the single-crystal face upon treatment with a dialkyl zinc vapor. *Angew. Chem. Int. Ed.* **2011**, *50*, 6796–6798. [CrossRef]
106. Caglioti, L.; Hajdu, C.; Holczknecht, O.; Zékány, L.; Zucchi, C.; Micskei, K.; Pályi, G. The concept of racemates and the Soai-reaction. *Viva Origino* **2006**, *34*, 62–80.
107. Maioli, M.; Varadi, G.; Kurdi, R.; Caglioti, L.; Palyi, G. Limits of the classical concept of concentration. *J. Phys. Chem. B* **2016**, *120*, 7438–7445. [CrossRef]
108. Barabas, B.; Caglioti, L.; Zucchi, C.; Maioli, M.; Gál, E.; Micskei, K.; Pályi, G. Violation of distribution symmetry in statistical evaluation of absolute enantioselective synthesis. *J. Phys. Chem. B* **2007**, *111*, 11506–11510. [CrossRef]
109. Soai, K.; Shibata, T.; Kowata, Y. Japan Kokai Tokkyo Koho. Patent No. JP1997–268179, 2 January 1997.
110. Soai, K.; Sato, I.; Shibata, T.; Komiya, S.; Hayashi, M.; Matsueda, Y.; Imamura, H.; Hayase, T.; Morioka, H.; Tabira, H.; Yamamoto, J.; Kowata, Y. Asymmetric synthesis of pyrimidyl alkanol without adding chiral substances by the addition of diisopropylzinc to pyrimidine-5-carbaldehyde in conjunction with asymmetric autocatalysis. *Tetrahedron Asymm.* **2003**, *14*, 185–188. [CrossRef]
111. Kawasaki, T.; Suzuki, K.; Shimizu, M.; Ishikawa, K.; Soai, K. Spontaneous absolute asymmetric synthesis in the presence of achiral silica gel in conjunction with asymmetric autocatalysis. *Chirality* **2006**, *18*, 479–482. [CrossRef]
112. Suzuki, K.; Hatase, K.; Nishiyama, D.; Kawasaki, T.; Soai, K. Spontaneous absolute asymmetric synthesis promoted by achiral amines in conjunction with asymmetric autocatalysis. *J. Syst. Chem.* **2010**, *1*, 5. [CrossRef]
113. Singleton, D.A.; Vo, L.K. A few molecules can control the enantiomeric outcome. Evidence supporting absolute asymmetric synthesis using the Soai asymmetric autocatalysis. *Org. Lett.* **2003**, *5*, 4337–4339. [CrossRef]
114. Lente, G. Stochastic kinetic models of chiral autocatalysis: A general tool for the quantitative interpretation of total asymmetric synthesis. *J. Phys. Chem. A* **2005**, *109*, 11058–11063. [CrossRef]
115. Islas, J.R.; Lavabre, D.; Grevy, J.-M.; Lamoneda, R.H.; Cabrera, H.R.; Micheau, J.-C.; Buhse, T. Mirror-symmetry breaking in the Soai reaction: A kinetic understanding. *Proc. Natl. Acad. Sci. USA* **2005**, *102*, 13743–13748. [CrossRef]
116. Lavabre, D.; Micheau, J.-C.; Rivera Islas, J.; Buhse, T. Kinetic insight into specific features of the autocatalytic Soai reaction. *Top. Curr. Chem.* **2008**, *284*, 67–96.
117. Saito, Y.; Hyuga, H. Rate equation approaches to amplification of enantiomeric excess and chiral symmetry breaking. *Top. Curr. Chem.* **2008**, *284*, 97–118.
118. Kaimori, Y.; Hiyoshi, Y.; Kawasaki, T.; Matsumoto, A.; Soai, K. Formation of enantioenriched alkanol with stochastic distribution of enantiomers in the absolute asymmetric synthesis under heterogeneous solid–vapor phase conditions. *Chem. Commun.* **2019**, *55*, 5223–5226. [CrossRef]
119. Kawasaki, T.; Matsumura, Y.; Tsutsumi, T.; Suzuki, K.; Ito, M.; Soai, K. Asymmetric autocatalysis triggered by carbon isotope ($^{13}C/^{12}C$) chirality. *Science* **2009**, *324*, 492–495. [CrossRef] [PubMed]

120. Matsumoto, A.; Ozaki, H.; Harada, S.; Tada, K.; Ayugase, T.; Ozawa, H.; Kawasaki, T.; Soai, K. Asymmetric induction by nitrogen $^{14}N/^{15}N$ isotopomer in conjunction with asymmetric autocatalysis. *Angew. Chem. Int. Ed.* **2016**, *55*, 15246–15249. [CrossRef]
121. Kawasaki, T.; Okano, Y.; Suzuki, E.; Takano, S.; Oji, S.; Soai, K. Asymmetric autocatalysis: Triggered by chiral isotopomer arising from oxygen isotope substitution. *Angew. Chem. Int. Ed.* **2011**, *50*, 8131–8133. [CrossRef]
122. Matsumoto, A.; Oji, S.; Takano, S.; Tada, K.; Kawasaki, T.; Soai, K. Asymmetric autocatalysis triggered by oxygen isotopically chiral glycerin. *Org. Biomol. Chem.* **2013**, *11*, 2928–2931. [CrossRef]
123. Horeau, A.; Nouaille, A.; Mislow, K. Secondary deuterium isotope effects in asymmetric syntheses and kinetic resolutions. *J. Am. Chem. Soc.* **1965**, *87*, 4957–4958. [CrossRef]
124. Pracejus, H. Ein sterischer isotopeneffekt als ursache einer katalytisch-asymmetrischen synthese. *Tetrahedron Lett.* **1966**, *7*, 3809–3813. [CrossRef]
125. Sato, I.; Omiya, D.; Saito, T.; Soai, K. Highly enantioselective synthesis induced by chiral primary alcohols due to deuterium substitution. *J. Am. Chem. Soc.* **2000**, *122*, 11739–11740. [CrossRef]
126. Kawasaki, T.; Ozawa, H.; Ito, M.; Soai, K. Enantioselective synthesis induced by compounds with chirality arising from partially deuterated methyl groups in conjunction with asymmetric autocatalysis. *Chem. Lett.* **2011**, *40*, 320–321. [CrossRef]
127. Kawasaki, T.; Shimizu, M.; Nishiyama, D.; Ito, M.; Ozawa, H.; Soai, K. Asymmetric autocatalysis induced by meteoritic amino acids with hydrogen isotope chirality. *Chem. Commun.* **2009**, 4396–4398. [CrossRef]
128. Sato, I.; Ohgo, Y.; Igarashi, H.; Nishiyama, D.; Kawasaki, T.; Soai, K. Determination of absolute configurations of amino acids by asymmetric autocatalysis of 2-alkynylpyrimidyl alkanol as a chiral sensor. *J. Organomet. Chem.* **2007**, *692*, 1783–1787. [CrossRef]
129. Sato, I.; Yamashima, R.; Kadowaki, K.; Yamamoto, J.; Shibata, T.; Soai, K. Asymmetric induction by helical hydrocarbons: [6]- and [5]helicenes. *Angew. Chem. Int. Ed.* **2001**, *40*, 1096–1098. [CrossRef]
130. Kawasaki, T.; Suzuki, K.; Licandro, E.; Bossi, A.; Maiorana, S.; Soai, K. Enantioselective synthesis induced by tetrathia-[7]-helicenes in conjunction with asymmetric autocatalysis. *Tetrahedron Asymm.* **2006**, *17*, 2050–2053. [CrossRef]
131. Matsumoto, A.; Yonemitsu, K.; Ozaki, H.; Míšek, J.; Starý, I.; Stará, I.G.; Soai, K. Reversal of the sense of enantioselectivity between 1- and 2-aza[6]helicenes used as chiral inducers of asymmetric autocatalysis. *Org. Biomol. Chem.* **2017**, *15*, 1321–1324. [CrossRef]
132. Kawasaki, T.; Tanaka, H.; Tsutsumi, T.; Kasahara, T.; Sato, I.; Soai, K. Chiral discrimination of cryptochiral saturated quaternary and tertiary hydrocarbons by asymmetric autocatalysis. *J. Am. Chem. Soc.* **2006**, *128*, 6032–6033. [CrossRef]
133. Kawasaki, T.; Hohberger, C.; Araki, Y.; Hatase, K.; Beckerle, K.; Okuda, J.; Soai, K. Discrimination of cryptochirality in chiral isotactic polystyrene by asymmetric autocatalysis. *Chem. Commun.* **2009**, 5621–5623. [CrossRef]
134. Sato, I.; Kadowaki, K.; Urabe, H.; Hwa Jung, J.; Ono, Y.; Shinkai, S.; Soai, K. Highly enantioselective synthesis of organic compound using right- and left-handed helical silica. *Tetrahedron Lett.* **2003**, *44*, 721–724. [CrossRef]
135. Kawasaki, T.; Araki, Y.; Hatase, K.; Suzuki, K.; Matsumoto, A.; Yokoi, T.; Kubota, Y.; Tatsumi, T.; Soai, K. Helical mesoporous silica as an inorganic heterogeneous chiral trigger for asymmetric autocatalysis with amplification of enantiomeric excess. *Chem. Commun.* **2015**, *51*, 8742–8744. [CrossRef] [PubMed]

Review

Biological Homochirality on the Earth, or in the Universe? A Selective Review

Vadim A. Davankov

Institute of Organo-Element Compounds, Russian Academy of Sciences, 119991 Moscow, Russia; davank@ineos.ac.ru

Received: 21 November 2018; Accepted: 10 December 2018; Published: 13 December 2018

Abstract: The discovery of meteoritic alpha-amino acids with significant enantiomeric excesses of the L-form has suggested that some cosmic factors could serve as the initial source for chiral imbalance of organic compounds delivered to the early Earth. The paper reviews major hypothesis considering the influence of chiral irradiation and chiral combinations of physical fields on the possible ways asymmetric synthesis and transformations of organics could take place within the solar system. They could result in a small enantiomeric imbalance of some groups of compounds. More attention is paid to the hypothesis on parity violation of weak interaction that was supposed to cause homochirality of all primary particles and a more significant homochirality of compounds directly synthesized from the latter in a plasma reactor. The first experiment with material synthesized in a plasma torch resulting from a super-high-velocity impact showed formation of alanine with the excess of L-form between 7 and 25%. The supposed conclusion is that L-amino acids could serve as a starting homochiral biomolecular pool for life to emerge all over the Universe.

Keywords: origin of life; biological homochirality; deracemization; super-high-velocity impact; plasma reactor; absolute asymmetric synthesis; amino acids

---

1. Introduction

The recently revised estimates of the age of our Earth amount to 4.54 billion years. Some 4.3 billion years old rocks already show traces of water withering, implying the existence of oceans and land. Surprisingly, isotopic analysis of some zircon crystals, 4.1 billion years old, reveals carbon-rich inclusions which allow suspecting the existence of primitive organic life at that time. In any case, it did not take long, just a few hundred million years, for life to emerge on the young Earth. This could only be possible if abundant organic matter with a predominant homochirality has been accumulated in its aqueous basins. Indeed, there is no doubt that numerous organic compounds have been transported to the planet's surface or were formed on it during the times of intense bombardment of planets with meteorites. Even now, the Mars rover "Curiosity" detects organic molecules in the Gale Crater, a depression that was most likely a freshwater lake about 3.5 billion years ago. Among such compounds are thiophenes and dimethyl sulfides, but also aromatic compounds, such as toluene, chlorobenzene, and naphthalene, as well as chain hydrocarbons. The soil samples examined by Curiosity come from the top 5 cm of the Martian surface and thus they survived in a chemically very aggressive and hostile environment exposed to cosmic rays and the ionizing and oxidizing conditions [1]. Where the molecules come from; whether they have been delivered from elsewhere or directly formed on the planet's surface is unclear.

As early as 1988, Gol'danskii and Kuz'min [2] convincingly proved that abundance of organic matter is the precondition that is required but not sufficient for life to emerge. Biopolymers present numerous high-molecular-weight compounds whose unique three-dimensional structures are absolutely essential for their productive functioning during their interaction with other components of a living cell. This uniqueness of each biopolymer structure is guaranteed not only by the unambiguous sequence of their building blocks—alpha amino acids, sugars, and nucleotides—but also by the homochirality of basic classes of compounds with asymmetric molecules, such as amino acids and sugars. The problem of the emergence of life thus cannot be solved without identifying mechanisms providing homochirality of the above series of compounds, since without preexisting homochirality, the self-replication characteristic of living matter could not occur in principle. No racemic primordial soup of organic compounds would ever give rise to any self-replicating system. Shandrasekhar [3] reasonably notes: "It is indeed sobering – if not depressing – to consider the fact that, to the extent that a firm answer to the question on the origin of molecular chirality does not emerge, the origin of life will remain a mystery."

Numerous literature exists dealing with the emergence or enhancement of chirality in a racemic system by a so-called *chance mechanism*: spontaneous resolution on sublimation or crystallization with or without Viedma ripening [4], spontaneous symmetry breaking via stereospecific autocatalysis, asymmetric adsorption, or asymmetric synthesis on chiral crystals. Some of the above processes can give, within a rather small location, individual products having very high enantiomeric purity of an undetermined sign, with a similar probability for creating the opposite stereochemical result in another small location. None of the chance processes could thus provide a *global* predominance of L-enantiomers for the whole series of amino acids and D-isomers for sugars. Besides, it is quite impossible to imagine the above processes to proceed efficiently in the complex matrix of the primordial soup of organic compounds. Therefore, if the chance scenario for the origin of abiotic chiral purity is not viable, then all arguments strongly favor theories of *induced extraterrestrial homochirality* of initial organic matter.

Searching for chiral organic matter in our solar system is a major challenge for the space research community and has attained increasing attention from many research groups. Review [5] analyzed functional possibilities generated by numerous sophisticated instrumentation and technologies available or being developed especially for space survey missions. One advanced space mission, namely, the Rosetta spacecraft, after a 10-years-long journey, detached in November 2014 the Philae lander on the surface of comet 67P/Churyumov-Gerasimenko (67P/C-G). The instrument was fully prepared to conduct for the first time an extraterrestrial stereochemical investigation of a soil extract by employing a multicolumn gas chromatograph and a time-of-flight mass spectrometer [6]. After methylation with *N,N*-dimethylformamide dimethyl acetal, a series of chiral molecules, such as amino acids, hydrocarbons, amines, alcohols, diols, and carboxylic acids, could be analyzed. Unfortunately, the experiment failed because of insufficient energy supply, so that no enantiomeric on-site analysis has been performed thus far on any organic extraterrestrial material. On the contrary, numerous laboratory works have been conducted in many countries simulating possible sources of extraterrestrial organic matter, including chiral compounds.

2. Origins of Extraterrestrial Non-Racemic Organic Matter

Since Miller's seminal works from 1953 and later, it has become clear that specific biological building blocks could well be formed in the atmosphere of any planet from a primitive reducing gas mixture of water, methane, and ammonia, but also from the later suggested gas composition consisting mostly of carbon dioxide, nitrogen, and water. For instance, both UV and charged particle irradiation of simple gas mixtures of $H_2O/CH_3OH/NH_3/CH_3CN$ were shown to produce up to 26 racemic amino acids, diamino acids, and N-(2-aminoethyl)glycine [7], but also purine and pyrimidine compounds, urea, and polyols, along with other prebiotic molecular structures [8].

Simultaneous formation of many more important organic compounds, namely, that of nucleobases (cytosine and uracil), various proteinogenic amino acids (glycine, alanine, serine, aspartic acid, glutamic acid, valine, leucine, isoleucine, and proline), non-proteinogenic amino acids, and aliphatic amines, was reported in experiments simulating reactions induced by extraterrestrial objects impacting on the early oceans [9]. The latter are expected to have had sufficient amounts of dissolved bicarbonates and inorganic nitrogen as the sours of required elements.

Another medium suitable for emergence of organic molecules is cosmic ice, even at extremely low temperatures. The latter are beneficial to the survival of molecules that were formed in space by one mechanism or another. In general, individual molecules and heavier atoms are assumed to tend to gather or adhere to already existing cold solid dust material. Since water is one of most abundant compounds in the interstellar space, formation of ice bodies with the incorporation of larger molecules and dust particles is a very common process. Indeed, astrophysical ices contain simple molecules (CO_2, H_2O, CH_4, HCN, NH_3, CO, and others). Once formed, the ices are exposed to complex radiation fields, e.g., UV, γ- and X-rays, stellar/solar wind particles, cosmic rays, and collisions with other cosmic bodies. It is only natural to assume that cosmic ices not only entrap various larger organic molecules but also present a suitable matrix for the formation of new ones, e.g., via radical mechanisms. Indeed, glycine (Gly), the simplest amino-acid building-block of proteins, has been identified on icy dust grains in the interstellar medium, icy comets, and ice-covered meteorites, as in the comet 81P/Wild-2 samples collected and returned to Earth by NASA's Stardust spacecraft.

Martins et al. [10] presented results of laboratory experiments in which ice mixtures analogous to those found in a comet were subjected to a shock with a steel projectile fired at a hypervelocity using a light gas gun. The hypervelocity impact shock of a typical comet ice mixture was found to produce several amino acids after hydrolysis. These included equal amounts of D- and L-alanine, and the non-protein amino acids α-aminoisobutyric acid and isovaline as well as their precursors. Simulation of astrophysical conditions by exposure of multilayer CO_2:CH_4:NH_3 ice films to 0–70 eV electrons also revealed formation of multiple products, among them glycine [11]. Other amino acids could well form too, though in smaller quantities.

The discovery of meteoritic α-amino acids with significant enantiomeric excesses of the L-form has suggested that some cosmic factors could serve as the initial source for chiral imbalance of organic compounds. In this case, the non-racemic extraterrestrial organic materials delivered to Earth by carbonaceous meteorites (chondrites) may have contributed to prebiotic chemistry and finally directed to homochirality of amino acids and carbohydrates on Earth. Chondrites present remnants of asteroids that survived extreme heating in the Earth's atmosphere and impact with the solid surface. Chondrites contain numerous extractable organic compounds and, in even larger amounts, insoluble crosslinked polymeric kerogen-like organic material. As summarized in the review paper by the group of d'Hendecourt and Meierhenrich [6], extensive studies of the Murchison and Murray meteorites "revealed the presence of a series of alkyl-substituted bicyclic and tricyclic aromatic compounds, aliphatic compounds ranging from C1 to C7, including both saturated and unsaturated hydrocarbons, more than 80 amino acids, including diamino acids, N-alkylated amino acids and iminodiacids, small amounts of aldehydes and ketones up to C5, a wide spectrum of carboxylic acids and hydroxycarboxylic acids, several nucleobases, and sugar acids." Thank to available sensitive chromatographic techniques for enantiomeric analysis, from the above diverse array of meteoritic organic molecules, four groups of compounds having asymmetric carbon atoms—namely, amino acids, hydroxycarboxylic acids, monocarboxylic acids, and amines—have been examined for their stereochemical composition. All amino acids were found to display the predominance of one type of configuration with a wide range of *ee* values. α-Substituted amino acids, which are known to be especially resistant to racemization, exhibited extremely high enantiomeric excess: up to 60% for D-alloisoleucine [12] and 18% for L-isovaline [13]. Surprisingly, meteoritic aliphatic amines and carboxylic acids that might share a common chemical origin with amino acids were found to be racemic with the exception of α-hydroxypropionic (lactic) acid [14]. According to Soai et al. [15], the insoluble and resistant to hydrolysis meteoritic material also incorporates chiral fragments.

Several mechanisms for the emergence of extraterrestrial chirality have been considered and subjected to experimental testing. Rather popular is the idea of *deracemization* of pre-formed racemic amino acids, i.e., distortion of the initial 1:1 proportion of their enantiomers. A number of deracemization scenario are discussed involving chiral irradiation, chiral combinations of physical fields, and beams of chiral particles, all of which are known to be emitted by nascent neutron stars, short supernova bursts, or growing black holes. Indeed, a supermassive black hole, billions of times heavier than the Sun, spinning at the galactic center shoots out two very intensive narrow jets of matter and chiral radiation in opposite directions [16].

As was shown, circularly polarized irradiation may cause slightly enantioselective destruction (deracemization) of racemic compounds. Numerous extensive papers and reviews analyze astronomical sources of circularly polarized light [17,18] and results of model experiments on deracemization of amino acids [19,20] in terms of possible origins of biological homochirality on the Earth [6,21].

Besides deracemization of pre-formed amino acids, direct asymmetric synthetic processes can take place on space objects exposed to circularly polarized irradiation. Thus, in circumstantial laboratory experiments on model achiral extraterrestrial ice [22,23], the asymmetrical synthesis of amino acids was found to take place under irradiation with visible and UV circularly polarized light (CPL) at two different photon energies (6.6 and 10.2 eV). Sixteen distinct amino acids were identified and the enantiomeric composition of five of them (α-alanine, 2,3-diaminopropionic acid, 2-aminobutyric acid, valine, and norvaline) was precisely measured using the enantioselective two dimensional gas chromatography-time of flight mass spectrometry technique technique. The results obtained on such irradiated ices showed certain symmetry breaking of amino acids with the *ee* values rising with time from *ee* L = $-0.20\% \pm 0.14\%$ to *ee* L = $-2.54\% \pm 0.28\%$. The sign of the induced *ee* (that depends on the helicity of the CPL) was the same for all five considered amino acids and did not depend on the evolutionary stage of the samples. Surprisingly, the chirality sign was also found to depend on the energy of the CPL.

Similar formation of slightly enantiomerically enriched amino acids was earlier considered by irradiation of their initially racemic pools with intensive antineutrino streams [24]. More recent works [25,26] suggested that the deracemization of the amino acids could be established due to the magnetic field of a nascent neutron star from a core-collapsed supernova via amino acid processing by the neutrinos that would be emitted. A theoretically derived supernova neutrino amino acid processing model, or SNAAP model, not only appears to produce a small chiral imbalance, but, importantly, always produces the same sign of the chirality for different amino acids [27].

Chirality also emerges when subjecting thin films of amino acids to polarized light from synchrotron radiation and a free electron laser. This process is thought to mimic the evolution of organics adsorbed on the surface of interstellar dust particles [28].

Finally, spin-polarized electrons emitted by the β-decay of radioactive nucleons were also shown to cause ionization of two enantiomeric molecules of amino acids at slightly different rate thus resulting in small enantiomeric enhancement of the last surviving portions of the initial racemic material. This effect known as "electronic circular dichroism" was applied to induce small optical activity in leucine [29], tryptophan [30], and alanine [31]. However, the kinetic energy of the investigated spin-polarized electrons was by orders of magnitude too high for efficient enantiomer-discriminating absorption, which results in almost total destruction of the initial portion of amino acids.

All the above considered *determinate mechanisms* of creating homochiral groups of organic compounds (which include symmetry breaking induced by combinations of circularly polarized irradiation, and magnetic or even gravitational [32] fields) expand the search area for origins of chirality and life from some locations on the Earth to the interplanetary matter existing in space regions as big as our Solar system.

Indeed, a more or less expressed enantioselective interaction between organic matter and various physical fields and elementary particle beams seems to be a much more common phenomenon than has been earlier expected. Just to mention that electron transmission through chiral molecules was found to depend on the electron spin, and thin films of chiral compounds, in particular nucleic acids and peptides, have been shown to act as spin filters, as summarized in the recent review [33].

Still, all the above hypotheses of extraterrestrial origins of biological homochirality on Earth suffer from several important weak points.

First of all, for life to start on Earth, it is critical that the building blocks of amino acids, sugars, and nucleosides *be created in space in homochiral form*, namely, with a predominant L-configuration for all amino acids and D-configuration for all sugars. This unique direction of enantioselectivity is not confirmed for the action of all types of neutrino, antineutrino, polarized electrons, or streams of other cosmic particles. As for the rays of electromagnetic waves, in different area of space, they may have different signs of circular polarization. Up to now, a relatively large (from +17% to −5%) circular polarization of irradiation (namely in the IR range of radiation) was reported from the Orion nebula OMC-1 star-forming region extending about 400 times the size of the solar system [34].

Notably, such CPL sources will not usually generate a net circular polarization because they will have regions of positive and negative sign that cancel when averaged over any galactic-sized area like the Milky Way. The discussed irradiation scenario for the origin of homochirality requires that the Sun system be formed during only one single chiral cosmic event dominating over a very large region of a few parsecs in size. The hypothesis involves development of the planetary system in a high-mass star-forming region where matter was subjected to the influence of an external source of CPL or chiral combination of physical fields of a given helicity and a favorable dominant energy during the protoplanetar phase of evolution, thus inducing one definite stereo-specific photochemistry in the system. That the solar system originated in such a massive star formation region is supported by isotopic studies of meteorites' compositions, such as those including ^{60}Fe, suggesting that a supernova explosion occurred near the Sun [35].

Another general problem is that the configuration of amino acids induced by circularly polarized irradiation depends on the energy of its quanta, i.e. on its wavelength [23]. Photolysis of amino acids requires UV radiation, rather than the infrared radiation observed in the Orion massive star forming area. Of course, UV radiation cannot be directly observed as it is unable to penetrate the dust that lies along the line-of-sight between the Earth and regions of highly polarized radiation activity. The general stereochemical outcome of irradiation with CPL of the whole pallet of amino acids was critically analyzed by Cerf and Jorissen [36]. When taking into account that asymmetric photolysis implies the preferential destruction of the enantiomer having the higher absorption coefficient, the highest stereodifferentiation efficiency of irradiation is directly related to the circular dichroism (CD) band of the compound. For amino acids, the carboxyl group bound to the alpha-carbon has a strong CD band centered at about 210 nm. The sign of this CD band is the same for all aliphatic L-amino acids. It is not the case for tryptophan [37], whose indole chromophore exhibits a strong CD band centered at about 195 nm, with the opposite sign to the carboxyl 210 nm band. Proline also has a strong CD band of opposite sign around 193 nm in a neutral solution. Therefore, these two amino acids would violate the homochirality of the pool as a whole if the asymmetry would arise from the enantioselective photolysis of all the racemates with CPL with the wave length around 200 nm. At the same time, there seems to be no spectral window where all the biogenic amino acids have a strong CD band of *one and the same sign*. Besides, the intensities and signs of CD bands depend on the properties of the medium, so that extrapolation of laboratory data obtained in liquid solutions to infer the CD properties of amino acids in space (where they are likely to be found in solid or gas phase) is not straightforward.

It follows from the above considerations that the energy of the circularly polarized UV irradiation or particle beams must fit into a very narrow range in order to cause efficient ionization and stereochemical transformations of organic molecules resulting in the deracemization of preformed compounds or asymmetric synthesis of new ones. Less energetic impacts are inefficient while impacts of higher energy will totally destroy larger organic molecules. In fact, no circularly polarized cosmic radiation in the UV region that is required to trigger desired chemical transformations has been detected thus far. As a rule, cosmic rays of accelerated particles and radiation are destructive to unprotected organics rather than stimulating efficient synthetic processes. Maybe, cosmic ice bodies present exclusion due to the existence of some layers with reduced energies of radiation and emergence of secondary electrons. On the other hand, the extent of circular polarization of their impacts still remains to be examined. Also, one should bear in mind that we deal with dynamic processes where surface layers of ice bodies partially evaporate while they experience bombardment with cosmic rays.

When taking into account the predominant destructive role of all cosmic rays, it is difficult to expect any significant accumulation of chiral biomolecules on unprotected supports, the more that a measurable degree of enantioenrichment is only attained for the last small portion of the initial material. Even then, *ee* levels attained for these last portions of amino acids could not exceed few per cent. These values are by orders of magnitude smaller than what is observed in chondritic meteorites. All these considerations lead to a suggestion that the major part of homochirally enriched organic molecules were synthesized directly on the atmosphere-protected planets, rather than delivered to them by radiation-exposed cosmic dust particles and ice bodies.

3. Parity Violation in the Weak Interaction and Homochirality of Matter

The above-discussed processes with determinate origins of chirality could possibly generate very small chirality within regions as large as the solar system, but only as a result of one single cosmic event, such as a supernova burst, since any second similar burst event could generate the opposite sign of chirality of all enantioselective processes. The *only universal and constant source of invariant chirality* thus remains the phenomenon called violation of parity in the weak interaction.

Parity conservation implying that nature is symmetrical under reflections in a mirror enjoyed the status of a fundamental law of physics along with those of conservation of energy, momentum, and electric charge. Yet, although there were many experiments that established parity conservation in strong interactions, the weak force, which was first postulated to explain disintegration of elementary particles, was shown to violate the parity conservation law. It was demonstrated on radioactive ^{60}Co that the β-decay of nuclei oriented with a strong magnetic field emits exclusively left-polarized electrons, which unequivocally violates parity expectations [38,39]. This discovery revolutionized our understanding of nature's fundamental laws. As one of its consequences, we must expect a small energy difference between any two molecular structures that we used to call enantiomers. The energy gap between them was calculated to have an order of 10^{-12}–10^{-15} Jmol^{-1} [40] with a magnitude and sign that depend on the particular molecule and, unfortunately, also on its conformation. This energy difference is too small to be measured and also too small to cause any measurable enantiomeric enrichment under conditions of thermodynamic equilibration of a racemate.

Nevertheless, the role of parity violation in the creation of initial partial homochirality of organic matter under kinetic conditions, *far from a thermodynamic equilibrium*, must be considered more precisely. According to Davankov [41], parity violation in β-decay processes may be interpreted in terms of the chirality of elementary particles and is closely related to the phenomenon of the experimentally observed chirality of atoms [42], assuming the difference in terminology used in chemistry and physics is neglected. Since the discovery of parity violation, physicists now and then, analyzed the idea of "how chiral symmetry, its pattern of breaking and restoration under extreme conditions manifest themselves in the nucleon, nuclei, nuclear matter and hadronic matter" [43]. Even now, asymmetry at the level of atoms and elementary particles remains both a wonder and a challenge. In principle, it may give a valuable hint toward something new in the standard model and reconsiders the role of neutrinos, which are elusive particles that are difficult to detect. Moreover, current measurements of violation with respect to simultaneous space reflection and charge conjugation (charge-parity (CP) violation) in the decay of B mesons, which is the phenomenon of asymmetry itself, appear to account for the excess of matter over antimatter [44,45].

Recently, the European Space Agency's Planck telescope finished examining the cosmic microwave background, the faint afterglow of the Big Bang, so that final maps of the early Universe could be adjusted and released. These data helped researchers to pin down the age of the Universe (about 13.8 billion years), and its composition (95% dark matter and dark energy). The observable 5% of mass, which is the Universe, is made of matter, not antimatter. This means, that soon after Big Bang, particles would decay somewhat slower than their counterparts, antiparticles, and "CP violation" in particle decays could be the reason for that. It is assumed that stability of particles and their clusters is related to their weak charges. Thus, the proton's weak charge defines the strength of certain interactions between protons and other particles. More recently, scattering of left-polarized and right-polarized electrons on protons was precisely measured. The difference between the two measurement results characterizes the weak current of protons, its asymmetry, and was found to amount to -226.5 ± 9.3 parts per billion, where the minus sign indicates that left-handed electrons are more likely to be scattered than their right-handed counterparts. The corresponding proton's weak charge is 0.0719 ± 0.0045, while the proton's electric charge is +1. To put the magnitude of the above asymmetry in perspective: if parity symmetry were violated for the height of mountains, Mount Everest and its mirror-image twin would differ in height by a mere 2 millimeters [46]. Nonetheless, the important message in terms of chemistry is: protons are chiral.

Earlier, Davankov [47] presented a more common argument for the chirality of another elementary particle, the electron. It is known that any electric current, that is propagation of electrons along a wire (or just in vacuum), is accompanied with circular magnetic field of *one definite direction*. This makes the whole system, moving electrons/circular magnetic field, invariantly chiral. The intensity of the field only depends on the number of electrons that pass through the wire cross section during a unit time, but not on the creep velocity of electrons along the wire that would depend on voltage applied. One could argue that the voltage drop may cause partial spin-polarization of electrons. However, the velocity of random movements of electrons within a metal at room temperature exceeds the creep velocity by ten orders of magnitude, such that any spin polarization at that large ratio of rates appears highly improbable. The only known source of the invariant chirality of the system is the inherent chirality of electrons. It is only natural that any moving charged particle, both negative and positive, combined with the corresponding magnetic field, would compose a system with one definite sign of chirality. This leads to a conclusion that every charged elementary particle is inherently chiral. Importantly, any antiparticle having an opposite charge, like in the pair electron/positron, will produce an opposite chirality. For this reason, Davankov's suggestion [47] is that charged particles and their antiparticles have the opposite inherent chirality sign. This suggestion of some kind of enantiomeric relationship was further extended to all pairs of neutral particles and antiparticles.

What is even more important, Davankov formulated a general hypothesis [41,47] that all elementary particles in our universe compose a *homochiral pool* of which all atoms and molecules are built, while all elementary antiparticles would compose a homochiral pool that formed antimatter. The reality is that the collision of all particles with corresponding antiparticles results in their annihilation with an emission of electromagnetic impulses, i.e. photons of varying energy up to gamma quanta and neutrino, all of which are chiral as well. (According to Davankov [47] each individual photon and each energy quantum are chiral, but, contrary to elementary particles, they can exist in both left-rotating and right-rotating forms). The early universe is theorized to have spawned the exact same amounts of normal matter and antimatter. Since we observe only matter and no consolidated antimatter, two kinds of particles must have played by slightly different rules, letting a fraction of normal particles outlast their bizarre enantiomeric twins. Therefore, scientists are looking for theoretical particles called the axions that could possibly explain the curious asymmetry of the Universe and the massive cosmic mystery of dark matter. Theory says that axions will be absurdly light and hardly interact with normal matter. Until above fundamental theoretical problems receive their explanation, our notion of homochirality of all elementary particles has its right to exist and should be seriously taken into account.

Basically, this hypotheses states that inherent homochirality of normal matter was established in the whole universe soon after the Big Bang, about 13.8 billion years ago, and invariantly subsists for the whole period of its existence. It is only natural to suggest that homochirality of elementary particles exerts certain invariable influence on the overall chirality of atoms. The influence of parity violation could be much less expressed in organic molecules. Chirality of molecular structures belongs to a completely different hierarchy of chirality phenomena along with morphological chirality of living organisms or that of crystals. Besides, organic molecules were constantly formed and destroyed during prebiological chemical evolution of matter according to laws of organic chemistry, rather than that of physics.

Still, one can speculate that the above postulated homochirality of elementary particles and atoms could reveal itself to the most significant extent in the processes of the direct synthesis of asymmetric organic molecules from ionized atoms and electrons, rather than from transformations of simple organic precursors. Such direct synthetic processes may proceed during dissipation and the fast cooling of plasma that contain electrons and all required atoms, such as C, N, H, and O, in their ionized form. In any event, all numerous simple organic molecules in the universe [5] have emerged in exactly these kinds of processes.

According to Managadze [48], the major part of the more complex organic matter in the universe was also synthesized by collisions between solid cosmic bodies at relative velocities above 15–20 km s^{-1}, so-called super-high-velocity impacts. The latter bring about a super-quick heating of the interacting regions of both the target and the projectile up to temperatures of 10^6 °C, which results in a total atomization and 100% ionization of the affected matter. The initially solid matter converts into high-density and high-temperature plasma. The latter dissipates in the form of a so-called plasma torch or burst and rapidly cools down. During the process of the adiabatic expansion of plasma, recombination of ionized atoms and electrons proceeds with the eventual combination of the organic-bearing elements C, H, O, N, and other abandoned elements of matter having a particularly high affinity for each other. They finally combine into both simple and more complex organic compounds including those having centers of asymmetry. A significant part of organic compounds could thus be formed directly on the early Earth's surface (and everywhere else), as well as delivered by meteorites, which would have been especially intense during the first 500 million years after formation of solar planetary system [48].

Davankov's idea is that the general inherent homochirality of elementary particles should manifest itself to the most obvious extent in a kind of direct absolute asymmetric synthesis of first organic molecules and result in the invariant predominant homochirality in all classes of similarly constructed molecules. Notably, the nonequilibrium synthetic processes considered proceed under unsteady flows of colliding primary components in open systems that are far from equilibrium. If this hypothesis is true, amino acids enriched in L-enantiomers and sugars enriched in D-enantiomers have existed in the universe since the times where sufficient solid material has been accumulated in space to form planetary systems around numerous suns. The extent of enantioenrichment of organic molecules on each particular spot should depend on the conditions of their formation in super-high-velocity impacts and their subsequent transformations.

It further follows from the assumption of general homochirality of matter and that of initially formed biomolecules that preconditions for life to emerge do exist everywhere in the universe and that everywhere proteins, sugars, and nucleic acids must have the same unequivocal homochirality sign as that in living organisms on the Earth. At that, the pallet of life-forming building blocks and hereditary codes selected for living matter could well be different on different planets. In other words, the homochirality sign of living matter is predetermined by the inherent universal parity violation on the level of elementary particles, whereas amino acid and sugar compositions of biopolymers could be a matter of chance.

Among the above several fundamental predictions that follow from the hypothesis on homochirality of elementary particles, one could be put to the test by using already existing experimental techniques. It reads that any direct self assembling of α-amino acid molecules from ionized atoms and electrons must result in a noticeable homochiral enantiomeric enrichment of the products with predominance of molecules having L-configuration. Arguments for this supposition have been outlined in details by Davankov [47,49] and Managadze [50] in their conceptual publications.

The main idea of the designed experiment consists in creating of a plasma torch reactor by a super-high-velocity impact of a projectile against a suitable target. Both of them must contain elements required for the formation of amino acids [51]. The dissipating and cooling-down plasma was expected to act as a reactor for the self-assembling of ions, radicals, and atoms, all of them being inherently homochiral, into enantiomerically enriched organic molecules. The plasma torch itself, as a flow of chiral electrons, followed by a flow of heavier, chiral, positively charged particles, presents a reactor with very strong chiral combinations of magnetic and electric fields, as well as intensive circularly polarized light beams. Since the chirality sign of all the components and fields within the plasma torch is invariantly determined by the homochirality of all primary particles, amino acids, if formed at all, are expected to be enriched in L-isomers, especially if the chirality of all building elements act in concert with the induction caused by the chirality of the fields.

As a substantial corroboration of the above-outlined expectation, the striking fact was considered that all amino acids, including nonnatural α-methylated amino acids, in meteorites (some of them being much older than Earth) are found to exhibit a predominance of L-isomers with *ee* as high as 15–18% (for the racemization-resistant isovaline).

A large international team of scientists [51] started experiments with high purity ^{13}C carbon black that was converted, via the action of high temperatures and high pressure, into diamond particles of about 1.5–2.5 mm in size. The latter served as projectiles for a super-high-velocity impact experiment and were expected to completely vaporize and label the plasma synthesis products with high concentration of the rare ^{13}C carbon isotope. A tablet of ammonium nitrate served as the O,H,N-providing impact target. Since its mechanical properties are week, a strong plate of pure ^{12}C graphite was positioned behind the target. To easily collect the products of the impact, the target was placed into a steel container lined with a titanium foil. The hypervelocity impact experiments were performed using a ballistic launcher equipped with a light-gas gun [52]. The gun accelerated a set of 20–50 diamond projectiles, about 0.4 g in total weight, to a velocity as high as 7 km/s. The impact products presented dark gray fine powder, composed of both plasma torch synthesis products and debris of the graphite plate. The latter was 5.6 g in weight, so that the $^{12}C/^{13}C$ ratio in any organic matter synthesized would be 14 if complete mixing of diamond and graphite carbon atoms occurred. One portion of organic material (2 mg) was extracted from the impact sample with 2 M hydrochloric acid at 150 °C for 24 h and subjected to careful mass-spectrometric investigation. Three separate analytical instruments with different strengths, limitations, and detection limits were used: a matrix assisted laser desorption ionization (MALDI) TOF instrument, a laser desorption TOF instrument, and a GC–MS instrument. In the mass-spectra obtained, the three simplest amino acids and their typical fragmentation ions could be safely observed: glycine, alanine, and small amounts of serine (Figure 1). A large proportion of the amino acids was found to contain heavy ^{13}C atoms, implying that both carbons from the diamond projectiles and ^{12}C target base plate participated in the plasma torch synthesis process. Mass-spectra also showed formation of much heavier organic products, among them, possibly, peptides composed of the above amino acids.

The most important results were generated by the numerous repeated chiral gas chromatography-mass-spectrometry experiments. To perform them, an aqueous soluble fraction of organic products was subjected to acidic hydrolysis and the amino acids were converted into volatile isopropyl esters of N-trifluoroacetyl derivatives. The chiral stationary phase in the 50 m x 0.25 mm capillary column was Chirosil-Val from Altech. Glycine and alanine, enriched in ^{13}C to an extent of 40%, showed up very clearly in the chromatogram. Mass spectra were recorded in both the single ion monitoring (SIM) and total ion current modes (TIC). The chiral imbalance of alanine was presented as the ratio of its L- and D-enantiomers. The fundamental result of the whole study was summed up by the authors as follows: "This measurement enabled us to calculate L/D values for alanine based on peak integrals and peak amplitudes in both the SIM and total ion modes. Using SIM data, the values for alanine are L/D = 1.15 based on the peak amplitudes and L/D = 1.68 from evaluating the integrals of peaks. For the total ion mode, the values for alanine are L/D = 2.5 for the peak amplitudes and L/D = 2.4 for integrals of peaks. Given the unavoidable background noise in the total ion mode, the L/D values obtained for the SIM mode should be considered more reliable. Thus, the violation of symmetry could be in the range from 1.15 to 1.68" (Figure 2). This range corresponds to enantiomeric excess values between 7 and 25%, which is similar to *ee* values found in meteorites and outperforms by far the values expected for all deracemization effects under action of cosmic polarized irradiations. Remarkably, alanine synthesized in the plasma torch reactor is noticeably enriched in the L-enantiomer, exactly as predicted from the hypothesis on homochirality of all elementary particles in Universe.

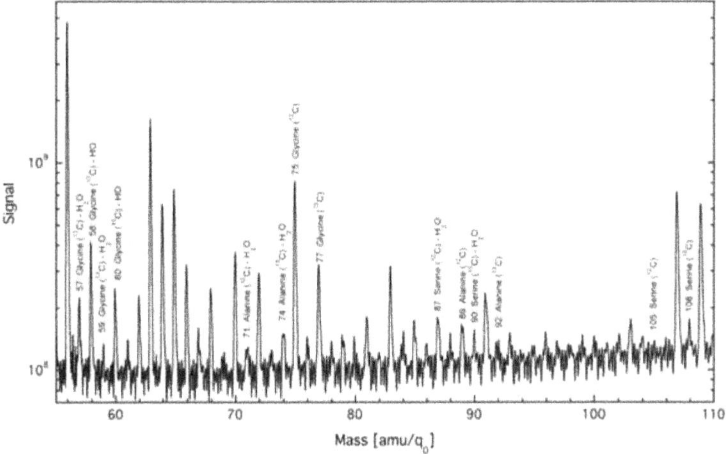

Figure 1. The amino acid-corresponding section of the MALDI mass spectra of impact products. The high content of ^{13}C isotope is evident. Notably, glycine peaks dominate over those of alanine (by 4.75 times) and serine, which is not characteristic of bioproteins. Source: Reference [51].

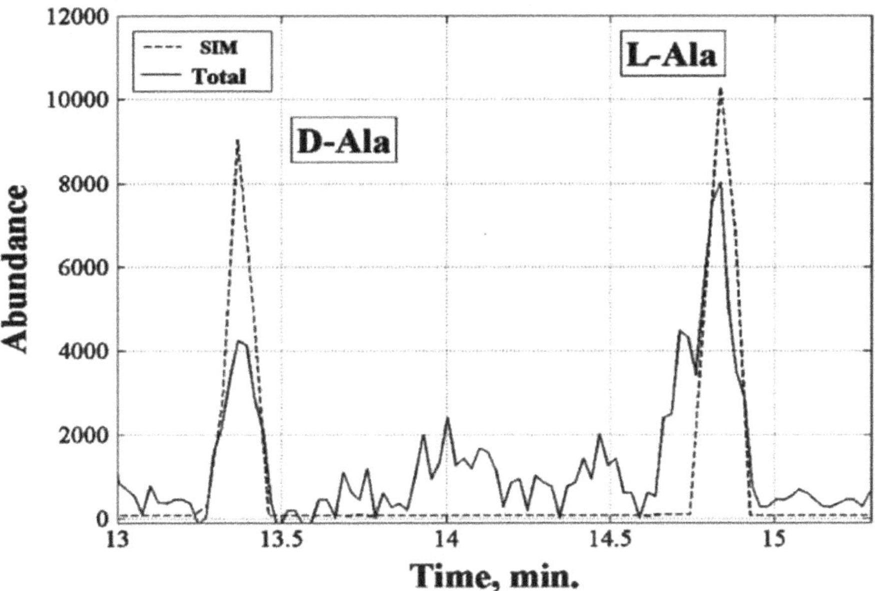

Figure 2. Section of the total chiral GC-MS profile with D-Ala and L-Ala peaks (retention times 13.3 and 14.7 min, respectively) shown in SIM (140 amu) and TIC modes. Source: Reference [51].

4. Conclusions

Based on the above reliable and extremely informative experimental findings, one can speculate that many other α-amino acids, including "non-natural" species, will be synthesized everywhere in the universe under conditions of much stronger hypervelocity collisions of cosmic bodies and that all of them will be characterized by the predominance of the L configuration. In a similar direct synthetic way or during subsequent chemical transformations of organic matter, D-isomers could become the predominating isomers of sugars. Mechanisms of further enantiomeric enrichment of some particular compounds [53] through the Soai-type asymmetric autocatalysis [15] and/or Viedma-type deracemization [4] are being intensively investigated but none of them could function efficiently and unidirectionally in a complex concentrated aqueous broth. Equally unknown remain the rules of selection of amino acids and sugars for creating first primitive self-reproducing organic systems. If one also takes into consideration the essential unique complex of physical conditions on the planet (such as the presence of water in its liquid form, a suitable atmosphere, sufficient magnetic field of the planet, and appropriate rotation rate of the latter around its axis and the central star, etc.) the fact of the emergence of life on the Earth appears a big mystery. Important is only that the general homochiral shift of organic building blocks and, hence, unique stereochemistry of living matter in the whole universe is predetermined by the parity violation of elementary particles, whereas realization of life forms on different planets may remain a matter of a single happy chance.

Funding: There is no extra funding.

Conflicts of Interest: The author declares no conflict of interest.

References

1. Eigenbrode, J.L.; Summons, R.E.; Steele, A.; Freissinet, C.; Millan, M.; Navarro-González, R.; Sutter, B.; McAdam, A.C.; Franz, H.B.; Glavin, D.P.; et al. Organic matter preserved in 3-billion-year-old mudstones at Gale crater, Mars. *Science* **2018**, *360*, 1096–1101. [CrossRef]
2. Goldanskii, V.I.; Kuz'min, V.V. Spontaneous mirror symmetry breaking in nature and the origin of life. *Z. Phys. Chem.* **1988**, *269*, 216–274. [CrossRef]
3. Chandrasekhar, S. Molecular homochirality and the parity-violating energy difference. A critique with new proposals. *Chirality* **2008**, *20*, 84–95. [CrossRef]
4. Viedma, C.; McBride, J.M.; Kahr, B.; Cintas, P. Enantiomer-specific oriented attachment: Formation of macroscopic homochiral crystal aggregates from a racemic system. *Angew. Chem.* **2013**, *52*, 10545–10548. [CrossRef]
5. Poinot, P.; Geffroy-Rodier, C. Searching for organic compounds in the Universe. *Trends Anal. Chem.* **2015**, *65*, 1–12. [CrossRef]
6. Myrgorodska, I.; Meinert, C.; Martins, Z.; Le Sergeant, L.; d'Hendecourt; Meierhenrich, U.J. Molecular chirality in meteorites and interstellar ices, and the chirality experiment on board the ESA Cometary Rosetta Mission. *Angew. Chem. Int. Ed.* **2015**, *54*, 1402–1412. [CrossRef]
7. Meinert, C.; Filippi, J.-J.; de Marcellus, P.; Le Sergeant, L.; d'Hendecourt; Meierhenrich, U.J. Why interstellar ices can be considered as precursors for prebiotic chemistry. *ChemPlusChem* **2012**, *77*, 186–191. [CrossRef]
8. Nuevo, M.; Milam, S.N.; Sandford, S.A. Molecules in organic residues produced from the ultraviolet photo-irradiation of pyrimidine in NH_3 and H_2O+NH_3 ices. *Astrobiology* **2012**, *12*, 295–314. [CrossRef]
9. Furukawa, Y.; Sekine, T.; Kobayashi, T.; Kakegawa, T. Nucleobase and amino acid formation through impacts of meteorites on the early ocean. *Earth Planet. Sci. Lett.* **2015**, *429*, 216–222. [CrossRef]
10. Martins, Z.; Price, M.C.; Goldman, N.; Sephton, M.A.; Burchell, M.J. Shock synthesis of amino acids from impacting cometary and icy planet surface analogues. *Nat. Geosci.* **2013**, *6*, 1045–1049. [CrossRef]
11. Esmaili, S.; Bass, A.D.; Cloutier, P.; Sanche, L.; Huels, M.A. Glycine formation in CO_2:CH_4:NH_3 ices induced by 0–70 eV electrons. *J. Chem. Phys.* **2018**, *148*, 164702. [CrossRef]
12. Pizzarello, S.; Schrader, D.L.; Monroe, A.A.; Lauretta, D.S. Large enantiomeric excesses in primitive meteorites and the diverse effects of water in cosmochemical evolution. *Proc. Natl. Acad. Sci. USA* **2012**, *109*, 11949–11954. [CrossRef]

13. Glavin, D.P.; Dworkin, J.P. Enrichment of the amino acid L-isovaline by aqueous alteration on CI and CM meteorite parent bodies. *Proc. Natl. Acad. Sci. USA* **2009**, *106*, 5487–5492. [CrossRef]
14. Pizzarello, S.; Wang, Y.; Chaban, G.M. A comparative study of the hydroxy acids from the Murchison GRA 95229 and LAP 02342 meteorites. *Geochim. Cosmochim. Acta* **2010**, *74*, 6206–6217. [CrossRef]
15. Soai, K.; Kawasaki, T.; Matsumoto, A. The origins of homochirality examined by using asymmetric autocatalysis. *Chem. Rec.* **2014**, *14*, 70–83. [CrossRef]
16. Gabuzda, D. Astrophysics: The MAD world of black holes. *Nature* **2014**, *510*, 42–43. [CrossRef]
17. Bailey, J.; Chrysostomou, A.; Hough, J.H.; Gledhill, T.M.; McCall, A.; Clark, S.; Menard, F.; Tamura, M. Circular polarization in star-formation regions: Implications for biomolecular homochirality. *Science* **1988**, *281*, 672–674. [CrossRef]
18. Bailey, J. Astronomical sources of circularly polarized light and the origin of homochirality. *Orig. Life Evol. Biosph.* **2001**, *31*, 167–183. [CrossRef]
19. Meinert, C.; Filippi, J.J.; Nahon, L.; Hoffmann, S.V.; d'Hendecourt, L.; de Marcellus, P.; Bredehöft, J.H.; Thiemann, W.H.P.; Meierhendrich, U.J. Photochirogenesis: Photochemical models on the origin of biomelecular homochirality. *Symmetry* **2010**, *2*, 1055–1080. [CrossRef]
20. Meinert, C.; de Marcellus, P.; d'Hendecourt, L.; Nahon, L.; Jones, N.C.; Hoffmann, S.V.; Bredehöft, J.H.; Thiemann, W.H.P.; Meierhendrich, U.J. Photochirogenesis: Photochemical models on the absolute asymmetric formation of amino acids in interstellar space. *Phys. Life Rev.* **2011**, *8*, 307–330. [CrossRef]
21. Bartmess, J.E.; Pagni, R.M. A photochemical mechanism for homochirogenesis. Part 2. *Chirality* **2013**, *25*, 16–21. [CrossRef]
22. De Marcellus, P.; Meinert, C.; Nuevo, M.; Filippi, J.-J.; Danger, G.; Deboffle, D.; Nahon, L.; d'Hendecourt, L.L.S.; Meierhenrich, U.J. Non-racemic amino acid production by ultraviolet irradiation of achiral interstellar ice analogs with circularly polarized light. *Astrophys. J.* **2011**, *727*, L1–L6. [CrossRef]
23. Modica, P.; Meinert, C.; De Marcellus, P.; Nahon, L.; Meierhenrich, U.J.; D'Hendecourt, L.L.S. Enantiomeric excesses induced in amino acids by ultraviolet circularly polarized light irradiation of extraterrestrial ice analogs: A possible source of asymmetry for prebiotic chemistry. *Astrophys. J.* **2014**, *788*, 79–90. [CrossRef]
24. Cline, D.B. Supernova antineutrino interactions cause chiral symmetry breaking and possible homochiral biomaterials for life. *Chirality* **2005**, *17*, S234–S239. [CrossRef]
25. Boyd, R.N.; Kajino, T.; Onaka, T. Supernovae and the chirality of the amino acids. *Astrobiology* **2010**, *10*, 561–568. [CrossRef]
26. Boyd, R.N.; Kajino, T.; Onaka, T. Supernovae, neutrinos, and the chirality of the amino acids. *Int. J. Mol. Sci.* **2011**, *12*, 3432–3444. [CrossRef]
27. Famiano, M.; Boyd, R.; Kajino, T.; Onaka, T.; Koehler, K.; Hulbert, S. Determining Amino Acid Chirality in the Supernova Neutrino Processing Model. *Symmetry* **2014**, *6*, 909–925. [CrossRef]
28. Takahashi, J.-I.; Shinojima, H.; Seyama, M.; Ueno, Y.; Kaneko, T.; Kobayashi, K.; Mita, H.; Adachi, M.; Hosaka, M.; Katoh, M. Chirality emergence in thin solid films of amino acids by polarized light from synchrotron radiation and free electron laser. *Int. J. Mol. Sci.* **2009**, *10*, 3044–3064. [CrossRef]
29. Bonner, W.A.; van Dort, M.A.; Yearian, M.R.; Zeman, H.D.; Li, G.C. Polarized Electrons and the Origin of Optical Activity. *Isr. J. Chem.* **1976**, *15*, 89–95. [CrossRef]
30. Darge, W.; Laczko, I.; Thiemann, W. Stereoselectivity of beta irradiation of D,L-tryptophan in aqueous solution. *Nature* **1976**, *261*, 522–524. [CrossRef]
31. Akaboshi, M.; Noda, M.; Kawai, K.; Maki, H.; Kawamoto, K. Asymmetrical radical formation of D- and L0alanins irradiated with tritium-rays. *Orig. Life* **1982**, *12*, 395–399. [CrossRef]
32. Klabunowskii, E.I.; Pavlov, V.A. Homochiraity origine in nature: Possible versions. *Curr. Org. Chem.* **2014**, *18*, 93–114.
33. Naaman, R. Chirality—Beyond the Structural Effects. *Isr. J. Chem.* **2016**, *56*, 1010–1015. [CrossRef]
34. Fukue, T.; Tamura, M.; Kandori, R.; Kusakabe, N.; Hough, J.H.; Bailey, J.; Whittet, D.C.B.; Lucas, P.W.; Nakajima, Y.; Hashimoto, J. Extended high circular polarization in the Orion massive star forming region: Implications for the origin of homochirality in the Solar system. *Orig. Life Evol. Biosph.* **2010**, *40*, 335–346. [CrossRef]
35. Tachibana, S.; Huss, G.R.; Kita, N.T.; Shimoda, G.; Morishita, Y. ^{60}Fe in Chondrites: debris from a nearby supernova in the early solar system? *Astrophys. J.* **2006**, *639*, L87–L90. [CrossRef]

36. Cerf, C.; Jorissen, A. Is amino-acid homochirality due to asymmetric photolysis in space? *Space Sci. Rev.* **2000**, *92*, 603–612. [CrossRef]
37. Myer, Y.P.; MacDonald, L.H. Circular dichroism of L-tryptophan by an improved dichrograph. *J. Am. Chem. Soc.* **1967**, *89*, 7142–7144. [CrossRef]
38. Lee, T.D.; Yang, C.N. Question of parity conservation in the weak interaction. *Phys. Rev.* **1956**, *104*, 254–258. [CrossRef]
39. Wu, C.S.; Ambler, A.; Hayward, R.W.; Hoppes, D.D.; Hudson, R.P. Experimental test of parity conservation in beta decay. *Phys. Rev.* **1957**, *105*, 1413–1415. [CrossRef]
40. Berger, R. *Theoretical and Computational Chemistry*; Peter, S., Ed.; Elsevier: Amsterdam, The Netherlands, 2004; Volume 14, pp. 188–288.
41. Davankov, V. Chirality as an inherent general property of matter. *Chirality* **2006**, *18*, 459–461. [CrossRef]
42. Emmons, T.P.; Reeves, J.M.; Forson, E.N. Parity-Nonconserving Optical Rotation in Atomic Lead. *Phys. Rev. Lett.* **1983**, *51*, 2089–2092. [CrossRef]
43. Brown, G.E.; Rho, M. On the manifestation of chiral symmetry in nuclei and dense nuclear matter. *Phys. Rep.* **2002**, *363*, 85–171. [CrossRef]
44. Avalos, M.; Babiano, R.; Cintas, P.; Jiménez, J.L.; Palacios, J.C. What does elementary chirality have to do with neutrinos? *ChemPhysChem.* **2002**, *3*, 1001–1003. [CrossRef]
45. Peskin, M. The matter with antimatter. *Nature* **2002**, *419*, 24–27. [CrossRef]
46. Zheng, X. Precision measurement of the weak charge of the proton. *Nature* **2018**, *557*, 171–172. [CrossRef]
47. Davankov, V.A. Inherent homochirality of primary particles and meteorite impacts as possible source of prebiotic molecular chirality. *Russ. J. Phys. Chem. A* **2009**, *83*, 1247–1256. [CrossRef]
48. Managadze, G. A new universal mechanism of organic compounds synthesis during prebiotic evolution. *Planet. Space Sci.* **2006**, *55*, 134–140. [CrossRef]
49. Davankov, V.A. Homochirality of organic matter—Objective law or curious incident? *Israel J. Chem.* **2016**, *56*, 1036–1041. [CrossRef]
50. Managadze, G. Plasma and collision processes of hypervelocity meteorite impact in the prehistory of life. *Int. J. Astrobiol.* **2010**, *9*, 157–174. [CrossRef]
51. Managadze, G.G.; Engel, M.H.; Getty, S.; Wurz, P.; Brinckerhoff, W.B.; Shokolov, A.G.; Sholin, G.V.; Terent'ev, S.A.; Chumikov, A.E.; Skalkin, A.S.; et al. Excessof L-alanine in amino acids synthesized in a plasma torch generated by a hypervelocity meteorite impact reproduced in the laboratory. *Planet. Space Sci.* **2016**, *131*, 70–78. [CrossRef]
52. Skalkin, A.S.; Suntsov, G.N.; Shokolov, A.G.; Yakhlakov, Y.V. Investigation of the crater-formation process in the case of a hypervelocity impact of an aluminium particle onto a massive barrier made of AMg-6 alloy. *Kosmonavtika i Raketostroenie* **2011**, 62.
53. Ribó, M.; Blanco, C.; Crusats, J.; El-Hachemi, Z.; Hochberg, D.; Moyano, A. Absolute asymmetric synthesis in enantioselective autocatalytic reaction networks: Theoretical games, speculations on chemical evolution and perhaps a synthetic option. *Chem. Eur. J.* **2014**, *20*, 17250–17271. [CrossRef]

Review

Chiral and Racemic Fields Concept for Understanding of the Homochirality Origin, Asymmetric Catalysis, Chiral Superstructure Formation from Achiral Molecules, and B-Z DNA Conformational Transition

Valerii A. Pavlov [1,*], Yaroslav V. Shushenachev [2] and Sergey G. Zlotin [1]

1 N. D. Zelinsky Institute of Organic Chemistry, Russian Academy of Sciences, 47 Leninsky prosp., Moscow 119991, Russia; zlotin@ioc.ac.ru

2 N. S. Kurnakov Institute of General and Inorganic Chemistry, Russian Academy of Sciences, 31 Leninsky prosp., Moscow 119991, Russia; slavash@mail.com

* Correspondence: pvlv69@mail.ru; Tel.: +(07)-495-750-2545

Received: 6 March 2019; Accepted: 23 April 2019; Published: 8 May 2019

Abstract: The four most important and well-studied phenomena of mirror symmetry breaking of molecules were analyzed for the first time in terms of available common features and regularities. Mirror symmetry breaking of the primary origin of biological homochirality requires the involvement of an external chiral inductor (environmental chirality). All reviewed mirror symmetry breaking phenomena were considered from that standpoint. A concept of chiral and racemic fields was highly helpful in this analysis. A chiral gravitational field in combination with a static magnetic field (Earth's environmental conditions) may be regarded as a hypothetical long-term chiral inductor. Experimental evidences suggest a possible effect of the environmental chiral inductor as a chiral trigger on the mirror symmetry breaking effect. Also, this effect explains a conformational transition of the right-handed double DNA helix to the left-handed double DNA helix (B-Z DNA transition) as possible DNA damage.

Keywords: environmental chirality; C_1- and C_2-symmetric catalysts; chiral field (memory); racemic field; Viedma ripening effect; Wallach's rule

---

1. Introduction

Curie [1] was convinced that "without asymmetric physical impact no asymmetric chemical effect arises". Modern experimental data support this criterion: asymmetric induction in asymmetric catalysis is only implemented through the asymmetric (C_1 symmetry axis) key intermediate [2].

If a substrate has two or three functional groups which can be coordinated in that intermediate, chiral C_2-symmetric catalyst loses C_2 symmetry in the substrate coordination stage. Nature has chosen such substrates (amino acids and sugars) as components of important macromolecules. However, the only variant of configuration ratios (L–D) of amino acids and sugars has been selected from the four possible: D–D, L–L, D–L, and L–D (Figure 1).

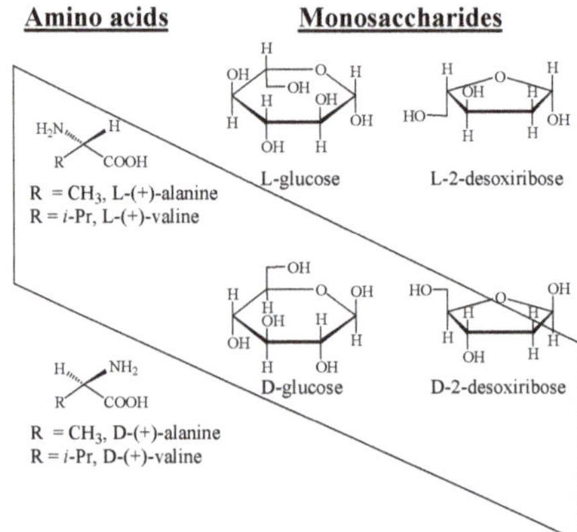

Figure 1. Schematic corridor of amino acid and sugar configurations of life support processes.

A reason for this choice has not received a generally recognized explanation up to now [3–17]. These references indicate a variety of scenarios for the emergence of homochirality and therefore origin of life [16]. Various scenarios are explained by external and internal reasons existing on the primary Earth. Possible scenarios of homochirality origin include Earth and exoterrestrial origins, mirror-symmetrical and non-mirror symmetrical forces, different ampllification mechanisms leading to L- or D-amino acids and sugars, and L-amino acids excess during meteorite impact.

We believe [3,4] that a possible basis for such a ratio of configurations is the right-handed helix (*P*) conformation of important biomacromolecules formed from amino acids and monosaccharides. (Scheme 1).

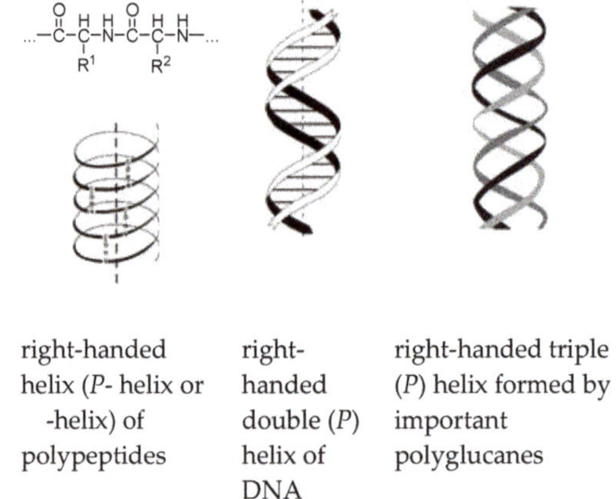

Scheme 1. The triad of right-handed important biomacromolicules.

The L–D ratio of amino acid and sugar monomer configurations may merely occur where there is a right-handed helix conformation. Mutual relations of single, double and triple right-handed (*P*) helixes are also not accidental. The single helix (α-polypeptide) protects amino acids from racemization by hydrogen bonds between helical turns.

Contradictory situation arises during the single helix formation from monocaccharide hexose. Indeed, α-(1→2)-D-mannan and β-(1→3)-D-glucan generate *M*- and *P*-helices, corresponddingly, in the same conditions [18]. A helical configuration of amylose relies on the crystallization conditions

(Table 1). A possible stabilizing factor of the P-helix formation is the double helix (Table 1, items 1 and 2).

Table 1. Structures of linear polysaccharide amylose as function of the solvent used for crystallization.

No	Amylose	Solvent	Helix	Reference
1	Amylose A [1]	water	*P*	[19]
2	Amylose B [1]	water	*P*	[20]
3	Amylose V	–	*M*	[21]
4	KOH-Amylose	water, KOH	*M*	[22]
5	Amylose $V_{propan\text{-}2\text{-}ol}$	propan-2-ol, water	*M*	[23]
6	Amylose	water	*M*	[24]

[1] Double helix.

Indeed, linear polysaccharides, viz. i-carrageenan [25] and xanthan [26], form double helices with the *P*-structure as well. Schisofillan, a nonlinear polysaccharide, also gives a double *P*-helix [27–29].

Thus, it is not by chance that nature has chosen the double helix structure to stabilize the right-handed conformation of DNA. In this case, the configuration error reducing factor is also important [27]. There is directionality to a right-handed helix in triple helical structures such as in fungi, mushrooms, and so on. For example, the simplest β-(1→3)-D-glucans, namely curdlan [27], lentinan [30], scleroglucan [31], xylan [32], etc., with side chains [33–38] form triple *P*-helices with the right-handed helical conformation. Hence, there is a triad pronounced in Scheme 1. That is why right-handed helix structures of important biomacromolecules are representatives of the regular trend in nature.

Since all natural processes occur in the open system, in previous work [39], we attemted to assess the external chiral effect on the homochirality origin on the early Earth. A possible influence of an external asymmetric inductor was analyzed for metastable (stochastic and spontaneous) reactions [39]. In this review, we continue the search for traces of the external chiral inductor, such as the chiral gravitational field, in the most studied reactions with mirror symmetry breaking.

2. Gravity as a Chirality Inductor

Davankov [40] suggested chirality to be an indispensable feature of different levels of matter. While evolving this idea, we [39] considered a possibility of chiral effects of various physical phenomena on chemical reactions (electric field [41], electric field (propeller effect) [42], a combination of electric and magnetic fields [4,43], magnetic field [4,39,44], circularly polarized light [45], plasma torch of meteorite impact [46,47], solar irradiation [48,49], parity violation energy difference [29], and similar effects [39]). Gravity is among such physical phenomena as asymmetric factors [39]. Basing on Barron's concept [50] of true and false chirality, we can assume that the mutual gravitational influence of a space object and its satellites is a chiral factor. In the schematic diagram (Figure 2), we tried to summarize information on mutual gravitational impact (moving in space) of Sun, Earth, Moon, and Venus. Their mutual movements create a combination of trajectories in the form of virtual chiral helices. The image below gives a view of the chiral gravitational environment that is probably strictly individual for Earth (The Earth–Moon (ratio of masses 81/1) is a double planet system unlike other planets and satellites of the Solar system, for example, Jupiter–Europe ($4{\cdot}10^4/1$), Mars–Phobos ($6{\cdot}10^7/1$), and so on). A hypothesis about the formal similarity and possible effect of the Earth's right-handed (spin) rotation near the Moon, alongside with the right-handed Earth's orbital motion around moving Sun and right-handed helix symmetry of biomacromolecules, was published earlier [3,29,39,51].

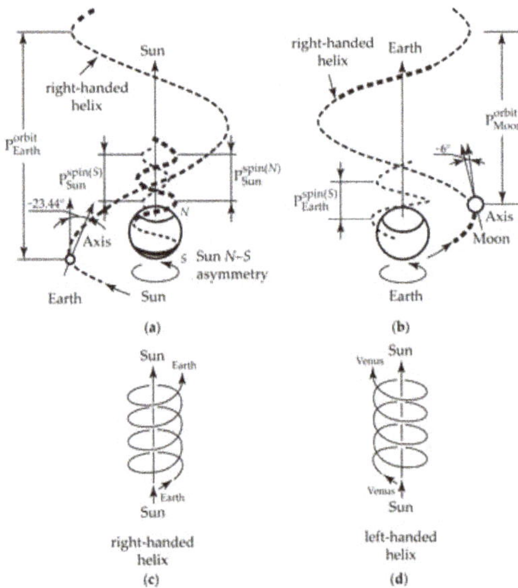

Figure 2. Schematic image of right-handed helices created by the orbital and spin rotation of moving Sun – Earth (**a**, **c**), Earth – Moon (**b**), and the left-handed helix as a result of the orbital rotation of Venus around Sun (**d**). Pitches (P) of the helices: P_E^{spin} =1 day, P_E^{orbit} =1 year, P_M^{spin} =P_M^{orbit} = 1 lunar (sunderic) month, $P_s^{spin(N)}$ = $P_s^{spin(S)}$ =~ 38 Earth's days (near polar caps). The rotation of Sun around the axis tilted 82°45′ to the plane of the Earth's orbit occurs in the same direction as the spin rotation of Earth (counterclockwise).

Perhaps, all spiral movments shown in Figure 2 should be considered as a manifestation of the unified structure of the gravitational field of our Galaxy. The Milky Way Galaxy is a double snail of flat structure (Figure 3a,b) [52–55]. A symmetry plane virtually divides the Galaxy into a mirror symmetric left-handed snail – "bottom" (Figure 4c), and a right-handed snail – "top" (Figure 3d). Hence, the Milky Way may be presented as a mesostructure (an inner racemate).

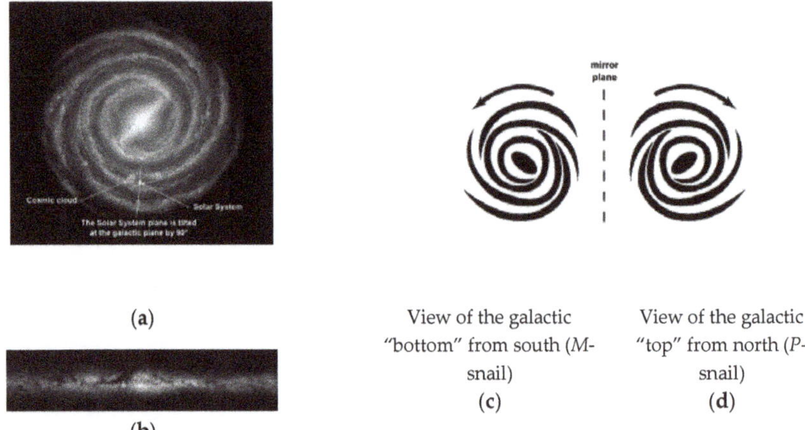

Figure 3. The Milky Way Galaxy (**a**) top (view from the galactic north) and (**b**) sideways. Schematic structure of the Galaxy: (**c**) "bottom" snail and (**d**) "top" snail.

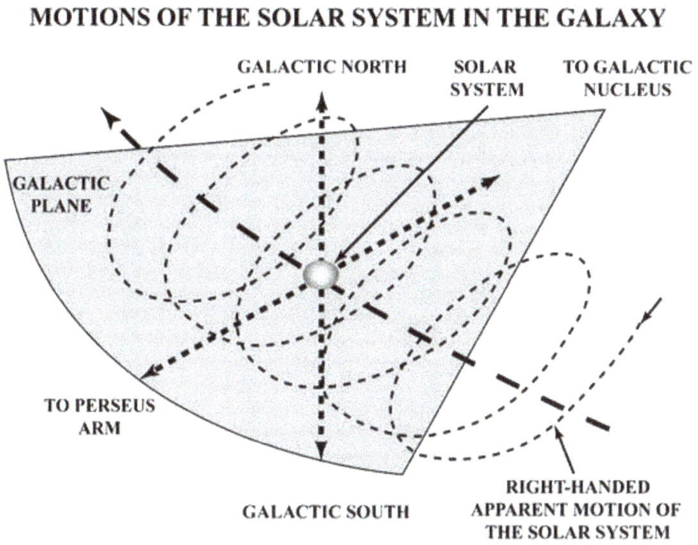

MOTIONS OF THE SOLAR SYSTEM IN THE GALAXY

GALACTIC NORTH
SOLAR SYSTEM
TO GALACTIC NUCLEUS
GALACTIC PLANE
TO PERSEUS ARM
GALACTIC SOUTH
RIGHT-HANDED APPARENT MOTION OF THE SOLAR SYSTEM

Figure 4. Motions of the the solar system in the Milky Way [51].

In terms of symmetry, the Galaxy bears a similarity to inner racemate (for example, meso–tartaric acid. It is conceivable that this mesostructure is confirmed by the anomalous rotation (counterclockwise) of Venus in the Solar system. This movement does not contradict with the complexity of the Galaxy structure. Indeed, the mirror symmetry plane divides "clockwise" and "counterclockwise" fields (Mirror reflection effect or a positive–negative photoeffect.) of the galactic space (*P*- and *M*-snails). Because the Solar system spins and moves along the right-handed trajectory [52–55] (Figure 4) in the *P*-snail field, Earth and other planets move identically (clockwise) with the exception of Venus.

An example of the chiral environment's (chiral gravitational field?) influence is found in configurational stability (mirror symmetry breaking) of sea snail shells [56,57] (Figure 5). All collectors of sea shells (conchiologists) evidence that the opposite left-handed structure is an extreme rarity [58].

Figure 5. The right-handed helix shape of arbitrary sampled shells of sea mollusks (about 70,000 species in total [54]).

How does gravity affect symmetry of molecules? In contrast to the influence of the electromagnetic field [59], the effect of gravity on chemical reactions is difficult to verify experimentally. It was previously believed that this effect was absent or very insignificant. Another point of view appeared possible with the beginning of the space flight era.

3. Gravitational Field Impact on Chemical Reactions

Analyzing an open system, we have to evaluate all possible physical inductors which may affect the system. Among the non-obvious physical factors that may affect chemical reactions is gravity. After the discovery of gravitational waves (Gravitational waves discovery (Abbott B.P. et. al. Observation of Gravitational Waves from a Binary Black Hole Merger. *Phys. Rev. Lett.*, **2016**. *116*. 061102) suggests the complex structure of the gravitational field.) the influence of this factor, especially in unstable asymmetric reactions, became more obvious. A nuclear decay has a reputation of a stable process that does not depend on external physical inductors. If an external inductor such as a gravitational field can alter the radioactive decay rate it is bound to affect the nucleus mass. Therefore, if such effect occurs, this phenomenon may be called a "mass resonator" or "mass resonance". Fischbach et al. [60–62] analyzed a ^{137}Cs decay sample onboard the Messenger spacecraft during its mission to Mercury and ^{54}Mn decay data during the solar flare on December 13 2006. The goal was to show the limits of a possible correlation between nuclear decay rates and solar activity. Such correlation was suggested not only on the basis of the ^{54}Mn decay during the solar flare but also by indications of annual and other periodic variations in the decay rates of ^{32}Si, ^{36}Cl and ^{226}Ra. Data from five measurements of the ^{137}Cs count rate over a period of approximately 5.4 years was fit to a formula which accounts for a typical exponential decrease in the count rate over time, alongside with the addition of a theoretical solar contribution varying with the Messinger—Sun distance [61]. These controversial data on nonexponential periodic decay rates drew attention and gave rise to discussion [63–65].

Ivanova et al. [48] observed North–South solar asymmetry and anisotropy of cosmic rays over solar polar caps. The measurements were taken onboard the Kosmos-480 satellite on 18 April 1972 over a 10-h period. In particular, the flow above the south pole was an order of magnitude higher than that above the north pole. Svirzhevsky et al. [49] examined N–S solar poles asymmetry relative to the solar wind speed, plasma density, and some other solar parameters. Asymmetry between the north and south solar fields was observed in the plasma density, solar radio flux, and geomagnetic indices [66]. Consequently, N–S solar asymmetry (Figure 3a) is an additional chirality trigger alongside with chirality of the right-handed helix orbital trajectory of Earth revolving around Sun (Figure 3c). Indeed, N–S solar asymmetric poles during the Sun's movement form a virtual double right-handed helix (Figure 3a). Gravitation field asymmetry throughout the entire life time of the Solar system can culminate in the chiral biomacromolecules origin as well as in biological objects (Figure 6). Nevertheless, the question of the noticeable effect of asymmetric gravity on symmetry of molecules remains unanswered. However, asymmetric gravity could affect metastable chiral reactions if assume the gravitational field affects the radioactive decay.

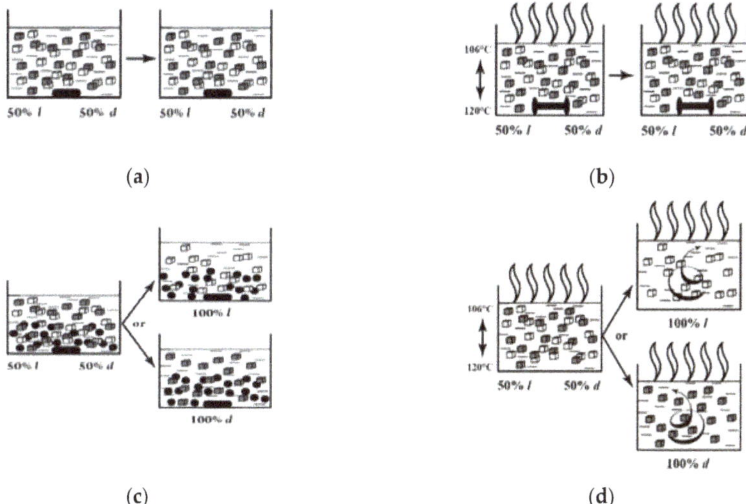

Figure 6. (**a**) Stirring of *l*- and *d*-crustals (50/50%) of $NaClO_3$ without glass beads and (**b**) boiling of the mixture with stirring. (**c**) Stirring of the mixture with glass beads and (**d**) deracemization of the mixture under possible convection fluxes by the temperature gradient during boiling.

4. Stirring (Helical Flux) of the Reaction Mass as a Chirality Trigger

The formation of a conglomerate (50/50% *l*- and *d*-crystals) during crystallization of $NaClO_3$ from a saturated aqueous solution proceeds over a short time interval. However, stirring of the $NaClO_3$ solution (for appreciable time) leads to mirror symmetry breaking (Scheme 2). The enantiomeric direction of crystallization during stirring was ascribed to the primary crystal as a crystallization germ ('secondary' crystal nuclei) [67,68].

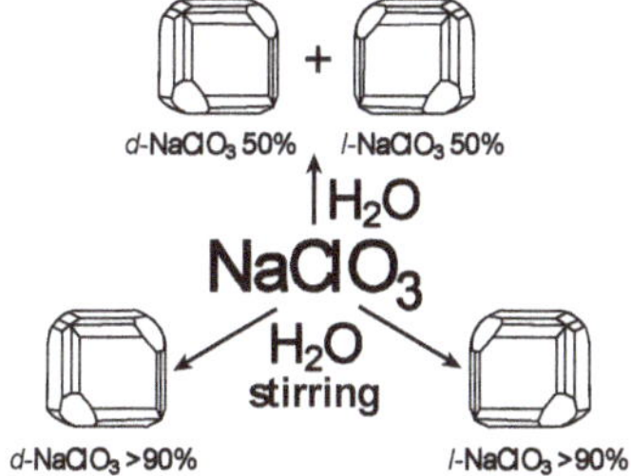

Scheme 2. Influence of solution helical flux.

Similar enantiomeric and racemic conglomerate crystallization was observed in stirred and unstirred 1,1′-binaphthyl melts [69,70]. The enantiomeric excess (*ee* up to 80%) in each stirred crystallization test varied randomly as well.

Stirring of *l*- and *d*-crystals mixture of $NaClO_3$ (50/50%) with glass beads led to 100% *l*- or *d*-crystals, randomly, whereas stirring without glass beads left the mixture unchanged [71] (Figure 7a,c). It is interesting to note that the "stirring time to achieve 100% *ee* depends on the number of glass beads and the stirring rate" [72]. In attempt to explain $NaClO_3$ crystallization data, Kondepudi et al. [66–70,73,74] and other researchers [75–78] attached importance to the experimental fact of stirring. To explain enantioselectivity of $NaClO_3$ *l*- and *d*-crystals stirring with and without glass beads (Figure 6a,c) many researchers [71,79–87] argued that attrition/grinding was the key factor.

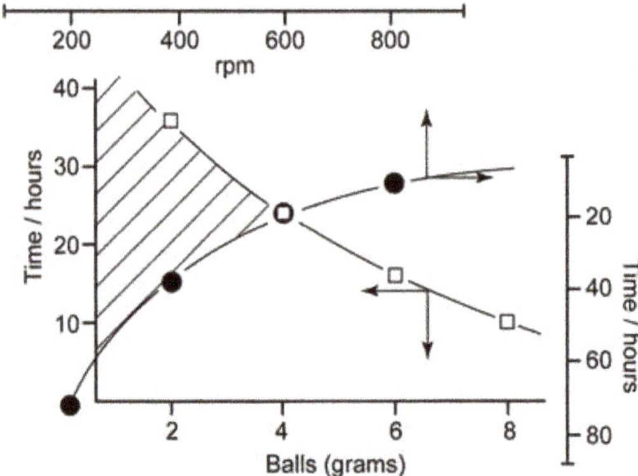

Figure 7. Stirring of the mixture of *l-/d*–$NaClO_3$ crystals and glass balls. Time necessary for achievement of chiral purity is plotted for the constant speed of its agitation (600 rpm)—□ and for the agitation speed, with the constant number of balls (4g of balls)—●.

There are data that argue against the role of stirring. In boiling solutions of the racemic mixture of $NaClO_3$ crystals, stirring (without abrasion) did not bring mirror symmetry breaking [88] (Figure 6b). However, the temperature gradient (120 °C lower layer–106 C upper layer of colution in the reaction vessel) [88] led to the formation of spiral flows inb the oiling solution (Figure 6d) (shown by the snail-shaped arrows). These flows could appear destroyed while stirring, which is why the enantioenrichment did not occur (Figure 6b).

Even more obvious evidence of the effect of temperature flows of the reaction mixture on enantioselectivity is sublimation. For example, valine, racemic amino acid, was converted *via* sublimation into a conglomerate [89]. During continued heating, the crystals underwent substantial chirality amplification (increase of initial *ee*). This phenomenon occurred both in the closed and in the open system [89]. The authors observed the appearance of three sublimation regions depending on their location in the form of rings on the conical flask walls. The most enantioselective region corresponded to the enantioenrichment of the valine sublimate with *ee* 80% (the closed system) and *ee* 70% (the open system). In our opinion, the intervention of the external chiral inductor in this experiment is reasonably evident.

We believe that the rotation of the flow energy (stirring) or temperature convectional flow of the helical structure can be a reason for the emergence of enantioselectivity. This scenario can also be triggered by an external chiral field.

The possible existence of this field is evidenced by statistical experiments while studying the Viedma ripening procedure (VRP). Viedma [90] discarded any explanation of VRP by the parity violation energy difference (PVED) effect (Table 2).

Table 2. The number of VRP experiments for $NaClO_3$ or $NaBrO_3$ with the initial racemic population of *l*- and *d*-crysals and their final chiral purity [90].

Starting Racemate in VPR	Number of VPR Experiments	Final Chirality Purity		
		l-Crystals	*d*-Crystals	% of *l*-Crystals
$NaClO_3$	200	160	40	79.5
	240	236	4	98.3
	200	102	98	51.0
	100	83	17	83.0
	100	49	51	49.0
	100	73	27	73.0
$NaBrO_3$	260	258	2	99.2
	280	274	6	97.8
	200	89	111	44.5
	100	80	20	80.0
	100	52	48	52.0
	100	68	32	68.0

"The experiments were performed with racemic mixtures obtained spontaneously from the same solutions and in the same competing conditions between *l*- and *d*-crystals" ("natural" or true population) [90]. Those data support Viedma's opinion that the handedness of chiral crystals remaining in the solution (Table 2) is not random. As seen from Table 2, there is a predominance of *l*-crystals (in some cases up to 99.5%), alongside with a few exceptions (*l*-crystals 49% and 44.5%). Thus, there is quite a definite trend expressed in the predominant formation of *l*-crystals. According to Viedma [90] these experimental results may be explained by the "cryptochiral environment in control".

To our knowledge, a thorough statistical analysis of changes in left-handed or right-handed chirality of reaction products, depending on the direction of reaction mixture stirring (clockwise or counterclockwise), has not been undertaken (there is an opinion that the stirring direction is not important in VRP expirements [76]). In addition, the nature and origin of this cryptochiral environment's effect have not been discussed [81–89,91]. About the same statistical likelihood of random (stochastic) signs of chirality is observed even with the same rotation direction (also, stochastic distribution of the enantiomer outcome was observed in crystallization under cooling [92]).

$NaClO_3$ crystallization and similar dissolution–crystallizations are metastable processes [93,94]. Therefore, even a small energy of the spiral flow during mixing can affect chirality as an inductor or trigger. It explains well the influence of product enantiopurity on the number of glass beads and the stirring rate [72]. It resembles a search of resonance such as the action of the ultrasonic radiation field (20 KHz) on enantiopurity of the product during threonine crystallization (5(D)→70–87(D)%*ee*) [95].

Indeed, Figure 7 shows that the maximal enantioenrichment of $NaClO_3$ crystallization occurs in the time corridor of 20–24 h of stirring at a speed of 600 rpm (rotation per minute) and with 4 g of glass balls (data from Reference [72]). This result is likely to be a reflection of a search of resonance with some external chiral inductor. The shaded area in this figure is the possible resonance region.

Moreover, chiral enrichment vectors are the same in time but are manifested as stochastically (random) reactions, while enantioselectivity vectors are not the same for different enantiomers (Figure 8).

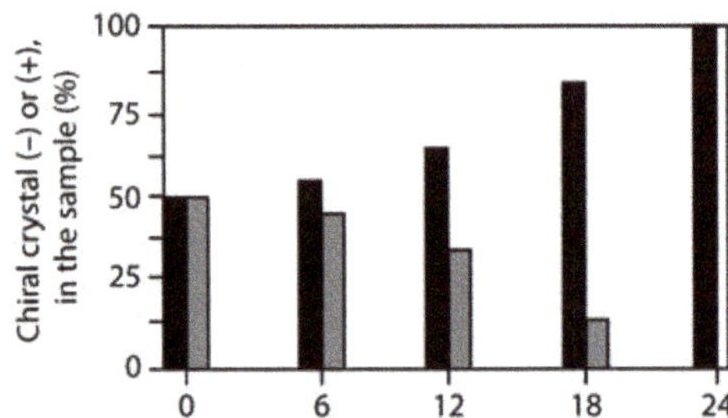

Figure 8. Stirring of 50/50% *l*- and *d*-crystals mixture of NaClO3 and 4g of balls lead to 100% chiral purity after 24 h (600 rpm) randomly.

The results given in Table 2 also testify to the vector of predominant "natural" enantioselectivity.

A striking case of a literally "mechanistic" embodiment of stirring in the chiral effect can be demonstrated by the example of antracene polymerization into a chiral nanofiber structure (Figure 9) [96].

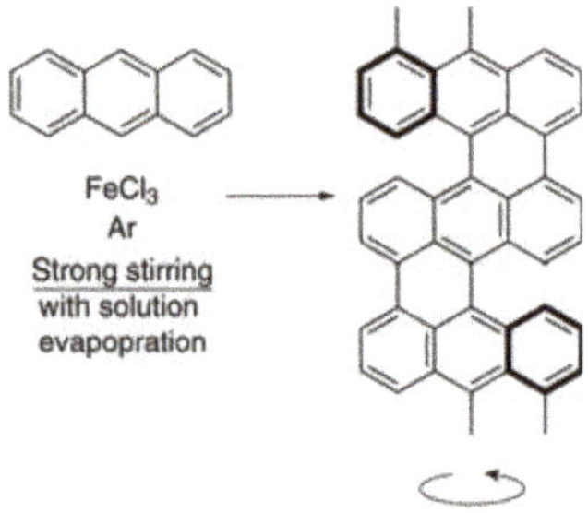

Figure 9. Schematic illustration of the formation of supermolecular polyantracene with left-handed helical nanoarchitectures.

An average diameter of the rotational helix flux of the reaction solution as a result of stirring and the supermolecular polyantracene helix cannot be directrly interdependent because of the incompatibility of helix sizes (the ratio of the diameter of the helices is ~$5{\cdot}10^9$–1 nm). Therefore, chirality of the reaction product could appear apparently only due to an external chiral inductor as a resonator or the rotation flux of the reaction mixture as a trigger.

Low temperature can also be a stabilization factor of chirality. Mirror symmetry breaking occurred upon low temperature spontaneous crystallization of achiral macrocyclic imines [97]. An absolute asymmetric synthesis was also carried out with a chiral reagent from chiral crystals with axial chirality [98]. Crystalline **1** was melted at 120 °C and then gradually solidified by lowering the temperature to 110 °C with vigorous stirring. Reactions (1) and (2) proceeded as 'frozen chirality' [98] at a low temperature ("frozen chirality") [92]. Low temperature monitoring data for the reaction (2) are given in Table 3.

Table 3. Cyclopropanation of **1** using sulfur ylide.

Entry	Temp./°C	Conv. of 1/%	Yield of 2/%	*ee* of 2/%
1	20	69	37	0
2	0	38	53	50
3	−20	27	65	67
4	−40	16	93	97

(R)-1 → [hv (365 nm), 2-methoxypropene, benzophenone; MeOH, −20°C, frozen chirality in cold solvent] → product (CONEt$_2$), 99% yield, 99% ee

$P2_12_12_1$ crystal system

(1)

$Me_3S^{\oplus}(O)\ I^{\ominus}$ → [NaH, DFM, r.t.] → $H_2C^{\ominus}–S^{\oplus}(O)Me_2$ → [powdered (R)-1; DMF, low temp. 48 h] → (R,R)-2

(2)

There are other examples of axially chiral reagents with "frozen chirality" at a low temperature [99].

Crystallization of racemic axially chiral pyrimidinethione led to the chiral product with up to 91% *ee* at a high temperature only [100].

Strong temperature stabilization of axial chirality suggests the existence of a chirality resonator. A similar assumption can be made in analyzing a dependence of the helicity sign from temperature. The helicity dependence from temperature was vividly exemplified by experimenting with poly {(S)-3,7-dimethyloctyl-3-methylbutulsilylene} **3** [101] (Scheme 3). Also, a similar $P \leftrightarrow M$ transition was observed for other polysilylenes and analogous polymers at different temperatures, with various solvents and varying other reaction conditions [102–107].

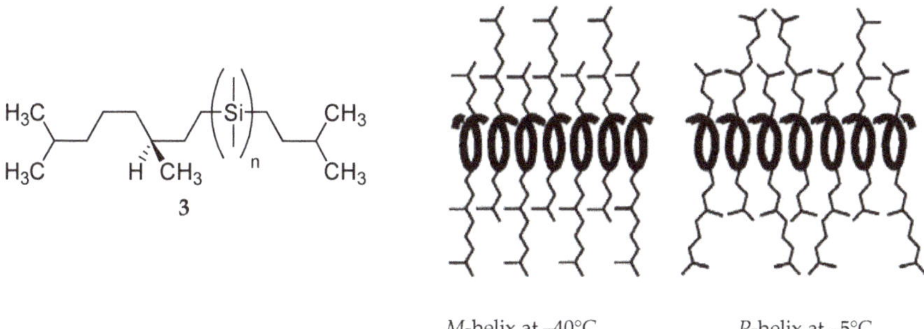

Scheme 3. Schemtic helicity dependence of **3** from temperature.

Reactions (3) and (4) resulted in helical products with different signs. When analyzing the differences in the energy of these product conformations, it was shown that small amount of energy (~1/600 of the ambient energy) is only distinguished *P*- and *M*-helices in these deuterated polyisocyanates [108–111]. Therefore, it is quite possible that even a weak external inductor may be capable of affecting the chiral conformation of helical nanostructures. This assumption should be verified with special care, for example, in the case of the $P \leftrightarrow M$ conversion of DNA (see Section 8).

NCO ... NaCN, DMF, −58°C ... (3)

NCO ... NaCN, DMF, −58°C ... (4)

A possible existence of the chiral environmental inductor is also evidenced by asymmetric reactions of achiral molecules adsorbed on the surface of metals [112–115]. Ernst rightly noted that product's chirality depends not only on the surface structure but also on the molecules orientation (Figure 10) [116]. Chirality in this case can occur, apart from under the action of the chiral surface structure, under the impact of the adsorbing molecule orientation. The environmental chiral inductor in turn can contribute to this chiral orientation. Problematical chirality of metal surfaces suggests that the chiral orientation is the main factor that provides enantioselectivity of the reaction of adsorbed achiral molecules.

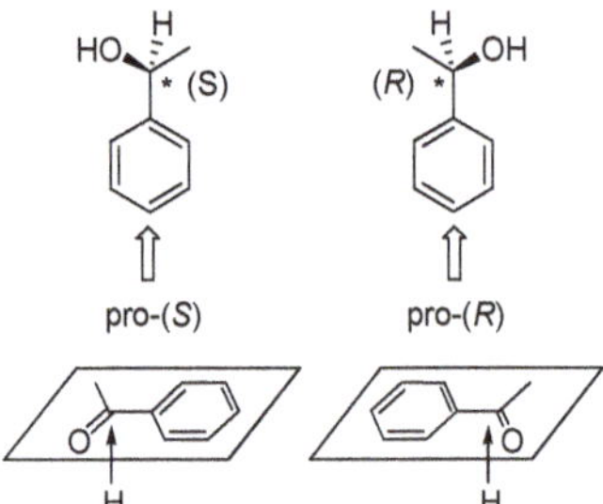

Figure 10. Mirror symmetry in planar molecules on a surface.

5. Chiral and Racemic Field in Asymmetric Catalysis and Nonlinear Effects

Racemic compounds can also exist in solution, e.g., in ongoing reactions. The formation of racemic intermediates of a reaction is observed in catalytic asymmetric reactions. The use of non-enantiopure chiral auxiliaries as asymmetric catalysts sometimes causes a deviation from the proportional linearity to *ee* of a catalytic reaction product (nonlinear effect NLE) in asymmetric catalysis [117] (Figure 11). Scheme 4 shows two models for explaining nonlinear effects: Kagan's and Noyori's models [118]. According to both models racemates and homochiral dimers function as real actors of the reaction mechanism. However, both schemes do not take into account the specifics of the reaction and substrate to predict the NLE occurance.

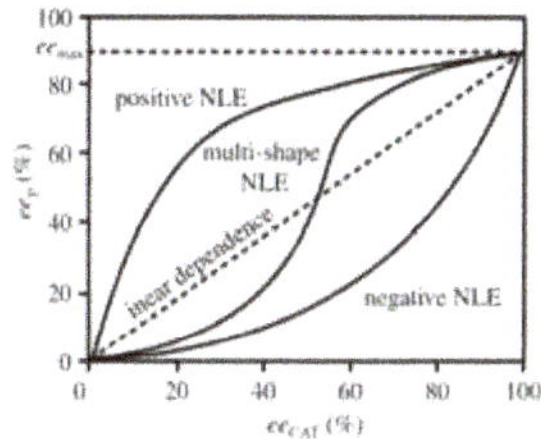

Figure 11. Examples of nonlinear effects in asymmetric catalysis $ee_{\text{max}} = ee$ provided by an enantiopure catalyst.

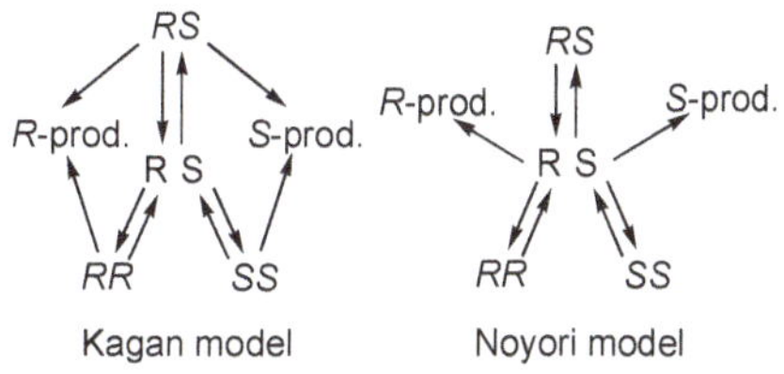

RS
R-prod. S-prod.
R S
RR SS
Kagan model

RS
R-prod. S-prod.
R S
RR SS
Noyori model

Scheme 4. Kagan and Noyori NLE models.

Steigelmann et al. [119] observed remarkable (–)-NLE in diethylzinc addition to benzaldehyde in the presence of (S)– and (R)–fenchols (5). Methylzinc dimeric C2-symmetrical complexes (*R,R*-**4**) and (*S,S*-**4**) were formed in this reaction (Scheme 5).

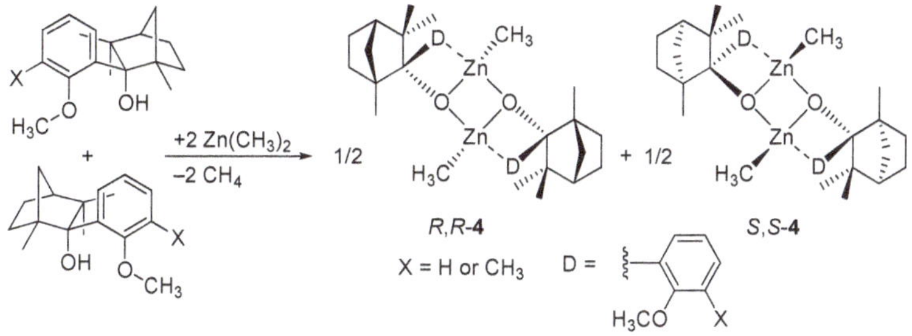

+2 $Zn(CH_3)_2$
−2 CH_4
1/2 *R,R*-**4** + 1/2 *S,S*-**4**
X = H or CH_3
D = 2-H_3CO-3-X-phenyl

Scheme 5. Racemic complex formation with (-)-NLE at X = H or CH_3.

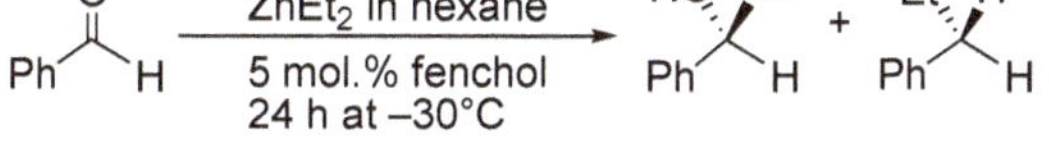

$$\text{PhCHO} \xrightarrow[\text{5 mol.\% fenchol, 24 h at } -30\,^{\circ}\text{C}]{\text{ZnEt}_2 \text{ in hexane}} \text{PhCH(OH)Et} + \text{PhCH(Et)H} \tag{5}$$

A racemic (heterochiral) complex was formffed where X = SMe_3 or *t*-Bu (Scheme 6).

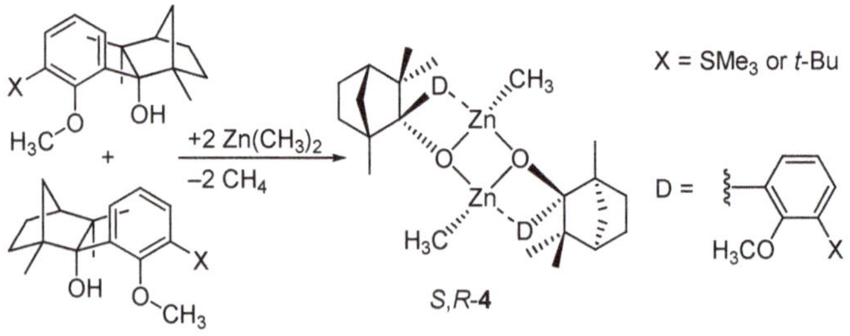

Scheme 6. Racemic complex formation with linear relationship at X = SMe_3 or *t*-Bu.

The use of fenchols with X = H and X = CH_3 in the reaction (5) resulted in (–)-NLE while fenchols with X = SMe_3 and X-*t*-Bu catalyzed this reaction with a linear relationship. Thus, the emergence or absence of (–)-NLE in the reaction (5) depends only on the structure of ligands. Therefore, in this case, the (–)-NLE emergence is formally defined by Kagan-Noyori models.

Chen et al. [120] discovered a substrate dependence of the nonlinear effect in diethylzinc addition to aromatic aldehydes over chiral auxiliary **5**.

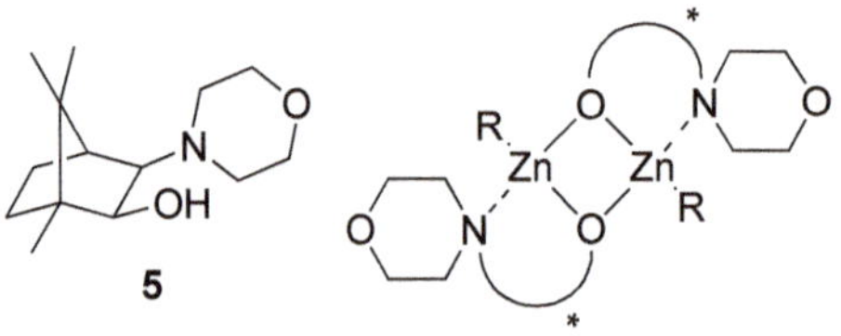

Reactions with electron-donating substituents on the aromatic ring of aldehyde exhibited a greater (+)–NLE than those with electron-withdrawing substituents. These data contradict with the Noyori-Kagan models. The observed substrate dependence can be explained by a reaction of aldehydes with diastereomeric homo- and hetero-chiral dimers of **6** *via* different pathways.

Dimethylzinc addition to aromatic aldehydes (6) using BINOLate–titanium complexes **7** as catalysts produced weak (–)-NLE [121].

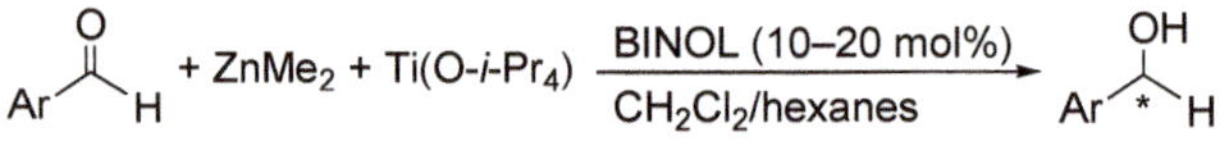

$$ArCHO + ZnMe_2 + Ti(O\text{-}i\text{-}Pr_4) \xrightarrow[CH_2Cl_2/hexanes]{BINOL\ (10–20\ mol\%)} ArC^{*}H(OH)H \quad (6)$$

Several BINOLate–titanium complexes (e.g., **7**—reaction (7)) were synthesized at a low temperature and characterized by crystallography X-ray crystallography. It was shown that (–)-NLE was only observed in stoichiometric experiments (Figure 12). Catalytic experiments replicated the proportional linearity of the *ee* product and the *ee* catalyst [122].

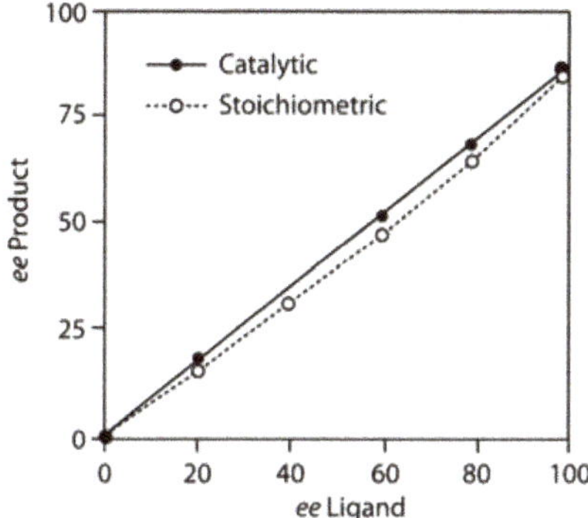

Figure 12. And stoichiometric experimental data (Ti/ non-enantiopure BINOL ligand).

$Ti(O\text{-}i\text{-}Pr)_4$ + BINOL-OR; CH_2Cl_2/hexanes, – HO-*i*-Pr; R = Me, ^{t}-Bu (*R*,*R*-**7**) (7)

Catalytic reactions were run with 20 mol% of (BINOLate)Ti(O-i-Pr)$_2$ and 80 mol% of titanium tetraisopropoxide. The stoichiometric reaction were performed using 100 mol% of (BINOLate)Ti(O-i-Pr)$_2$.

Such a difference between the stoichiometric and catalytic experiments is true if homochiral dimers catalyze the reaction path to enantiomerically pure products (no NLE). The stoichiometric reaction results in non-enantiomerically pure products. Given the observed difference the NLE formation mechanism remains unclear. The heterochiral dimer was shown to be significantly more stable than homochiral dimers [123]. It is safe to assume that the concentration of homochiral dimers is higher in the stochiometric experiments than in the catalytic ones. On the basis of this ratio, we may hypothesize a more pronounced role of homochiral dimers (C$_2$-symmetrical) in the occurance of (–)-NLE during the stochiometric experiments.

The assumption about the leading role of C_2–symmetric homochiral dimers in the appearance of (–)-NLE gained experimental justification in other reactions. For example, pronounced (–)-NLE was observed in Friedel–Crafts alkylation of pyrrole with chalcones (8) catalyzed by C$_2$-symmetric chiral dinuclear zinc catalyst **8** [124] and in the hetero-Diels–Alder reaction of Danishefsky's diene and aromatic aldehydes (9) over BINOLate–Zn complex **9** [125].

(*S*,*S*)-**8**/$ZnEt_2$ (1/2), 15 mol%; THF, 20 °C, 24 h (8)

1) (*R*)-**9**/$ZnEt_2$ (10 mol%), toluene, −25°C
2) CF_3CO_2H

(9)

(*S*,*S*)-**8** (*S*,*S*)-**9**

Thus, the deviation from the linearity in favor of the racemic product ((−)-NLE) occurs in case aromatic substrates in reactions catalyzed by homochiral C_2–symmetric dimeres in situ formed from dialkylzinc. This observation resembles catalytic reactions of aromatic substrates under the action of C_2–symmetric chiral catalysts that occur with anomalous low *ee*, i.e., an almost racemic product. Addition of diethylzinc to trifluoromethyl ketones (aromatic) over C_2-symmetric chiral auxiliaries (10) pertains to such reactions (Table 4) [126].

Table 4. Results of chiral ligand **10–15** screening in the reaction (10).

Ligand	*ee* %
10	37
11	8
12	1
13	7
14	0
15	0

$ZnEt_2$ (1/2 equiv.)
ligand (10 mol%)

(10)

It can be seen that all complexes with C_2-symmetric ligands in the reaction (10) lead to racemate-like products.

Addition of diethylzinc to aromatic aldehydes (11) in the presence of chiral auxiliaries manifests other interesting features: differences in enantioselectivity of C_2– and C_1-symmetric chiral auxiliaries of similar structures are sometimes greater than an order of magnitude (Table 5).

Table 5. Results of chiral ligand **16–23** screening in the reaction (11).

L	*ee* %	Config Product	Ref.
16	97	R	[127]
17	100	R	[128]
18	99	R	[129]
19	8	S	[130]
21	26	*S*	[130]
20	8	*R*	[131]
22	20	*R*	[130]
23	4	*S*	[132]

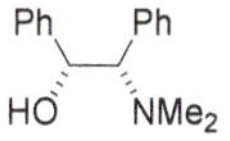

16 (C_1) 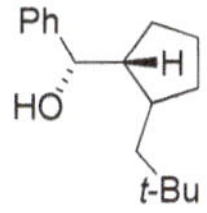 **17** (C_1) 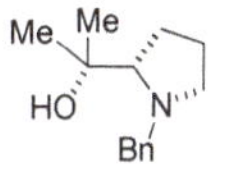 **18** (C_1)

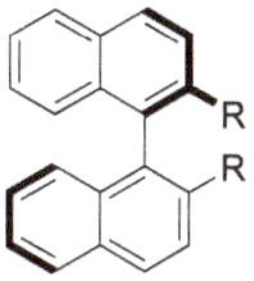

19 R = OH (C_2) **20** R = NMePh (C_2)

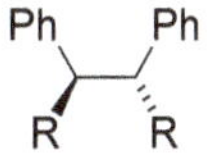

21 R = NH_2 (C_2) **22** R = OH (C_2)

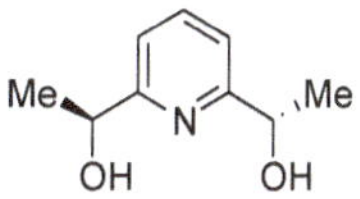

23 (C_2)

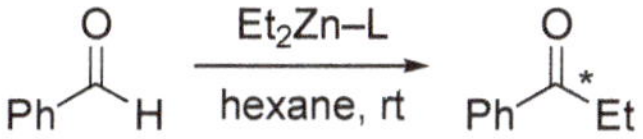

$$PhCHO \xrightarrow[\text{hexane, rt}]{Et_2Zn\text{–}L} PhC(O)^{*}Et \quad (11)$$

The catalytic cycle of this reaction in the presence of asymmetric (C_1) ligands **16–18** was suggested on the basis of a calorimetric study of kinetics [133]. The decrease in enantioselectivity in the reaction (11) over C_2-symmetric chiral auxiliaries can be explained on the grounds of a similar catalytic cycle (Scheme 7) [134].

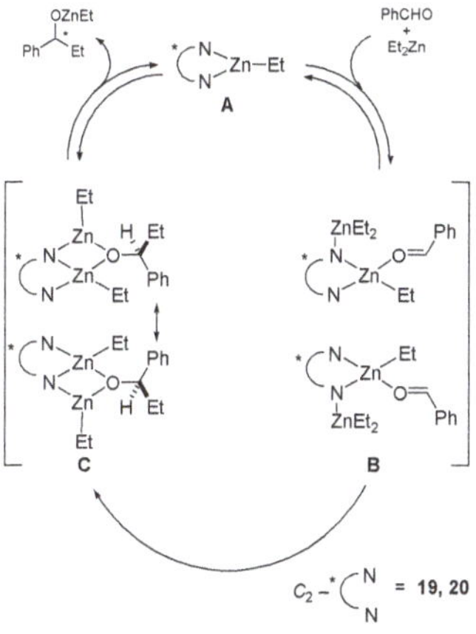

Scheme 7. Possible catalytic cycle of the reaction (11) in the presence of C_2-symmetric ligands **19, 20**.

Key intermediate A in this scheme is also C_2-symmetric. Both intermediate C and B are racemic dimers. A possible $\pi \rightarrow \pi^*$ conjugation (shown by the arrow in intermediate C) reduces the energy barrier and facilitates this reaction path. Therefore, the reaction (11), in accordance with the mechanism, can proceed with the formation of a product of low optical purity, in accordance to the experiment.

The deviation from the linearity to a racemic product ((–)-NLE) was observed not only in the reactions of dialkylzinc addition to aromatic substrates catalyzed by homochiral dimers on the basis of zinc. Remarkable (–)-NLEs were observed in reactions (12) and (13) [135,136] and similar reactions [137] of aromatic substrates over C_2-symmetric chiral catalysts:

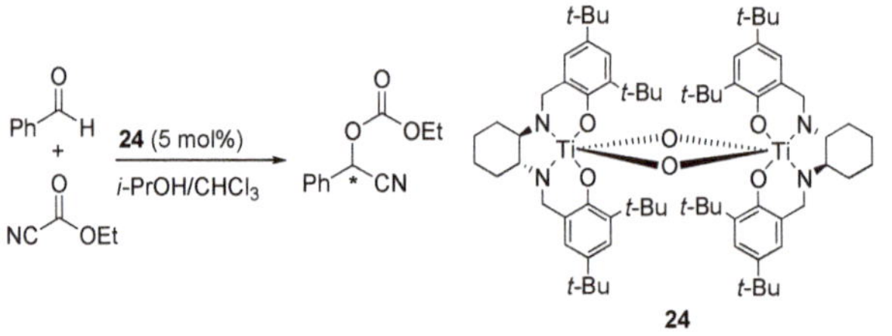

(12)

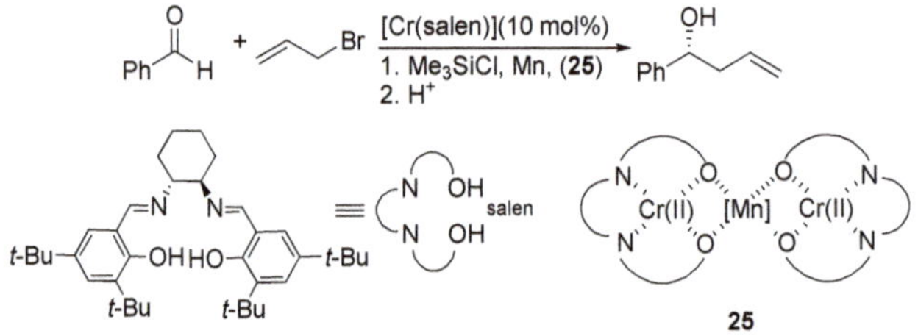

(13)

Thus, all reactions with (–)-NLE occur on the basis of aromatic substrates under the action of C_2-symmetric chiral complexes. A similar effect was mentioned above for the reactions over in situ formed catalysts with the dialkylzinc participation. The deviation from the linearity with the formation of the low *ee* product in asymmetric catalytic reactions occurs where there is an abnormally low *ee* product in the presence of C_2-symmetric chiral complexes as compared to C_1 complexes of similar structure [2].

Noticeable distinctions between enantioselectivities of identical catalysts with ligands having a similar structure but different symmetry (C_2-**21,26,27** and C_1-**28**) were observed in reactions (14) and (15) [138–141] (Scheme 8) where aromatic prochiral compounds were used as substrates.

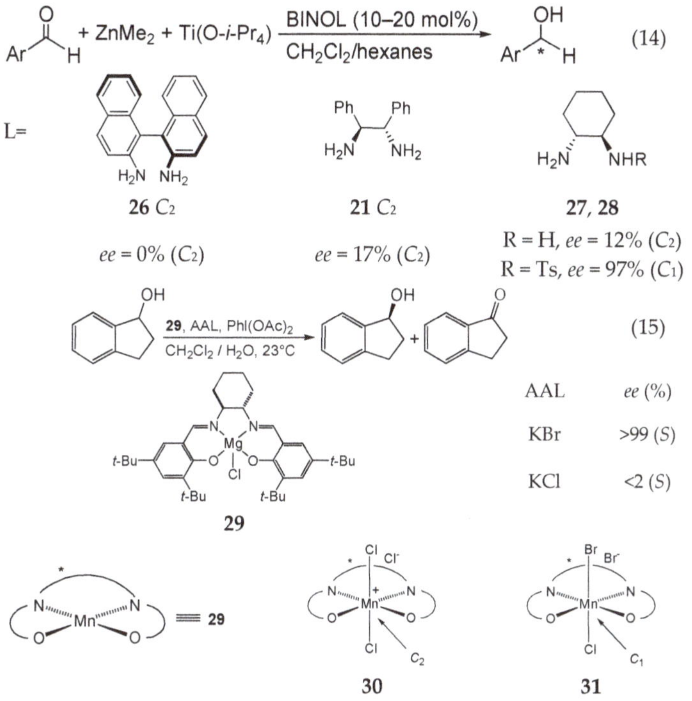

$$ArCHO + ZnMe_2 + Ti(O\text{-}i\text{-}Pr_4) \xrightarrow[CH_2Cl_2/\text{hexanes}]{\text{BINOL (10–20 mol\%)}} Ar\overset{*}{C}H(OH) \quad (14)$$

L=

26 C_2 — ee = 0% (C_2)

21 C_2 — ee = 17% (C_2)

27, 28 — R = H, ee = 12% (C_2); R = Ts, ee = 97% (C_1)

$$\xrightarrow[CH_2Cl_2\,/\,H_2O,\ 23^{\circ}C]{\mathbf{29},\ \text{AAL},\ PhI(OAc)_2} \quad (15)$$

AAL	ee (%)
KBr	>99 (*S*)
KCl	<2 (*S*)

29

30 C_2 **31** C_1

Scheme 8. Dependence of the reaction (14) and (15) enantioselectivity upon C_1 and C_2 symmetry of the ligand **21, 26–29**.

Thus, simple aromatic substrates (acetophenone, etc.) exhibit similar behavior with respect to C_2 or C_2 inductor (catalysts). These monofunctional substrates which coordinate with a catalyst by the reacting group (single–point coordination) only receive stereospecific information directly from the asymmetric catalyst. A racemic field formed by the $\pi\rightarrow\pi^*$ coordination of aromatic substrates (Scheme 7) can induce a decrease in enantioselectivity. A stage of intermediate complexing with C_2-symmetric ligand may be responsible for the product *ee* reduction. This mechanism, perhaps, is similar to the mechanism of the (–)-NLE occurrence established for aromatic substrates.

Since asymmetric induction occurs at the catalyst–substrate stage, symmetry of this intermediate determines enantioselectivity. If the substrate has two or three functional groups, the chiral C_2-symmetric catalyst loses C_2 symmetry at the stage of coordination with such substrate. Indeed, hydrogenation of amino acid precursors on C_2-symmetric chiral complexes (three–point coordination) involves the formation of the C_1 (asymmetric) intermediate [2].

All asymmetric reactions on chiral metal complexes as catalysts support this concept [2]. The experimental data are entirely consistent with Curie–Pasteur's doctrine: only asymmetric factors are responsible for products with asymmetric carbon atoms [1].

A similar situation is observed in asymmetric reactions over chiral organocatalysts. A typical example of such reactions is two-component reactions such as aldol reactions of acetone with α-ketoesters, reactions of cyclohexanone derivatives and β-nitrostyrenes (Michael reactions), nitroaldol (Henry) reactions [39,142–148], etc. Two plausible transition states TS-*R* and TS-*S* (Figure 13) reflect the structure and bonds of the key intermediate of the reaction (16) over C_2 organocatalyst **32** [149]. As seen from the figure, each of the possible transition states (key intermediates) do already not have C_2 symmetry like the basicorganocatalyst. These key intermediates are asymmetric structures in the reaction (16).

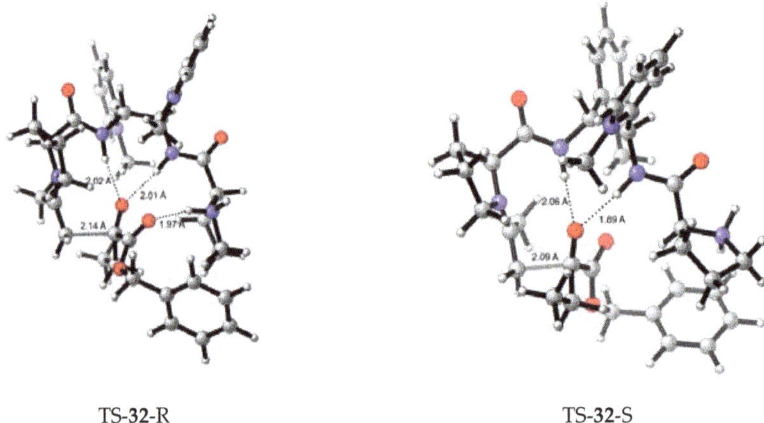

Figure 13. Transition states of the reaction (16) on C_2 organocatalyst **32**.

Me(C=O)Me + 3-ClC$_6$H$_4$C(=O)CO$_2$Me —(**32**, r.t.)→ MeC(=O)CH$_2$C(OH)(3-ClC$_6$H$_4$)CO$_2$Me

ee = 92 (*R*) (16)

2OTf$^-$ + 2TFA

Therefore, asymmetric induction proceeds through the C_1-symmetric (asymmetric) key intermediate in two-component reactions catalyzed by C_2 organocatalysts. Another example in support of this conclusion is Michael's reaction (17) on novel C_2-symmetric *N,N'*-bis-[(pyrrolidin-2-yl)methyl-squaramide] TFA ((*R,R*)-**33** or (*S,S*)-**33**) salts [150].

4-methylcyclohexanone + Ph–CH=CH–NO$_2$ —((*R,R*)33 (10 mmol%), TEA (35 mmol), Solvent, r.t.)→ (2*S*,2'*R*,4*S*) + diastereomer

(86/14%) *ee* = 93% (THF) (17)

Figure 14 shows two possible favoured and two disfavoured intermediates which are different from the viewpoint of an approach of the reagents to C_2 **33** (a front or rear approach). Both favoured-1 and favoured-2 intermediates, alongside with those disfavoured-1 and disfavoured-2, are formed as a result of the reactants coordination with different side nitrogen atoms of C_2- symmetrical **33**. Therefore, they are similar in terms of the structure and symmetry of both **1** and **2** favoured or disfavoured **1** and **2** key intermediates (C_2 symmetry of **33** and C_1 symmetry of key intermediates).

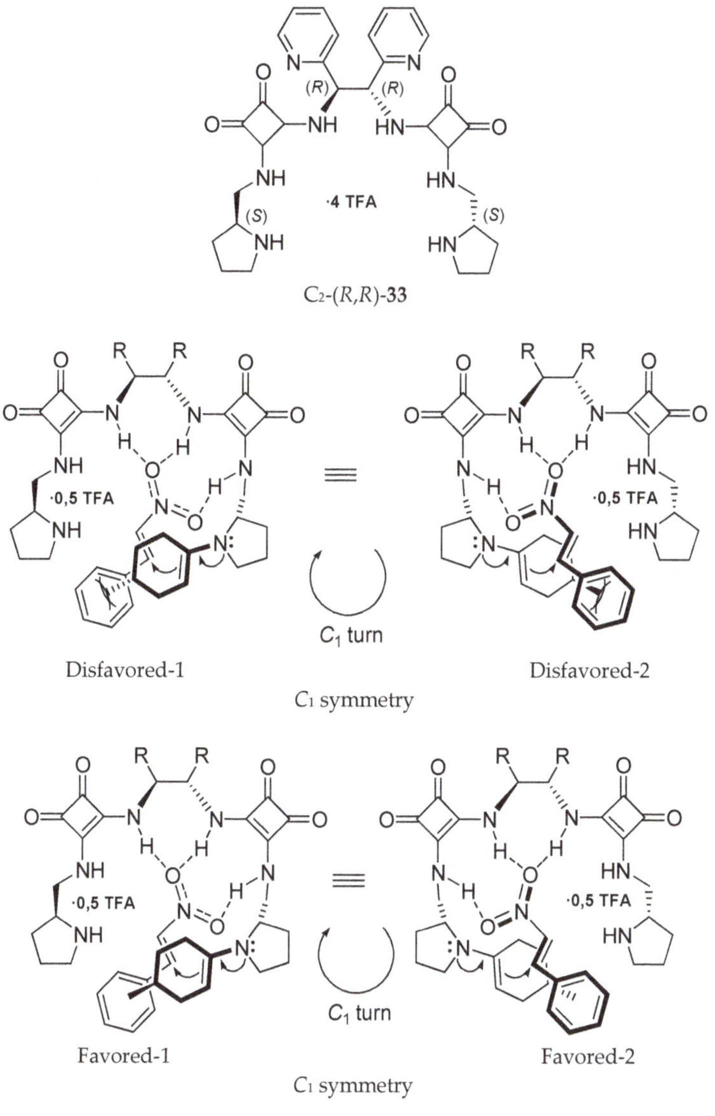

C_2-(R,R)-**33**

Disfavored-1

Disfavored-2

C_1 symmetry

Favored-1

Favored-2

C_1 symmetry

Figure 14. Proposed transition states for the asymmetric Michael reaction (17) over (*R*,*R*)-**33**.

A similar loss of C_2 symmetry of an organocatalyst was observed at the key intermediate stage in the aldol reaction (18) of β-carbonyl acids with trifluoroacetalaldehyde over the C_2-symmetric bisoxazoline catalyst [151] and analogou reactions [151–156].

Nonetheless, the C_2 symmetry loss by the key intermediate does not guarantee high enantioselectivity. An example is a Henry two-component reaction catalyzed by $Cu(OAc)_2$ complexes with C_2 ligands **34–36** and **37–39** (Table 6) [157]. Supposedly, the proposed structure of key intermediate disfavoured and favoured mechanisms of the reaction (19) on complexes with **35** (**34**,**36**) only corresponds to high enantioselectivity (Scheme 9).

Table 6. Henry reaction of *p*-nitrobenzaldehyde with nitromethane in the presence of different ligands.

Ligand	Product *ee*%
34	69
35	71
36	67
37	0
38	2
39	0

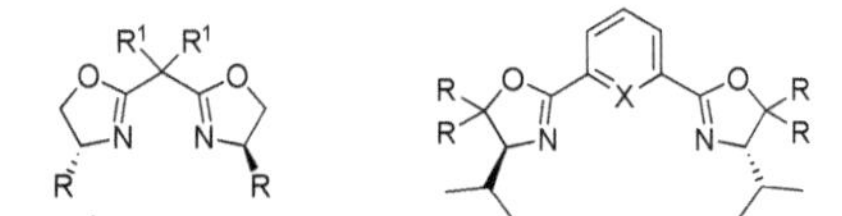

34 R^1 = Et, R = Bn
35 R^1 = Et, R = Ph
36 R^1 = Me, R = ^{i}Pr

37 X = N, R = H
38 X = N, R = Ph
39 X = C, R = ^{i}Pr

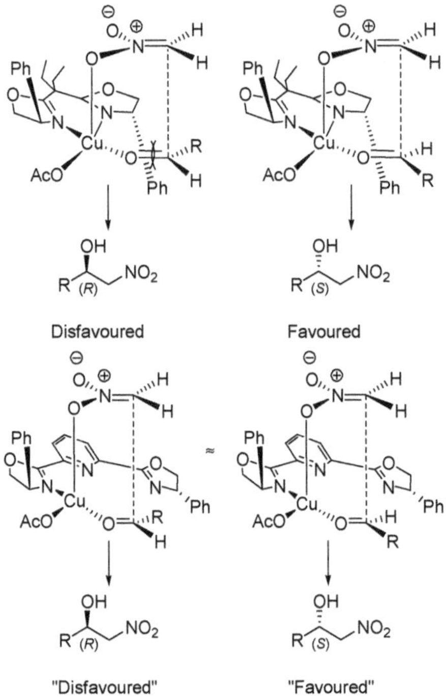

Disfavoured

Favoured

"Disfavoured"

"Favoured"

Scheme 9. Key intermediate strctures of reaction (19).

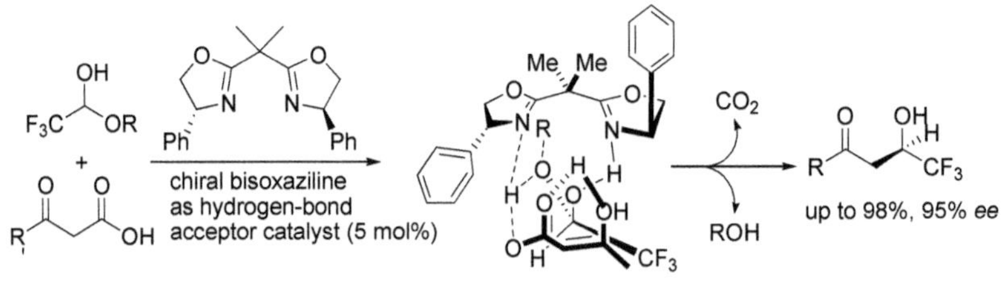

chiral bisoxaziline as hydrogen-bond acceptor catalyst (5 mol%)

CO_2

ROH

up to 98%, 95% *ee*

(18)

$$O_2N\text{–}C_6H_4\text{–}CHO + CH_3NO_2 \xrightarrow[\text{MeOH, 25°C}]{L^*,\ Cu(OAc)_2,} O_2N\text{–}C_6H_4\text{–}CH(OH)\text{–}CH_2NO_2\ (S) \tag{19}$$

Favoured and disfavoured routes of the reaction pathway are equally likely (low enantioselectivity) in the caseof the reaction (19) on $Cu(OAc)_2$/**37** (**38**,**39**). Therefore, this mechanism is in agreement with the experiment.

Thus, all the above-mentioned reactions of asymmetric catalysis proceed by the standard scheme: asymmetric catalyst (an asymmetry inductor) – asymmetric product. C_2 chiral catalysts (a C_2 chiral metallocomplex and a C_2 chiral organocatalyst) lose C_2 symmetry of the key intermediate that determines asymmetric induction. Therefore, asymmetric catalytic reactions run in the asymmetric field and do not occur in the chiral field (asymmetry-(C_1), chirality-(C_1, C_2, C_n, D_2).

6. Spontaneous Chiral Ordering of Achiral Molecules in Liquid Crystals

A common feature of achiral molecules involved in chiral ordering is the ability to bind to each other due to $\pi - \pi^*$ staking, hydrogen bonds or hydrophobic interactions. Linked by those bonds, they shift relative to each other around the central axis on the right or left screw. This synchronization of achiral molecules in chiral superstructures occurs in smectic (lamellar) or nematic liquid crystal (LC) phases. Numerous studies [158–165] of these nanostructures with a "banana", "hat" or crown–like shape show that a ratio of layers with chiral superstructures of different signs in the LC phase is 50:50% or so. It is viewed in photomicrographs as chiral domains (dark/bright spots of equal areas) between slightly uncrossed polarizers. Therefore, the definition of such spontaneous crystallization of achiral molecules as "mirror symmetry breaking" [158–165] can only refer to one individual domain or monolayer with superstructures of similar chirality. Since the ratio of domains of different signs is close to racemic, this definition is not correct for crystallization of the entire mass of molecules.

Thus, the formation of chiral superstructures from achiral molecules in LC and distribution of these chiral structures of different signs in the LC matrix is strikingly similar to the formation of chiral crystals from achiral inorganic compounds/salt molecules (Section 4). For example, a similar pattern of the equal ratio (*l/d* = 1:1) of chiral inorganic crystals is observed in deposits of quartz [166–169] (Figure 15) as well as in the formation of a conglomerate (a mixture of *l*- and *d*–crystals) during $NaClO_3$ crystallization (Scheme 2).

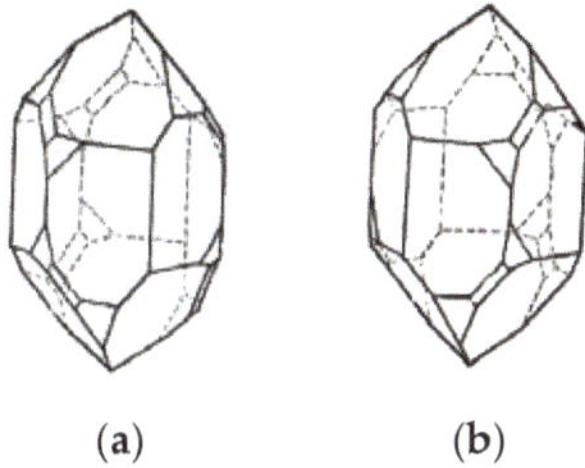

Figure 15. Natural quartz (**a**) left-handed crystals versus (**b**) right-handed ones [167].

Powdering of natural chiral quartz crystals blends left- and right-handed domains [166–169]. Chiral surfaces of centric calcite ($CaCO_3$,) were observed as well [170–172]. Asymmetric crystals of gypsum ($CaSO_4·2H_2O$) were also found [173] and used for asymmetric adsorption [174]. It is believed that natural processes of nucleation and growth of these crystals last for a very long time under the influence of the chiral environment.

Crystallization of achiral molecules in LC and crystallization of quartz and other minerals and salts (see Scheme 8) in nature are likely to proceed according to the laws same as of chirality occurrence. This regularity was also confirmed in the conditions of a standard organic reaction (20) [175].

OMe
H_2N
+
O
OMe
OMe
achiral solution
crist. *S*-**40** as dopant
MeO
NH O
MeO
MeO
S-**40**, 50%
OMe
O HN
OMe
OMe
R-**40**, 50%
MeO
NH O
MeO
MeO
S-**40**, 100% (Solids)
in high yield

(20)

Achiral molecules crystallize in the LC matrix with an approximately equal area of domains of chiral superstructures with different chirality signs. It is difficult to measure the exact ratio of superstructures with different signs of chirality using a polarizing microscope technique (domains) or scanning electron microscopy (twisted ribbons). Therefore, the (1:1) ratio (a racemic mixture) can be assumed valid on the grounds of many publications [158–165].

Spontaneous formation of chiral nanostructures from achiral molecules in the LC matrix has a high inertia depending on heating or cooling to reach a certain temperature. For example, a paradoxal situation may arise when chiral domains of different chirality signs can be formed at the same temperature depending on its achievement by heating or cooling (Figures 16 and 17) [57,176].

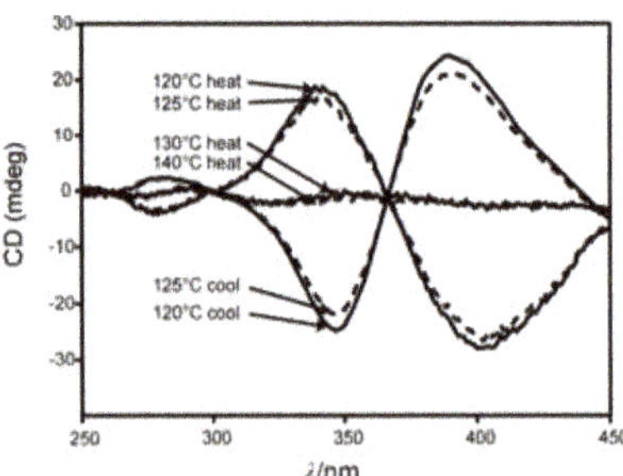

Figure 16. Temperature dependence of circular dichroism (CD) spectra (ellipticity) of **41** [176].

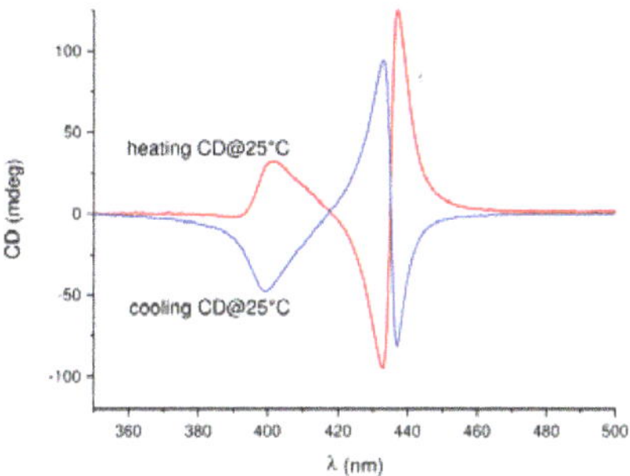

Figure 17. Temperature dependence of CD spectra of **42** [57].

41

43

42

$R^1 = -(CH_2-CH_2-)_2CH_3$
$R^2 = -(CH_2-CH_2-)_nCH_3$

Similar effects were observed in the ongoing formation of other chiral superstructures from achiral molecules [57,177–180]. The dependence of chiral nanostructures from various mesophase temperatures of LC (maximal order) suggests that this regularity is of general nature. The temperature inertia of the formation of chiral superstructures is comparable with the time inertia of crystallization of calcite, gypsum or quartz in nature.

Mineo et al. [57] suggested that chiral superstructures on the basis of "achiral porphyrin–based molecules **42** can be induced and controlled by means of a weak asymmetric thermal gradient" or "asymmetric heat flow." It can be considered as evidence of "very weak forces having an important role in natural chiral selective processes" [57]. May it be regarded as a manifestation of resonance with the external chiral field?

Another feature of the reaction of the chiral superstructure formation from achiral molecules is a dependence of the conformation of *P* or *M* helical structures from irradiation of different energy. The ratio of *P*⇔*M* enantiomeres was found to be reproducible depending on irradiation [158,162] (Figure 18).

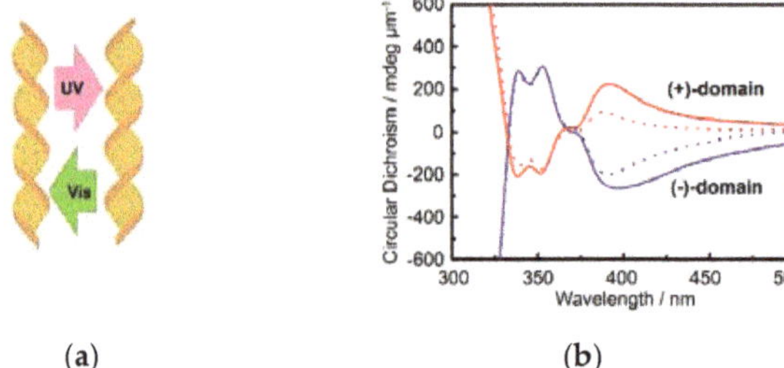

(a) (b)

Figure 18. (**a**) Photoresponsive azobenzene **43**. (**b**) CD spectra for (+) and (–) domains of sample **43** [162].

If enantiomers are an energy equivalent, the ratio of enantiomers should be random and non-reproducible. The reproducibility of $M \rightarrow P$ ratio UV and the opposite $P \rightarrow M$ sequence after Vis irradiation (Figure 18) could result from the action of the external chiral inductor as a chiral trigger.

Another strange feature of crystallization of achiral molecules in the LC phase was found during crystallization of pyridinum–tailored antracene **44** [181] in the presence of iodine and other pseudo-halogen anions with a similar anionic radius.

I^- (or OCN^-, SCN^-, $SeCN^-$)

44

A specific role of the anionic radius to form chiral supromolecular assemblies from achiral molecule **44** testifies to the influence of the helix pitch size in the ongoing formation of a helix structure. An external chiral inductor may also be responsible for this experimental data. It is difficult to explain a similar sensitivity of the organic helical framework to the metallic core radius by internal reasons [182].

An extremely interesting effect was observed in a supamolecular self-assembly of achiral tetraphenylethylene **45** (Figures 19 and 20) [183] due to solvent evaporation. (A) R = C_7, (B) R = C_8, (C) R = C_9, (D) R = C_{10}, *M*, *P*, *M*, *P*, correspondingly.

45

R = n-Heptyl C_7, n-Octyl C_8, n-Nonyl C_9, n-Decyl C_{10}

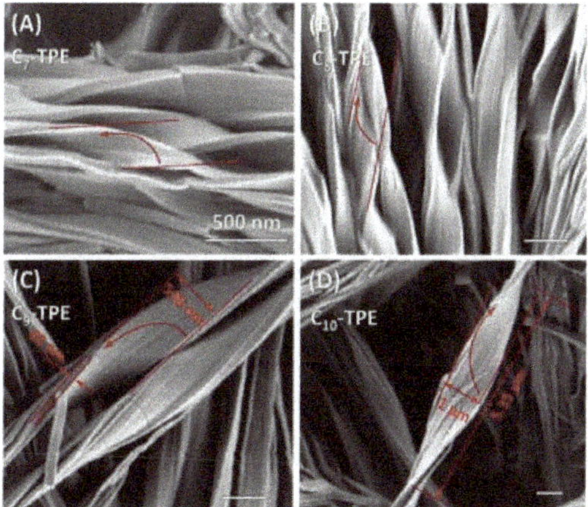

Figure 19. Twisted ribbons of **45** on silicon wafers from MeCN/THF (9:1 v/v). (**A**) R = C_7. (**B**) R = C_8. (**C**) R = C_9. (**D**) R = C_{10} [183].

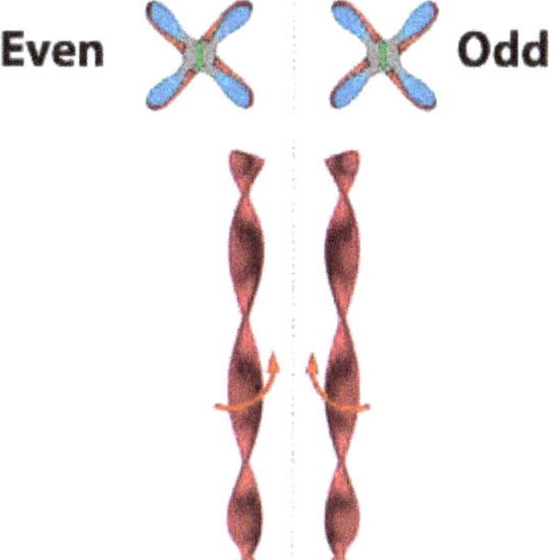

Figure 20. Graphical illustration of the **45** self-assembly effect [183].

Hence, **45** with the even number of carbon atoms in alkyl chains was produced by right-handed superstructures, whereas the odd number led to the left-handed supramolecular structure (Figure 20).

No physical rationale for this phenomenon has been discussed. We believe that a possible explanation for this effect may be the influence of an external chiral inductor as a resonator.

The sensitivity of this resonator to the structure of achiral molecules could be illustrated by the following examples (Figures 21–24).

Figure 21. Schematic illustration of the chiral superstructure on the basis of **46** [148,164].

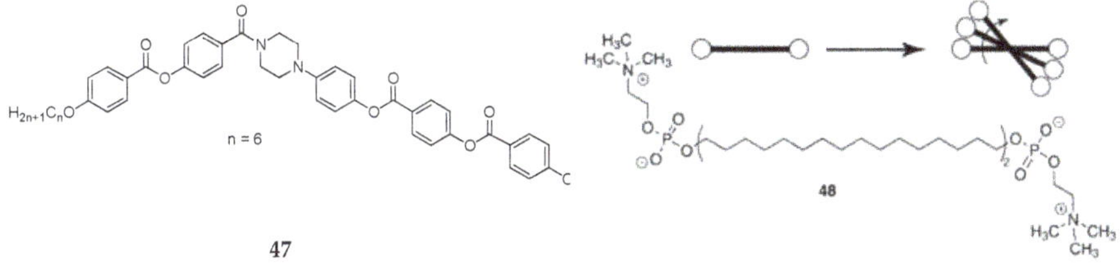

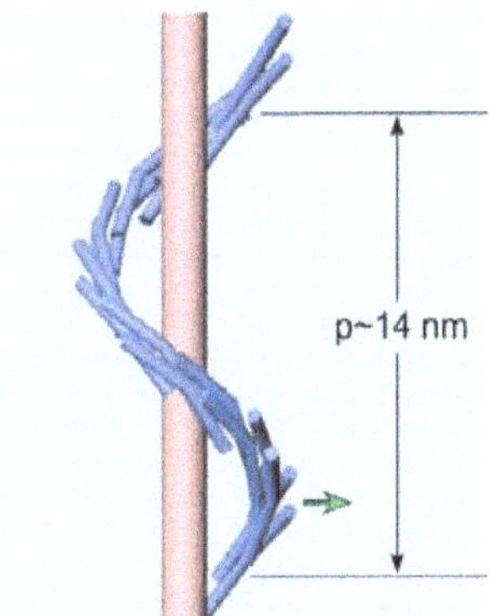

Figure 22. Schematic representation of the chiral superstructure on the basis of **47** [183].

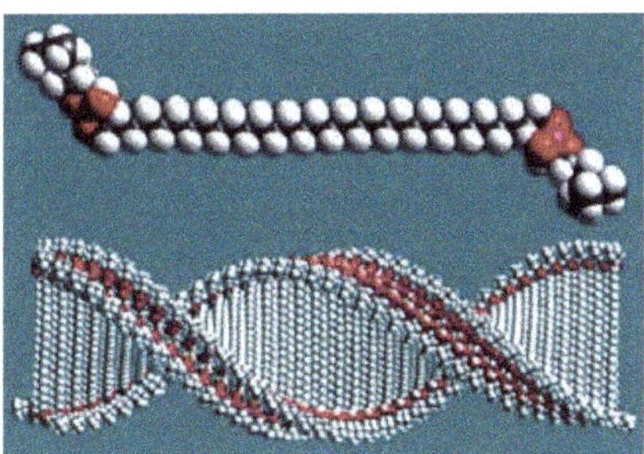

Figure 23. Schematic model of the chiral superstructure based on propeller-shaped molecule **48** [184,185].

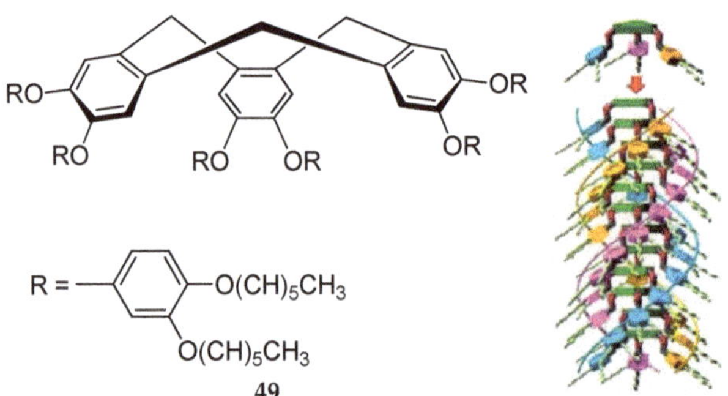

Figure 24. Schematic illustration of the chiral superstructure on the basis of crown-shaped molecule **49** [186,187].

As can be seen, symmetry (C_2, C_4) of corresponding chiral superstructures coincides with symmetry of the underlying achiral molecules (**46**,**45**,**48**). Crown-shaped molecule **49** with 12 peripheral alkyl chains (C_3) form single columns with 12-fold triple helices (C_3) (Figure 24) [186,187]. Asymmetric molecule **47** (C_1 symmetry axis) with the center of gravity in the side chain forms a helix from the linear sequence of these molecules (tail to head) with C_1 symmetry (Figure 22).

Thus, the centers of gravity (symmetry) of achiral molecules and relevant chiral superstructures are similar. This coincidence could be possible evidence of the gravitational influence in the construction of chiral ensembles from achiral molecules.

Not necessarily organic compounds or salts solely participate in the formation of chiral superstructures. Singh et al. [188] observed that cubic nanocrystals (~13.4 nm) of magnetite Fe_3O_4 as dipoles self-assembled into arrays of helical superstructures. Fe_3O_4 dipoles are oriented along the applied magnetic field (H = 700g). The chains of single magnetite particles aggregated as the solvent evaporated (Figure 25). Examples of self-assembly of one-dimensional Fe_3O_4 nanocube belts can be seen in Figure 26.

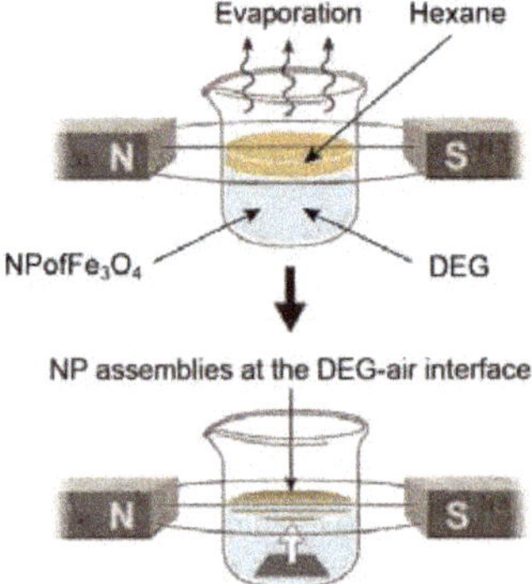

Figure 25. Assemblies of the formation of Fe_3O_4 nanoparticles (NP) in diethylene glycol (DEG)-air interface in the magnetic field [188].

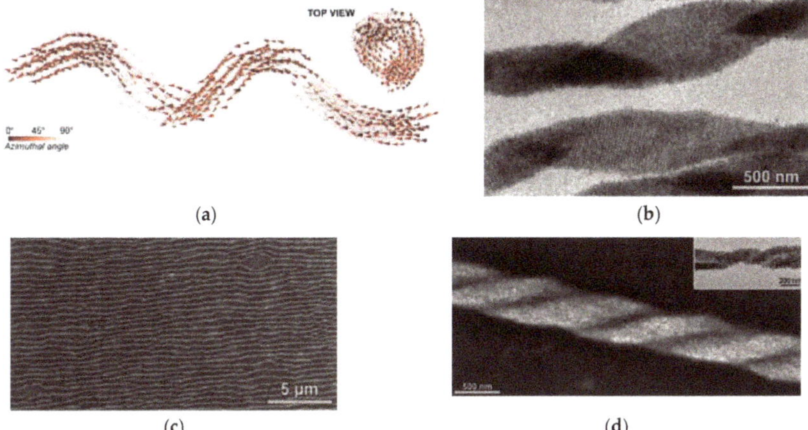

(**a**) (**b**)

(**c**) (**d**)

Figure 26. Of Fe_3O_4 nanoparticle chiral arrays. (**a**) One-dimensional belt folding into a left-handed helix. (**b**) Transmission electron microscopy (TEM) image of an individual left-handed helix. (**c**) Scanning electron microscopy (SEM) of single-stranded helices. (**d**) SEM image of (left-handed) double helix [188].

The authors concluded that "there was no intrinsic preference for helices of either handedness; each experiment began with the nucleation of either right- or left-handed helices with equal probability". At the same time, chiral arrays during a single experiment retained the identical handedness. Double or triple helices were observed during some experiments (Figure 27a,b). Figure 27c,d also shows chirality inversion cases in some single experiments (denoted by arrows). Consequently, chirality was preserved during the experiments, except for some cases of partial inversion which reduced enantioselectivity. Whereas crystallization of chiral superstructures from achiral molecules in LC can be compared with $NaClO_3$ crystallization without stirring (Scheme 2 top), chiral arrays of Fe_3O_4 nanoparticles in the magnetic field can be compared with $NaClO_3$ crystallization with stirring (Scheme 2, bottom). Indeed, the images (Figures 26 and 27a) show domains with homochiral superstructures (single handedness).

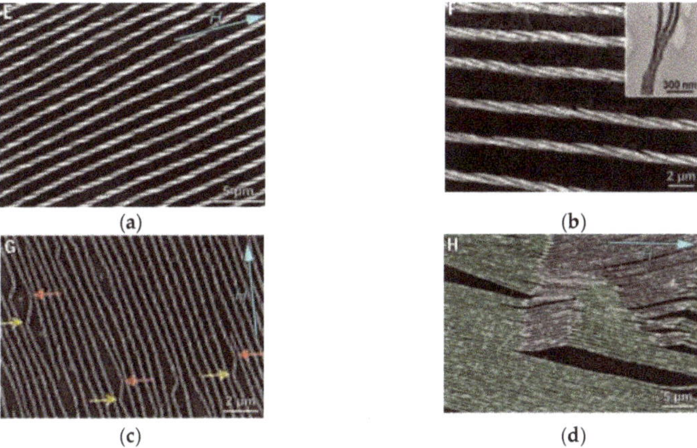

(**a**) (**b**)

(**c**) (**d**)

Figure 27. (**a**) Self-assembly of helical Fe_3O_4 nanocrystal superstructures. (**b**) SEM image of the right–handed array of double helices. (B) Array of right-handed triple helices and the end of a triple helix (inset). (**c**) SEM image of self-healing (handedness inversion or damage). The arrows indicate chirality inversion sites. (**d**) SEM Image of left-handed double or triple helices (Fe_3O_4Ag) with sites of helices with the sign reversal [188].

A comparison to $NaClO_3$ crystallization with stirring implies a presence of a hypothetical chiral inductor. Hence, the magnetic field alone cannot be a chiral inductor in the formation of the chiral order of magnetite crystals. It is believed that a combination of the magnetic field only with another field [189] (electrical or gravitational) can be an asymmetric inductor.

A number of nanotubes were synthesized (BN, WS_2, MoS_2, $NbSe_2$, $NiCl_2$, SiO_2, TiO_2, MoO_3, and V_2O_5) [189]. Celik-Aktas et al. [190] studing boron nitride nanotubes by transmission electron microscopy (TEM), observed "regular, zigzag, dark and bright spots on the side walls of the nanotubes."

These spots moved in a regular fashion around the tube. Basing on this evidence, the authors [191,192] suggested a double-helix structural model (Figure 28) "as a result of a stronger wall–wall interaction associated with the ionic bonding in boron nitride". However, stabilization of the right-handed helix structure (Figure 28) is opposite to that of the left-handed. It is possible that the preference of right–handed helix symmetry can be explained as the result of an external chiral inductor (chiral environment).

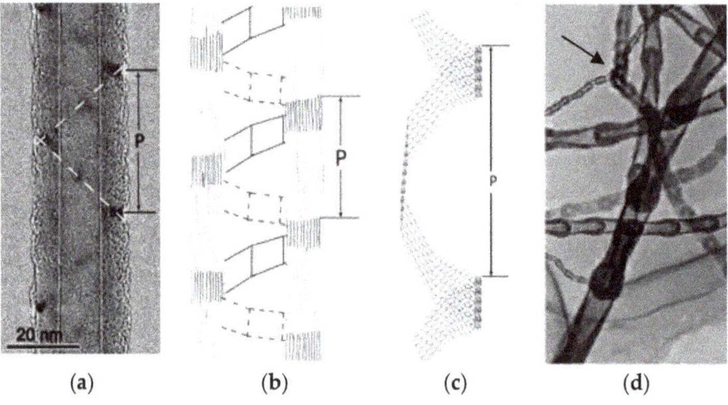

Figure 28. TEM image of BN nanotubes. (**a**) "Dark contrast regions in the middle of the tube and on the side walls are visible." (**b**) Schematic helix structure of BN nanotubes. (**c**) Atomic structure of BN hexagonal helices. (**d**) BN nanotubes (the change point is indicated by an arrow) [191–193].

The sharp turning point of the BN nanotube at an angle of about 30–40° (Figure 28) [190–193] is extremely similar to the inversion point of the chirality sign of the magnetite nanocrystals self-assembly (Figure 27c,d). Therefore, it can be assumed that this BN point is the turning point of the transition of the right–handed helix to the left-handed one (or the contrary).

7. Chiral Field (Chiral Memory) and Racemic Field

An example of the chiral field existence can be an enantioselective reaction (21) of prochiral substrate hydrogenation on an achiral catalyst in the cholesteric liquid crystal (ChLC) [189].

$$\text{PhCH=C(COOH)NHCOMe} + H_2 \xrightarrow[\text{ChLC}]{RhCl(PPh_3)_3} \text{PhCH}_2\text{-}^{*}\text{CH(COOH)NHCOMe} \tag{21}$$

The temperature maximum of hydrogenation enantioselectivity (*ee* = 16% at 60–70 °C) coincides with the temperature maximum of ChLC (cholesteryl tridecanoate) helical ordering. Mirror symmetry breaking in this case is contrary to common sense. Indeed, molecule sizes of α-acetamidocinnamic acid and the $RhCl(PPh_3)_3$ catalyst are much smaller than the ChLC helical pitch (300–400 nm). Therefore, asymmetric induction at the stage of coordination of substrate and catalyst molecules cannot only occur under the influence of ChLC helical chirality. A possible explanation of mirror symmetry breaking in this experiment is the effect of the chiral field on individual molecules of the substrate and catalyst [194,195]. Indeed, it was shown by the method of induced circular dichroism that helical ordering was built by achiral reactant molecules as well as from achiral Wilkinson's catalyst molecules in the conditions of the hydrogenation reaction in the ChLC medium. Thus, the chiral structure of the key intermediate in this reaction was provided by the action of the chiral field [194,195].

A strong chiral field was shown to exist between two flat aromatic "pincers" (tweezers) [196,197] separated by some kind of a more or less rigid "teter" with a chiral moiety as a chirality inductor. Porphyrin or metalloporphyrin groups at the chain ends with the chiral inductor in the middle of this host molecule, for example **50**, can be used as tweezers [198–201]. The length of the chain the inductor and porphyrin group separating can exceed 13 single and double bonds[173]. In the absence of a chiral inductor, such structure is used to study molecular chirality of chiral bidentate guest molecules by circular dichroism of metalloporphyrin hosts as a sensor of chirality [202–207]. End porphyrin groups can be located at a considerable distance from each other in formulations such as dibenzo30-crown-10 skeleton **51** [208].

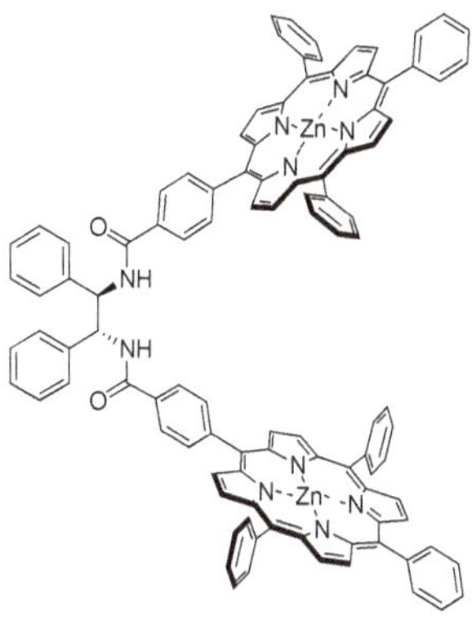

50

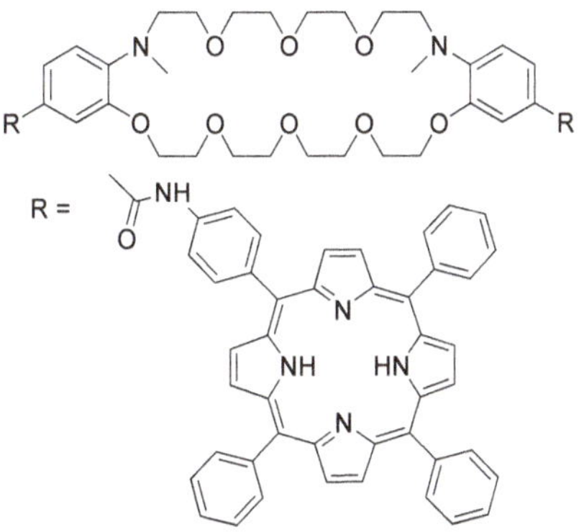

51

There is an assumption that tweezer's architecture of **51** lead to the mesostructure formation. Mmirror symmetry of N-atoms in **51** is realized through the π–π^* interaction of porphirin groups. This structure provides a strong racemic field between flat porphirin groups. Chirality of self-assembled achiral porphyrins (imprinted chirality) in host–chiral guest structures persists after a removal of the chiral guest–inductor which caused this chirality [209]. This is a new example of chiral memory.

A property of silica gel to retain information ("remember"it) about complex or chiral molecules dissolved in it during molding was noticed long ago [210–213]. After their removal, silica gel could selectively adsorb these molecules or enantiomers. This selectivity of silica gel could be attributed individually to silylium ions [214]. However, the experimental variety of the ability to memorize structural information, which is characteristic of different gels, allows us to attribute this property to physical gels [215–223].

The discotic trisamides and trisureas of the **52** and **53** type form fibers and organic gels [224–226].

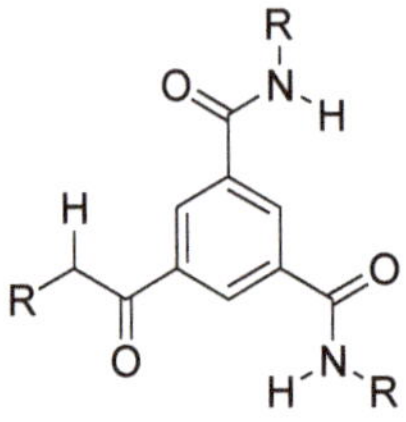

52 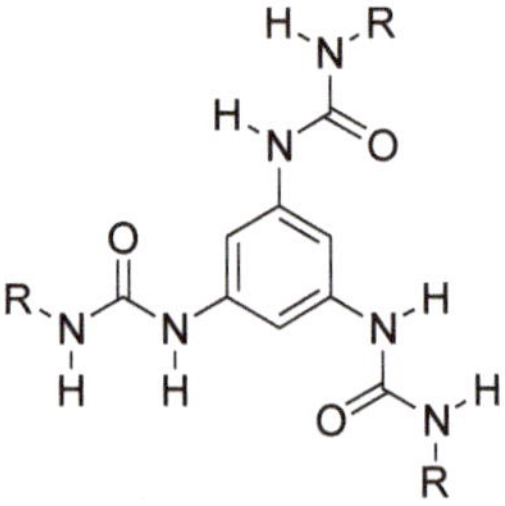 53

A minor structural variation in this type molecules may dramatically disturb helical columnar superstructures built on the basis of these molecules. The structure and properties variability of these helical columnar aggregates may be related to the gel structure as well as π–π* stacking and H-bonding. It is believed that the gel formation leads to reaction medium structuring. Together with π–π* stacking [227,228], H-bonding [229], and donor–acceptor interaction [230], this structuring facilitates the chiral information transfer. Polymers **54** and **55** form co-gels with the *N*,*N*′-bis(octadecyl)L(D)-Boc-glutamic gelator [218].

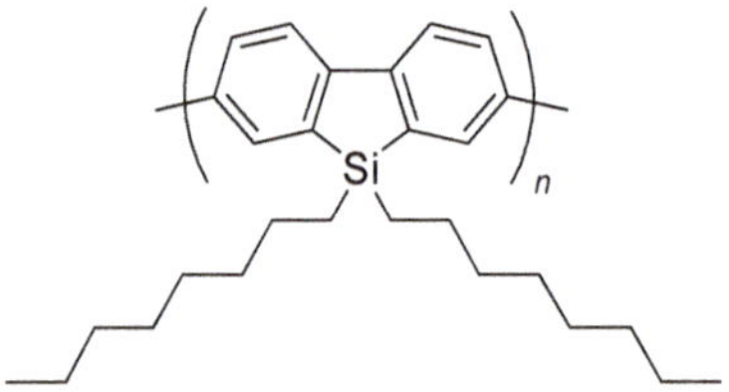

54 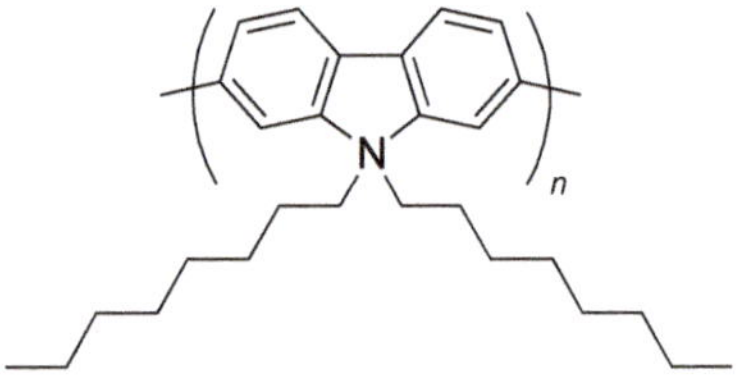 55

Supramolecular chirality of polymers **54** and **55** follows the chirality type of the gelator. Helicity of the polymer assemblies can be memorized even after a removal of gelator molecules. Reaction mixture structuring by means of physical gel and π–π* stacking or H-bonding relationships of interacting molecules is necessary for displaying chiral memory in the formation of chiral superstructures. These relations and the chiral memory effect were observed for self-assembly of molecules **56** [231] and **57** [216], and other aromatic molecules in the presence of molecular low weight gelators [217,229–231]. The idea of reaction mixture media structuring with gel has been further evolved experimentally [232–235].

56 **57**

Numerous examples of the chiral memory effect have been observed in polymerization [236–243]. Polymerization creates higher-order structural ordering than the structural arrangement by gel. Therefore, π–π* stacking, in addition to structural ordering during polymerization, creates conditions for the emergence of chiral memory. The chiral memory effect in polymerization of oligomers is also observed with the participation of hydrogen bonding [244]. It is also possible that other factors may contribute to chiral structuring of the medium during polymerization, e.g., vicinal chirality (Figure 29) [245].

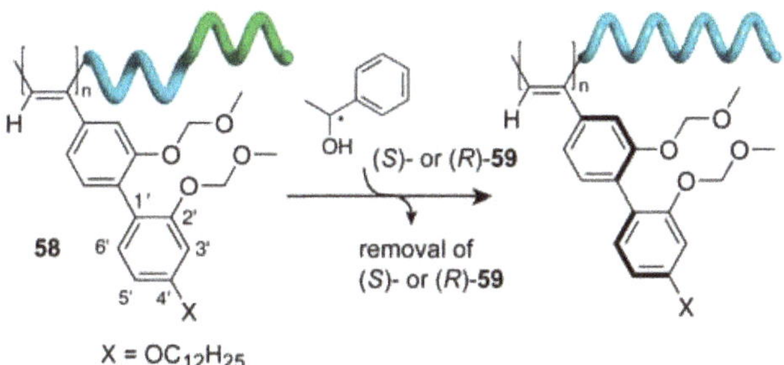

Figure 29. Illustration of the chiral memory effect in poly(biphenylacetylene)s.

We believe that all above-mentioned examples of chiral memory can be explained by the existence of a chiral field. The conditions of the chiral field existence and preservation correspond to the necessity of structural ordering in the overall variety of chiral memory examples.

Chiral effects are known to have only a small energy barrier (see, e.g., reactions (3) and (4) with the *P–M* transition) [107–111]). Therefore, for example, the *P–M* conversion occurs when the temperature changes within 30 °C [101–111]. The chiral field effect may be compared with the chiral effect of other fields or inductors. An example is the chiral effect of ultrasound radiation on chiral amplification [95] in crystallization. It is of interest that the chiral memory motif is also observed in chirality amplification during porphyrin self-assembling [246].

A racemic field is misappreciated in comparison with a chiral field. Perhaps, a reason for this is a kinetic dependence, in which an asymmetric reaction (without an asymmetric inductor) produces the same number of "right" and "left" products. With a high energy of this reaction, the racemic field may not be noticed. Therefore, the comparison between the racemic field and the chiral field can be valid only for not high-energy reactions such as crystallization [247]. Thomas and Tor [248] synthesized a novel 1.10-phenanthroline ligand with branched multifunctional dentritic groups. When an octahedral metal ion self-assembled with these ligands two enantiomers, Δ and Λ, were formed (Figure 30). This can apparently be considered as an example of racemic field impact.

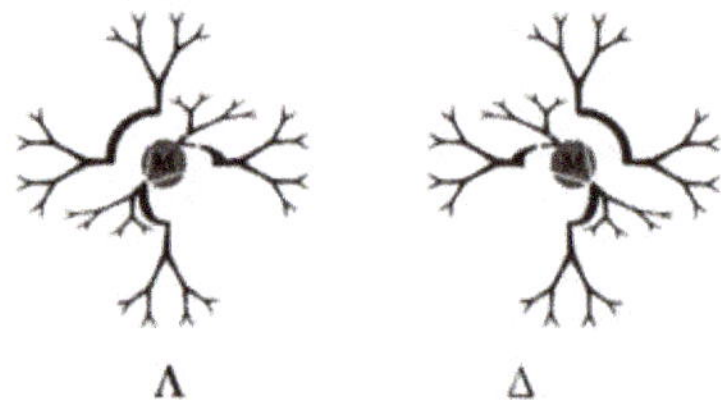

Figure 30. Dendritic fragments assembled around an octahedral metal ion from Λ and Δ enantiomers [248].

A colourless crystal of helical $[Cu_4Cl_4(ally)_4]_\infty$ was formed after neat triallylamine (ally) addition to copper (I) chloride at ambient temperature [249] (Figure 31). Out of the five crystallizations three resulted in the predominantly *P*-helix and two in the predominantly *M*-helix. Those examples can be considered as a manifestation of the stochastic action of the chiral fieid (Figure 31) as well as the racemic field (Figure 30).

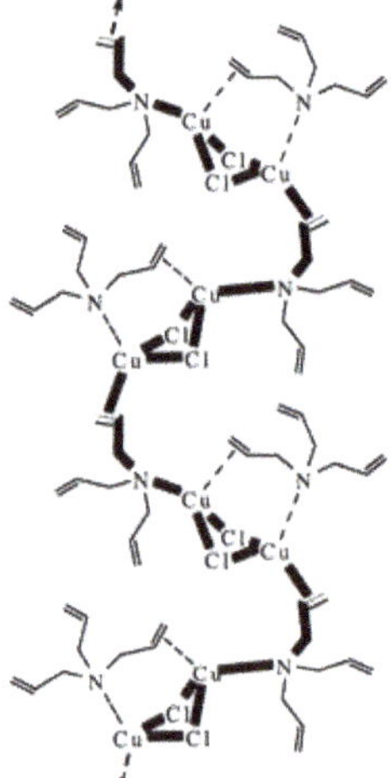

Figure 31. Schematical view of the $[Cu_4Cl_4(ally)_4]_\infty$ helix [249].

These cases demonstrate an analogy with crystallization of sodium chlorate (Scheme 2) and the stochastic helicity formation (Section 6).

There is abundant evidence that a collection of right- and left-handed enantiomers possesses other physical properties than a collection of pure enantiomers [39]. It is reflected in tighter packing (more than 4% denser) or higher melting points of racemates than those of their enantiomers (Wallach's rule) [250–252]. Also, data on solubility and some other physical properties show that racemates tend to be slightly more stable than pure enantiomers. The simplest explanation of Wallach's phenomenon is associated with a difference in hydrogen bonding energies of "true" racemates and pure enantiomers [253–260]. Opponents of this concept validity believe that "true" racemates and pure enantiomers with analogous H-bonding are not rare in occurrence [261–266]. The discussion is complicated by the fact that there are the examples of anti-Wallach's rule [267]. Therefore, the cause of Wallach's phenomenon has not been completely understood up to now. Some researchers designated Wallach's rule as "mutually exclusive binding" in "true" racemates [268].

Thus, the differences in the physical properties of racemate and enantiomer crystals as well as mirror symmetry breaking or conservation in crystallization of sodium chlorate (Scheme 2) or dendritic complexes (Figure 30) and triallylamine complexes (Figure 31) suggest the existence of a racemic field alongside with a chiral field. It is possible that the existence of mirror symmetry in the form of a racemic field is global (mirror symmetry space). After all, our Galaxy has a mesostructure (Figure 3).

There are direct evidences of sophisticated structures of racemates. For example, collagen peptides form tight ridges-in-grooves packing of right- and left-handed triple helices (Figure 32) [269].

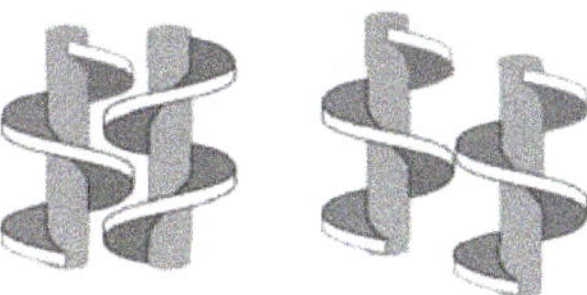

Figure 32. Illustration of triple helices of collagen LD (left) and LL (right) conformations [269].

It seems there is a force which provides an energy advantage for the synthesis of a denser racemate structure. The difference between racemic and chiral fields could explain this experimental observation.

Racemic compounds exist in solution as well (Scheme 4). According to Kagan and Noyori's models racemates and homochiral dimers function as real actors in catalytic asymmetric reactions with nonlinear effects (see Section 5).

Racemic superstructures formed by *P*- and *M*-helices represent a large group of racemates [39,270–272]. There is no consensus about a reason for the structural difference between heterochiral (racemic) and homochiral dimers [273–275].

8. *B–Z* DNA Conformational Transition

Having a seemingly stable molecular structure, the DNA molecule is a vulnerable object. For example, there are more than a hundred of oxidative damages of the DNA molecular structure. Alongside with the structural damages, the DNA molecule may be subject to conformational distortions (Table 7) [276]. All these conformations of canonical *B*-DNA (right-handed double helix) lead to genetic instability and genetic diseases [277–283]. From this viewpoint, the conversion of right-handed *B*-DNA into left-handed *Z*-DNA attracts widespread attention.

Table 7. DNA conformations [276–278].

Name	Conformation	Name	Conformation
Cruciform		Slipped (hairpin) structures	
Triplex		Left-handed Z-DNA	

Harvey [284] presented a schematic structure (Figure 33) of *B*- and *Z*-DNA molecules after the inversion point. The base pairs in *Z*-DNA are directed opposite (arrows) from those in *B*-DNA.

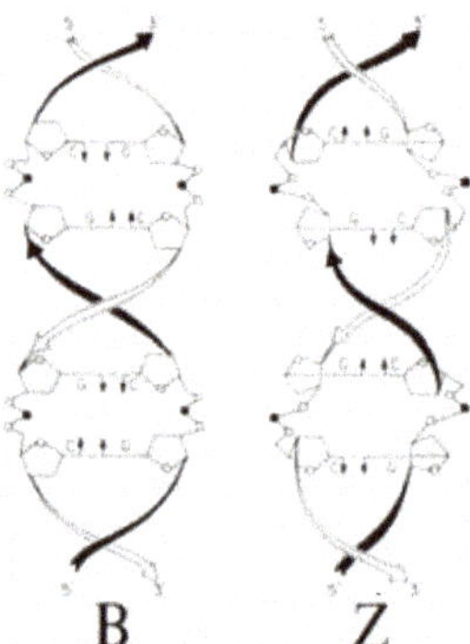

Figure 33. *B*→*Z* DNA after the inversion point.

It is unlikely that a transition of right-handed *B*-DNA into left-handed *Z*-DNA shown in Table 7 occurs linearly without changing the angle of the DNA thread. It is known that the angle between the polymer helices of opposite senses is about 130° (Figure 34) [285,286], which has been confirmed by other experimental data (see Figure 27c,d).

Figure 34. Calculated helical reversal of poly(ethyl isocyanate) [285].

Lee et al. [287] calculated a free energy difference between *B*-DNA and *Z*-DNA which is 0.9 kcal/mol per dinucleotide unit. The low energy of this transition was confirmed by the influence of relatively weak chemical factors on this transition. Indeed, ions, especially cations, strongly affect *B*–*Z* DNA transitions [288,289]. This influence can be explained by reducing phosphate–phosphate interactions between phosphate groups on opposite strands. Phosphate groups got closer to each other in *Z*-DNA than in *B*-DNA (7.7 Å in *Z*-DNA compared to 11.7 Å in *B*-DNA) [290]. Therefore, cations clustering around the negatively charged phosphate group affected *B*-DNA and *Z*-DNA in a different manner. It is not surprising that ions with higher valencies appeared more effective than monovalent ions [291–293]. The agents that change the dielectric constant of water (or alcohol) were found to stabilize *Z*-DNA [294–296]. Small molecules affected *B*↔*Z* equilibrium as well [297,298]. The *B*–*Z* DNA transition was also influenced by Co, Mn, Ru and Pt complexes [299–301]. Nevertheless, chiral metal complexes failed to convert *B*-DNA to *Z*-DNA (see also [302–306]).

Xu et al. [307] first reported that the (*P*) and (*M*) helicene **60** helix molecule (Figure 35) displayed structural selectivity in binding to DNA. The circular dichroism (CD) spectra of the (*P*)-**60**/Z-DNA mixture showed the 70% decrease in intensity of CD whereas no change occurred in binding of (*P*)-**60** to B-DNA. No discrimination was seen in the CD spectra of the (*M*)-**60**/B-DNA and (*M*)-**60**/Z-DNA mixture. There was only the 20% decrease in CD intensity.

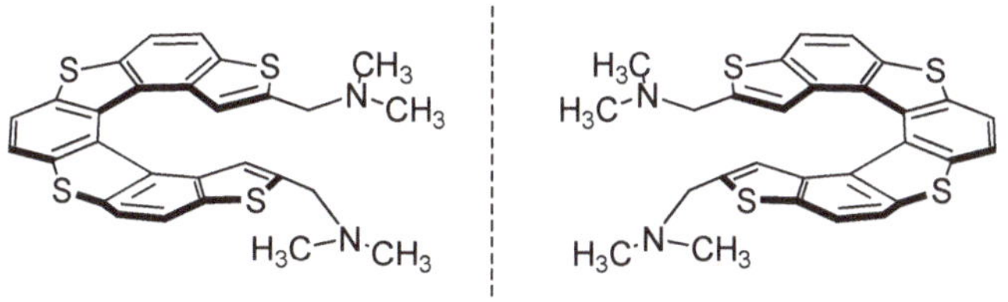

Figure 35. Enantiomeric helical pair of helicene **60**.

This observation allows a very important and fundamental conclusion: an external chiral inductor can affect B–Z DNA conformations or their equilibrium. The influence of this inductor is apparently not related to the chemical interaction. Optically active hexahelicen and its derivatives are known for their huge optical rotational ability (CD activity and chiral field). Therefore, their effect on chiral conformations of DNA is due to a greater degree of the helicene chiral field.

By analogy, Tsuji et al. [308] showed that an optically active helicene-spermine conjugate (Figure 36) might discriminate *B–Z* DNA conformations as well. The authors proposed a schematic illustration of the intermediate complex of the B-DNA and Z-DNA interaction with (*P*)-**64** and (*M*)-**64**, respectively (Scheme 10).

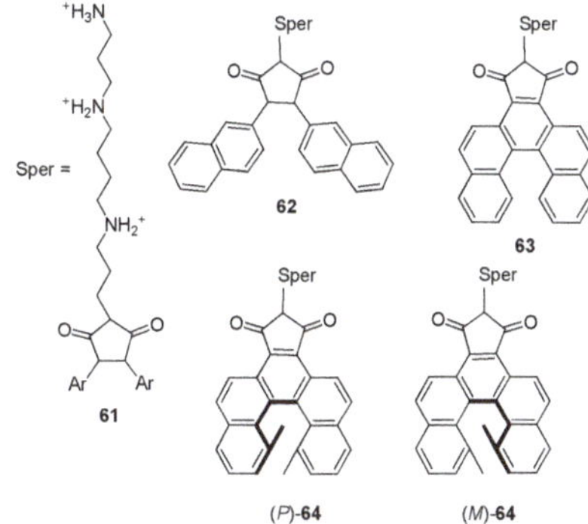

Figure 36. Structures of several spermine-conjugated ligands.

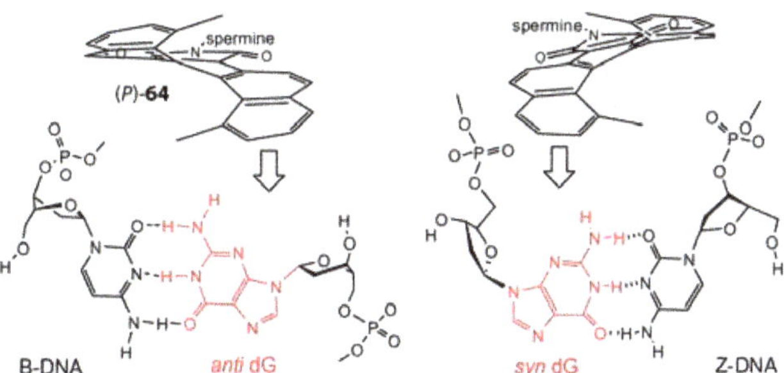

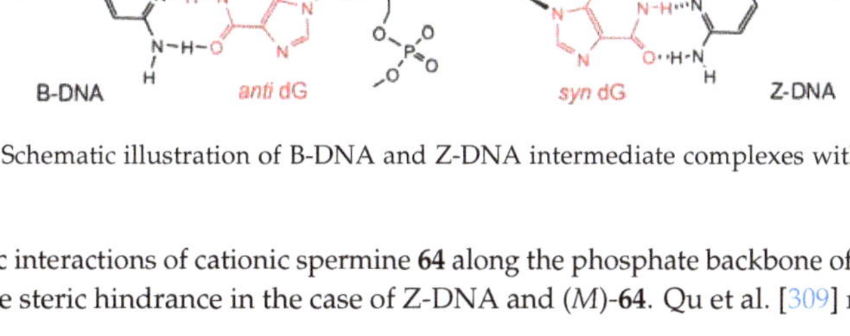

Scheme 10. Schematic illustration of B-DNA and Z-DNA intermediate complexes with (*P*)-64 and (*M*)-**64**.

Electrostatic interactions of cationic spermine **64** along the phosphate backbone of the DNA minor groove led to the steric hindrance in the case of Z-DNA and (*M*)-**64**. Qu et al. [309] reported that an anticancer agent (+)-dunorubicin and its novel (-)-enantiomer (WP 900) exhibited enantioselectivity in binding to DNA.

Apart from the above-mentioned chiral inductors, complex structures containing large aromatic ensembles can exert a discriminatory effect on DNA. Doi et al. [310] found that spermine achiral conjugate **65** affected the B→Z transition of d(CGCGCG)$_2$ at a low salt concentration. Haque et al. [311] observed an opposite transition of left-handed Z-DNA into *B*-DNA in the presence of benzophenanthridine plant alkaloid chelerythrine **66**.

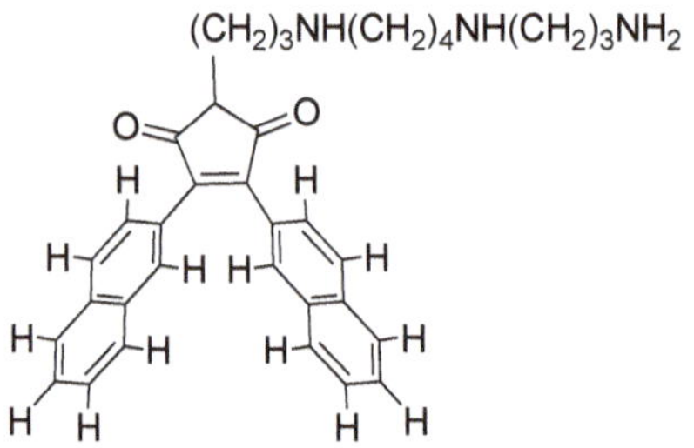

$(CH_2)_3NH(CH_2)_4NH(CH_2)_3NH_2$

65

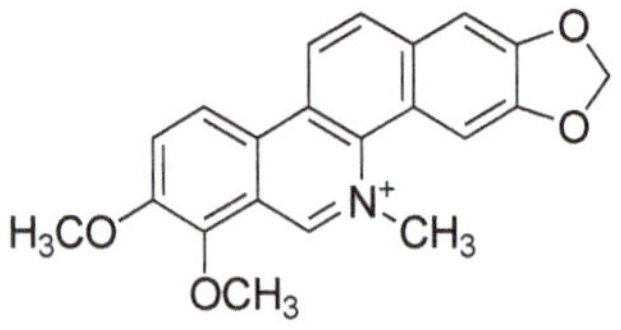

H_3CO N^+ CH_3 OCH_3

66

An analysis of the discriminatory ability of irradiations and electromagnetic fields to initiate DNA degradation shows that almost each of them exerts impact on DNA, alone or jointly with some other factors. For example, a magnetic field with an extremely low frequency can induce a DNA double-strand break (the most potent form of DNA damage and genomic instability) [312]. A similar effect was observed for DNA marker exposure under a pulsed magnetic field (25 Hz) [313]. The pulsed magnetic field increased a spontaneous genomic DNA degradation in this case. The pulsed magnetic field enhanced the cell-killing effect of UV radiation [314].

Impact of the electromagnetic field on DNA obeys the same pattern as that of the magnetic field. It appears that a stable electromagnetic field does not affect DNA. These are only low frequency electromagnetic fields that exert a genotoxic effect on DNA [315]. For example, 50 Hz low frequency electromagnetic fields enhanced cell proliferation and DNA damage [316]. Genotoxic effects were observed in human fibroblasts after intermittent exposure to 50 Hz electromagnetic fields [317] (see also [318,319]). Regretfully, a combination of two factors (ionizing radiation and a presence of organic salts) is required to protect DNA from damage. Zheng and Sanche [320] marked that organic salts were efficient in protecting DNA from damage by electrons of 1 eV to 60 keV. The authors suggest that anions of organic salts create additional electric fields within the DNA groove which protect the molecule (see also [321]).

Therefore, usually, a low frequency electromagnetic or a low frequency magnetic field (alone or in combination with other factors) can affect DNA; however, some scientists disagree with this opinion [322,323].

Investigations of the static magnetic field effect on DNA have led to less definite conclusions. Li et al. [324] showed that a "magnetic field could potentiate the activity of oxidant radicals" and could bring about "both stabilizing and destabilizing effects to DNA". According to Ruiz-Gomez et al. [325] the magnetic field effect on DNA is not certain. Aydin et al. [326] believe that a low intensity static magnetic field may trigger genomic instability. "But this genotoxic effect of the magnetic field, however, is minimized in living organisms due to the presence of protectic cellular responses" [327] (see, however [328]).

This assumption has sound grounds. It is now known that each type of DNA damage corresponds to a certain repair mechanism in a living cell such as nucleotide excision repair, base excision repair, mismatch repair, and so on [329]. It is logical to assume that impact of a stable factor of the Earth's magnetic field over millions of years of evolution has led to the development of a living cell protective mechanism, alongside with a corresponding DNA repair mechanism. However, since the pulsed magnetic field is not found in nature, DNA has no protective mechanism from it. Probably, that is why the static magnetic field has little effect on DNA and the cell in vivo whereas the pulsated magnetic field affects DNA and the cell radically [320–322,330]. According to the same logic we can suggest that periodic long-term gravitational pulsations (lunar tides), as a common phenomenon on Earth, have a protective mechanism against DNA damage (corresponding DNA repair mechanism) whereas static gravitation is uncommon for living organisms on Earth. Indeed, gravitational impact on the Moon (by Earth) has no pulsation. This impact is constant because only one side of the Moon faces the Earth. Therefore, living organisms may have difficulties with the *B*–*Z* DNA transition as there is no protection from static gravity.

9. Conclusions

An obvious advantage of the right-handed helical structure of the most important biological macromolecules is possible evidence of the existence of an external chiral inductor with similar symmetry.

Effects of chiral and racemic fields as possible inductors and conductors of the corresponding influence explain chemical processes with mirror symmetry breaking or retention.

It is possible that the mirror symmetry effect of a racemic field generates additional energy for material objects. This effect may probably explain the mesostructure of galaxies or the formation of chiral crystals of different signs in equal proportions in Earth's deposits. Also, this effect may account for the difference in the physical properties of racemic and enantiomeric crystals (Wallach's rule) or spontaneous chiral ordering of achiral molecules in the form of domains of different chirality signs in approximately equal proportions. The physical basis of mirror symmetry breaking and mirror symmetry retention effects can be found in stochastic chemical reactions running by two routes: with the formation either of racemic or chiral products.

External inductor's impact on these reactions and processes has been also discussed. This influence corresponds to the mesostructure of our Galaxy as a source of chiral and racemic gravitational fields. The fields can serve as a chiral (chiral environment) or racemic inductor and trigger.

These fields, in turn, can affect all biological processes in living organisms. This is especially relevant for Z-DNA areas of native DNA damage with the left-handed conformational conversion. Indeed, there are mechanisms in Nature that protect DNA from destructive factors (mismatch repair, nucleotide or base excision repair, oxidative defects repair, and so on). It is highly likely that Nature has a corresponding DNA repair mechanism against DNA conformational damage caused by the pulsed gravitational field of Earth (Moon tides). Changing of this status quo in lunar settlements (stable gravity and zero magnetic field) can be dangerous for living organisms. Indeed, it is possible that DNA molecules (especially Z-DNA fragments) in organisms placed in the other chiral gravitational environment could behave *via* a different mechanism and become more vulnerable. Moreover, the corresponding DNA repair mechanism will function in a different manner.

Author Contributions: The contribution of the authors is equivalent and approximately proportional to their order in order.

Funding: Work on the review received no external funding.

Conflicts of Interest: The authors declare no conflict of interest, financial or otherwise.

References

1. Curie, P. Sur la symétrie dans les phénomènes physiques, symétrie d'un champ électrique et d'un champ magnétique. *J. Phys. Theor. Appl.* **1894**, *3*, 393–415. [CrossRef]
2. Pavlov, V.A. C2 and C1 Symmetry of chiral auxiliaries in catalytic reactions on metal complexes. *Tetrahedron* **2008**, *64*, 1147–1179. [CrossRef]

3. Pavlov, V.A.; Klabunovskii, E.I. The origin of homochirality in nature: A possible version. *Russ. Chem. Rev.* **2015**, *84*, 121–133. [CrossRef]
4. Pavlov, V.; Klabunovskii, E. Homochirality Origin in Nature: Possible Versions. *Curr. Org. Chem.* **2014**, *18*, 93–114. [CrossRef]
5. Breslow, R. Formation of L Amino Acids and D Sugars, and Amplification of their Enantioexcesses in Aqueous Solutions, Under Simulated Prebiotic Conditions. *Isr. J. Chem.* **2011**, *51*, 990–996. [CrossRef]
6. Weissbuch, I.; Lahav, M. Crystalline architectures as templates of relevance to the origins of homochirality. *Chem. Rev.* **2011**, *111*, 3236–3267. [CrossRef] [PubMed]
7. Meierhenrich, U.J. Amino Acids and the Asymmetry of Life. *Eur. Rev.* **2013**, *21*, 190–199. [CrossRef]
8. Lente, G. Open system approaches in deterministic models of the emergence of homochirality. *Chirality* **2010**, *22*, 907–913. [CrossRef]
9. Toxvaerd, S. Origin of homochirality in biosystems. *Int. J. Mol. Sci.* **2009**, *10*, 1290–1299. [CrossRef] [PubMed]
10. Plasson, R.; Kondepudi, D.K.; Bersini, H.; Commeyras, A.; Asakura, K. Emergence of homochirality in far-from-equilibrium systems: Mechanisms and role in prebiotic chemistry. *Chirality* **2007**, *19*, 589–600. [CrossRef]
11. Pizzarello, S. The chemistry of life's origin: A carbonaceous meteorite perspective. *Acc. Chem. Res.* **2006**, *39*, 231–237. [CrossRef]
12. Jorissen, A.; Cerf, C. Asymmetric Photoreactions as the Origin of Biomolecular Homochirality: A Critical Review. *Orig. Life Evol. Biosph.* **2002**, *32*, 129–142. [CrossRef]
13. Podlech, J. Origin of organic molecules and biomolecular homochirality. *Cell Mol. Life Sci.* **2001**, *58*, 44–60. [CrossRef]
14. Feringa, B.L.; van Delden, R.A. Absolute Asymmetric Synthesis: The Origin, Control, and Amplification of Chirality. *Angew. Chem. Int. Ed.* **1999**, *38*, 3418–3438. [CrossRef]
15. Prelog, V. Chirality in chemistry. *Science* **1976**, *193*, 17–24. [CrossRef]
16. Chandrasekhar, S. Molecular homochirality and the parity-violating energy difference. A critique with new proposals. *Chirality* **2008**, *20*, 84–95. [CrossRef]
17. Carroll, J.D. A new definition of life. *Chirality* **2009**, *21*, 354–358. [CrossRef] [PubMed]
18. Kuttel, M.; Ravenscroft, N.; Foschiatti, M.; Cescutti, P.; Rizzo, R. Conformational properties of two exopolysaccharides produced by Inquilinus limosus, a cystic fibrosis lung pathogen. *Carbohydr. Res.* **2012**, *350*, 40–48. [CrossRef]
19. Hsien-Chih, H.W.; Sarko, A. The double-helical molecular structure of crystalline a-amylose. *Carbohydr. Res.* **1978**, *61*, 27–40. [CrossRef]
20. Hsein-Chih, H.W.; Sarko, A. The double-helical molecular structure of crystalline b-amylose. *Carbohydr. Res.* **1978**, *61*, 7–25. [CrossRef]
21. French, A.; Zaslow, B. Conformation of the "V" amylose helix. *J. Chem. Soc. Chem. Commun.* **1972**, *0*, 41–42. [CrossRef]
22. Sarko, A.; Biloski, A. Crystal structure of the koh-amylose complex. *Carbohydr. Res.* **1980**, *79*, 11–21. [CrossRef]
23. Nishiyama, Y.; Mazeau, K.; Morin, M.; Cardoso, M.B.; Chanzy, H.; Putaux, J.-L. Molecular and Crystal Structure of 7-Fold V-Amylose Complexed with 2-Propanol. *Macromolecules* **2010**, *43*, 8628–8636. [CrossRef]
24. Lopez, C.A.; de Vries, A.H.; Marrink, S.J. Amylose folding under the influence of lipids. *Carbohydr. Res.* **2012**, *364*, 1–7. [CrossRef]
25. Janaswamy, S.; Chandrasekaran, R. Heterogeneity in iota-carrageenan molecular structure: Insights for polymorph II–>III transition in the presence of calcium ions. *Carbohydr. Res.* **2008**, *343*, 364–373. [CrossRef]
26. Matsuda, Y.; Biyajima, Y.; Sato, T. Thermal Denaturation, Renaturation, and Aggregation of a Double-Helical Polysaccharide Xanthan in Aqueous Solution. *Polym. J.* **2009**, *41*, 526. [CrossRef]
27. Sheehan, J.K.; Gardner, K.H.; Atkins, E.D.T. Hyaluronic acid: A double-helical structure in the presence of potassium at low pH and found also with the cations ammonium, rubidium and caesium. *J. Mol. Biol.* **1977**, *117*, 113–135. [CrossRef]
28. Numata, M.; Shinkai, S. 'Supramolecular wrapping chemistry' by helix-forming polysaccharides: A powerful strategy for generating diverse polymeric nano-architectures. *Chem. Commun.* **2011**, *47*, 1961–1975. [CrossRef]
29. Pavlov, V.; Pavlova, T. Paradoxes of Symmetry: Homochirality; Cryptochiral Reactions; Chiral Field, Memory, and Induction; Chiral and Racemic Environment. *Curr. Org. Chem.* **2017**, *21*, 872–888. [CrossRef]

30. Wang, X.; Zhang, Y.; Zhang, L.; Ding, Y. Multiple conformation transitions of triple helical lentinan in DMSO/water by microcalorimetry. *J. Phys. Chem. B* **2009**, *113*, 9915–9923. [CrossRef]
31. Bocchinfuso, G.; Mazzuca, C.; Sandolo, C.; Margheritelli, S.; Alhaique, F.; Coviello, T.; Palleschi, A. Guar gum and scleroglucan interactions with borax: Experimental and theoretical studies of an unexpected similarity. *J. Phys. Chem. B* **2010**, *114*, 13059–13068. [CrossRef] [PubMed]
32. Miyoshi, K.; Uezu, K.; Sakurai, K.; Shinkai, S. Inter-chain and arrayed hydrogen bonds in β-1,3-d-xylan triple helix predicted by quantum mechanics calculation. *Carbohydr. Polym.* **2006**, *66*, 352–356. [CrossRef]
33. Okobira, T.; Miyoshi, K.; Uezu, K.; Sakurai, K.; Shinkai, S. Molecular dynamics studies of side chain effect on the beta-1,3-D-glucan triple helix in aqueous solution. *Biomacromolecules* **2008**, *9*, 783–788. [CrossRef] [PubMed]
34. Villares, A. Polysaccharides from the edible mushroom Calocybe gambosa: Structure and chain conformation of a (1–>4),(1–>6)-linked glucan. *Carbohydr. Res.* **2013**, *375*, 153–157. [CrossRef]
35. Gagnon, M.A.; Lafleur, M. From curdlan powder to the triple helix gel structure: An attenuated total reflection-infrared study of the gelation process. *Appl. Spectrosc.* **2007**, *61*, 374–378. [CrossRef]
36. Harrington, J.C.; Morris, E.R. Conformational ordering and gelation of gelatin in mixtures with soluble polysaccharides. *Food Hydrocoll.* **2009**, *23*, 327–336. [CrossRef]
37. Yanaki, T.; Norisuye, T.; Fujita, H. Triple Helix of Schizophyllum commune Polysaccharide in Dilute Solution. 3. Hydrodynamic Properties in Water. *Macromolecules* **1980**, *13*, 1462–1466. [CrossRef]
38. Yanaki, T.; Ito, W.; Tabata, K.; Kojima, T.; Norisuye, T.; Takano, N.; Fujita, H. Correlation between the antitumor activity of a polysaccharide schizophyllan and its triple-helical conformation in dilute aqueous solution. *Biophys. Chem.* **1983**, *17*, 337–342. [CrossRef]
39. Pavlov, V.A.; Zlotin, S.G. Homochirality, Stochastic Chiral Reactions, Spontaneous Chiral Ordering of Achiral Molecules, and Similar Chiral Effects. Is there a Physical Basis for these Mirror Symmetry Breaking Phenomena? *Curr. Org. Chem.* **2018**, *22*, 2029–2054. [CrossRef]
40. Davankov, V. Chirality as an inherent general property of matter. *Chirality* **2006**, *18*, 459–461. [CrossRef]
41. Kane, A.; Shao, R.-F.; Maclennan, J.E.; Wang, L.; Walba, D.M.; Clark, N.A. Cover Picture: Electric Field-Driven Deracemization (ChemPhysChem 1/2007). *ChemPhysChem* **2007**, *8*, 170–174. [CrossRef] [PubMed]
42. Baranova, N.B.; Zel'dovich, B.Y. Separation of mirror isomeric molecules by radio-frequency electric field of rotating polarization. *Chem. Phys. Lett.* **1978**, *57*, 435–437. [CrossRef]
43. Pavlov, V.A. Mechanisms of asymmetric induction in catalytic hydrogenation, hydrosilylation and cross-coupling on metal complexes. *Russ. Chem. Rev.* **2002**, *71*, 33–48. [CrossRef]
44. Klabunovskii, E.I. Homochirality and its significance for biosphere and the origin of life theory. *Russ. J. Org. Chem.* **2012**, *48*, 881–901. [CrossRef]
45. Takano, Y.; Takahashi, J.-I.; Kaneko, T.; Marumo, K.; Kobayashi, K. Asymmetric synthesis of amino acid precursors in interstellar complex organics by circularly polarized light. *Earth Planet. Sci. Lett.* **2007**, *254*, 106–114. [CrossRef]
46. Managadze, G.G.; Engel, M.H.; Getty, S.; Wurz, P.; Brinckerhoff, W.B.; Shokolov, A.G.; Sholin, G.V.; Terent'ev, S.A.; Chumikov, A.E.; Skalkin, A.S.; et al. Excess of L-alanine in amino acids synthesized in a plasma torch generated by a hypervelocity meteorite impact reproduced in the laboratory. *Planet. Space Sci.* **2016**, *131*, 70–78. [CrossRef]
47. Davankov, V.A. Inherent homochirality of primary particles and meteorite impacts as possible source of prebiotic molecular chirality. *Russ. J. Phys. Chem. A* **2009**, *83*, 1247–1256. [CrossRef]
48. Ivanova, T.A.; Kuznettsov, S.N.; Logachev, Y.I.; Sosnovetc, E.N. North-south asymmetry and anisotropy of solar cosmic rays during the flare of April18, 1972. *Kosmicheskie Issledovaniya* **1976**, *14*, 235–238.
49. Svirzhevsky, N.S.; Svirzhevskaya, A.K.; Bazilevskaya, G.A.; Stozhkov, Y.I. North–South asymmetry in cosmic ray fluxes as measured in the stratosphere and in selected solar wind parameters in the near-Earth space. *Adv. Space Res.* **2005**, *35*, 671–676. [CrossRef]
50. Barron, L.D. True and false chirality and absolute asymmetric synthesis. *J. Am. Chem. Soc.* **1986**, *108*, 5539–5542. [CrossRef]
51. He, Y.J.; Qi, F.; Qi, S.C. Earth's orbital chirality and driving force of biomolecular evolution. *Med. Hypotheses* **2001**, *56*, 493–496. [CrossRef]
52. Stone, E.C.; Cummings, A.C.; McDonald, F.B.; Heikkila, B.C.; Lal, N.; Webber, W.R. Voyager 1 explores the termination shock region and the heliosheath beyond. *Science* **2005**, *309*, 2017–2020. [CrossRef]

53. Shu, F.H. *The Physical Universe: An Introduction to Astronomy*; University Science Books, Mill Valley, CA: Sasalito, CA, USA, 1982.
54. Decker, R.B.; Krimigis, S.M.; Roelof, E.C.; Hill, M.E.; Armstrong, T.P.; Gloeckler, G.; Hamilton, D.C.; Lanzerotti, L.J. Voyager 1 in the foreshock, termination shock, and heliosheath. *Science* **2005**, *309*, 2020–2024. [CrossRef]
55. Vidal-Madjar, A.; Laurent, C.; Bruston, P.; Audouze, J. Is the solar system entering a nearby interstellar cloud. *Astrophys. J.* **1978**, *223*. [CrossRef]
56. Grande, C.; Patel, N.H. Nodal signalling is involved in left-right asymmetry in snails. *Nature* **2009**, *457*, 1007–1011. [CrossRef]
57. Mineo, P.; Villari, V.; Scamporrino, E.; Micali, N. New Evidence about the Spontaneous Symmetry Breaking: Action of an Asymmetric Weak Heat Source. *J. Phys. Chem. B* **2015**, *119*, 12345–12353. [CrossRef]
58. Erschov, V.E.; Cantor, J.I. *Sea Shells. Concise Guide*; "Kursiv": Moscow, Russia, 2008.
59. Reiss, H.R. Nuclear beta decay induced by intense electromagnetic fields: Basic theory. *Phys. Rev. C* **1983**, *27*, 1199–1228. [CrossRef]
60. Fischbach, E.; Buncher, J.B.; Gruenwald, J.T.; Jenkins, J.H.; Krause, D.E.; Mattes, J.J.; Newport, J.R. Time-Dependent Nuclear Decay Parameters: New Evidence for New Forces? *Space Sci. Rev.* **2009**, *145*, 285–335. [CrossRef]
61. Fischbach, E.; Chen, K.J.; Gold, R.E.; Goldsten, J.O.; Lawrence, D.J.; McNutt, R.J.; Rhodes, E.A.; Jenkins, J.H.; Longuski, J. Solar influence on nuclear decay rates: Constraints from the MESSENGER mission. *Astrophys. Space Sci.* **2011**, *337*, 39–45. [CrossRef]
62. Sturrock, P.A.; Buncher, J.B.; Fischbach, E.; Javorsek, D., II; Jenkins, J.H.; Mattes, J.J. Concerning the Phases of the Annual Variations of Nuclear Decay Rates. *Astrophys. J.* **2011**, *737*. [CrossRef]
63. Silverman, M.P. Search for anomalies in the decay of radioactive Mn-54. *EPL* **2016**, *114*, 62001. [CrossRef]
64. O'Keefe, D.; Morreale, B.L.; Lee, R.H.; Buncher, J.B.; Jenkins, J.H.; Fischbach, E.; Gruenwald, T.; Javorsek, D.; Sturrock, P.A. Spectral content of 22Na/44Ti decay data: Implications for a solar influence. *Astrophys. Space Sci.* **2013**, *344*, 297–303. [CrossRef]
65. Norman, E.B.; Browne, E.; Shugart, H.A.; Joshi, T.H.; Firestone, R.B. Evidence against correlations between nuclear decay rates and Earth–Sun distance. *Astropart. Phys.* **2009**, *31*, 135–137. [CrossRef]
66. El-Borie, M.A.; El-Abshehy, M.; Talaat, S.; Taleb, W.M.A. North-south asymmetry in solar, interplanetary, and geomagnetic indices. *Astrophysics* **2012**, *55*, 127–139. [CrossRef]
67. Kondepudi, D.K.; Kaufman, R.J.; Singh, N. Chiral symmetry breaking in sodium chlorate crystallizaton. *Science* **1990**, *250*, 975–976. [CrossRef]
68. Kondepudi, D.K.; Asakura, K. Chiral Autocatalysis, Spontaneous Symmetry Breaking, and Stochastic Behavior. *Acc. Chem. Res.* **2001**, *34*, 946–954. [CrossRef]
69. Kondepudi, D.K.; Laudadio, J.; Asakura, K. Chiral Symmetry Breaking in Stirred Crystallization of 1,1′-Binaphthyl Melt. *J. Am. Chem. Soc.* **1999**, *121*, 1448–1451. [CrossRef]
70. Asakura, K.; Soga, T.; Uchida, T.; Osanai, S.; Kondepudi, D.K. Probability distributions of enantiomeric excess in unstirred and stirred crystallization of 1,1′-binaphthyl melt. *Chirality* **2002**, *14*, 85–89. [CrossRef]
71. Cintas, P.; Viedma, C. On the physical basis of asymmetry and homochirality. *Chirality* **2012**, *24*, 894–908. [CrossRef]
72. Viedma, C. Chiral symmetry breaking during crystallization: Complete chiral purity induced by nonlinear autocatalysis and recycling. *Phys. Rev. Lett.* **2005**, *94*, 065504. [CrossRef]
73. Kondepudi, D.K.; Bullock, K.L.; Digits, J.A.; Hall, J.K.; Miller, J.M. Kinetics of chiral symmetry breaking in crystallization. *J. Am. Chem. Soc.* **1993**, *115*, 10211–10216. [CrossRef]
74. McBride, J.M.; Carter, R.L. Spontaneous Resolution by Stirred Crystallization. *Angew. Chem. Int. Ed. Engl.* **1991**, *30*, 293–295. [CrossRef]
75. Tsogoeva, S.B.; Wei, S.; Freund, M.; Mauksch, M. Generation of highly enantioenriched crystalline products in reversible asymmetric reactions with racemic or achiral catalysts. *Angew. Chem. Int. Ed. Engl.* **2009**, *48*, 590–594. [CrossRef]
76. Wei, S.; Mauksch, M.; Tsogoeva, S.B. Autocatalytic enantiomerisation at the crystal surface in deracemisation of scalemic conglomerates. *Chemistry* **2009**, *15*, 10255–10262. [CrossRef] [PubMed]
77. Kondepudi, D.K.; Bullock, K.L.; Digits, J.A.; Yarborough, P.D. Stirring Rate as a Critical Parameter in Chiral Symmetry Breaking Crystallization. *J. Am. Chem. Soc.* **1995**, *117*, 401–404. [CrossRef]

78. Osuna-Esteban, S.; Zorzano, M.P.; Menor-Salvan, C.; Ruiz-Bermejo, M.; Veintemillas-Verdaguer, S. Asymmetric chiral growth of micron-size NaClO3 crystals in water aerosols. *Phys. Rev. Lett.* **2008**, *100*, 146102. [CrossRef] [PubMed]
79. Levilain, G.; Rougeot, C.; Guillen, F.; Plaquevent, J.-C.; Coquerel, G. Attrition-enhanced preferential crystallization combined with racemization leading to redissolution of the antipode nuclei. *Tetrahedron Asymmetry* **2009**, *20*, 2769–2771. [CrossRef]
80. Cartwright, J.H.; Piro, O.; Tuval, I. Ostwald ripening, chiral crystallization, and the common-ancestor effect. *Phys. Rev. Lett.* **2007**, *98*, 165501. [CrossRef]
81. Noorduin, W.L.; Meekes, H.; van Enckevort, W.J.; Millemaggi, A.; Leeman, M.; Kaptein, B.; Kellogg, R.M.; Vlieg, E. Complete deracemization by attrition-enhanced ostwald ripening elucidated. *Angew. Chem. Int. Ed. Engl.* **2008**, *47*, 6445–6447. [CrossRef] [PubMed]
82. Cartwright, J.H.; Garcia-Ruiz, J.M.; Piro, O.; Sainz-Diaz, C.I.; Tuval, I. Chiral symmetry breaking during crystallization: An advection-mediated nonlinear autocatalytic process. *Phys. Rev. Lett.* **2004**, *93*, 035502. [CrossRef]
83. Noorduin, W.L.; van Enckevort, W.J.; Meekes, H.; Kaptein, B.; Kellogg, R.M.; Tully, J.C.; McBride, J.M.; Vlieg, E. The driving mechanism behind attrition-enhanced deracemization. *Angew. Chem. Int. Ed. Engl.* **2010**, *49*, 8435–8438. [CrossRef]
84. Plasson, R.; Kondepudi, D.K.; Asakura, K. Three-dimensional description of the spontaneous onset of homochirality on the surface of a conglomerate crystal phase. *J. Phys. Chem. B* **2006**, *110*, 8481–8487. [CrossRef]
85. El-Hachemi, Z.; Arteaga, O.; Canillas, A.; Crusats, J.; Llorens, J.; Ribo, J.M. Chirality generated by flows in pseudocyanine dye J-aggregates: Revisiting 40 years old reports. *Chirality* **2011**, *23*, 585–592. [CrossRef]
86. Noorduin, W.L.; van der Asdonk, P.; Meekes, H.; van Enckevort, W.J.; Kaptein, B.; Leeman, M.; Kellogg, R.M.; Vlieg, E. Complete chiral resolution using additive-induced crystal size bifurcation during grinding. *Angew. Chem. Int. Ed. Engl.* **2009**, *48*, 3278–3280. [CrossRef]
87. Hein, J.E.; Cao, B.H.; Viedma, C.; Kellogg, R.M.; Blackmond, D.G. Pasteur's tweezers revisited: On the mechanism of attrition-enhanced deracemization and resolution of chiral conglomerate solids. *J. Am. Chem. Soc.* **2012**, *134*, 12629–12636. [CrossRef]
88. Viedma, C.; Cintas, P. Homochirality beyond grinding: Deracemizing chiral crystals by temperature gradient under boiling. *Chem. Commun.* **2011**, *47*, 12786–12788. [CrossRef]
89. Viedma, C.; Noorduin, W.L.; Ortiz, J.E.; de Torres, T.; Cintas, P. Asymmetric amplification in amino acid sublimation involving racemic compound to conglomerate conversion. *Chem. Commun.* **2011**, *47*, 671–673. [CrossRef]
90. Viedma, C. Selective Chiral Symmetry Breaking during Crystallization: Parity Violation or Cryptochiral Environment in Control? *Cryst. Growth Des.* **2007**, *7*, 553–556. [CrossRef]
91. Steendam, R.R.E.; Harmsen, B.; Meekes, H.; Enckevort, W.J.P.v.; Kaptein, B.; Kellogg, R.M.; Raap, J.; Rutjes, F.P.J.T.; Vlieg, E. Controlling the Effect of Chiral Impurities on Viedma Ripening. *Cryst. Growth Des.* **2013**, *13*, 4776–4780. [CrossRef]
92. Zhang, Q.; Jia, L.; Wang, J.-R.; Mei, X. Absolute asymmetric synthesis of a sanguinarine derivative through crystal–solution interactions. *CrystEngComm* **2016**, *18*, 8834–8837. [CrossRef]
93. Spix, L.; Meekes, H.; Blaauw, R.H.; van Enckevort, W.J.P.; Vlieg, E. Complete Deracemization of Proteinogenic Glutamic Acid Using Viedma Ripening on a Metastable Conglomerate. *Cryst. Growth Des.* **2012**, *12*, 5796–5799. [CrossRef]
94. El-Hachemi, Z.; Crusats, J.; Ribo, J.M.; McBride, J.M.; Veintemillas-Verdaguer, S. Metastability in supersaturated solution and transition towards chirality in the crystallization of $NaClO_3$. *Angew. Chem. Int. Ed. Engl.* **2011**, *50*, 2359–2363. [CrossRef]
95. Medina, D.D.; Gedanken, A.; Mastai, Y. Chiral amplification in crystallization under ultrasound radiation. *Chemistry* **2011**, *17*, 11139–11142. [CrossRef]
96. Han, B.; Shen, F.; Su, H.; Zhang, X.; Shen, Y.; Zhang, T. Self-assembly of achiral monomer into left-handed helical polyanthracene nanofibers. *Mater. Express* **2016**, *6*, 88–92. [CrossRef]
97. Ziach, K.; Jurczak, J. Mirror symmetry breaking upon spontaneous crystallization from a dynamic combinatorial library of macrocyclic imines. *Chem. Commun.* **2015**, *51*, 4306–4309. [CrossRef]

98. Yagishita, F.; Kato, M.; Uemura, N.; Ishikawa, H.; Yoshida, Y.; Mino, T.; Kasashima, Y.; Sakamoto, M. Asymmetric Synthesis Using Chiral Crystals of Coumarin-3-carboxamides and Carbenoids. *Chem. Lett.* **2016**, *45*, 1310–1312. [CrossRef]
99. Sakamoto, M.; Kato, M.; Aida, Y.; Fujita, K.; Mino, T.; Fujita, T. Photosensitized 2 + 2 cycloaddition reaction using homochirality generated by spontaneous crystallization. *J. Am. Chem. Soc.* **2008**, *130*, 1132–1133. [CrossRef]
100. Sakamoto, M.; Yagishita, F.; Ando, M.; Sasahara, Y.; Kamataki, N.; Ohta, M.; Mino, T.; Kasashima, Y.; Fujita, T. Generation and amplification of optical activity of axially chiral *N*-(1-naphthyl)-2(1H) pyrimidinethione by crystallization. *Org. Biomol. Chem.* **2010**, *8*, 5418–5422. [CrossRef]
101. Fujiki, M. Helix Magic. Thermo-Driven Chiroptical Switching and Screw-Sense Inversion of Flexible Rod Helical Polysilylenes. *J. Am. Chem. Soc.* **2000**, *122*, 3336–3343. [CrossRef]
102. Fujiki, M. Optically Active Polysilylenes: State-of-the-Art Chiroptical Polymers. *Macromol. Rapid Commun.* **2001**, *22*, 539–563. [CrossRef]
103. Fujiki, M. Helix Generation, Amplification, Switching, and Memory of Chromophoric Polymers. In *Amplification of Chirality*; Soai, K., Ed.; Springer: Berlin/Heidelberg, Germany, 2008; pp. 119–186. [CrossRef]
104. Tabei, J.; Nomura, R.; Sanda, F.; Masuda, T. Design of Helical Poly(N-propargylamides) that Switch the Helix Sense with Thermal Stimuli. *Macromolecules* **2004**, *37*, 1175–1179. [CrossRef]
105. Koe, J.R.; Fujiki, M.; Nakashima, H.; Motonaga, M. Tempera ture-dependent helix–helix transition of an optically active poly(diarylsilylene). *Chem. Commun.* **2000**, 389–390. [CrossRef]
106. Watanabe, J.; Okamoto, S.; Satoh, K.; Sakajiri, K.; Furuya, H.; Abe, A. Reversible Helix−Helix Transition of Poly(β-phenylpropyllaspartate) Involving a Screw-Sense Inversion in the Solid State. *Macromolecules* **1996**, *29*, 7084–7088. [CrossRef]
107. Yashima, E.; Maeda, K.; Sato, O. Switching of a Macromolecular Helicity for Visual Distinction of Molecular Recognition Events. *J. Am. Chem. Soc.* **2001**, *123*, 8159–8160. [CrossRef]
108. Green, M.M.; Park, J.-W.; Sato, T.; Teramoto, A.; Lifson, S.; Selinger, R.L.B.; Selinger, J.V. The Macromolecular Route to Chiral Amplification. *Angew. Chem. Int. Ed.* **1999**, *38*, 3138–3154. [CrossRef]
109. Okamoto, Y.; Nakano, T.; Ono, E.; Hatada, K. Synthesis and Reversible Stereomutation of Optically Active Poly[(S)-diphenyl(1-methylpyrrolidin-2-yl)methyl methacrylate]. *Chemistry Letters* **1991**, *20*, 525–528. [CrossRef]
110. Gu, H.; Nakamura, Y.; Sato, T.; Teramoto, A.; Green, M.M.; Andreola, C.; Peterson, N.C.; Lifson, S. Molecular-Weight Dependence of the Optical Rotation of Poly((R)-2-deuterio-n-hexyl isocyanate). *Macromolecules* **1995**, *28*, 1016–1024. [CrossRef]
111. Okamoto, N.; Mukaida, F.; Gu, H.; Nakamura, Y.; Sato, T.; Teramoto, A.; Green, M.M.; Andreola, C.; Peterson, N.C.; Lifson, S. Molecular Weight Dependence of the Optical Rotation of Poly((R)-1-deuterio-n-hexyl isocyanate) in Dilute Solution. *Macromolecules* **1996**, *29*, 2878–2884. [CrossRef]
112. Cai, Y.; Bernasek, S.L. Adsorption-induced asymmetric assembly from an achiral adsorbate. *J. Am. Chem. Soc.* **2004**, *126*, 14234–14238. [CrossRef]
113. Humblot, V.; Lorenzo, M.O.; Baddeley, C.J.; Haq, S.; Raval, R. Local and global chirality at surfaces: Succinic acid versus tartaric acid on Cu110. *J. Am. Chem. Soc.* **2004**, *126*, 6460–6469. [CrossRef]
114. Raval, R. Chiral expression from molecular assemblies at metal surfaces: Insights from surface science techniques. *Chem. Soc. Rev.* **2009**, *38*, 707–721. [CrossRef]
115. Elemans, J.A.; De Cat, I.; Xu, H.; De Feyter, S. Two-dimensional chirality at liquid-solid interfaces. *Chem. Soc. Rev.* **2009**, *38*, 722–736. [CrossRef]
116. Ernst, K.H. Surface chemistry: Single handedness in flatland. *Nat. Chem.* **2017**, *9*, 195–196. [CrossRef]
117. Satyanarayana, T.; Abraham, S.; Kagan, H.B. Nonlinear effects in asymmetric catalysis. *Angew. Chem. Int. Ed. Engl.* **2009**, *48*, 456–494. [CrossRef]
118. Ercolani, G. Principles for designing an achiral receptor promoting asymmetric autocatalysis with amplification of chirality. *Tetrahedron Asymmetry* **2014**, *25*, 405–410. [CrossRef]
119. Steigelmann, M.; Nisar, Y.; Rominger, F.; Goldfuss, B. Homo- and Heterochiral Alkylzinc Fencholates: Linear or Nonlinear Effects in Dialkylzinc Additions to Benzaldehyde. *Chem. Eur. J.* **2002**, *8*, 5211–5218. [CrossRef]
120. Chen, Y.K.; Costa, A.M.; Walsh, P.J. Substrate Dependence of Nonlinear Effects: Mechanistic Probe and Practical Applications. *J. Am. Chem. Soc.* **2001**, *123*, 5378–5379. [CrossRef]

121. Balsells, J.; Davis, T.J.; Carroll, P.; Walsh, P.J. Insight into the Mechanism of the Asymmetric Addition of Alkyl Groups to Aldehydes Catalyzed by Titanium−BINOLate Species. *J. Am. Chem. Soc.* **2002**, *124*, 10336–10348. [CrossRef]
122. Mori, M.; Imma, H.; Nakai, T. Asymmetric Catalytic Cyanosilylation of Aldehydes Using a Chiral Binaphthol-Titanium Complex. *Tetrahedron Lett.* **1997**, *38*, 6229–6232. [CrossRef]
123. Kitamura, M.; Suga, S.; Oka, H.; Noyori, R. Quantitative Analysis of the Chiral Amplification in the Amino Alcohol-Promoted Asymmetric Alkylation of Aldehydes with Dialkylzincs. *J. Am. Chem. Soc.* **1998**, *120*, 9800–9809. [CrossRef]
124. Hua, Y.Z.; Han, X.W.; Yang, X.C.; Song, X.; Wang, M.C.; Chang, J.B. Enantioselective Friedel-Crafts alkylation of pyrrole with chalcones catalyzed by a dinuclear zinc catalyst. *J. Org. Chem.* **2014**, *79*, 11690–11699. [CrossRef]
125. Du, H.; Long, J.; Hu, J.; Li, X.; Ding, K. 3,3′-Br2-BINOL-Zn complex: A highly efficient catalyst for the enantioselective hetero-Diels-Alder reaction. *Org. Lett.* **2002**, *4*, 4349–4352. [CrossRef]
126. Yearick, K.; Wolf, C. Catalytic enantioselective addition of diethylzinc to trifluoromethyl ketones. *Org. Lett.* **2008**, *10*, 3915–3918. [CrossRef]
127. ShengJian, L.; Yaozhong, J.; Aiqiao, M. Asymmetric synthesis XVII. New chiral catalysts for the stereocontrolled addition of benzaldehyde by diethylzinc. *Tetrahedron Asymmetry* **1992**, *3*, 1467–1474. [CrossRef]
128. Soai, K.; Ookawa, A.; Kaba, T.; Ogawa, K. Catalytic asymmetric induction. Highly enantioselective addition of dialkylzincs to aldehydes using chiral pyrrolidinylmethanols and their metal salts. *J. Am. Chem. Soc.* **1987**, *109*, 7111–7115. [CrossRef]
129. Yang, X.; Shen, J.; Da, C.; Wang, R.; Choi, M.C.K.; Yang, L.; Wong, K.-Y. Chiral pyrrolidine derivatives as catalysts in the enantioselective addition of diethylzinc to aldehydes. *Tetrahedron Asymmetry* **1999**, *10*, 133–138. [CrossRef]
130. Ding, K.; Ishii, A.; Mikami, K. Super High Throughput Screening (SHTS) of Chiral Ligands and Activators: Asymmetric Activation of Chiral Diol-Zinc Catalysts by Chiral Nitrogen Activators for the Enantioselective Addition of Diethylzinc to Aldehydes. *Angew. Chem. Int. Ed.* **1999**, *38*, 497–501. [CrossRef]
131. Vyskočil, Š.; Jaracz, S.; Smrčina, M.; Štícha, M.; Hanuš, V.; Polášek, M.; Kočovský, P. Synthesis of N-Alkylated and N-Arylated Derivatives of 2-Amino-2′-hydroxy-1,1′-binaphthyl (NOBIN) and 2,2′-Diamino-1,1′-binaphthyl and Their Application in the Enantioselective Addition of Diethylzinc to Aromatic Aldehydes†. *J. Org. Chem.* **1998**, *63*, 7727–7737. [CrossRef]
132. Le Goanvic, D.; Holler, M.; Pale, P. Chiral tridentate versus bidentate pyridines as catalysts in the enantioselective alkylation of benzaldehyde with diethylzinc. *Tetrahedron Asymmetry* **2002**, *13*, 119–121. [CrossRef]
133. Rosner, T.; Sears, P.J.; Nugent, W.A.; Blackmond, D.G. Kinetic Investigations of Product Inhibition in the Amino Alcohol-Catalyzed Asymmetric Alkylation of Benzaldehyde with Diethylzinc. *Org. Lett.* **2000**, *2*, 2511–2513. [CrossRef]
134. Chen, S.-K.; Peng, D.; Zhou, H.; Wang, L.-W.; Chen, F.-X.; Feng, X.-M. Highly Enantioselective Cyanoformylation of Aldehydes Catalyzed by a Mononuclear Salen-Ti(OiPr)4 Complex Produced In Situ. *Eur. J. Org. Chem.* **2007**, *2007*, 639–644. [CrossRef]
135. Bandini, M.; Cozzi, P.G.; Umani-Ronchi, A. Enantioselective catalytic addition of allyl organometallic reagents to aldehydes promoted by [Cr(Salen)]: The hidden role played by weak Lewis acids in metallo-Salen promoted reactions. *Tetrahedron* **2001**, *57*, 835–843. [CrossRef]
136. Mikami, K.A.; Yamanaka, M. Negative nonlinear effect in aquo palladium catalysis depending on tropos biphenylphosphine ligand chirality controlled by chiral diaminobinaphthyl activator. *Pure Appl. Chem.* **2004**, *76*, 537–540. [CrossRef]
137. Feng, X.; Liu, Y.; Liu, X.; Xin, J. Asymmetric Cyanosilylation of Aldehydes Catalyzed by Novel Chiral Tetraaza-Titanium Complexes. *Synlett* **2006**, *2006*, 1085–1089. [CrossRef]
138. Gamez, P.; Fache, F.; Mangeney, P.; Lamaire, M. Enantioselective catalytic reduction of ketones using C2-symmetric diamines as chiral ligands. *Tetrahedron Lett.* **1993**, *34*, 6897–6898. [CrossRef]
139. Spogliarich, R.; Kašpar, J.; Graziani, M.; Morandini, F. Asymmetric transfer hydrogenation of ketones catalyzed by phosphine-rhodium(I) and -iridium(I) complexes. *J. Organomet. Chem.* **1986**, *306*, 407–412. [CrossRef]

140. Gamez, P.; Fache, F.; Lemaire, M. Asymmetric catalytic reduction of carbonyl compounds using C2 symmetric diamines as chiral ligands. *Tetrahedron Asymmetry* **1995**, *6*, 705–718. [CrossRef]
141. Brown, M.K.; Blewett, M.M.; Colombe, J.R.; Corey, E.J. Mechanism of the enantioselective oxidation of racemic secondary alcohols catalyzed by chiral Mn(III)-salen complexes. *J. Am. Chem. Soc.* **2010**, *132*, 11165–11170. [CrossRef]
142. Zlotin, S.G.; Kochetkov, S.V. C2-Symmetric diamines and their derivatives as promising organocatalysts for asymmetric synthesis. *Russ. Chem. Rev.* **2015**, *84*, 1077–1099. [CrossRef]
143. Kucherenko, A.S.; Siyutkin, D.E.; Nigmatov, A.G.; Chizhov, A.O.; Zlotin, S.G. Chiral Primary Amine Tagged to Ionic Group as Reusable Organocatalyst for Asymmetric Michael Reactions of C-Nucleophiles with α,β-Unsaturated Ketones. *Adv. Synth. Catal.* **2012**, *354*, 3078–3086. [CrossRef]
144. Kucherenko, A.S.; Kostenko, A.A.; Gerasimchuk, V.V.; Zlotin, S.G. Stereospecific diaza-Cope rearrangement as an efficient tool for the synthesis of DPEDA pyridine analogs and related C2-symmetric organo catalysts. *Org. Biomol. Chem.* **2017**, *15*, 7028–7033. [CrossRef]
145. Kucherenko, A.S.; Kostenko, A.A.; Zhdankina, G.M.; Kuznetsova, O.Y.; Zlotin, S.G. Green asymmetric synthesis of Warfarin and Coumachlor in pure water catalyzed by quinoline-derived 1,2-diamines. *Green Chem.* **2018**, *20*, 754–759. [CrossRef]
146. Boucherif, A.; Duan, S.-W.; Yuan, Z.-G.; Lu, L.-Q.; Xiao, W.-J. Catalytic Asymmetric Allylation of 3-Aryloxindoles by Merging Palladium Catalysis and Asymmetric H-Bonding Catalysis. *Adv. Synth. Catal.* **2016**, *358*, 2594–2598. [CrossRef]
147. Nájera, C.; Yus, M. Chiral benzimidazoles as hydrogen bonding organocatalysts. *Tetrahedron Lett.* **2015**, *56*, 2623–2633. [CrossRef]
148. Palomo, C.; Oiarbide, M.; Laso, A. Recent Advances in the Catalytic Asymmetric Nitroaldol (Henry) Reaction. *Eur. J. Org. Chem.* **2007**, *2007*, 2561–2574. [CrossRef]
149. Lisnyak, V.G.; Kucherenko, A.S.; Valeev, E.F.; Zlotin, S.G. (1,2-Diaminoethane-1,2-diyl)bis(N-methylpyridinium) Salts as a Prospective Platform for Designing Recyclable Prolinamide-Based Organocatalysts. *J. Org. Chem.* **2015**, *80*, 9570–9577. [CrossRef]
150. Kucherenko, A.S.; Lisnyak, V.G.; Kostenko, A.A.; Kochetkov, S.V.; Zlotin, S.G. C2-Symmetric pyrrolidine-derived squaramides as recyclable organocatalysts for asymmetric Michael reactions. *Org. Biomol. Chem.* **2016**, *14*, 9751–9759. [CrossRef]
151. Yang, Z.Y.; Zeng, J.L.; Ren, N.; Meng, W.; Nie, J.; Ma, J.A. C2-Symmetric Chiral Bisoxazolines as Hydrogen-Bond-Acceptor Catalysts in Enantioselective Aldol Reaction of beta-Carbonyl Acids with Trifluoroacetaldehyde Hemiacetals. *Org. Lett.* **2016**, *18*, 6364–6367. [CrossRef]
152. Servín, F.A.; Madrigal, D.; Romero, J.A.; Chávez, D.; Aguirre, G.; Anaya de Parrodi, C.; Somanathan, R. Synthesis of C_2-symmetric 1,2-diamine-functionalized organocatalysts: Mimicking enzymes in enantioselective Michael addition reactions. *Tetrahedron Lett.* **2015**, *56*, 2355–2358. [CrossRef]
153. Ogasawara, M.; Kotani, S.; Nakajima, H.; Furusho, H.; Miyasaka, M.; Shimoda, Y.; Wu, W.Y.; Sugiura, M.; Takahashi, T.; Nakajima, M. Atropisomeric chiral dienes in asymmetric catalysis: C(2)-symmetric (Z,Z)-2,3-bis[1-(diphenylphosphinyl)ethylidene]tetralin as a highly active Lewis base organocatalyst. *Angew. Chem. Int. Ed. Engl.* **2013**, *52*, 13798–13802. [CrossRef]
154. Gomez-Torres, E.; Alonso, D.A.; Gomez-Bengoa, E.; Najera, C. Conjugate addition of 1,3-dicarbonyl compounds to maleimides using a chiral C2-symmetric bis(2-aminobenzimidazole) as recyclable organocatalyst. *Org. Lett.* **2011**, *13*, 6106–6109. [CrossRef]
155. Delaney, J.P.; Brozinski, H.L.; Henderson, L.C. Synergistic effects within a C2-symmetric organocatalyst: The potential formation of a chiral catalytic pocket. *Org. Biomol. Chem.* **2013**, *11*, 2951–2960. [CrossRef] [PubMed]
156. Sohtome, Y.; Hashimoto, Y.; Nagasawa, K. Guanidine-Thiourea Bifunctional Organocatalyst for the Asymmetric Henry (Nitroaldol) Reaction. *Adv. Synth. Catal.* **2005**, *347*, 1643–1648. [CrossRef]
157. Ginotra, S.K.; Singh, V.K. Enantioselective Henry reaction catalyzed by a C_2-symmetric bis(oxazoline)Cu(OAc)$_2$. H_2O complex. *Org. Biomol. Chem.* **2007**, *5*, 3932–3937. [CrossRef] [PubMed]
158. Le, K.V.; Takezoe, H.; Araoka, F. Chiral Superstructure Mesophases of Achiral Bent-Shaped Molecules—Hierarchical Chirality Amplification and Physical Properties. *Adv. Mater.* **2017**, *29*, 1602737. [CrossRef] [PubMed]

159. Otani, T.; Araoka, F.; Ishikawa, K.; Takezoe, H. Enhanced optical activity by achiral rod-like molecules nanosegregated in the B4 structure of achiral bent-core molecules. *J. Am. Chem. Soc.* **2009**, *131*, 12368–12372. [CrossRef] [PubMed]
160. Hough, L.E.; Spannuth, M.; Nakata, M.; Coleman, D.A.; Jones, C.D.; Dantlgraber, G.; Tschierske, C.; Watanabe, J.; Korblova, E.; Walba, D.M.; et al. Chiral isotropic liquids from achiral molecules. *Science* **2009**, *325*, 452–456. [CrossRef]
161. Takanishi, Y.; Shin, G.J.; Jung, J.C.; Choi, S.-W.; Ishikawa, K.; Watanabe, J.; Takezoe, H.; Toledano, P. Observation of very large chiral domains in a liquid crystal phase formed by mixtures of achiral bent-core and rod molecules. *J. Mater. Chem.* **2005**, *15*. [CrossRef]
162. Kim, K.; Kim, H.; Jo, S.Y.; Araoka, F.; Yoon, D.K.; Choi, S.W. Photomodulated Supramolecular Chirality in Achiral Photoresponsive Rodlike Compounds Nanosegregated from the Helical Nanofilaments of Achiral Bent-Core Molecules. *ACS Appl. Mater. Interfaces* **2015**, *7*, 22686–22691. [CrossRef]
163. Nagayama, H.; Varshney, S.K.; Goto, M.; Araoka, F.; Ishikawa, K.; Prasad, V.; Takezoe, H. Spontaneous deracemization of disc-like molecules in the columnar phase. *Angew. Chem. Int. Ed. Engl.* **2010**, *49*, 445–448. [CrossRef]
164. Gortz, V.; Goodby, J.W. Enantioselective segregation in achiral nematic liquid crystals. *Chem. Commun.* **2005**, 3262–3264. [CrossRef]
165. Zhang, C.; Diorio, N.; Lavrentovich, O.D.; Jakli, A. Helical nanofilaments of bent-core liquid crystals with a second twist. *Nat. Commun.* **2014**, *5*, 3302. [CrossRef]
166. Hazen, R.M.; Sholl, D.S. Chiral selection on inorganic crystalline surfaces. *Nat. Mater.* **2003**, *2*, 367–374. [CrossRef]
167. Weissbuch, I.; Addadi, L.; Leiserowitz, L. Molecular recognition at crystal interfaces. *Science* **1991**, *253*, 637–645. [CrossRef]
168. Koretsky, C.M.; Sverjensky, D.A.; Sahai, N. A model of surface site types on oxide and silicate minerals based on crystal chemistry; implications for site types and densities, multi-site adsorption, surface infrared spectroscopy, and dissolution kinetics. *Am. J. Sci.* **1998**, *298*, 349–438. [CrossRef]
169. Dana, E.S. *A Text-Book of Mineralogy: With an Extended Treatise on Crystallography and Physical Mineralogy*; Wiley: Hoboken, NJ, USA, 1898.
170. Hazen, R.M.; Filley, T.R.; Goodfriend, G.A. Selective adsorption of L- and D-amino acids on calcite: Implications for biochemical homochirality. *Proc. Natl. Acad. Sci. USA* **2001**, *98*, 5487–5490. [CrossRef]
171. Van Cappellen, P.; Charlet, L.; Stumm, W.; Wersin, P. A surface complexation model of the carbonate mineral-aqueous solution interface. *Geochim. Cosmochim. Acta* **1993**, *57*, 3505–3518. [CrossRef]
172. Stipp, S.L.; Hochella, M.F. Structure and bonding environments at the calcite surface as observed with X-ray photoelectron spectroscopy (XPS) and low energy electron diffraction (LEED). *Geochim. Cosmochim. Acta* **1991**, *55*, 1723–1736. [CrossRef]
173. Bąbel, M. Crystal lography and genesis of the giant intergrowths of gypsum from the Miocene evaporites of Poland. *Arch. Miner.* **1990**, *44*, 103–135.
174. Cody, A.M.; Cody, R.D. Chiral habit modifications of gypsum from epitaxial-like adsorption of stereospecific growth inhibitors. *J. Cryst. Growth* **1991**, *113*, 508–519. [CrossRef]
175. Steendam, R.R.; Verkade, J.M.; van Benthem, T.J.; Meekes, H.; van Enckevort, W.J.; Raap, J.; Rutjes, F.P.; Vlieg, E. Emergence of single-molecular chirality from achiral reactants. *Nat. Commun.* **2014**, *5*, 5543. [CrossRef]
176. Dressel, C.; Liu, F.; Prehm, M.; Zeng, X.; Ungar, G.; Tschierske, C. Dynamic mirror-symmetry breaking in bicontinuous cubic phases. *Angew. Chem. Int. Ed. Engl.* **2014**, *53*, 13115–13120. [CrossRef]
177. Alaasar, M.; Poppe, S.; Dong, Q.; Liu, F.; Tschierske, C. Isothermal Chirality Switching in Liquid-Crystalline Azobenzene Compounds with Non-Polarized Light. *Angew. Chem. Int. Ed. Engl.* **2017**, *56*, 10801–10805. [CrossRef]
178. Alaasar, M.; Prehm, M.; Tschierske, C. Helical Nano-crystallite (HNC) Phases: Chirality Synchronization of Achiral Bent-Core Mesogens in a New Type of Dark Conglomerates. *Chemistry* **2016**, *22*, 6583–6597. [CrossRef]
179. Dressel, C.; Reppe, T.; Prehm, M.; Brautzsch, M.; Tschierske, C. Chiral self-sorting and amplification in isotropic liquids of achiral molecules. *Nat. Chem.* **2014**, *6*, 971–977. [CrossRef]

180. Ueda, T.; Masuko, S.; Araoka, F.; Ishikawa, K.; Takezoe, H. A General Method for the Enantioselective Formation of Helical Nanofilaments. *Angew. Chem.* **2013**, *125*, 7001–7004. [CrossRef]
181. Hu, J.; Gao, L.; Zhu, Y.; Wang, P.; Lin, Y.; Sun, Z.; Yang, S.; Wang, Q. Chiral Assemblies from an Achiral Pyridinium-Tailored Anthracene. *Chemistry* **2017**, *23*, 1422–1426. [CrossRef]
182. Akine, S.; Sairenji, S.; Taniguchi, T.; Nabeshima, T. Stepwise helicity inversions by multisequential metal exchange. *J. Am. Chem. Soc.* **2013**, *135*, 12948–12951. [CrossRef]
183. La, D.D.; Al Kobaisi, M.; Gupta, A.; Bhosale, S.V. Chiral Assembly of AIE-Active Achiral Molecules: An Odd Effect in Self-Assembly. *Chem. Eur. J.* **2017**, *23*, 3950–3956. [CrossRef]
184. Tschierske, C.; Ungar, G. Mirror Symmetry Breaking by Chirality Synchronisation in Liquids and Liquid Crystals of Achiral Molecules. *ChemPhysChem* **2016**, *17*, 9–26. [CrossRef]
185. Chen, D.; Nakata, M.; Shao, R.; Tuchband, M.R.; Shual, M.; Baumeister, U.; Weissflog, W.; Walba, D.M.; Glaser, M.A.; Maclennan, J.E.; Clark, N.A. Twist-bend heliconical chiral nematic liquid crystal phase of an achiral rigid bent-core mesogen. *Phys. Rev. E* **2014**, *89*, 022506. [CrossRef]
186. Kohler, K.; Forster, G.; Hauser, A.; Dobner, B.; Heiser, U.F.; Ziethe, F.; Richter, W.; Steiniger, F.; Drechsler, M.; Stettin, H.; et al. Temperature-dependent behavior of a symmetric long-chain bolaamphiphile with phosphocholine headgroups in water: From hydrogel to nanoparticles. *J. Am. Chem. Soc.* **2004**, *126*, 16804–16813. [CrossRef]
187. Roche, C.; Sun, H.J.; Prendergast, M.E.; Leowanawat, P.; Partridge, B.E.; Heiney, P.A.; Araoka, F.; Graf, R.; Spiess, H.W.; Zeng, X.; et al. Homochiral columns constructed by chiral self-sorting during supramolecular helical organization of hat-shaped molecules. *J. Am. Chem. Soc.* **2014**, *136*, 7169–7185. [CrossRef]
188. Singh, G.; Chan, H.; Baskin, A.; Gelman, E.; Repnin, N.; Kral, P.; Klajn, R. Self-assembly of magnetite n5anocubes into helical superstructures. *Science* **2014**, *345*, 1149–1153. [CrossRef]
189. Pokropivny, V.V. Powder. *Metal. Met. Ceram.* **2001**, *40*, 582–594. [CrossRef]
190. Celik-Aktas, A.; Zuo, J.M.; Stubbins, J.F.; Tang, C.; Bando, Y. Double-helix structure in multiwall boron nitride nanotubes. *Acta Crystallogr. A* **2005**, *61*, 533–541. [CrossRef]
191. Zhi, C.; Bando, Y.; Tang, C.; Golberg, D. Boron nitride nanotubes. *Mater. Sci. Eng. R* **2010**, *70*, 92–111. [CrossRef]
192. Wang, Z.L. Zinc oxide nanostructures: Growth, properties and applications. *J. Phys. Condens. Matter* **2004**, *16*, R829–R858. [CrossRef]
193. Ma, R.; Bando, Y.; Sato, T. Controlled Synthesis of BN Nanotubes, Nanobamboos, and Nanocables. *Adv. Mater.* **2002**, *14*, 366–368. [CrossRef]
194. Pavlov, V.A.; Spitsina, N.I.; Klabunovsky, E.I. Enantioselective hydrogenation in a cholesteric liquid crystal as a chiral matrix. *Bull. Acad. Sci. USSR Div. Chem. Sci.* **1982**, *31*, 2509. [CrossRef]
195. Pavlov, V.A.; Spitsina, N.I.; Klabunovsky, E.I. Enantioselective hydrogenation in cholesteryl tridecanoate as a chiral liquid-crystalline matrix. *Bull. Acad. Sci. USSR Div. Chem. Sci.* **1983**, *32*, 1501–1503. [CrossRef]
196. Chen, C.W.; Whitlock, H.W. Molecular tweezers: A simple model of bifunctional intercalation. *J. Am. Chem. Soc.* **1978**, *100*, 4921–4922. [CrossRef]
197. Zimmerman, S.C. *Rigid Molecular Tweezers as Hosts for the Complexation of Neutral Guests;* Springer: Berlin/Heidelberg, Germany, 1993; Volume 165, pp. 70–102.
198. Harmata, M. Chiral molecular tweezers. *Acc. Chem. Res.* **2004**, *37*, 862–873. [CrossRef] [PubMed]
199. Saha, B.; Ikbal, S.A.; Petrovic, A.G.; Berova, N.; Rath, S.P. Complexation of Chiral Zinc-Porphyrin Tweezer with Achiral Diamines: Induction and Two-Step Inversion of Interporphyrin Helicity Monitored by ECD. *Inorg. Chem.* **2017**, *56*, 3849–3860. [CrossRef] [PubMed]
200. Ouyang, Q.; Zhu, Y.Z.; Li, Y.C.; Wei, H.B.; Zheng, J.Y. Diastereoselective synthesis of chiral diporphyrins via intramolecular meso-meso oxidative coupling. *J. Org. Chem.* **2009**, *74*, 3164–3167. [CrossRef]
201. Ema, T.; Misawa, S.; Nemugaki, S.; Sakai, T.; Utaka, M. New Optically Active Diporphyrin Having a Chiral Cyclophane as a Spacer. *Chem. Lett.* **1997**, *26*, 487–488. [CrossRef]
202. Berova, N.; Pescitelli, G.; Petrovic, A.G.; Proni, G. Probing molecular chirality by CD-sensitive dimeric metalloporphyrin hosts. *Chem. Commun.* **2009**, 5958–5980. [CrossRef]
203. Borovkov, V.V.; Hembury, G.A.; Inoue, Y. Origin, control, and application of supra molecular chirogenesis in bisporphyrin-based systems. *Acc. Chem. Res.* **2004**, *37*, 449–459. [CrossRef]
204. Huang, X.; Nakanishi, K.; Berova, N. Porphyrins and metalloporphyrins: Versatile circular dichroic reporter groups for structural studies. *Chirality* **2000**, *12*, 237–255. [CrossRef]

205. D'Urso, A.; Nicotra, P.F.; Centonze, G.; Fragala, M.E.; Gattuso, G.; Notti, A.; Pappalardo, A.; Pappalardo, S.; Parisi, M.F.; Purrello, R. Induction of chirality in porphyrin-(bis)calixarene assemblies: A mixed covalent-non-covalent vs a fully non-covalent approach. *Chem. Commun.* **2012**, *48*, 4046–4048. [CrossRef]
206. Gholami, H.; Anyika, M.; Zhang, J.; Vasileiou, C.; Borhan, B. Host-Guest Assembly of a Molecular Reporter with Chiral Cyanohydrins for Assignment of Absolute Stereochemistry. *Chemistry* **2016**, *22*, 9235–9239. [CrossRef]
207. Ikbal, S.A.; Dhamija, A.; Brahma, S.; Rath, S.P. A Nonempirical Approach for Direct Determination of the Absolute Configuration of 1,2-Diols and Amino Alcohols Using Mg(II)bisporphyrin. *J. Org. Chem.* **2016**, *81*, 5440–5449. [CrossRef]
208. Ishii, Y.; Yoshizawa, T.; Kubo, Y. Dibenzodiaza-30-crown-10-appended bis(zinc porphyrin) tweezers: Synthesis and crown-assisted chiroptical behaviour. *Org. Biomol. Chem.* **2007**, *5*, 1210–1217. [CrossRef]
209. Tanasova, M.; Anyika, M.; Borhan, B. Sensing remote chirality: Stereochemical determination of beta-, gamma-, and delta-chiral carboxylic acids. *Angew. Chem. Int. Ed. Engl.* **2015**, *54*, 4274–4278. [CrossRef]
210. Beckett, A.H.; Anderson, P. A Method for the Determination of the Configuration of Organic Molecules using 'Stereo-selective Adsorbents'. *Nature* **1957**, *179*, 1074–1075. [CrossRef]
211. Bartels, H.; Prijs, B.; Erlenmeyer, H. Über spezifisch adsorbierende Silicagele V. *Helv. Chim. Acta* **1966**, *49*, 1621–1625. [CrossRef]
212. Erlenmeyer, H.; Bartels, H. Über das Problem der Ähnlichkeit in der Chemie Über spezifisch adsorbierende Silikagele II [1]. *Helv. Chim. Acta* **1964**, *47*, 1285–1288. [CrossRef]
213. Bartels, H.; Erlenmeyer, H. Über das Problem der Ähnlichkeit in der Chemie Über spezifisch adsorbierende Silicagele III. *Helv. Chim. Acta* **1965**, *48*, 285–290. [CrossRef]
214. Ducos, P.; Liautard, V.; Robert, F.; Landais, Y. Chiral Memory in Silylium Ions. *Chemistry* **2015**, *21*, 11573–11578. [CrossRef]
215. Miyabe, T.; Iida, H.; Ohnishi, A.; Yashima, E. Enantioseparation on poly(phenyl isocyanide)s with macromolecular helicity memory as chiral stationary phases for HPLC. *Chem. Sci.* **2012**, *3*, 863–867. [CrossRef]
216. Huang, H.; Deng, J.; Shi, Y. Optically Active Physical Gels with Chiral Memory Ability: Directly Prepared by Helix-Sense-Selective Polymerization. *Macromolecules* **2016**, *49*, 2948–2956. [CrossRef]
217. Zhao, Y.; Abdul Rahim, N.A.; Xia, Y.; Fujiki, M.; Song, B.; Zhang, Z.; Zhang, W.; Zhu, X. Supramolecular Chirality in Achiral Polyfluorene: Chiral Gelation, Memory of Chirality, and Chiral Sensing Property. *Macromolecules* **2016**, *49*, 3214–3221. [CrossRef]
218. Yang, D.; Zhao, Y.; Lv, K.; Wang, X.; Zhang, W.; Zhang, L.; Liu, M. A strategy for tuning achiral mainchain polymers into helical assemblies and chiral memory systems. *Soft. Matter.* **2016**, *12*, 1170–1175. [CrossRef]
219. Sobczuk, A.A.; Tsuchiya, Y.; Shiraki, T.; Tamaru, S.; Shinkai, S. Creation of chiral thixotropic gels through a crown-ammonium interaction and their application to a memory-erasing recycle system. *Chemistry* **2012**, *18*, 2832–2838. [CrossRef]
220. De Jong, J.J.; Lucas, L.N.; Kellogg, R.M.; van Esch, J.H.; Feringa, B.L. Reversible optical transcription of supramolecular chirality into molecular chirality. *Science* **2004**, *304*, 278–281. [CrossRef]
221. Inoue, K.; Ono, Y.; Kanekiyo, Y.; Ishii, T.; Yoshihara, K.; Shinkai, S. Chiroselective re-binding of saccharides to the fibrous aggregates prepared from organic gels of cholesterylphenylboronic acid. *Tetrahedron Lett.* **1998**, *39*, 2981–2984. [CrossRef]
222. Gural'skiy, I.y.A.; Reshetnikov, V.A.; Szebesczyk, A.; Gumienna-Kontecka, E.; Marynin, A.I.; Shylin, S.I.; Ksenofontov, V.; Fritsky, I.O. Chiral spin crossover nanoparticles and gels with switchable circular dichroism. *J. Mater. Chem. C* **2015**, *3*, 4737–4741. [CrossRef]
223. Duan, P.; Zhu, X.; Liu, M. Isomeric effect in the self-assembly of pyridine-containing L-glutamic lipid: Substituent position controlled morphology and supramolecular chirality. *Chem. Commun.* **2011**, *47*, 5569–5571. [CrossRef]
224. Van Gorp, J.J.; Vekemans, J.A.J.M.; Meijer, E.W. C3-Symmetrical Supramolecular Architectures: Fibers and Organic Gels from Discotic Trisamides and Trisureas. *J. Am. Chem. Soc.* **2002**, *124*, 14759–14769. [CrossRef]
225. Tobe, Y.; Utsumi, N.; Kawabata, K.; Nagano, A.; Adachi, K.; Araki, S.; Sonoda, M.; Hirose, K.; Naemura, K. m-Diethynylbenzene Macrocycles: Syntheses and Self-Association Behavior in Solution. *J. Am. Chem. Soc.* **2002**, *124*, 5350–5364. [CrossRef]

226. Lahiri, S.; Thompson, J.L.; Moore, J.S. Solvophobically Driven π-Stacking of Phenylene Ethynylene Macrocycles and Oligomers. *J. Am. Chem. Soc.* **2000**, *122*, 11315–11319. [CrossRef]
227. Xing, P.; Zhao, Y. Controlling Supramolecular Chirality in Multicomponent Self-Assembled Systems. *Acc. Chem. Res.* **2018**, *51*, 2324–2334. [CrossRef]
228. Haino, T.; Tanaka, M.; Fukazawa, Y. Self-assembly of tris(phenylisoxazolyl)benzene and its asymmetric induction of supramolecular chirality. *Chem. Commun.* **2008**, 468–470. [CrossRef]
229. Edwards, W.; Smith, D.K. Enantioselective component selection in multicomponent supramolecular gels. *J. Am. Chem. Soc.* **2014**, *136*, 1116–1124. [CrossRef]
230. Molla, M.R.; Das, A.; Ghosh, S. Chiral induction by helical neighbour: Spectroscopic visualization of macroscopic-interaction among self-sorted donor and acceptor pi-stacks. *Chem. Commun.* **2011**, *47*, 8934–8936. [CrossRef]
231. Tanaka, M.; Ikeda, T.; Mack, J.; Kobayashi, N.; Haino, T. Self-assembly and gelation behavior of tris(phenylisoxazolyl)benzenes. *J. Org. Chem.* **2011**, *76*, 5082–5091. [CrossRef]
232. Smith, D.K. Lost in translation? Chirality effects in the self-assembly of nanostructured gel-phase materials. *Chem. Soc. Rev.* **2009**, *38*, 684–694. [CrossRef]
233. Shen, Z.; Wang, T.; Liu, M. Macroscopic chirality of supramolecular gels formed from achiral tris(ethyl cinnamate) benzene-1,3,5-tricarboxamides. *Angew. Chem. Int. Ed. Engl.* **2014**, *53*, 13424–13428. [CrossRef]
234. Yu, X.; Wang, Z.; Li, Y.; Geng, L.; Ren, J.; Feng, G. Fluorescent and Electrochemical Supramolecular Coordination Polymer Hydrogels Formed from Ion-Tuned Self-Assembly of Small Bis-Terpyridine Monomer. *Inorg. Chem.* **2017**, *56*, 7512–7518. [CrossRef]
235. Jin, W.; Fukushima, T.; Niki, M.; Kosaka, A.; Ishii, N.; Aida, T. Self-assembled graphitic nanotubes with one-handed helical arrays of a chiral amphiphilic molecular graphene. *Proc. Natl. Acad. Sci. USA* **2005**, *102*, 10801–10806. [CrossRef]
236. Onouchi, H.; Miyagawa, T.; Morino, K.; Yashima, E. Assisted formation of chiral porphyrin homoaggregates by an induced helical poly(phenylacetylene) template and their chiral memory. *Angew. Chem. Int. Ed. Engl.* **2006**, *45*, 2381–2384. [CrossRef]
237. Mammana, A.; D'Urso, A.; Lauceri, R.; Purrello, R. Switching off and on the supramolecular chiral memory in porphyrin assemblies. *J. Am. Chem. Soc.* **2007**, *129*, 8062–8063. [CrossRef]
238. Randazzo, R.; Mammana, A.; D'Urso, A.; Lauceri, R.; Purrello, R. Reversible "chiral memory" in ruthenium tris(phenanthroline)-anionic porphyrin complexes. *Angew. Chem. Int. Ed. Engl.* **2008**, *47*, 9879–9882. [CrossRef]
239. Gaeta, M.; Oliveri, I.P.; Fragala, M.E.; Failla, S.; D'Urso, A.; Di Bella, S.; Purrello, R. Chirality of self-assembled achiral porphyrins induced by chiral Zn(ii) Schiff-base complexes and maintained after spontaneous dissociation of the templates: A new case of chiral memory. *Chem. Commun.* **2016**, *52*, 8518–8521. [CrossRef]
240. Lauceri, R.; Raudino, A.; Scolaro, L.M.; Micali, N.; Purrello, R. From Achiral Porphyrins to Template-Imprinted Chiral Aggregates and Further. Self-Replication of Chiral Memory from Scratch. *J. Am. Chem. Soc.* **2002**, *124*, 894–895. [CrossRef]
241. Yashima, E.; Maeda, K.; Okamoto, Y. Memory of macromolecular helicity assisted by interaction with achiral small molecules. *Nature* **1999**, *399*, 449. [CrossRef]
242. Hase, Y.; Mitsutsuji, Y.; Ishikawa, M.; Maeda, K.; Okoshi, K.; Yashima, E. Unexpected thermally stable, cholesteric liquid-crystalline helical polyisocyanides with memory of macromolecular helicity. *Chem. Asian J.* **2007**, *2*, 755–763. [CrossRef]
243. Ishikawa, M.; Maeda, K.; Mitsutsuji, Y.; Yashima, E. An unprecedented memory of macromolecular helicity induced in an achiral polyisocyanide in water. *J. Am. Chem. Soc.* **2004**, *126*, 732–733. [CrossRef]
244. Takashima, S.; Abe, H.; Inouye, M. Unexpected chain length dependence on a chiral memory effect of 'meta-ethynylpyridine' oligomers. *Tetrahedron Asymmetry* **2013**, *24*, 527–531. [CrossRef]
245. Ishidate, R.; Shimomura, K.; Ikai, T.; Kanoh, S.; Maeda, K. Macromolecular helicity induction and memory in a poly(biphenylylacetylene) bearing an ester group and its application to a chiral stationary phase for high-performance liquid chromatography. *Chem. Lett.* **2015**, *44*, 946–948. [CrossRef]
246. Rosaria, L.; D'Urso, A.; Mammana, A.; Purrello, R. Chiral memory: Induction, amplification, and switching in porphyrin assemblies. *Chirality* **2008**, *20*, 411–419. [CrossRef]
247. Sung, B.; de la Cotte, A.; Grelet, E. Chirality-controlled crystallization via screw dislocations. *Nat. Commun.* **2018**, *9*, 1405. [CrossRef]

248. Thomas, C.W.; Tor, Y. Dendrimers and chirality. *Chirality* **1998**, *10*, 53–59. [CrossRef]
249. Vestergren, M.; Johansson, A.; Lennartson, A.; Håkansson, M. Non-stochastic homochiral helix crystallization: Cryptochirality in control? *Mendeleev Commun.* **2004**, *14*, 258–260. [CrossRef]
250. Wallach, O. Zur Kenntniss der Terpene und der ätherischen Oele. *Justus Liebig's Ann. Chem.* **1895**, *286*, 90–118. [CrossRef]
251. Dunitz, J.D.; Gavezzotti, A. Proteogenic amino acids: Chiral and racemic crystal packings and stabilities. *J. Phys. Chem. B* **2012**, *116*, 6740–6750. [CrossRef]
252. Brock, C.P.; Schweizer, W.B.; Dunitz, J.D. On the validity of Wallach's rule: On the density and stability of racemic crystals compared with their chiral counterparts. *J. Am. Chem. Soc.* **1991**, *113*, 9811–9820. [CrossRef]
253. Slepukhin, P.A.; Gruzdev, D.A.; Chulakov, E.N.; Levit, G.L.; Krasnov, V.P.; Charushin, V.N. Structures of the racemate and (S)-enantiomer of 7,8-difluoro-3-methyl-2,3-dihydro-4H-[1,4]benzoxazine. *Russ. Chem. Bull.* **2011**, *60*, 955–960. [CrossRef]
254. Navare, P.S.; MacDonald, J.C. Investigation of Stability and Structure in Three Homochiral and Heterochiral Crystalline Forms of 3-Phenyllactic Acid. *Cryst. Growth Des.* **2011**, *11*, 2422–2428. [CrossRef]
255. Sørensen, H.O.; Larsen, S. Hydrogen bonding in enantiomericversusracemic mono-carboxylic acids; a case study of 2-phenoxypropionic acid. *Acta Crystallogr. B Struct. Sci.* **2003**, *59*, 132–140. [CrossRef]
256. Assaad, T.; Rukiah, M. Powder X-ray study of racemic (2RS,3RS)-5-amino-3-[4-(3-methoxyphenyl)piperazin-1-yl]-1,2,3,4-tetrahydronaphtha len-2-ol. *Acta Crystallogr. C* **2011**, *67*, o469–o472. [CrossRef]
257. Husin, H.; Leong, Y.-K.; Liu, J. Molecular attributes of an effective steric agent: Yield stress of dispersions in the presence of pure enantiomeric and racemate malic acids. *Adv. Powder Technol.* **2012**, *23*, 459–464. [CrossRef]
258. Sanabria, C.M.; Gomez, S.L.; Palma, A.; Cobo, J.; Glidewell, C. Four 1-naphthyl-substituted tetrahydro-1,4-epoxy-1-benzazepines: hydrogen bonded structures in one, two and three dimensions. *Acta Crystallogr. C* **2010**, *66*, o540–o546. [CrossRef]
259. Marthi, K.; Larsen, S.; Ács, M.; Fogassy, E. Enantiomer associations in the crystal structures of racemic and (2S,3S)-(+)-3-hydroxy-2-(4-methoxyphenyl)-2,3-dihydro-1,5-benzothiazepin-4(5H)-one. *J. Mol. Struct.* **1996**, *374*, 347–355. [CrossRef]
260. Luger, P.; Weber, M. DL-Cysteine at 298K. *Acta Crystallogr. C Cryst. Struct. Commun.* **1999**, *55*, 1882–1885. [CrossRef]
261. Krishnaswamy, S.; Patil, M.T.; Shashidhar, M.S. Comparison of racemic epi-inosose and (-)-epi-inosose. *Acta Crystallogr. C* **2011**, *67*, o435–o438. [CrossRef]
262. Kitoh, S.-I.; Kunimoto, K.-K.; Funaki, N.; Senda, H.; Kuwae, A.; Hanai, K.J. Crystal structures and vibrational spectra of racemic and chiral 4-phenyl-1,3-oxazolidine-2-thione. *Chem. Crystallogr.* **2002**, *32*, 547–553. [CrossRef]
263. Pella, E.; Restelli, R. Binary phase diagram of the enantiomers of indoprofen. *Mikrochim. Acta* **1983**, *79*, 65–74. [CrossRef]
264. Xie, S.; Nusbaum, D.A.; Stein, H.J.; Pink, M. 4-(3-Methoxy-phen-yl)-2,6-dimethyl-cyclo-hex3-enecarboxylic acid. *Acta Crystallogr. E Struct. Rep. Online* **2010**, *66*, o1443–o1449. [CrossRef]
265. Blazis, V.J.; Koeller, K.J.; Rath, N.P.; Spilling, C.D. Application of Wallach's Rule in a Comparison of the X-ray Crystal Structures of the Racemate and the (S) Enantiomer of (1-Hydroxy3-phenyl-2-propenyl) Dimethylphosphonate. *Acta Crystallogr. B Struct. Sci.* **1997**, *53*, 838–842. [CrossRef]
266. Simonov, Y.; Bourosh, P.; Kravtsov, V.; Gdanets, M.; Semenishyna, K.; Pavlovsky, V.; Kabanova, T.; Khalimova, O.; Andronati, S. A comparative analysis of the crystal structure of r,s-racemate and r-enantiomer of 7-bromo-3-(2-methoxy)ethoxy-5-phenyl-1,2-dihydro- h-1,4-benzodiazepine-2-one exhibiting a high analgesic activity. *Zhurnal Organichnoi ta Farmatsevtichnoi Khimi* **2011**, *9*, 70–73.
267. Patrick, B.O.; Brock, C.P. S,S-1,2-Dicyclohexylethane-1,2-diol and its racemic compound: A striking exception to Wallach's rule. *Acta Crystallogr. B* **2006**, *62*, 488–497. [CrossRef]
268. Benson, N.; Snelder, N.; Ploeger, B.; Napier, C.; Sale, H.; Birdsall, N.J.; Butt, R.P.; van der Graaf, P.H. Estimation of binding rate constants using a simultaneous mixed-effects method: Application to monoamine transporter reuptake inhibitor reboxetine. *Br. J. Pharmacol.* **2010**, *160*, 389–398. [CrossRef]
269. Xu, F.; Khan, I.J.; McGuinness, K.; Parmar, A.S.; Silva, T.; Murthy, N.S.; Nanda, V. Self-assembly of left- and right-handed molecular screws. *J. Am. Chem. Soc.* **2013**, *135*, 18762–18765. [CrossRef]

270. Kwon, S.; Shin, H.S.; Gong, J.; Eom, J.H.; Jeon, A.; Yoo, S.H.; Chung, I.S.; Cho, S.J.; Lee, H.S. Self-assembled peptide architecture with a tooth shape: Folding into shape. *J. Am. Chem. Soc.* **2011**, *133*, 17618–17621. [CrossRef]
271. Chen, C.C.; Hsu, W.; Hwang, K.C.; Hwu, J.R.; Lin, C.C.; Horng, J.C. Contributions of cation-pi interactions to the collagen triple helix stability. *Arch. Biochem. Biophys.* **2011**, *508*, 46–53. [CrossRef]
272. Brown, E.M. Development and utilization of a bovine type I collagen microfibril model. *Int. J. Biol. Macromol.* **2013**, *53*, 20–25. [CrossRef]
273. Miyoshi, K.; Uezu, K.; Sakurai, K.; Shinkai, S. Proposal of a new hydrogen-bonding form to maintain curdlan triple helix. *Chem. Biodivers.* **2004**, *1*, 916–924. [CrossRef]
274. Ishikawa, T.; Morita, T.; Kimura, S. Unique Helical Triangle Molecular Geometry Induced by Dipole–Dipole Interactions. *Bull. Chem. Soc. Jpn.* **2007**, *80*, 1483–1491. [CrossRef]
275. Kony, D.B.; Damm, W.; Stoll, S.; van Gunsteren, W.F.; Hunenberger, P.H. Explicit-solvent molecular dynamics simulations of the polysaccharide schizophyllan in water. *Biophys. J.* **2007**, *93*, 442–455. [CrossRef]
276. Wells, R.D. Non-B DNA conformations, mutagenesis and disease. *Trends Biochem. Sci.* **2007**, *32*, 271–278. [CrossRef] [PubMed]
277. Wells, R.D.; Dere, R.; Hebert, M.L.; Napierala, M.; Son, L.S. Advances in mechanisms of genetic instability related to hereditary neurological diseases. *Nucleic Acids. Res.* **2005**, *33*, 3785–3798. [CrossRef]
278. Zhang, H.; Yu, H.; Ren, J.; Qu, X. Reversible B/Z-DNA transition under the low salt condition and non-B-form polydApolydT selectivity by a cubane-like europium-L-aspartic acid complex. *Biophys. J.* **2006**, *90*, 3203–3207. [CrossRef]
279. Mirkin, S.M. DNA structures, repeat expansions and human hereditary disorders. *Curr. Opin. Struct. Biol.* **2006**, *16*, 351–358. [CrossRef]
280. Bacolla, A.; Wells, R.D. Non-B DNA conformations, genomic rearrangements, and human disease. *J. Biol. Chem.* **2004**, *279*, 47411–47414. [CrossRef]
281. Lupski, J.R. Genomic disorders: Structural features of the genome can lead to DNA rearrangements and human disease traits. *Trends Genet.* **1998**, *14*, 417–422. [CrossRef]
282. Lupski, J.R.; Stankiewicz, P. *Genomic Disorders: The Genomic BAsis of Desease*; Humana Press: New York, NY, USA, 2006.
283. *Genetic Instabilities and Neurological Diseases*; Academic Press: Houston, TX, USA, 2006.
284. Harvey, S.C. DNA structural dynamics: Longitudinal breathing as a possible mechanism for the B $\rightleftarrows$ Z transition. *Nucleic Acids Res.* **1983**, *11*, 4867–4878. [CrossRef]
285. Green, M.M.; Peterson, N.C.; Sato, T.; Teramoto, A.; Cook, R.; Lifson, S. A helical polymer with a cooperative response to chiral information. *Science* **1995**, *268*, 1860–1866. [CrossRef] [PubMed]
286. Lifson, S.; Felder, C.E.; Green, M.M. Helical conformations, internal motion, and helix sense reversal in polyisocyanates and the preferred helix sense of an optically active polyisocyanate. *Macromolecules* **1992**, *25*, 4142–4148. [CrossRef]
287. Lee, J.; Kim, Y.G.; Kim, K.K.; Seok, C. Transition between B-DNA and Z-DNA: Free energy landscape for the B-Z junction propagation. *J. Phys. Chem. B* **2010**, *114*, 9872–9881. [CrossRef]
288. Pohl, F.M.; Jovin, T.M. Salt-induced co-operative conformational change of a synthetic DNA: Equilibrium and kinetic studies with poly(dG-dC). *J. Mol. Biol.* **1972**, *67*, 375–396. [CrossRef]
289. Zacharias, W.; Martin, J.C.; Wells, R.D. A condensed form of (dG-dC)n.cntdot.(dG-dC)n as an intermediate between the B- and Z-conformations induced by sodium acetate. *Biochemistry* **2002**, *22*, 2398–2405. [CrossRef]
290. Wang, A.J.; Quigley, G.J.; Kolpak, F.J.; van der Marel, G.; van Boom, J.H.; Rich, A. Left-handed double helical DNA: Variations in the backbone conformation. *Science* **1981**, *211*, 171–176. [CrossRef] [PubMed]
291. Wang, A.H.; Quigley, G.J.; Kolpak, F.J.; Crawford, J.L.; van Boom, J.H.; van der Marel, G.; Rich, A. Molecular structure of a left-handed double helical DNA fragment at atomic resolution. *Nature* **1979**, *282*, 680–686. [CrossRef] [PubMed]
292. Behe, M.; Felsenfeld, G. Effects of methylation on a synthetic polynucleotide: The B–Z transition in poly(dG-m5dC).poly(dG-m5dC). *Proc. Natl. Acad. Sci. USA* **1981**, *78*, 1619–1623. [CrossRef]
293. Russell, W.C.; Precious, B.; Martin, S.R.; Bayley, P.M. Differential promotion and suppression of Z leads to B transitions in poly[d(G-C)] by histone subclasses, polyamino acids and polyamines. *EMBO J.* **1983**, *2*, 1647–1653. [CrossRef] [PubMed]
294. Pohl, F.M. Polymorphism of a synthetic DNA in solution. *Nature* **1976**, *260*, 365–366. [CrossRef]

295. Feigon, J.; Wang, A.H.; van der Marel, G.A.; van Boom, J.H.; Rich, A. A one- and two-dimensional NMR study of the B to Z transition of (m5dC-dG)3 in methanolic solution. *Nucleic Acids Res.* **1984**, *12*, 1243–1263. [CrossRef]
296. Zimmer, C.; Tymen, S.; Marck, C.; Guschlbauer, W. Conformational transitions of poly(dA-dC). poly(dG-dT) induced by high salt or in ethanolic solution. *Nucleic Acids Res.* **1982**, *10*, 1081–1091. [CrossRef]
297. Van de Sande, J.H.; McIntosh, L.P.; Jovin, T.M. Mn^{2+} and other transition metals at low concentration induce the right-to-left helical transformation of poly[d(G-C)]. *EMBO J.* **1982**, *1*, 777–782. [CrossRef]
298. Pohl, F.M.; Jovin, T.M.; Baehr, W.; Holbrook, J.J. Ethidium Bromide as a Cooperative Effector of a DNA Structure. *Proc. Natl. Acad. Sci. USA* **1972**, *69*, 3805–3809. [CrossRef] [PubMed]
299. Mirau, P.A.; Kearns, D.R. The effect of Intercalating drugs on the kinetics of the B to Z transition of poly(dG–dC). *Nucleic Acids Res.* **1983**, *11*, 1931–1941. [CrossRef] [PubMed]
300. Zacharias, W.; Larson, J.E.; Klysik, J.; Stirdivant, S.M.; Wells, R.D. Conditions which cause the right-handed to left-handed DNA conformational transitions. Evidence for several types of left-handed DNA structures in solution. *J. Biol. Chem.* **1982**, *257*, 2775–2782.
301. Wu, Z.; Tian, T.; Yu, J.; Weng, X.; Liu, Y.; Zhou, X. Formation of sequence-independent Z-DNA induced by a ruthenium complex at low salt concentrations. *Angew. Chem. Int. Ed. Engl.* **2011**, *50*, 11962–11967. [CrossRef]
302. Johnson, A.; Qu, Y.; Van Houten, B.; Farrell, N. B↑ Z DNA conformational changes induced by a family of dinuclear bis(platinum) complexes. *Nucleic Acids Res.* **1992**, *20*, 1697–1703. [CrossRef] [PubMed]
303. Nordén, B.; Lincoln, P.; Akerman, B.; Tuite, E. DNA interactions with substitution-inert transition metal ion complexes. *Met. Ions. Biol. Syst.* **1996**, *33*, 177–252. [PubMed]
304. Nordén, B.; Tjerneld, F. Binding of inert metal complexes to deoxyribonucleic acid detected by linear dichroism. *FEBS Lett.* **1976**, *67*, 368–370. [CrossRef]
305. Barton, J.K. Tris (phenanthroline) metal complexes: Probes for DNA helicity. *J. Biomol. Struct. Dyn.* **1983**, *1*, 621–632. [CrossRef]
306. Chow, C.S.; Barton, J.K. Transition metal complexes as probes of nucleic acids. *Methods Enzymol.* **1992**, *212*, 219–242.
307. Xu, Y.; Zhang, Y.X.; Sugiyama, H.; Umano, T.; Osuga, H.; Tanaka, K. (P)-helicene displays chiral selection in binding to Z-DNA. *J. Am. Chem. Soc.* **2004**, *126*, 6566–6567. [CrossRef]
308. Tsuji, G.; Kawakami, K.; Sasaki, S. Enantioselective binding of chiral 1,14-dimethyl[5]helicene-spermine ligands with B- and Z-DNA. *Bioorg. Med. Chem.* **2013**, *21*, 6063–6068. [CrossRef]
309. Qu, X.; Trent, J.O.; Fokt, I.; Priebe, W.; Chaires, J.B. Allosteric, chiral-selective drug binding to DNA. *Proc. Natl. Acad. Sci. USA* **2000**, *97*, 12032–12037. [CrossRef]
310. Doi, I.; Tsuji, G.; Kawakami, K.; Nakagawa, O.; Taniguchi, Y.; Sasaki, S. The spermine-bisaryl conjugate as a potent inducer of B- to Z-DNA transition. *Chemistry* **2010**, *16*, 11993–11999. [CrossRef]
311. Haque, L.; Pradhan, A.B.; Bhuiya, S.; Das, S. Structural conversion of left handed protonated form of calf thymus DNA to right handed B-DNA by the alkaloid chelerythrine. *J. Lumin.* **2016**, *173*, 44–51. [CrossRef]
312. Kim, J.; Ha, C.S.; Lee, H.J.; Song, K. Repetitive exposure to a 60-Hz time-varying magnetic field induces DNA double-strand breaks and apoptosis in human cells. *Biochem. Biophys. Res. Commun.* **2010**, *400*, 739–744. [CrossRef]
313. Lopez-Diaz, B.; Mercado-Saenz, S.; Martinez-Morillo, M.; Sendra-Portero, F.; Ruiz-Gomez, M.J. Long-term exposure to a pulsed magnetic field (1.5 mT, 25 Hz) increases genomic DNA spontaneous degradation. *Electromagn. Biol. Med.* **2014**, *33*, 228–235. [CrossRef]
314. Ruiz-Gomez, M.J.; Martinez-Morillo, M. Enhancement of the cell-killing effect of ultraviolet-C radiation by short-term exposure to a pulsed magnetic field. *Int. J. Radiat. Biol.* **2005**, *81*, 483–490. [CrossRef]
315. Ivancsits, S.; Diem, E.; Pilger, A.; Rüdiger, H.W.; Jahn, O. Induction of DNA strand breaks by intermittent exposure to extremely-low-frequency electromagnetic fields in human diploid fibroblasts. *Mutat. Res. Genet. Toxicol. Environ. Mutagen.* **2002**, *519*, 1–13. [CrossRef]
316. Wolf, F.I.; Torsello, A.; Tedesco, B.; Fasanella, S.; Boninsegna, A.; D'Ascenzo, M.; Grassi, C.; Azzena, G.B.; Cittadini, A. 50-Hz extremely low frequency electromagnetic fields enhance cell proliferation and DNA damage: Possible involvement of a redox mechanism. *Biochim. Biophys. Acta.* **2005**, *1743*, 120–129. [CrossRef]
317. Scarfí, M.R.; Sannino, A.; Perrotta, A.; Sarti, M.; Mesirca, P.; Bersani, F. Evaluation of Genotoxic Effects in Human Fibroblasts after Intermittent Exposure to 50 Hz Electromagnetic Fields: A Confirmatory Study. *Radiat. Res.* **2005**, *164*, 270–276. [CrossRef]

318. Igarashi, A.; Kobayashi, K.; Matsuki, H.; Endo, G.; Haga, A. Evaluation of damage in DNA molecules resulting from very-low-frequency magnetic fields by using bacterial mutation repairing genetic system. *IEEE Trans. Magn.* **2005**, *41*, 4368–4370. [CrossRef]
319. Lai, H.; Singh, N.P. Magnetic-field-induced DNA strand breaks in brain cells of the rat. *Environ. Health Perspect.* **2004**, *112*, 687–694. [CrossRef]
320. Zheng, Y.; Sanche, L. Influence of organic ions on DNA damage induced by 1 eV to 60 keV electrons. *J. Chem. Phys.* **2010**, *133*, 155102. [CrossRef]
321. Zmyslony, M.; Palus, J.; Dziubaltowska, E.; Politanski, P.; Mamrot, P.; Rajkowska, E.; Kamedula, M. Effects of in vitro exposure to power frequency magnetic fields on UV-induced DNA damage of rat lymphocytes. *Bioelectromagnetics* **2004**, *25*, 560–562. [CrossRef]
322. Williams, P.A.; Ingebretsen, R.J.; Dawson, R.J. 14.6 mT ELF magnetic field exposure yields no DNA breaks in model system Salmonella, but provides evidence of heat stress protection. *Bioelectromagnetics* **2006**, *27*, 445–450. [CrossRef]
323. McNamee, J.P.; Bellier, P.V.; Chauhan, V.; Gajda, G.B.; Lemay, E.; Thansandote, A. Evaluating DNA Damage in Rodent Brain after Acute 60 Hz Magnetic-Field Exposure. *Radiat. Res.* **2005**, *164*, 791–797. [CrossRef]
324. Li, S.H.; Chow, K.C. Magnetic field exposure induces DNA degradation. *Biochem. Biophys. Res. Commun.* **2001**, *280*, 1385–1388. [CrossRef]
325. Ruiz-Gómez, M.J.; Martínez-Morillo, M. Electromagnetic fields and the induction of DNA strand breaks. *Electromagn. Biol. Med.* **2009**, *28*, 201–214. [CrossRef]
326. Aydin, M.; Taspinar, M.S.; Cakmak, Z.E.; Dumlupinar, R.; Agar, G. Static magnetic field induced epigenetic changes in wheat callus. *Bioelectromagnetics* **2016**, *37*, 504–511. [CrossRef]
327. Potenza, L.; Cucchiarini, L.; Piatti, E.; Angelini, U.; Dacha, M. Effects of high static magnetic field exposure on different DNAs. *Bioelectromagnetics* **2004**, *25*, 352–355. [CrossRef]
328. Zhang, Q.M.; Tokiwa, M.; Doi, T.; Nakahara, T.; Chang, P.W.; Nakamura, N.; Hori, M.; Miyakoshi, J.; Yonei, S. Strong static magnetic field and the induction of mutations through elevated production of reactive oxygen species in Escherichia coli soxR. *Int. J. Radiat. Biol.* **2003**, *79*, 281–286. [CrossRef]
329. Boulikas, T. Evolutionary consequences of nonrandom damage and repair of chromatin domains. *J. Mol. Evol.* **1992**, *35*, 156–180. [CrossRef]
330. Villarini, M.; Moretti, M.; Scassellati-Sforzolini, G.; Boccioli, B.; Pasquini, R. Effects of co-exposure to extremely low frequency (50 Hz) magnetic fields and xenobiotics determined in vitro by the alkaline comet assay. *Sci. Total Environ.* **2006**, *361*, 208–219. [CrossRef]

Review

Possible Roles of Amphiphilic Molecules in the Origin of Biological Homochirality

Nozomu Suzuki [1,2,*] and Yutaka Itabashi [3,4,*]

1 Department of Chemistry, College of Science, Rikkyo University, Toshima-ku, Tokyo 171-8501, Japan
2 Department of Molecular Design and Engineering, Graduate School of Engineering, Nagoya University, Chikusa-ku, Nagoya 464-8603, Japan
3 Faculty of Fisheries Sciences, Hokkaido University, Hakodate 041-8611, Japan
4 Japan Association for Inspection and Investigation of Foods Including Fats and Oils, 3-27-8, Nihonbashi-Hamacho, Chuo-ku, Tokyo 103-0007, Japan
* Correspondence: nsuzuki@chembio.nagoya-u.ac.jp (N.S.); yutaka@fish.hokudai.ac.jp (Y.I.)

Received: 12 July 2019; Accepted: 25 July 2019; Published: 1 August 2019

Abstract: A review. The question of homochirality is an intriguing problem in the field of chemistry, and is deeply related to the origin of life. Though amphiphiles and their supramolecular assembly have attracted less attention compared to biomacromolecules such as RNA and proteins, the lipid world hypothesis sheds new light on the origin of life. This review describes how amphiphilic molecules are possibly involved in the scenario of homochirality. Some prebiotic conditions relevant to amphiphilic molecules will also be described. It could be said that the chiral properties of amphiphilic molecules have various interesting features such as compositional information, spontaneous formation, the ability to exchange components, fission and fusion, adsorption, and permeation. This review aims to clarify the roles of amphiphiles regarding homochirality, and to determine what kinds of physical properties of amphiphilic molecules could have played a role in the scenario of homochirality.

Keywords: lipid; supramolecular assembly; symmetry breaking; homochirality

1. Introduction

The question of homochirality arises from the simple fact that living organisms are composed of chiral molecules. Functional biomolecules, such as nucleic acids, proteins, and lipids, are made of enantiomerically-pure chiral building blocks (e.g., sugars, amino acids, and glycerolphosphates). Typical questions regarding homochirality are "How did life choose chirality?", "Why does the enantiopurity of chiral molecules need to be high in living organisms?", "How and when did the homochirality occur?". One specific question about lipids is "Why is the chiral moiety of archaea (*sn*-glycerol-1-phosphate) the mirror image of that of bacteria and eukaryotes (*sn*-glycerol-3-phosphate) [1–3]?". One reason that makes it difficult to answer these questions is that we do not know how life appeared, nor when it emerged. Some researchers assume that life emerged when functional biopolymers, such as RNA or proteins, emerged from a soup of their monomers and oligomers, represented by the RNA world [4–6], protenoid microsphere [7,8], and the protein world [9]. Others postulate that it was not these polymers, but the formation of a supramolecular assembly that came first (e.g., micelle, vesicles, coacervate, oil droplets, etc.), represented by a lipid world hypothesis [10–12]. The latter view is held by relatively few scientists, but it is gaining attention and could provide new insight into the problem of homochirality.

To understand how amphiphilic molecules may have affected homochirality, approaches from a wide range of sciences are required. The importance of amphiphilic molecules, from the view point of the origin of life, has been largely recognized by researchers in system chemistry [13]. This field provides a bird's eye view of the origin of life, including the roles of lipids, as well as RNA and proteins.

Computational methods in this field do not require chemicals and can explore the key features of a supramolecular assembly, including compositional information, reproduction, and the evolution of life [14]. Supramolecular chemistry is key in understanding the roles of cell membranes in a prebiotic system [15], and also provides profound insight into homochirality [16]. The field may be able to answer the basic questions of homochirality: "How are chiral properties connected with the various features of a supramolecular assembly, such as critical aggregate concentration, chemical reaction, and the permeation of chemicals through the boundary of the assembly, etc.?". The amplification of the chirality is an important phenomena, and a mechanistic study could provide a general framework with which to understand the origin of homochirality [17]. Organic chemistry provides knowledge of the synthetic routes of prebiotic chemicals taking prebiotic reaction conditions into account. In addition to inorganic catalysts, the growing field of organocatalysis may drastically change the view of stereoselective chemical reactions under prebiotic conditions [18]. Analytical chemistry provides basic information about lipids under prebiotic conditions by establishing analytical methods to separate and identify lipids. One surprising fact is that vesicles can be formed from lipids extracted from meteorites [19], and identifying the chemical structures of the components by analytical methods may uncover possible prebiotic molecules, since meteorites are recognized as carbon sources on earth.

This review focuses on amphiphilic molecules in the context of homochirality and aims to (1) show how could amphiphiles be emerged prior to biopolymers, (2) review prebiotic conditions and candidates of prebiotic molecules, and (3) clarify the connection between the features of amphiphilic molecules and homochirality. The main interest lies in (3), but before going into depth, it is worth considering (1), and understand why amphiphilic molecules, specifically lipids, are attracting attention from scientists who are studying the origin of life. By knowing what kinds of prebiotic conditions and prebiotic molecules are probable with the aim of (2), researchers will be able to construct plausible experimental systems or devise theories to solve the problems of homochirality.

2. Amphiphilic Molecules and the Origin of Life

In the following sections, the authors attempt to briefly summarize the relevance of amphiphilic molecules in the context of the origin of life. The emphasis is placed on amphiphiles rather than metabolism in this review (see ref. [13] for a broader view encompassing the metabolic system).

2.1. *Theories of Origin of Life*

It could be said that the importance of amphiphiles was gradually admitted after Dyson [20–22] revisited Oparin's droplet theory (coacervate theory), though his emphasis was on a metabolic system rather than the compartment of the system. In his lecture in 1984 [22], Dyson categorizes theories about the origin of life into three groups, i.e., droplet theories by Oparin [23], genetic theories by Eigen [24], and clay theories by Cairns-Smith [25]. The criteria to classify the theories are in the order of events (Figure 1). Cells are assumed to form a boundary to confine the content (enzymes and genes), enzymes to form a self-sustaining metabolic cycle, and genes to form biopolymers containing genetic information (e.g., RNA and DNA). The droplet theories assume that the formation of the cell comes first, enzymes come second, and genes last. Genetic theories are represented by the RNA world, which arose from the striking fact that RNA can replicate itself if the oligomer is provided for ligation. The theories place genes first, enzymes second, and cells last. The last theory assumes that inorganic crystals contributed to the emergence of life before nucleic acids.

Based on the droplet theory, Dyson describes how the first living organism proliferated as follows [22]: "I propose that the original living creatures were cells with a metabolic apparatus directed by protein enzymes but with no genetic apparatus. Such cells would lack the capacity for exact replication but could grow and divide and reproduce themselves in an approximate statistical fashion." He carefully distinguished replication and reproduction in the text. The former indicates that it is only possible for a molecule to construct an exact copy of itself, and the latter that cells are able to divide and inherit approximately the same composition (see review [26] for more details about replication and reproduction). The mechanism of replication requires a very precise copy of a molecule; otherwise, that molecule loses the ability to copy itself.

Interestingly, he also suggested that RNA emerges as a parasite of the prebiotic system that utilizes adenosine triphosphate (ATP) as an energy source. ATP has high energy phosphate bonds, and ATP is used as an energy carrier by the present living cells. The prebiotic system possibly discovered a way to synthesize ATP and other nucleoside triphosphates. RNA could have emerged as a parasite of the prebiotic system, since RNA is synthesized from nucleoside triphosphates (ATP, GTP, CTP, and UTP). This idea was inspired by Lynn Margulis, who proposed that mitochondria and the photosynthetic plastids in eukaryotic cells were symbiotically acquired [27]. Her idea is that the evolution of cellular complexity was often caused by parasitism and symbiosis.

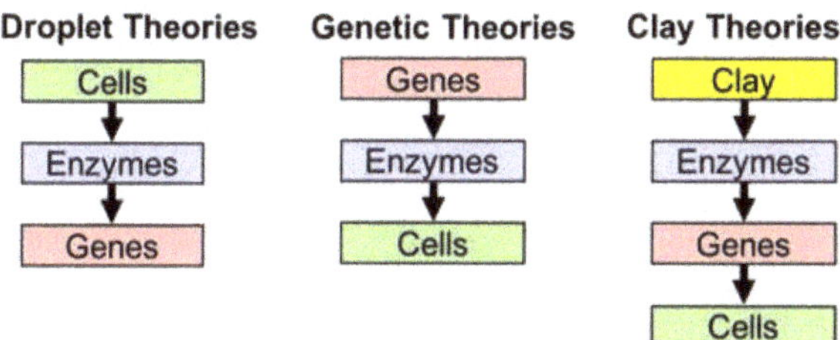

Figure 1. The three theories described by Dyson [21].

Dyson raises three other notions in support of droplet theories. The first is based on von Neumann's idea of hardware (process information) and software (embodies information) [22] (p. 7) [28]. They have an exact analogue in living cells, mainly proteins (component of metabolism) and nucleic acids (component of replication), respectively. Hardware comes prior to software, and can process information without the software. On the other hand, software has a parasitic character, and it needs a host to make a copy of itself. Therefore, the hardware characteristics of proteins is thought to come prior to nucleic acids with software characteristics. The second reason is that amino-acid synthesis is simpler than nucleotide synthesis. Dyson dealt with Miller's experiment that shows how an electric discharge in reducing gas mixtures can generate amino acids [29,30]. Whether the atmosphere was reductive or neutral has been discussed after the original report by Miller. We recognize that Miller himself admitted that the atmosphere is more likely to be neutral than reductive in his posthumous work in 2008, yet the fact that an amino acid can be formed relatively easily is unchanged [31]. Nucleotides have far more complex structures compared to amino acids, and it is normally thought that the spontaneous formation of RNA under prebiotic conditions is difficult [13] (p. 286) [32,33]. The possible synthetic pathway of RNA under prebiotic conditions is still debated and has not been settled [34]. It is likely that the synthetic route of ATP was gradually developed prior to the appearance of RNA in the metabolic system of the prebiotic cell. The last reason is that the hypothesis may be testable. He says that a geochemical approach may determine whether or not a primitive cell in a microfossil contains a clue of RNA (e.g., high content of phosphorus). Geochemical research on the origin of life is on its way; yet, the scarcity and poor preservation conditions of Archean rocks on Earth hampers the investigation [35] (p. 9). Other problems specific to the RNA world are described in the literature [32,36,37].

In addition to these reasonings, it is reasonable to put the emergence of cells before enzymes and genes when one considers the problem of the concentration threshold and permeation of the cell membrane [38–40]. First, since the replication or enzymatic reaction takes places via an intermolecular interaction, the reaction takes a very long time in very diluted conditions. Considering that the destruction and deactivation of the biopolymers take place at the same time, there should be a certain concentration threshold for the biopolymers to increase. Secondly, the permeability of a biopolymer is relatively low compared to that of small molecules, and if one assumes that those biopolymers obtained a cell membrane in the later stage than the emergence of the biopolymers, then it will be difficult to permeate through the membrane. Even if one assumes that biopolymers get into the cell, they can also be permeated through the membrane to exit from the membrane using the same mechanism by which they entered the cell.

Regarding the problem of the concentration threshold and permeation, however, a clue to overcoming this difficulty was found by Luisi and coworkers. They found that ferritin and a green fluorescent protein in a solution are overcrowded in the cell compartment when liposomes are formed in the solution [40,41]. The concentration within a compartment is high, contrary to the expectation of Poisson statistics (they call it an all-or-nothing situation). The mechanism is under investigation, but this fact is interesting, since it may make it possible to cross the concentration threshold of metabolism or replication. However, some difficulties still remain regarding the assumption that biopolymers emerged prior to the cell membrane.

2.2. Lipid World Hypothesis

Lancet and coworkers proposed the concept of the "lipid world" in 1999 [10], and emphasized the importance of the amphiphilic molecules in the context of the origin of life (see ref. [14] for a review written in 2018). It seems that Dyson considers the role of the cell as an inert compartment of the metabolic system, and he was focused more on the cell content that dominates metabolism and reproduction (i.e., protein enzymes), rather than the cell boundary itself (i.e., assembly of lipid and amphiphilic molecules). However, Lancet proposed that the lipid and amphiphilic molecules could also play an important role in information carriage, reproduction, catalysis, selection, and evolution.

Lancet developed a graded autocatalysis replication domain (GARD) model (Figure 2) to investigate the possibility of the lipid world hypothesis. The model assumes that amphiphilic molecules are formed from high energy monomers, while some amphiphiles show weak catalytic activity. Examples of the catalytic reaction with a supramolecular assembly are reported with the help of metal ions [42,43]. The catalytic activity is described by parameter β derived from a receptor affinity distribution (RAD) model [10] (p. 132) [44,45]. The simulation can describe the growth and splitting of the amphiphilic assembly and monitor the compositional changes during growth and fission. Fission generates progeny assemblies, and it may be problematic if one follows all of the generated assemblies, since the population exponentially increases. For the GARD simulation, one of the two progeny assemblies is discarded in the simulation [10] (p. 4), so the model basically focuses on one assembly. This simulation enables one to investigate the relationship among the compositional information, reproduction, catalysis, selection, and evolution. Unlike the experimental approach, the model is not necessarily restricted by the availability of molecules, but it can introduce kinetic constants derived from experiments which may be more plausible [46,47]. The model is based on the idea of the autocatalytic set by Kauffman, and he defines the term as follows: "*By autocatalytic set we mean that each member is the product of at least one reaction catalyzed by at least one other member*" [48] (p. 50). The meaning of the autocatalytic set, and the similar terms, autocatalytic reaction and autocatalytic cycle, are sometimes confusing; the difference was recently clarified by Hordijk [49]. A comparison among the models proposed by Lancet, Dyson, and Kauffman is reported in reference [50]. Other cell replication models are also proposed by Solé based on a dissipative particle dynamics approach [51].

The stated ideas of Dyson and Lancet are examples of the recently described metabolism first claim, and are often contrasted with the genetic first claim represented by the RNA world. Both approval [14,52] and disapproval [53] can be found for the metabolism first claim, but it is recognized that there is no decisive evidence to choose between the two, since each has its own shortcomings [13] (p. 287).

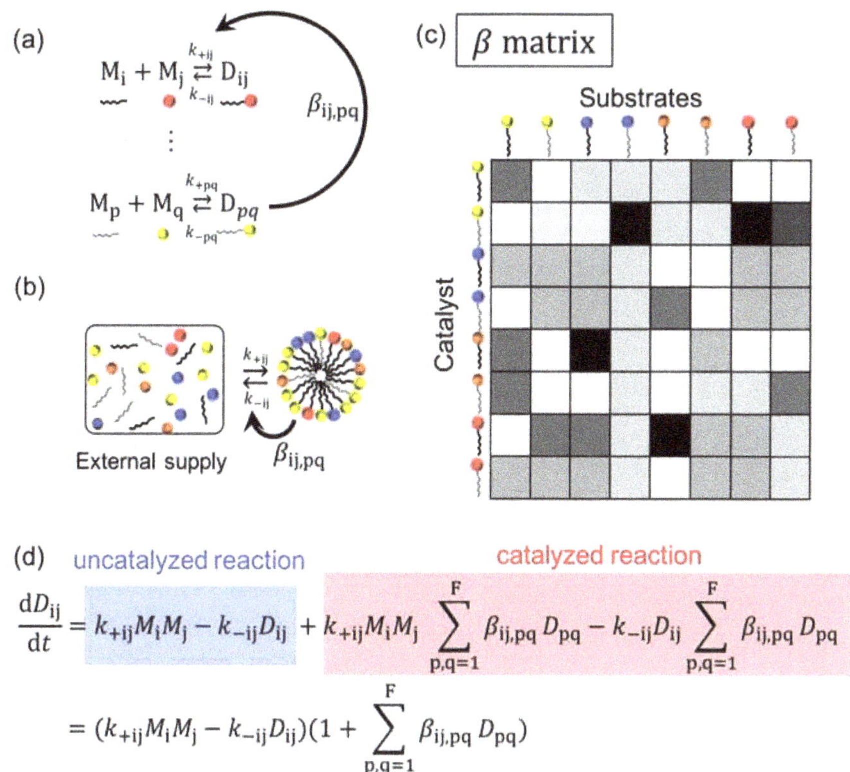

$$\frac{dD_{ij}}{dt} = k_{+ij}M_iM_j - k_{-ij}D_{ij} + k_{+ij}M_iM_j \sum_{p,q=1}^{F} \beta_{ij,pq} D_{pq} - k_{-ij}D_{ij} \sum_{p,q=1}^{F} \beta_{ij,pq} D_{pq}$$

$$= (k_{+ij}M_iM_j - k_{-ij}D_{ij})(1 + \sum_{p,q=1}^{F} \beta_{ij,pq} D_{pq})$$

Figure 2. Description of GARD model based on ref [50,54]. (**a**) Chemical reaction of amphiphilic dimer, D, from monomers, M. Constants k_{+ij} and k_{-ij} indicates association and dissociation rate constants for a reaction forming a dimer (D_{ij}) from monomers (M_i and M_j), respectively. The semicircle arrow indicates the catalysis of the formation and degradation reactions of D_{ij} by D_{pq} with the factor $\beta_{ij,pq}$. (**b**) Assumed condition of GARD model. Monomers are supplied from the outside and the components of assembly catalyze the reaction. (**c**) β matrix of the reaction. The gray scaled color indicates the value of β. (**d**) Rate equation of the system.

2.3. Amphiphiles and Definition of Life

Another piece of key research seeking to unvover the relation between the origin of life and amphiphiles was carried out by Luisi and coworkers (see reviews [15,39,55–57]). In the context of the origin of life, he points out that the criterion of life and criterion of evolution are often confused [39]. Having the question "What is life?" in mind, he introduced the word "autopoiesis" (i.e., self-reproduction) to supramolecular chemistry [15]. The autopoietic system is defined as "*a system which continuously produces the components that specify it, while at the same time realizing it (the system) as a concrete unity in space and time, which makes the network of production of components possible*" [58]. The definition can differentiate the criteria of evolution from those of the living or dead. In addition, it helps to think of a minimal living unit based on the supramolecular assembly. Figure 3 describes this situation. A unit of assembly S (surfactant) is formed from a precursor A. S forms the decay product P, which is eliminated from the system. The kinetics of formation and destruction of S determines whether the assembly grows, maintains (being the state of homeostasis), or decays.

The actual self-reproducible supramolecular system of reversed micelles, micelles, and vesicles are reported [15] (p. 3642). The process of self-reproduction is often referred to as autocatalytic, meaning "the catalysis of a reaction by the products" in this context. For example, in the case of autocatalytic micelles [59], the system is composed of two phases (water and oil), and micelles of sodium caprylates are formed in the aqueous phase by hydrolysis from ethyl caprylate (insoluble in water and forms an oil phase). The rate of the reaction is initially slow, but it exponentially increases after the formation of a certain number of micelles.

It could be interesting to ask whether the finding of the autocatalytic micelle means the discovery of the model system in which the surfactants show a catalytic activity as described in the GARD model. However, the mechanism of the autocatalytic micelle was attributed to a transport phenomenon rather than micellar catalysis [60–62]. This indicated that the catalytic property of the autocatalytic micelle is not the identical definition of catalytic activity assumed in the previously described GARD model. Even if this is true, finding a surfactant with catalytic activity or establishing a method to measure the low catalytic activity in the experimental system would be a big challenge, and it enables a dialog between the experiment and theory. It will benefit experimentalists to have guidance by a theoretical model, and it also benefits theoreticians to use the significant phenomena observed by experiment in the model.

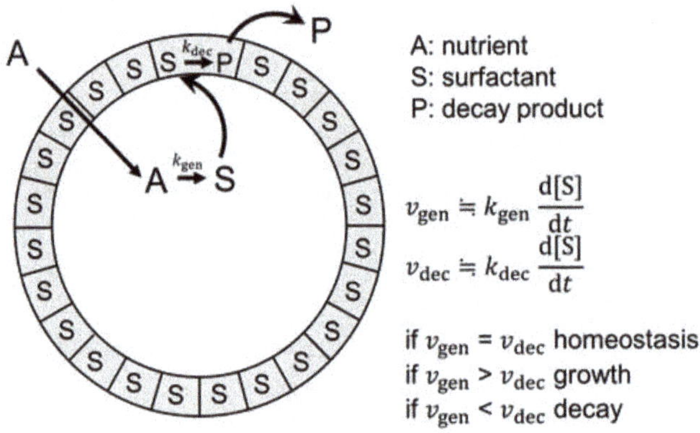

Figure 3. The concept of autopoiesis applied to the supramolecular assembly from reference [55].

3. Prebiotic Condition and Amphiphiles

3.1. Sources of Amphiphiles in Prebiotic Condition

Is it plausible to think that amphiphiles existed under prebiotic conditions? Among the sources of organic compounds (Table 1), the two main sources of amphiphiles are suggested to be extraterrestrial comets and thermal vents. The first source is a meteorite represented by carbonaceous chondrite (Table 1). Carbonaceous chondrite is a type of meteorite mainly composed of silicates, but containing 1.5–4% of carbons in organic forms [63]. Deamer and coworkers reported the striking fact that non-polar molecules extracted from the Murchison carbonaceous chondrite form a vesicle structure [19]. The components of the vesicles are thought to be mainly monocarboxylic acids. Later, analysis of the carbonaceous chondrite revealed that various kinds of monocarboxylic acids are present in meteorites [64,65] (Table 2). These findings led to the idea that a membrane-like structure was formed from lipids such as monocarboxylic acids [11]. The second source of organic compounds is synthesized via the Fischer-Tropsch-type reaction in thermal vents [66–68]. For instance, in aqueous solutions of formic acid or oxalic acid, various compounds, such as *n*-alkanols, *n*-alkanoic acids, *n*-alkenes, *n*-alkanes and alkanones ranging from C2 to over C35 are synthesized [67]. In addition to the Fischer-Tropsch-type reaction, the production of organic molecules catalyzed by various kinds of minerals was also reported [67,68]. It is notable that acyl glycerols can be formed under hydrothermal conditions, since acyl glycerols are a substructure of phospholipids at present, and important candidates for prebiotic supramolecular assembly [69]. The hydrothermal condition is not only preferable for abiotic organic synthesis; it is also proposed as a good condition for the selection and accumulation of amphiphilic molecules [70]. The lifetime of each hydrothermal vent is typically less than 100 years, but ranges from 1–10,000 years [1,71]. Therefore, some questions like "Is the time span enough for prebiotic cells to become independent from the hydrothermal condition?", and "Could a prebiotic cell increase the region of habitat before the living chimney becomes inactive?" will be important in evaluating whether or not prebiotic cells could have been produced in hydrothermal vents.

Table 1. Major sources (kg yr^{-1}) of prebiotic organic compounds in the early earth (from ref. [72] (p. 1459)).

Source	Amount [a]/kg yr^{-1}
Terrestrial Sources	
UV photolysis [b]	3×10^8
Electric discharge [c]	3×10^7
Shocks from impacts [d]	4×10^2
Hydrothermal vents [e]	1×10^8
Extraterrestrial Sources [f]	
Interplanetary dust particles	2×10^8
Comets	1×10^{11}
Total	10^{11}

Modified from Chyba and Sagan [73]. [a] Assumes neutral atmosphere, defined as $[H_2]/[CO_2] = 0.1$. [b] Synthesis of the Miller-Urey type. [c] Such as that caused by lightning interacting with a volcanic discharge. [d] An estimate for compounds created from the interaction between falling objects and the Earth's atmosphere. [e] Based on present-day estimates for total organic matter in hydrothermal vent effluent [74,75]. [f] Conservative estimate based on possible cumulative input calculated assuming flux of 10^{22} kg of cometary material during first Ga (10^9 yr) of Earth's history. If comets contain ~15 wt% organic material [76], and if ≈10% of this material survives, it will comprise approximately 10^{11}/kg yr^{-1} average flux via comets during the first 10^9 yr.

Table 2. Molecular abundances of main organic compounds found in the Murchison, Bells, and Ivuna meteorites from ref [77].

Compound(s)	Murchison [a] nmol/g [b]	n [c]	Bells nmol/g	n	Ivuna nmol/g	n
Ammonia	1100		280		5300	
Amines	130	20	nf [d]		38	5
Amino acids	600	>85	93	13	156 [e]	12
Aldehydes/ketones	200 [f]	18	134	14	1369	23
Hydroxy acids	455 [g]	17	1231	11	2136	10
Di-carboxylic acids	300	26	43	15	857	15
Carboxylic acids	3000	48	495	11	937	14
Hydrocarbons	1850	237	265	82	221	30
Alkanes	350	140	32	32	221	30
Aromatic	300	87	250	27	489 [h]	34
Polar	1200	10	ne [i]		ne	

[a] From ref [78] and references therein unless otherwise noted. [b] Nanomoles/gram of meteorite, indicates total weight of compounds identified with reference standards. [c] Number of species in the group. [d] Not found. [e] Ref [79] also reported for Ivuna amino acids, amounts in the table are new to relate all compounds' quantitative data to the same meteorite fragment. [f] From ref [80]. [g] From ref [81]. [h] Phenanthrene-subtracted. [i] Not estimated.

The amphiphiles described above are mainly lipids, but peptides could also have been components of the cell membrane [82]. Szostak and coworkers proposed a system of vesicle membranes encapsulating a dipeptide catalyst [83]. This dipeptide catalyzes the reaction that forms new dipeptides, and the product takes part in the vesicle membranes because of an enhanced affinity for fatty acids, thus promoting vesicle growth. Another peptide/lipid system was reported by Pascal and Ruiz-Mirazo [84]. They used the reaction of a 5(4H)-oxazolone with leucinamide to form a dipeptide product. The presence of a lipid vesicle increases the yield of the peptide product, leading to the stereoselective reversal of the dipeptide product above the critical aggregation concentration. They also demonstrated that the produced dipeptide showed an affinity for the lipid phase. Since amino acids are synthesized by Miller's experiment [31] and are found in meteorites [85], it is natural to think that a certain amount of peptides may be involved in the formation of the prebiotic cell membrane.

N-carboxyanhydride (NCA)-amino acids are a possible precursor of peptides, and could have been prebiotic compounds involved in the molecular evolution on the early Earth [86,87]. The present protein is synthesized by reactions with the help of the ribosome (composed of rRNA and proteins), mRNA, and tRNA. Under prebiotic conditions, a more primitive synthesis of proteins without such an elaborate reaction should take place. In the case of NCA, it undergoes a simple chemical reaction to form peptides activated by certain chemical species (e.g., CO_2 and carbonyl diimidazole). A particularly important finding is the case of activation by carbonyl diimidazole. It was found that α-amino acids can be efficiently oligomerized from NCA using carbonyl diimidazole, while β-amino acids do not oligomerize [88]. Proteins at the present time are composed from only α-amino acids, but as described below, the source of organic compounds (e.g., meteorites) contains not only α-amino acids, but also β-amino acids. If the polymerization of NCA took place with the help of carbonyl diimidazole, the selectivity would naturally form proteins made of α-amino acids.

Many studies of the origin of life utilize phospholipids to form vesicles, since they are the essential building blocks of current cell membranes. Their chiral configuration shows an interesting chiral recognition of amino-acid-related compounds. However, it should be noted that the formation of complex phospholipids (e.g., phosphatidylcholine) from fatty acids, glycerol, and phosphate would have been difficult under prebiotic conditions [89–91]. Phospholipids that can be found in the present cell membranes are products of highly-evolved metabolic pathways incorporated by multiple enzymes. However, the complexity of phospholipids does not mean that the study of the origin of life using phospholipids is wrong, but that the synthetic pathway of phospholipids was gradually formed in the prebiotic system [89]. Interplay between the phospholipids and other biomolecules (e.g., amino acids or nucleic acids) will provide a possible pathway to form more complex biopolymers, and to the interaction between lipids and other biomolecules and polymers [33,83,92–97].

The critical aggregation concentration is usually high for short-chain lipids or peptides, and doubts could be cast over whether or not a boundary structure can be formed with such short-chain lipids under prebiotic conditions. The pseudo-phase separation approach has been reported for micelles [98,99], and this model predicts the critical aggregation concentration of mixed micelles based on the parameter $\beta_{ij} = N(W_{ii} + W_{jj} - 2W_{ij})/RT$. W_{ii}, W_{jj}, and W_{ij} are the pair-wise interaction of the surfactants i and j, N is Avogadro's number, R is the gas constant and T is the temperature. When β_{ij} is close to 0, it will behave as an ideal system, but when the value is negative (e.g., in the case of a mixture of cationic and anionic surfactants), the critical aggregation concentration becomes lower than that of each component. Szostak and coworkers revealed that a similar model works not only for micelles, but also for vesicles composed of prebiotically-relevant lipid mixtures [100]. This result supports the fact that the boundary structure can be formed even if short chain amphiphiles take part in the vesicle structure.

3.2. *Chiral Amphiphiles in Prebiotic Condition*

It has been reported that carbonaceous chondrites show an enantiomeric excess of organic compounds. Amino acid and sugar are thought to be important chiral organic molecules, since they are chiral building blocks of proteins and nucleic acids. It is interesting that many of the amino acids and sugars found in meteorites have excess L and D configurations, respectively [101,102]. The chirality coincides with the handedness of the building blocks of biopolymers of present life and the fact supports that the source of the organic molecules are extraterrestrial.

It is also interesting that some of the amphiphiles extracted from a carbonaceous chondrite have an asymmetric carbon in their structures (e.g., branched monocarboxylic acids in the Murchison meteorite [64], and hydroxy acids found in the Murchison, GRA 95229 and LAP 02342 meteorites [81]). Most of these chemicals may have a very high solubility in water to form a membrane structure, but some could be seen as model chiral amphiphiles to investigate how the chiral structure affects the properties of a self-assembled entity.

4. Amphiphilic Molecule and Homochirality

The question of "How amphiphiles and homochirality are related?" is less understood compared to the relationship between polymers and homochirality. One reason for this is that it has been thought that the first biological entities were biopolymers, such as proteins and RNA, rather than lipids; thus, the discussion of homochirality has been mainly limited to biopolymers. The question was examined relatively recently [13,103] (p. 303). The relation between a supramolecular assembly and chirality is complicated compared to biopolymers, and it is necessary to clarify the connection between the features of an amphiphilic molecule and homochirality. Lancet and coworkers listed the advantages of lipid assemblies compared to biopolymers in the context of the origin of life [14]. In Figure 4, the physical properties in the list and two other physical properties (adsorption and permeation) are shown. Each physical property could be influenced by the chiral properties. The following sections are devoted to review studies related to chirality, and to some of the physical properties of the amphiphilic molecules.

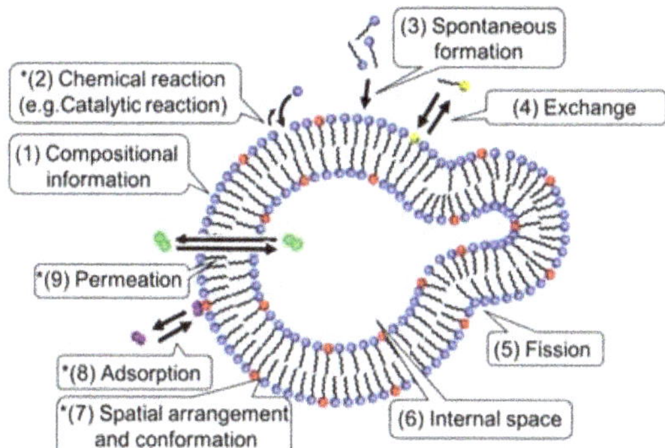

Figure 4. Functions of amphiphiles that could be important for the origin of life and homochirality. The properties (1)–(6) were taken from a table in reference [14] (p. 19) and slightly modified. (7)–(9) were added in this review to discuss the possible roles of chiral amphiphiles. In this article, only the properties with an asterisk are reviewed in detail.

4.1. Chemical Reactions

4.1.1. Advantage Factors

The developing process of homochirality can be described as shown in Figure 5 [104,105]. Without any asymmetric perturbation of chirality (i.e., the source of chirality), constituents of the pre-biotic system can be racemic on average (1). However, perturbation of the chiral purity (2a) or statistical fluctuation of the enantiopurity of chiral molecules (2b) can influence the asymmetrical initial conditions (3). Examples of the chiral perturbation are listed in Table 3. The enantiomeric excess of organic molecules in meteorites could have affected the initial conditions of the racemic soup on earth.

The process of (2a) and (2b) also affects the enhancement of asymmetry (4). The examples of the enhancement of asymmetry and the effect of the advantage factor in chemical reactions are described in the following section. Lastly, the system reaches chiral purity of the biological entity at the present time (5). The description of the advantage factors and chemical reactions described below was originally considered for biopolymers, but is general enough to also be applied to a supramolecular assembly.

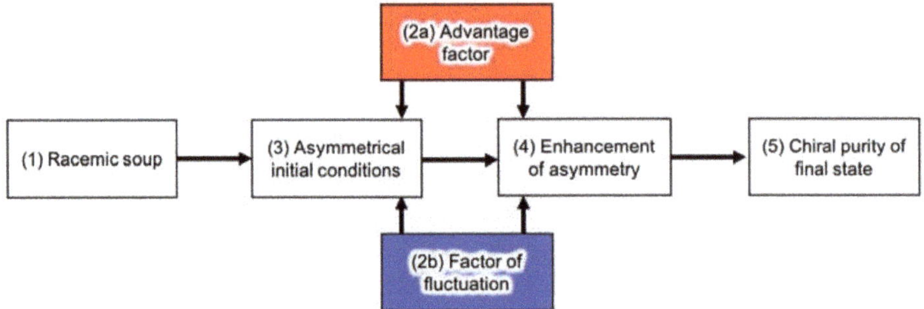

Figure 5. Process of developing homochirality.

Various types of physical advantage factors to induce chiral asymmetry of the system was summarized by Goldanskii and coworkers [105,106]. It was stated that the combinations of the static field are not a "true" advantage factor, and do not lead to inducing asymmetry. Quantified values of the asymmetry factor g are defined as follows.

$$g = \frac{k_L - k_D}{k_L + k_D} \tag{1}$$

where k_L and k_D are the rate constants for the mirror-image reactions. This value quantitatively evaluates how much each advantage factor can contribute to the chiral asymmetry. Table 3 lists this value for each reaction.

Table 3. Physical advantage factors [105,106].

Type of Advantage Factors	True (+) or Imaginary (−)	g [a]
Local advantage factors		
Circularly polarized light	+	$10^{-4} - 10^{-2}$
Static magnetic field (SMF)	−	
Static electric field (SEF)	−	
Gravitational field (GF)	−	
SMF + SEF	−	$\chi_i(EB)$
Rotation (Coriolis force) + GF	−	$\chi_j(\omega_v G) \simeq \chi_k(\Omega G)$
SMF + GF	−	$\chi_l(BG)$
Rotation + SMF + SEF	+	$\chi_m(\omega[EB]) < 10^{-4}$
Rotation + SMF + GF	+	$\chi_n(B[\omega G]) < 10^{-4}$
SMF + Linearly polarized light	+	$\chi_p(Bk) < 10^{-4}$
Global advantage factors		
Weak neutral currents	+	$10^{-20}\chi_r Z^5 \frac{1}{k_B T} \simeq 10^{-17}$
Longitudinally polarized β particles	+	$\chi_q h_e \frac{\sigma^L - \sigma^D}{\sigma^L + \sigma^D} \simeq 10^{-9} - 10^{-11}$

From reference [106]. [a] χ is a factor determined by molecular structure. E is electric field. B is magnetic field. Z is atomic number. k_B is Boltzmann's constant. h_e, helicity of β particles <sp>, where the operators s and p represent the spin and momentum of the particle. σ^L and σ^D are cross sections for the interaction of the polarized β particles with the molecules.

4.1.2. Reaction Building Blocks

Considering that the living organism at the present time consists of chirally-pure biopolymers, such as proteins, RNA, and DNA, it is natural to think that the g values of these advantage factors could be too weak, and that another mechanism to enhance or amplify the chirality of the system is necessary. One model that describes the spontaneous mirror-symmetry breaking was proposed by F. Frank in 1953 [107]. He proposed a model reaction of autocatalysis that results in products with a high enantiopurity, even if the starting chemicals are racemic. This approach was generalized by Morozov by describing the dynamic reaction of chiral polarization (η) [104,108].

$$\eta = \frac{(x_L - x_D)}{(x_L + x_D)} \tag{2}$$

where x_L and x_D are the concentrations of the L and D chiral molecules. $|\eta| = 1$ corresponds to the chirally-pure state while $|\eta| = 0$ corresponds to a racemic state. Based on the approach by Morozov, a broad class of kinetic diagrams and their dynamic equations are described in detail by Goldanskii and Kuzmin (Table 4) [105].

$$\theta = (x_L + x_D) \tag{3}$$

where θ is the concentration of antipodes in the system. The parameters η and θ can be replaced by x_L and x_D. The table has several important features:

(1) The table provides "reaction building blocks" which can construct complex models for the reaction process in an extremely simple way.
(2) The classification enables one to identify the type of process that efficiently leads to breaking the mirror symmetry.
(3) The table takes g into account to evaluate the contribution of the advantage factors to the process.

Figure 6 qualitatively describes the time dependency of the chiral polarization for each type of process. The racemization process decreases $|\eta|$ with time. The neutral process does not change $|\eta|$. The deracemization process increases the value of $|\eta|$ with time. The deracemization process is the key to increasing the chiral asymmetry of the system.

Lancet and coworkers investigated the homochirality based on the GARD model [103] (see Figure 2). Some models for the mirror symmetry breaking designed a reaction network based on the reaction building blocks shown in Table 4. However, for the GARD model, it randomly generated a large variety of chemical species including amphiphiles and catalytic networks without designing the reaction system. The model assumed that the N_G types of the chiral molecules form a catalytic network with equal amounts of the D and L optical isomers. The parity-violating energy difference between enantiomers is excluded and pairs of enantiomers having the same properties. A key chiral property is the ability of the chiral compound L_i to distinguish an enantiomer of other chiral compounds, L_j and D_j, described by the enantio-discrimination factor $\alpha_{ij} = \beta_{Li,Lj}/\beta_{Li,Dj}$, where β_{ij} is the catalytic intensity. It was observed that some chemical species in the stationary compositional states were enriched, with one of the two enantiomers indicating spontaneous chiral symmetry breaking. A considerable degree of chiral selection was observed when the values of α_{ij} are highly relative to the typical values. This indicates that assembly-based enantioselection could not have occurred during the early stage of the origin of life, since such a high discrimination factor is expected only for large molecules. Lancet and coworkers suggested that the homochirality could be a consequence of catalytic networks, rather than a prerequisite for the initiation of primeval life processes.

Table 4. Basic type of process in chiral system from reference [105].

Block	Name of the Reaction [a]	Reaction Formula [b]	Type of Process	Dynamic Equation ($d\eta/dt$) [c]		$\eta_{max}(t\to\infty)$	
				g=0	g≠0	g=0	g≠0
I	Synthesis	$A \xrightarrow{k_L} L$ $A \xrightarrow{k_D} D$	Racemizing	$-2\left(\frac{kx_A}{\theta}\right)\eta$	$\frac{(k_L+k_D)x_A}{\theta}\times(g-\eta)$	0	g
II	Racemization	$L \xrightarrow{k_L} D$ $D \xrightarrow{k_D} L$	"	$-2k\eta$	$-(k_L+k_D)(g+\eta)$	0	$-g$
III	Accidental autocatalysis	$A+L \xrightarrow{k_L} L+D$ $A+D \xrightarrow{k_D} L+D$	"	$-2kx_A\eta$	$-\frac{1}{2}(k_L+k_D)x_A\times\left(g+2\eta+g\eta^2\right)$	0	-
IV	Binary racemization	$2L \xrightarrow{k_L} L+D$ $2D \xrightarrow{k_D} L+D$	"	$-2k\theta\eta$	$-\frac{1}{2}(k_L+k_D)\times\left(g+2\eta+g\eta^2\right)$	0	-
V	Binary destruction	$2L \xrightarrow{k_L} A+B$ $2D \xrightarrow{k_D} A+B$	"	$-k\theta\eta\left(1-\eta^2\right)$	$-\frac{1}{2}(k_L+k_D)\theta\times(g+\eta)\left(1-\eta^2\right)$	0	-
VI	Accidental superautocatalysis	$A+2L \xrightarrow{k_L} 2L+D$ $A+2D \xrightarrow{k_D} L+2D$	"	$-\frac{k}{2}x_A\theta\times\eta\left(3+\eta^2\right)$	$-\frac{1}{4}(k_L+k_D)x_A\theta\times\left(g+3\eta+3g\eta^2+\eta^3\right)$	0	-
VII	Destruction	$L \xrightarrow{k_L} A$ $D \xrightarrow{k_D} A$	Neutral	0	$-\frac{1}{2}(k_L+k_D)\times g\left(1-\eta^2\right)$	η_0	$\lvert 1\rvert$
VIII	Autocatalysis	$A+L \xrightarrow{k_L} 2L$ $A+D \xrightarrow{k_D} 2D$	"	0	$\frac{1}{2}(k_L+k_D)x_A\times g\left(1-\eta^2\right)$	η_0	$\lvert 1\rvert$
IX	Cross-inversion	$L+D \xrightarrow{k_L} 2L$ $L+D \xrightarrow{k_D} 2D$	"	0	$\frac{1}{2}(k_L+k_D)\theta\times g\left(1-\eta^2\right)$	η_0	$\lvert 1\rvert$
X	Annihilation	$L+D \xrightarrow{k} A+B$	Deracemizing	$\frac{k}{2}\theta\eta\left(1-\eta^2\right)$	-	$\lvert 1\rvert$	-
XI	Superautocatalysis	$A+2L \xrightarrow{k_L} 3L$ $A+2D \xrightarrow{k_D} 3D$	"	$\frac{k}{2}x_A\theta\eta\left(1-\eta^2\right)$	$\frac{1}{4}(k_L+k_D)x_A\theta\times(g+\eta)\left(1-\eta^2\right)$	$\lvert 1\rvert$	$\lvert 1\rvert$

[a] The table in the original paper does not attribute Block XI to "Superautocatalysis" but it is evident from the main text. [b] The formula of block XI was corrected from "A + 2L → 3D" in the original paper to "A + 2D → 3D" in this paper. [c] x_A is the concentration of chemical A. $\theta = x_L + x_D$. For the case of $g = 0$, $k = k_L = k_D$. For $g \neq 0$, Block II, III, IV, and VI, were not given in the original paper which were completed in this paper (See supporting information). Block V of the original paper had a contradiction with the dynamic equation of $g = 0$ when g is substituted by 0. The equation of block V was revised in this paper.

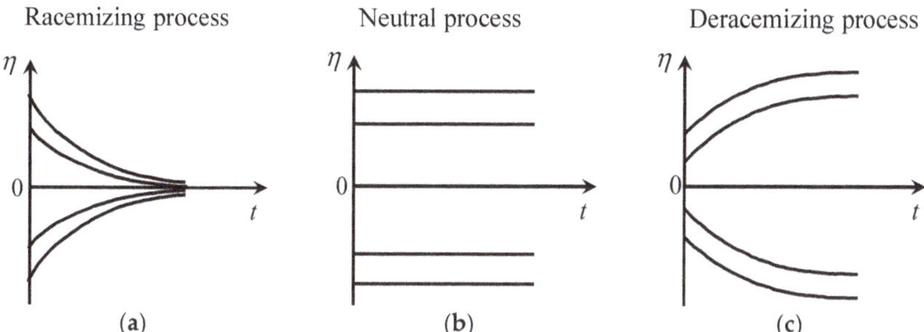

Figure 6. Function of $\eta(t)$ from reference [105]. (**a**) Racemizing process. (**b**) Neutral process. (**c**) Deracemizing process.

4.2. Spatial Arrangement and Conformation

When a chiral unit and racemic or achiral unit are confined in a polymer or supramolecular assembly, these units act cooperatively, which can result in non-linear relationships between the mole fraction of a chiral unit and chiral properties (e.g., optical activity and enantioselective catalytic activity). Such phenomena were first found for helical polymers, and similar phenomena were also found later for supramolecular assemblies, mainly in the form of one-dimensional helical fibers. The Italian researcher Pino and coworkers studied stereo-regular vinyl polymers and determined the cooperative phenomena in the 1960s [109–111]. However, understanding the phenomena was hindered by a weak cooperativity and the absence of a chromophore to observe the chiroptical properties that reflect the chiral conformational properties. Green and coworkers studied poly(isocyanates) (Nylon-1), which showed evident non-linear relationships between the mole fraction of the chiral unit and optical activity [112,113]. The mechanism behind the phenomena was described by one-dimensional Ising models utilizing four main parameters, i.e., temperature, energy of a helix sense reversal, the energy difference between the units of opposite helical senses, and the degree of polymerization [114–118]. The "optical activity" is often measured as a physical property that responds non-linearly to the mole fraction of the chiral unit, because the optical activity indicates the helical conformation of the polymer. However, this does not mean that those amplifications are limited to the chiroptical properties. Suginome and coworkers have shown that for the poly(quinoxaline-2,3-diyl)s, the majority rule [119,120] and sergeants-and-soldiers principal [121] can be coupled with asymmetric catalysis. The reported catalytic reaction of poly(quinoxaline-2,3-diyl)s may only have a slight relation with the prebiotic catalytic reaction, but implies possible roles of those phenomena in the homochirality if such a catalytic activity is observed in the supramolecular assembly. Many reviews are available for the amplification of chirality in helical polymers [113,122–127].

Cooperative phenomena are not limited to polymers whose units are covalently connected, but also for supramolecular assemblies whose units are non-covalently connected. A wide range of helical supramolecular assemblies are reviewed by Yashima and coworkers [128] and Meijer and coworkers [17,129–131]. Figure 7 depicts a model of cooperative supramolecular copolymerization of a disk-like monomer unit [17], a case of majority rule (Figure 7a), and a sergeants-and-soldiers reaction (Figure 7b). For the majority rules, a mismatch between the handedness of the helical structure ($\Delta H^0{}_{MMP}$) and chirality of the monomer and nucleation ($\Delta H^0{}_{NP}$) are assumed to reduce the enthalpy of the elongation process ($\Delta H^0{}_{ELO}$). For the sergeants-and-soldiers principal, the possible reactions are similar as those in the majority-rules model, but the principal distinguishes the enthalpy of chiral units from achiral units, since they have different achiral physical properties. The model is limited to a supramolecular assembly that undergoes one-dimensional growth, but it implies the possibility of amplification in the other chiral supramolecular system. It is an open question whether or not majority rules and sergeants-and-soldiers principal are involved in the emergence of homochirality in the context of the lipid world.

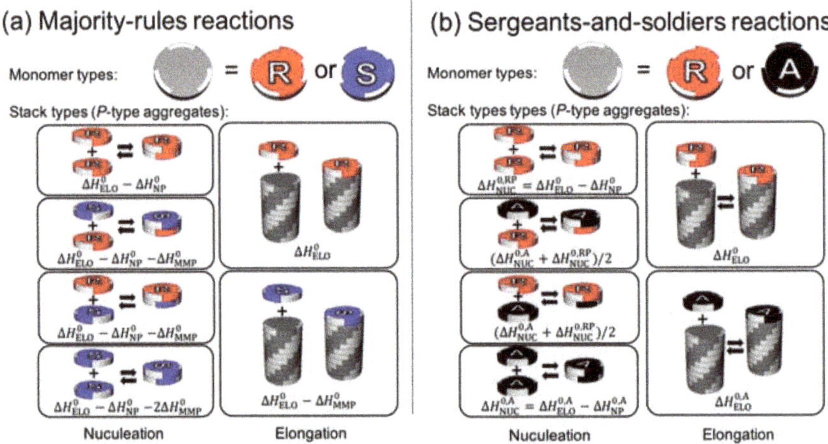

Figure 7. Theoretical models of cooperative supramolecular copolymerization from reference [17]. Formation of *M*- and *P*-type helical aggregates are possible but only the *P*-type aggregates are shown in this figure for the sake of simplicity. The *M*-type helical aggregate can be described in a similar manner by switching the helicity of the aggregates from *P* to *M* and switching the (*S*) and (*R*)-isomers. (**a**) Majority-rules reactions. Two types of monomers, (*R*)-isomer (red) and (*S*)-isomer (blue), forming *P*-type helical aggregate. The model is described by $\Delta H^0{}_{ELO}$, $\Delta H^0{}_{NP}$, $\Delta H^0{}_{MMP}$, and ΔS^0, where $\Delta H^0{}_{ELO}$ is the elongation enthalpy, $\Delta H^0{}_{NP}$ is the nucleation penalty, $\Delta H^0{}_{MMP}$ is the mismatch penalty, and ΔS^0 is the entropy. The (*R*)-isomer prefer *P*-helicity while (*S*)-isomer prefer *M*-helicity. This preference is taken in account by $\Delta H^0{}_{MMP}$. (**b**) Sergeants-and-soldiers reactions. Mismatch does not happen for this case but additional parameters, $\Delta H^{0,A}{}_{ELO}$, $\Delta H^{0,A}{}_{NP}$, and $\Delta S^{0,A}$, are introduced to consider the dependence of the energy on the helicity for achiral monomers.

4.3. Adsorption

It was reported that the membrane of liposomes assists in the formation of the amino acid sequence of the peptide chain by the adsorption of monomers. For instance, Luisi and coworkers reported that the 1-palmitoyl-2-oleoyl-*sn*-glycero-3-phosphocholine (POPC) membrane facilitates the NCA polycondensation of hydrophobic tryptophan-containing peptides up to 29 mer [132]. They also reported that 1,2-dipalmitoyl-*sn*-glycero-3-phosphocholine (DPPC) assisted the NCA-D/L-Tryptophan (L-Trp) condensation [97]. They concluded that the chiral structure of the lipid and the corresponding liposome phase transition temperature do not significantly affect the distribution of the enantiomeric composition of the resulting Trp oligomers.

Umakoshi and coworkers, however, reported that the liposome of DPPC enantioselectively adsorbs L-amino acids after 48 h of coincubation [133]. They tested 10 different amino acids (histidine, serine, aspartic acid, tyrosine, proline, cysteine, tryptophan, phenylalanine, leucine, and valine) and found that the L-stereoisomer of all the amino acids (except for serine) were preferentially adsorbed on the DPPC surface. The extreme cases were L-tryptophan and L-histidine, and their $S_{L/D}$ values (adsorbed concentration ratio of L-amino acid to D-amino acid) exceeded 1000. The adsorption isotherms were attributed to the Langmuir type. This finding implies the possibility of the condensation of the L-amino acid on the surface of the DPPC liposome. They also reported the enantioselective adsorption of ibuprofen [134].

Taking advantage of the enantioselective adsorption, the group showed that the chiral liposome can support homochiral oligomerization of an amino acid [135]. They first allowed L- or D-Histidine to be adsorbed on L-POPC for 72 h, then performed the oligomerization at 25 °C for 48 h by adding 1-hydroxybenzotriazole and 1-ethyl-3-(3-dimethylaminopropyl)carbodiimide hydrochloride to a 1-palmitoyl-2-oleoyl-*sn*-glycero-3-phosphocholine (L-POPC) suspension. They concluded that the homochiral oligomerization of L-amino acids was efficiently achieved compared to the D-amino acids. This result shows that the chirality of the lipid helps to induce the homochirality of peptides. Questions as to whether or not the opposite chiral transfer (peptides to phospholipid) happens, and whether such a reaction occurs with prebiotic molecules, such as NCA, may be interesting, and further investigation is anticipated.

4.4. Permeation

Though there are almost no studies on the enantioselective permeation from the point of view of homochirality, some studies from a pharmaceutical point of view have been reported. The blood-brain barrier protects our central nervous system against external aggressions, but it also prevents therapeutics from reaching targets in the brain. Teixido and Giralt developed a phenylproline tetrapeptide that can cross the blood-brain barrier by paracellular hydrophilic diffusion [136]. They synthesized a library of the 16 phenylproline tetrapeptides stereoisomers, and revealed the relationship among their stereochemistry, transport properties, and permeability. They categorized the tetrapeptides into two groups depending on the stereochemistry; (Group 1) peptides with a higher symmetry (i.e., containing an even number of L or D phenylproline units) and (Group 2) peptides with less symmetric peptides (i.e., containing an odd number of L or D phenylproline units). The enantiomers of the former group did not show similar transport and permeation properties. The latter showed a significant difference between the enantiomers, and basically, the L-rich enantiomer showed higher transport and permeation properties compared to the D-rich enantiomers. Other studies of stereochemical effects on the permeability of human and rat skin are reported based on transdermal drug delivery research [137,138].

5. Conclusions

Though it may not be a common view compared to the RNA world, the lipid world hypothesis provides new insight into the scenario of the origin of life. It emphasizes that an amphiphilic molecule could have played a key role in the origin of life prior to the emergence of the biopolymers. This area of research provides new knowledge about the supramolecular assembly and its interaction with other biomolecules and biopolymers. Far less is known about the relationship between the lipid world and homochirality, and it seems that there are various undiscovered features that chiral amphiphiles have when one considers the physical properties that are unique to supramolecular assemblies (compositional information, spontaneous formation, fission, permeation, etc.). Considering the scenario of homochirality, it is of great interest to determine how those properties could have connected with mirror symmetry breaking, and to elucidate the cooperative phenomena of chiral molecules, such as the sergeants-and-soldiers principal and majority rules.

Supplementary Materials: The following are available online at http://www.mdpi.com/2073-8994/11/8/966/s1: derivation of dynamic equations (S1–S54).

Author Contributions: Conceptualization, N.S. and Y.I.; Writing-Original Draft Preparation, N.S. and Y.I.; Writing—Review and Editing, N.S. and Y.I., Supervision, N.S. and Y.I.

Funding: This study was supported by JSPS KAKENHI Grant Number JP17K14083.

Acknowledgments: N.S. thanks E. Yashima, T. Ikai, and D. Taura, and the group members for the fruitful discussion on the amplification of chirality in helical polymers and supramolecular assembly.

Conflicts of Interest: The authors declare no conflict of interest.

References

1. Lombard, J.; López-García, P.; Moreira, D. The early evolution of lipid membranes and the three domains of life. *Nat. Rev. Microbiol.* **2012**, *10*, 507–515. [CrossRef] [PubMed]
2. Koga, Y. Early evolution of membrane lipids: How did the lipid divide occur? *J. Mol. Evol.* **2011**, *72*, 274–282. [CrossRef] [PubMed]
3. Caforio, A.; Siliakus, M.F.; Exterkate, M.; Jain, S.; Jumde, V.R.; Andringa, R.L.H.; Kengen, S.W.M.; Minnaard, A.J.; Driessen, A.J.M.; van der Oost, J. Converting escherichia coli into an archaebacterium with a hybrid heterochiral membrane. *Proc. Natl. Acad. Sci. USA* **2018**, *115*, 3704–3709. [CrossRef] [PubMed]
4. Forterre, P. The two ages of the RNA world, and the transition to the DNA world: A story of viruses and cells. *Biochimie* **2005**, *87*, 793–803. [CrossRef] [PubMed]
5. Orgel, L.E. Molecular replication. *Nature* **1992**, *358*, 203–209. [CrossRef] [PubMed]
6. Leslie, E.O. Prebiotic chemistry and the origin of the RNA world. *Crit. Rev. Biochem. Mol. Biol.* **2010**, *39*, 99–123. [CrossRef] [PubMed]
7. Fox, S.W. Self-sequencing of amino acids and origins of polyfunctional protocells. *Orig. Life* **1984**, *14*, 485–488. [CrossRef]
8. Fox, S.W.; Jungck, J.R.; Nakashima, T. From proteinoid microsphere to contemporary cell: Formation of internucleotide and peptide bonds by proteinoid particles. *Orig. Life* **1974**, *5*, 227–237. [CrossRef]
9. Ikehara, K. Possible steps to the emergence of life: The [gadv]-protein world hypothesis. *Chem. Rec.* **2005**, *5*, 107–118. [CrossRef]
10. Segré, D.; Ben-Eli, D.; Deamer, D.W.; Lancet, D. The lipid world. *Orig. Life Evol. Biospheres* **2001**, *31*, 119–145. [CrossRef]
11. Deamer, D. The role of lipid membranes in life's origin. *Life* **2017**, *7*, 5. [CrossRef] [PubMed]
12. Walde, P. Surfactant assemblies and their various possible roles for the origin(s) of life. *Orig. Life Evol. Biosph.* **2006**, *36*, 109–150. [CrossRef] [PubMed]
13. Ruiz-Mirazo, K.; Briones, C.; de la Escosura, A. Prebiotic systems chemistry: New perspectives for the origins of life. *Chem. Rev.* **2014**, *114*, 285–366. [CrossRef] [PubMed]
14. Lancet, D.; Zidovetzki, R.; Markovitch, O. Systems protobiology: Origin of life in lipid catalytic networks. *J. R. Soc. Interface* **2018**, *15*, 20180159. [CrossRef] [PubMed]
15. Stano, P.; Luisi, P.L. Achievements and open questions in the self-reproduction of vesicles and synthetic minimal cells. *Chem. Commun.* **2010**, *46*, 3639–3653. [CrossRef] [PubMed]
16. Morigaki, K.; Dallavalle, S.; Walde, P.; Colonna, S.; Luisi, P.L. Autopoietic self-reproduction of chiral fatty acid vesicles. *J. Am. Chem. Soc.* **1997**, *119*, 292–301. [CrossRef]
17. Markvoort, A.J.; ten Eikelder, H.M.M.; Hilbers, P.A.J.; de Greef, T.F.A.; Meijer, E.W. Theoretical models of nonlinear effects in two-component cooperative supramolecular copolymerizations. *Nat. Commun.* **2011**, *2*, 509. [CrossRef] [PubMed]
18. Dalko, P.I.; Moisan, L. In the golden age of organocatalysis. *Angew. Chem. Int. Ed.* **2004**, *43*, 5138–5175. [CrossRef]
19. Deamer, D.W. Boundary structures are formed by organic components of the murchison carbonaceous chondrite. *Nature* **1985**, *317*, 792–794. [CrossRef]
20. Dyson, F.J. A model for the origin of life. *J. Mol. Evol.* **1982**, *18*, 344–350. [CrossRef]
21. Dyson, F. *Origins of Life*; Cambridge University Press: Cambridge, UK, 1999.
22. Dyson, F. Origins of life. In *Nishina Memorial Lectures: Creators of Modern Physics*; Springer: Berlin/Heidelberg, Germany, 2007; Volume 746, pp. 71–97.
23. Oparin, A.I. Biochemical processes in the simplest structures. In *The Origin of Life on the Earth*; Elsevier: Amsterdam, The Netherlands, 1959; pp. 428–436.
24. Eigen, M.; Gardiner, W.; Schuster, P.; Winkler-Oswatitsch, R. The origin of genetic information. *Sci. Am.* **1981**, *244*, 88–118. [CrossRef]
25. Cairns-Smith, A.G. The origin of life and the nature of the primitive gene. *J. Theor. Biol.* **1965**, *10*, 53–88. [CrossRef]
26. Szathmáry, E. The origin of replicators and reproducers. *Philos. Trans. Royal. Soc. B* **2006**, *361*, 1761–1776. [CrossRef]
27. Sagan, L. On the origin of mitosing cells. *J. Theor. Biol.* **1967**, *14*, 225–274. [CrossRef]

28. Von Neumann, J. The general and logical theory of automata. In *Cerebral Mechanisms in Behavior; the Hixon Symposium*; Wiley: Oxford, UK, 1951; pp. 1–41.
29. Miller, S.L. A production of amino acids under possible primitive earth conditions. *Science* **1953**, *117*, 528–529. [CrossRef]
30. Schlesinger, G.; Miller, S.L. Prebiotic synthesis in atmospheres containing CH_4, CO, and CO_2. *J. Mol. Evol.* **1983**, *19*, 376–382. [CrossRef]
31. Cleaves, H.J.; Chalmers, J.H.; Lazcano, A.; Miller, S.L.; Bada, J.L. A reassessment of prebiotic organic synthesis in neutral planetary atmospheres. *Orig. Life Evol. Biosph.* **2008**, *38*, 105–115. [CrossRef]
32. Bernhardt, H.S. The RNA world hypothesis: The worst theory of the early evolution of life (except for all the others)[a]. *Biol. Direct* **2012**, *7*, 23. [CrossRef]
33. Mallik, S.; Kundu, S. The lipid-RNA world. *arXiv* **2012**, arXiv:1211.0413.
34. Biscans, A. Exploring the emergence of RNA nucleosides and nucleotides on the early earth. *Life* **2018**, *8*, 57. [CrossRef]
35. Nakashima, S.; Kebukawa, Y.; Kitadai, N.; Igisu, M.; Matsuoka, N. Geochemistry and the origin of life: From extraterrestrial processes, chemical evolution on earth, fossilized life's records, to natures of the extant life. *Life* **2018**, *8*, 39. [CrossRef]
36. Robertson, M.P.; Joyce, G.F. The origins of the RNA world. *Cold Spring Harb. Perspect. Biol.* **2010**, *4*, a003608. [CrossRef]
37. Hud, N.V.; Cafferty, B.J.; Krishnamurthy, R.; Williams, L.D. The origin of RNA and "my grandfather's axe". *Chem. Biol.* **2013**, *20*, 466–474. [CrossRef]
38. D'Aguanno, E.; Altamura, E.; Mavelli, F.; Fahr, A.; Stano, P.; Luisi, P. Physical routes to primitive cells: An experimental model based on the spontaneous entrapment of enzymes inside micrometer-sized liposomes. *Life* **2015**, *5*, 969–996. [CrossRef]
39. Luisi, P.L. Chemistry constraints on the origin of life. *Isr. J. Chem.* **2015**, *55*, 906–918. [CrossRef]
40. Luisi, P.L.; Stano, P.; de Souza, T. Spontaneous overcrowding in liposomes as possible origin of metabolism. *Orig. Life Evol. Biosph.* **2015**, *44*, 313–317. [CrossRef]
41. Luisi, P.L.; Allegretti, M.; Pereira de Souza, T.; Steiniger, F.; Fahr, A.; Stano, P. Spontaneous protein crowding in liposomes: A new vista for the origin of cellular metabolism. *ChemBioChem* **2010**, *11*, 1989–1992. [CrossRef]
42. Dwars, T.; Paetzold, E.; Oehme, G. Reactions in micellar systems. *Angew. Chem. Int. Ed.* **2005**, *44*, 7174–7199. [CrossRef]
43. Fendler, J.H. Atomic and molecular clusters in membrane mimetic chemistry. *Chem. Rev.* **1987**, *87*, 877–899. [CrossRef]
44. Lancet, D.; Sadovsky, E.; Seidemann, E. Probability model for molecular recognition in biological receptor repertoires: Significance to the olfactory system. *Proc. Natl. Acad. Sci. USA* **1993**, *90*, 3715–3719. [CrossRef]
45. Rosenwald, S.; Kafri, R.A.N.; Lancet, D. Test of a statistical model for molecular recognition in biological repertoires. *J. Theor. Biol.* **2002**, *216*, 327–336. [CrossRef]
46. Armstrong, D.L.; Markovitch, O.; Zidovetzki, R.; Lancet, D. Replication of simulated prebiotic amphiphile vesicles controlled by experimental lipid physicochemical properties. *Phys. Biol.* **2011**, *8*, 066001. [CrossRef]
47. Armstrong, D.L.; Lancet, D.; Zidovetzki, R. Replication of simulated prebiotic amphiphilic vesicles in a finite environment exhibits complex behavior that includes high progeny variability and competition. *Astrobiology* **2018**, *18*, 419–430. [CrossRef]
48. Farmer, J.D.; Kauffman, S.A.; Packard, N.H. Autocatalytic replication of polymers. *Physica D* **1986**, *22*, 50–67. [CrossRef]
49. Hordijk, W. Autocatalytic confusion clarified. *J. Theor. Biol.* **2017**, *435*, 22–28. [CrossRef]
50. Segré, D. A statistical chemistry approach to the origin of life. *Chemtracts–Biochem. Mol. Biol.* **1999**, *12*, 382–397.
51. Solé, R.V. Evolution and self-assembly of protocells. *Int. J. Biochem. Cell Biol.* **2009**, *41*, 274–284. [CrossRef]
52. Shapiro, R. A replicator was not involved in the origin of life. *IUBMB Life* **2000**, *49*, 173–176. [CrossRef]
53. Anet, F.A.L. The place of metabolism in the origin of life. *Curr. Opin. Chem. Biol.* **2004**, *8*, 654–659. [CrossRef]
54. Segré, D.; Lancet, D.; Kedem, O.; Pilpel, Y. Graded autocatalysis replication domain (gard): Kinetic analysis of self-replication in mutually catalytic sets. *Orig. Life Evol. Biospheres* **1998**, *28*, 501–514. [CrossRef]
55. Luisi, P.L. The minimal autopoietic unit. *Orig. Life Evol. Biospheres* **2015**, *44*, 335–338. [CrossRef]

56. Stano, P.; Luisi, P.L. Semi-synthetic minimal cells: Origin and recent developments. *Curr. Opin. Biotechnol.* **2013**, *24*, 633–638. [CrossRef]
57. Stano, P.; Luigi Luisi, P. Self-reproduction of micelles, reverse micelles, and vesicles: Compartments disclose a general transformation pattern. In *Advances in Planar Lipid Bilayers and Liposomes*; LiuLiu, A.L., Ed.; Elsevier Inc.: Amsterdam, The Netherlands, 2008; Volume 7, pp. 221–263.
58. Luisi, P.L.; Varela, F.J. Self-replicating micelles—A chemical version of a minimal autopoietic system. *Orig. Life Evol. Biospheres* **1989**, *19*, 633–643. [CrossRef]
59. Bachmann, P.A.; Luisi, P.L.; Lang, J. Autocatalytic self-replicating micelles as models for prebiotic structures. *Nature* **1992**, *357*, 57–59. [CrossRef]
60. Buhse, T.; Nagarajan, R.; Lavabre, D.; Micheau, J.C. Phase-transfer model for the dynamics of "micellar autocatalysis". *J. Phys. Chem. A* **1997**, *101*, 3910–3917. [CrossRef]
61. Buhse, T.; Pimienta, V.; Lavabre, D.; Micheau, J.C. Experimental evidence of kinetic bistability in a biphasic surfactant system. *J. Phys. Chem. A* **1997**, *101*, 5215–5217. [CrossRef]
62. Buhse, T.; Lavabre, D.; Nagarajan, R.; Micheau, J.C. Origin of autocatalysis in the biphasic alkaline hydrolysis of c-4 to c-8 ethyl alkanoates. *J. Phys. Chem. A* **1998**, *102*, 10552–10559. [CrossRef]
63. Pizzarello, S.; Shock, E. The organic composition of carbonaceous meteorites: The evolutionary story ahead of biochemistry. *Cold Spring Harb. Perspect. Biol.* **2010**, *2*, a002105. [CrossRef]
64. Huang, Y.; Wang, Y.; Alexandre, M.R.; Lee, T.; Rose-Petruck, C.; Fuller, M.; Pizzarello, S. Molecular and compound-specific isotopic characterization of monocarboxylic acids in carbonaceous meteorites. *Geochim. Cosmochim. Acta* **2005**, *69*, 1073–1084. [CrossRef]
65. Sephton, M.A. Organic compounds in carbonaceous meteorites. *Nat. Prod. Rep.* **2002**, *19*, 292–311. [CrossRef]
66. Olah, G.A.; Mathew, T.; Prakash, G.K.S. Chemical formation of methanol and hydrocarbon ("organic") derivatives from co2 and h2—carbon sources for subsequent biological cell evolution and life's origin. *J. Am. Chem. Soc.* **2016**, *139*, 566–570. [CrossRef]
67. Rushdi, A.I.; Simoneit, B.R.T. Lipid formation by aqueous fischer-tropsch-type synthesis over a temperature range of 100 to 400 degrees c. *Orig. Life Evol. Biospheres* **2001**, *31*, 103–118. [CrossRef]
68. McCollom, T.M.; Ritter, G.; Simoneit, B.R.T. Lipid synthesis under hydrothermal conditions by fischer-tropsch-type reactions. *Orig. Life Evol. Biospheres* **1999**, *29*, 153–166. [CrossRef]
69. Rushdi, A.I.; Simoneit, B.R.T. Abiotic condensation synthesis of glyceride lipids and wax esters under simulated hydrothermal conditions. *Orig. Life Evol. Biospheres* **2006**, *36*, 93–108. [CrossRef]
70. Mayer, C.; Schreiber, U.; Dávila, M. Selection of prebiotic molecules in amphiphilic environments. *Life* **2017**, *7*, 3. [CrossRef]
71. Lowell, R.P.; Rona, P.A.; Von Herzen, R.P. Seafloor hydrothermal systems. *J. Geophys. Res. Solid Earth* **1995**, *100*, 327–352. [CrossRef]
72. Ehrenfreund, P.; Irvine, W.; Becker, L.; Blank, J.; Brucato, J.R.; Colangeli, L.; Derenne, S.; Despois, D.; Dutrey, A.; Fraaije, H.; et al. Astrophysical and astrochemical insights into the origin of life. *Rep. Prog. Phys.* **2002**, *65*, 1427–1487. [CrossRef]
73. Chyba, C.; Sagan, C. Endogenous production, exogenous delivery and impact-shock synthesis of organic molecules: An inventory for the origins of life. *Nature* **1992**, *355*, 125–132. [CrossRef]
74. Kadko, D.; Baker, E.; Alt, J.; Baross, J. Global impact of submarine hydrothermal processes: Ridge. In *Vents Workshop, NSF RIDGE Initiative and NOAA Vents Program*; Durham Ridge Office: Durham, NH, USA, 1995.
75. Elderfield, H.; Schultz, A. Mid-ocean ridge hydrothermal fluxes and the chemical composition of the ocean. *Annu. Rev. Earth Planet. Sci.* **1996**, *24*, 191–224. [CrossRef]
76. Delsemme, A.H. Cometary origin of carbon, nitrogen and water on the earth. *Orig. Life Evol. Biospheres* **1991**, *21*, 279–298. [CrossRef]
77. Monroe, A.A.; Pizzarello, S. The soluble organic compounds of the bells meteorite: Not a unique or unusual composition. *Geochim. Cosmochim. Acta* **2011**, *75*, 7585–7595. [CrossRef]
78. Pizzarello, S. The chemistry of life's origin: A carbonaceous meteorite perspective. *Acc. Chem. Res.* **2006**, *39*, 231–237. [CrossRef]
79. Ehrenfreund, P.; Glavin, D.P.; Botta, O.; Cooper, G.; Bada, J.L. Extraterrestrial amino acids in orgueil and ivuna: Tracing the parent body of ci type carbonaceous chondrites. *Proc. Natl. Acad. Sci. USA* **2001**, *98*, 2138–2141. [CrossRef]

80. Pizzarello, S.; Holmes, W. Nitrogen-containing compounds in two CR_2 meteorites: 15N composition, molecular distribution and precursor molecules. *Geochim. Cosmochim. Acta* **2009**, *73*, 2150–2162. [CrossRef]
81. Pizzarello, S.; Wang, Y.; Chaban, G.M. A comparative study of the hydroxy acids from the murchison, gra 95229 and lap 02342 meteorites. *Geochim. Cosmochim. Acta* **2010**, *74*, 6206–6217. [CrossRef]
82. Fishkis, M. Steps towards the formation of a protocell: The possible role of short peptides. *Orig. Life Evol. Biosph.* **2007**, *37*, 537–553. [CrossRef]
83. Adamala, K.; Szostak, J.W. Competition between model protocells driven by an encapsulated catalyst. *Nat. Chem.* **2013**, *5*, 495–501. [CrossRef]
84. Murillo-Sánchez, S.; Beaufils, D.; González Mañas, J.M.; Pascal, R.; Ruiz-Mirazo, K. Fatty acids' double role in the prebiotic formation of a hydrophobic dipeptide. *Chem. Sci.* **2016**, *7*, 3406–3413. [CrossRef]
85. Pizzarello, S.; Schrader, D.L.; Monroe, A.A.; Lauretta, D.S. Large enantiomeric excesses in primitive meteorites and the diverse effects of water in cosmochemical evolution. *Proc. Natl. Acad. Sci. USA* **2012**, *109*, 11949–11954. [CrossRef]
86. Kricheldorf, H.R. Polypeptides and 100 years of chemistry of α-amino acidn-carboxyanhydrides. *Angew. Chem. Int. Ed.* **2006**, *45*, 5752–5784. [CrossRef]
87. Ehler, K.W.; Orgel, L.E. N, N′-carbonyldiimidazole-induced peptide formation in aqueous solution. *Biochim. Biophys. Acta Protein Struct. Mol. Enzymol.* **1976**, *434*, 233–243. [CrossRef]
88. Liu, R.; Orgel, L.E. Polymerization of β-amino acids in aqueous solution. *Orig. Life Evol. Biospheres* **1998**, *28*, 47–60. [CrossRef]
89. Monnard, P.A.; Deamer, D.W. Membrane self-assembly processes: Steps toward the first cellular life. *Anat. Rec.* **2002**, *268*, 196–207. [CrossRef]
90. Namani, T.; Deamer, D.W. Stability of model membranes in extreme environments. *Orig. Life Evol. Biospheres* **2008**, *38*, 329–341. [CrossRef]
91. Budin, I.; Szostak, J.W. Physical effects underlying the transition from primitive to modern cell membranes. *Proc. Natl. Acad. Sci. USA* **2011**, *108*, 5249–5254. [CrossRef]
92. Szostak, J.W.; Bartel, D.P.; Luisi, P.L. Synthesizing life. *Nature* **2001**, *409*, 387–390. [CrossRef]
93. Anella, F.; Danelon, C. Reconciling ligase ribozyme activity with fatty acid vesicle stability. *Life* **2014**, *4*, 929–943. [CrossRef]
94. Anella, F.; Danelon, C. Prebiotic factors influencing the activity of a ligase ribozyme. *Life* **2017**, *7*, 17. [CrossRef]
95. Olasagasti, F.; Kim, H.J.; Pourmand, N.; Deamer, D.W. Non-enzymatic transfer of sequence information under plausible prebiotic conditions. *Biochimie* **2011**, *93*, 556–561. [CrossRef]
96. Wilson, M.A.; Wei, C.; Pohorille, A. Towards co-evolution of membrane proteins and metabolism. *Orig. Life Evol. Biospheres* **2014**, *44*, 357–361. [CrossRef]
97. Hitz, T.; Blocher, M.; Walde, P.; Luisi, P.L. Stereoselectivity aspects in the condensation of racemic nca−amino acids in the presence and absence of liposomes. *Macromolecules* **2001**, *34*, 2443–2449. [CrossRef]
98. Holland, P.M.; Rubingh, D.N. Nonideal multicomponent mixed micelle model. *J. Phys. Chem.* **1983**, *87*, 1984–1990. [CrossRef]
99. Vora, S.; George, A.; Desai, H.; Bahadur, P. Mixed micelles of some anionic-anionic, cationic-cationic, and ionic-nonionic surfactants in aqueous media. *J. Surfactants Deterg.* **1999**, *2*, 213–221. [CrossRef]
100. Budin, I.; Prywes, N.; Zhang, N.; Szostak, J.W. Chain-length heterogeneity allows for the assembly of fatty acid vesicles in dilute solutions. *Biophys. J.* **2014**, *107*, 1582–1590. [CrossRef]
101. Burton, A.; Berger, E. Insights into abiotically-generated amino acid enantiomeric excesses found in meteorites. *Life* **2018**, *8*, 14. [CrossRef]
102. Cooper, G.; Rios, A.; Nuevo, M. Monosaccharides and their derivatives in carbonaceous meteorites: A scenario for their synthesis and onset of enantiomeric excesses. *Life* **2018**, *8*, 36. [CrossRef]
103. Kafri, R.; Markovitch, O.; Lancet, D. Spontaneous chiral symmetry breaking in early molecular networks. *Biol. Direct* **2010**, *5*, 38. [CrossRef]
104. Morozov, L.L.; Kuz Min, V.V.; Goldanskii, V.I. Comparative analysis of the role of statistical fluctuations and factor of advantage (parity nonconservation) in the origins of optical activity. *Orig. Life* **1983**, *13*, 119–138. [CrossRef]
105. Gol'danskiĭ, V.I.; Kuz'min, V.V. Spontaneous breaking of mirror symmetry in nature and the origin of life. *Sov. Phys. Usp.* **1989**, *32*, 1–29. [CrossRef]

106. Avetisov, V.A.; Kuz'min, V.V.; Goldanskii, V.I. Handedness, origin of life and evolution. *Phys. Today* **1991**, *44*, 33–41. [CrossRef]
107. Frank, F.C. On spontaneous asymmetric synthesis. *Biochim. Biophys. Acta* **1953**, *11*, 459–463. [CrossRef]
108. Morozov, L. Mirror symmetry breaking in biochemical evolution. *Orig. Life* **1979**, *9*, 187–217. [CrossRef]
109. Pino, P.; Lorenzi, G.P. Optically active vinyl polymers. Ii. The optical activity of isotactic and block polymers of optically active α-olefins in dilute hydrocarbon solution. *J. Am. Chem. Soc.* **1960**, *82*, 4745–4747. [CrossRef]
110. Pino, P.; Ciardelli, F.; Lorenzi, G.P.; Montagnoli, G. Optically active vinyl polymers. Ix. Optical activity and conformation in dilute solution of isotactic poly-α-olefins. *Makromol. Chem.* **1963**, *61*, 207–224. [CrossRef]
111. Luisi, P.L.; Pino, P. Conformational properties of optically active poly-Alpha-olefins in solution. *J. Phys. Chem.* **1968**, *72*, 2400–2405. [CrossRef]
112. Green, M.M.; Peterson, N.C.; Sato, T.; Teramoto, A.; Cook, R.; Lifson, S. A helical polymer with a cooperative response to chiral information. *Science* **1995**, *268*, 1860–1866. [CrossRef]
113. Green, M.M.; Park, J.W.; Sato, T.; Teramoto, A.; Lifson, S.; Selinger, R.L.B.; Selinger, J.V. The macromolecular route to chiral amplification. *Angew. Chem. Int. Ed.* **1999**, *38*, 3138–3154. [CrossRef]
114. Lifson, S.; Green, M.M.; Andreola, C.; Peterson, N.C. Macromolecular stereochemistry: Helical sense preference in optically active polyisocyanates. Amplification of a conformational equilibrium deuterium isotope effect. *J. Am. Chem. Soc.* **1989**, *111*, 8850–8858. [CrossRef]
115. Selinger, J.V.; Selinger, R.L.B. Theory of chiral order in random copolymers. *Phys. Rev. Lett.* **1996**, *76*, 58–61. [CrossRef]
116. Selinger, J.V.; Selinger, R.L.B. Cooperative chiral order in copolymers of chiral and achiral units. *Phys. Rev. E* **1997**, *55*, 1728–1731. [CrossRef]
117. Selinger, J.V.; Selinger, R.L.B. Cooperative chiral order in polyisocyanates: New statistical problems. *Macromolecules* **1998**, *31*, 2488–2492. [CrossRef]
118. Gu, H.; Sato, T.; Teramoto, A.; Varichon, L.; Green, M.M. Molecular mechanisms for the optical activities of polyisocyanates induced by intramolecular chiral perturbations. *Polym. J.* **1997**, *29*, 77–84. [CrossRef]
119. Ke, Y.Z.; Nagata, Y.; Yamada, T.; Suginome, M. Majority-rules-type helical poly(quinoxaline-2,3-diyl)s as highly efficient chirality-amplification systems for asymmetric catalysis. *Angew. Chem. Int. Ed.* **2015**, *54*, 9333–9337. [CrossRef]
120. Yamamoto, T.; Murakami, R.; Komatsu, S.; Suginome, M. Chirality-amplifying, dynamic induction of single-handed helix by chiral guests to macromolecular chiral catalysts bearing boronyl pendants as receptor sites. *J. Am. Chem. Soc.* **2018**, *140*, 3867–3870. [CrossRef]
121. Nagata, Y.; Nishikawa, T.; Suginome, M. Exerting control over the helical chirality in the main chain of sergeants-and-soldiers-type poly(quinoxaline-2,3-diyl)s by changing from random to block copolymerization protocols. *J. Am. Chem. Soc.* **2015**, *137*, 4070–4073. [CrossRef]
122. Green, M.M.; Cheon, K.S.; Yang, S.Y.; Park, J.W.; Swansburg, S.; Liu, W. Chiral studies across the spectrum of polymer science. *Acc. Chem. Res.* **2001**, *34*, 672–680. [CrossRef]
123. Yashima, E.; Maeda, K.; Nishimura, T. Detection and amplification of chirality by helical polymers. *Chem. Eur. J.* **2004**, *10*, 42–51. [CrossRef]
124. Fujiki, M. Helix generation, amplification, switching, and memory of chromophoric polymers. *Top. Curr. Chem.* **2008**, *284*, 119–186.
125. Yashima, E.; Maeda, K.; Furusho, Y. Single and double-stranded helical polymers: Synthesis, structures, and functions. *Acc. Chem. Res.* **2008**, *41*, 1166–1180. [CrossRef]
126. Yashima, E.; Maeda, K.; Iida, H.; Furusho, Y.; Nagai, K. Helical polymers: Synthesis, structures, and functions. *Chem. Rev.* **2009**, *109*, 6102–6211. [CrossRef]
127. Maeda, K.; Yashima, E. Helical polyacetylenes induced via noncovalent chiral interactions and their applications as chiral materials. *Top. Curr. Chem.* **2017**, *375*, 72. [CrossRef]
128. Yashima, E.; Ousaka, N.; Taura, D.; Shimomura, K.; Ikai, T.; Maeda, K. Supramolecular helical systems: Helical assemblies of small molecules, foldamers, and polymers with chiral amplification and their functions. *Chem. Rev.* **2016**, *116*, 13752–13990. [CrossRef]
129. Brunsveld, L.; Folmer, B.J.B.; Meijer, E.W.; Sijbesma, R.P. Supramolecular polymers. *Chem. Rev.* **2001**, *101*, 4071–4097. [CrossRef]
130. Palmans, A.R.A.; Meijer, E.W. Amplification of chirality in dynamic supramolecular aggregates. *Angew. Chem. Int. Ed.* **2007**, *46*, 8948–8968. [CrossRef]

131. De Greef, T.F.A.; Smulders, M.M.J.; Wolffs, M.; Schenning, A.P.H.J.; Sijbesma, R.P.; Meijer, E.W. Supramolecular polymerization. *Chem. Rev.* **2009**, *109*, 5687–5754. [CrossRef]
132. Blocher, M.; Liu, D.; Walde, P.; Luisi, P.L. Liposome-assisted selective polycondensation of α-amino acids and peptides. *Macromolecules* **1999**, *32*, 7332–7334. [CrossRef]
133. Ishigami, T.; Suga, K.; Umakoshi, H. Chiral recognition of l-amino acids on liposomes prepared with l-phospholipid. *ACS Appl. Mater. Interfaces* **2015**, *7*, 21065–21072. [CrossRef]
134. Okamoto, Y.; Kishi, Y.; Ishigami, T.; Suga, K.; Umakoshi, H. Chiral selective adsorption of ibuprofen on a liposome membrane. *J. Phys. Chem. B* **2016**, *120*, 2790–2795. [CrossRef]
135. Ishigami, T.; Kaneko, Y.; Suga, K.; Okamoto, Y.; Umakoshi, H. Homochiral oligomerization of l-histidine in the presence of liposome membranes. *Colloid. Polym. Sci.* **2015**, *293*, 3649–3653. [CrossRef]
136. Arranz-Gibert, P.; Guixer, B.; Malakoutikhah, M.; Muttenthaler, M.; Guzmán, F.; Teixidó, M.; Giralt, E. Lipid bilayer crossing—The gate of symmetry. Water-soluble phenylproline-based blood-brain barrier shuttles. *J. Am. Chem. Soc.* **2015**, *137*, 7357–7364. [CrossRef]
137. Valentová, J.; Bauerová, K.; Farah, L.; Devínsky, F. Does stereochemistry influence transdermal permeation of flurbiprofen through the rat skin? *Arch. Dermatol. Res.* **2010**, *302*, 635–638. [CrossRef]
138. Touitou, E.; Godin, B.; Kommuru, T.; Afouna, M.; Reddy, I. Transport of chiral molecules across the skin. In *Chirality in Drug Design and Development*; Marcel Dekker: New York, NY, USA, 2004.

Article

Characterization of Hidden Chirality: Two-Fold Helicity in β-Strands

Toshiyuki Sasaki [1,*] and **Mikiji Miyata** [2,*]

1 Department of Materials System Science, Graduate School of Nanobioscience, Yokohama City University, 22-2 Seto, Kanazawa-ku, Yokohama, Kanagawa 236-0027, Japan
2 The Institute of Scientific and Industrial Research, Osaka University, 8-1 Mihogaoka, Ibaraki, Osaka 567-0047, Japan
* Correspondence: tsasaki@yokohama-cu.ac.jp (T.S.); miyata@mls.eng.osaka-u.ac.jp (M.M.); Tel.: +81-45-787-2184 (T.S.); +81-6-6879-8496 (M.M.)

Received: 27 March 2019; Accepted: 4 April 2019; Published: 5 April 2019

Abstract: A β-strand is a component of a β-sheet and is an important structural motif in biomolecules. An α-helix has clear helicity, while chirality of a β-strand had been discussed on the basis of molecular twists generated by forming hydrogen bonds in parallel or non-parallel β-sheets. Herein we describe handedness determination of two-fold helicity in a zig-zag β-strand structure. Left- (*M*) and right-handedness (*P*) of the two-fold helicity was defined by application of two concepts: tilt-chirality and multi-point approximation. We call the two-fold helicity in a β-strand, whose handedness has been unrecognized and unclarified, as hidden chirality. Such hidden chirality enables us to clarify precise chiral characteristics of biopolymers. It is also noteworthy that characterization of chirality of high dimensional structures like a β-strand and α-helix, referred to as high dimensional chirality (HDC) in the present study, will contribute to elucidation of the possible origins of chirality and homochirality in nature because such HDC originates from not only asymmetric centers but also conformations in a polypeptide chain.

Keywords: β-strand; hidden chirality; two-fold helix; multi-point approximation; tilt-chirality; high dimensional chirality

---

1. Introduction

Chirality bestows variety and complexity to functions of substances and is one of the most important and fundamental properties in nature. In this context, elucidation of the origins of chirality and homochirality is a challenge of great importance for scientists, and there have been several theories [1,2] such as mechanical stirring [3] and photo reaction [4]. One of the possible answers to the challenge is chirality generation by chiral crystallization [5–8] with enantiomeric excess, or chiral symmetry breaking [9], followed by amplification [10,11] and transcription [12,13] of the excessed chirality.

In chiral crystallization, chirality is generated by assembling achiral components in chiral manners and is fixed in crystals. Such a phenomenon is observed in both inorganic [5] and organic materials [7,14–18]. A two-fold helix is an especially important structural motif [19] because a large number of chiral crystals belong to $P2_1$ and $P2_12_12_1$ Sohncke groups having two-fold helices, according to space group statistics of the Cambridge Structural Database [20]. It is surprising, however, that a two-fold helix is achiral as a symmetry operation from the viewpoint of 'mathematical' crystallography [21]. This fact brought us a question "what is the origin of chirality in two-fold helix-based crystals?" The supramolecular-tilt-chirality method based on two concepts, tilt chirality and multi-point approximation, answered the question as described below (Figure 1) [22–26].

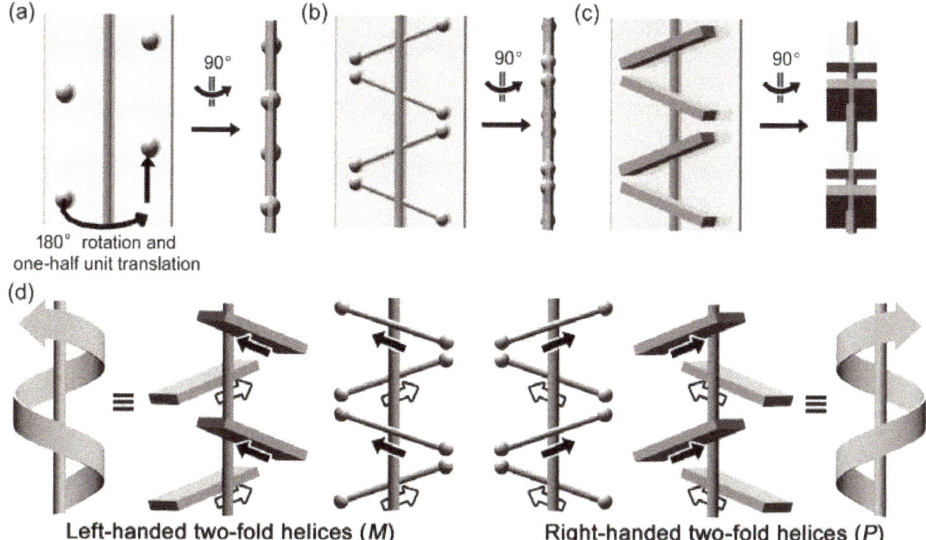

Figure 1. Concepts of supramolecular-tilt-chirality method and multi-point approximation method for chirality characterization of two-fold helices. Achiral two-fold helices represented by approximating components as (**a**) one-point (sphere); (**b**) two-point (line); (**c**) three-point (face) on a mirror plane; (**d**) Chirality of two-fold helices in two-point and three-point approximation methods. The lines or faces in front of the helical axes are tilted to the left or right in the left-(*M*) or right-handed (*P*) two-fold helices, respectively.

A two-fold helix is a symmetry operation of 180° rotation around a helical axis combined with one-half unit translation and is achiral in 'mathematical' crystallography (Figure 1a–c) [21]. On the other hand, a two-fold helix is chiral and has clear left- and right-handedness in 'chemical/material' crystallography [18,27] by considering molecular shapes (Figure 1d). This difference is attributed to an approximation method of molecules: one-point and multi-point approximation in the former and the latter, respectively. We call the chirality, whose handedness cannot be recognized in a general way by mathematical crystallography but definitely exists in materials, as hidden chirality. Not only clear helices, including three-fold, four-fold, and six-fold helices in crystals as well as α-helices in proteins, but also helices having hidden chirality like two-fold helices, may serve as the origin of chirality [17,18,28].

Inspired by the discovery of hidden chirality in a two-fold helix, we have been seeking for further hidden chirality not only in supermolecules [29] but also in a wide range of materials, including polymers. Consequently, we noticed the fact that chirality of a β-strand has been discussed according to twist of a β-strand generated by forming hydrogen bonds in a β-sheet [30–32], while the structure of a β-strand is recognized as a two-fold helix [33]. We can say that the fundamental chirality of a β-strand is two-fold helicity rather than the twist, which is a kind of secondary chirality. Herein we describe two-fold helicity as hidden chirality in a β-strand, which is a part of a β-sheet in biopolymers. Handedness of the two-fold helix is clarified and characterized by applying concepts of tilt-chirality, which is handedness determination based on the tilt of molecules, and multi-point approximation, which is consideration of molecules as multi-point rather than one-point. The unveiled fundamental chirality in a β-strand will give new clues for the origins of chirality generation, homochirality, and chiral properties of proteins.

2. Materials and Methods

Model molecules, pentaglycine and pentaalanine, having various dihedral angles, φ and ψ according to the Ramachandran plot [34], were constructed by using Gauss View5.0.8 [35] and Mercury CSD 4.0.0 [36].

3. Results

Polypeptides construct high dimensional structures, e.g., α-helices [37] and β-sheets [38], by changing dihedral angles of φ and ψ in the main chain [39]. Due to the simplicity of having no chiral centers, pentaglycine was firstly focused on as a model polypeptide. In addition, pentaalanine was also used to clarify effects of chiral centers and side chains by comparing with pentaglycine.

3.1. *Two-Fold Helicity of Extended Linear Pentaglycine*

Pentaglycine with all *trans* (φ = 180° and ψ = −180° (or φ = −180° and ψ = 180°)) formed an extended linear structure (Figure 2). The linear structure had two-fold helical symmetry, i.e., a glycine moiety corresponds to the neighboring one by the operation: 180° rotation around the two-fold axis chain followed by translation along the main chain. Such a two-fold helix was achiral because there was a mirror plane along the two-fold helical axis.

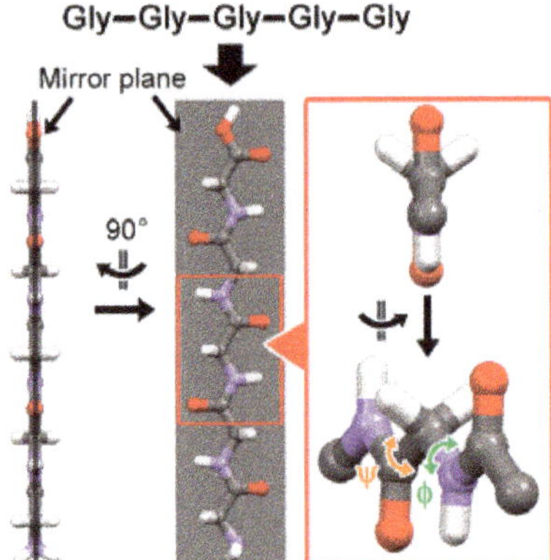

Figure 2. A structure of extended linear pentaglycine. Dihedral angles are φ = 180° and ψ = −180° (or φ = −180° and ψ = 180°).

3.2. *Two-Fold Helicity of Zig-Zag Pentaglycine*

When the dihedral angles became φ = 120° and ψ = −120° (or φ = −120° and ψ = 120°), the linear pentaglycine transformed into a zig-zag structure (Figure 3 left (or right)). Each chain was a so-called β-strand, which is a substructure of a β-sheet. The β-strand had no mirror plane along the two-fold helical axis and constituted a two-fold helix having handedness. Handedness of the two-fold helicity was defined by focusing on lines or faces in the β-strand. Here we defined a face by a peptide bond, which is a well-known planar structure in polypeptides and neighboring alpha carbons of glycine residues.

Handedness of the two-fold helicity in the β-strand was defined by application of the same concept of the supramolecular-tilt-chirality method, i.e., helical handedness was defined based on tilt of lines or faces in front of helical axes. Relative positions of the faces against the helical axis, front and back, were distinguished based on the position of an oxygen atom in the face when the β-strand was viewed from the direction that was vertical to the helical axis and parallel to the faces. When the faces in front of the helical axis were tilted to the left, the β-strand was defined as a left-handed (*M*) two-fold helix (φ = 120° and ψ = −120°, Figure 3 left). Its mirror-imaged one was a right-handed (*P*) two-fold helix having faces tilted to the right in front of the helical axis (Figure 3 right). In this case, dihedral angles of the β-strand were φ = −120° and ψ = 120°.

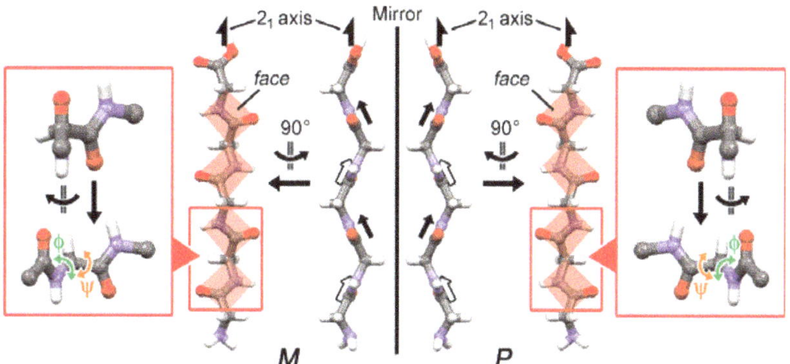

Figure 3. Left- (*M*) and right-handed (*P*) two-fold helicity in a zig-zag β-strand of which dihedral angles are φ = 120°, ψ = −120° and φ = −120°, ψ = 120°, respectively. Faces are comprised of peptide bonds and neighboring alpha carbons of glycine residues. Relative positions of the faces, front, and back are defined based on the relative position of an oxygen atom in each face against the two-fold helical axis.

4. Discussion

4.1. Two-Fold Helicity and Twists of β-Strands in Real Polypeptides

The φ and ψ values of the above mentioned linear structure (Ln) and zig-zag β-strand (Zg) are shown on the Ramachandran plot in addition to the representative secondary structures: β-strands in a parallel β-sheet (↑↑), antiparallel β-sheet (↑↓), α-helix (α), 3_{10}-helix (3_{10}), and π-helix (π) (Figure 4). The blue and red circles are used to distinguish observed structures in real polypeptides composed of *L*-amino acids and their mirror-imaged structures, respectively. In real polypeptides, β-strands exhibit similar structures to the two-fold helical linear and zig-zag chains. From the viewpoint of symmetry, however, β-strands do not correspond to two-fold helices, with the exception of the case |φ| = |ψ| (dotted line in Figure 4).

Instead, they have pseudo-two-fold helicity with twists along the two-fold helical axes. For example, a β-strand shows a left- or right-handed twist when its dihedral angles are φ = −150° and ψ = 120° or φ = −120° and ψ = 150°, respectively (Figure 4c,d). Handedness of pseudo-two-fold helicity, or tilt of the faces, is unchanged by twists of polypeptide chains. Conventionally, on the other hand, a zig-zag β-strand having a left- (|φ| > |ψ|) or right-handed twist (|φ| < |ψ|) is defined as *P* or *M* helices because neighboring amino acid residues in a polypeptide chain are related by less than 180° rotation about the helical axis combined with a unit translation [31].

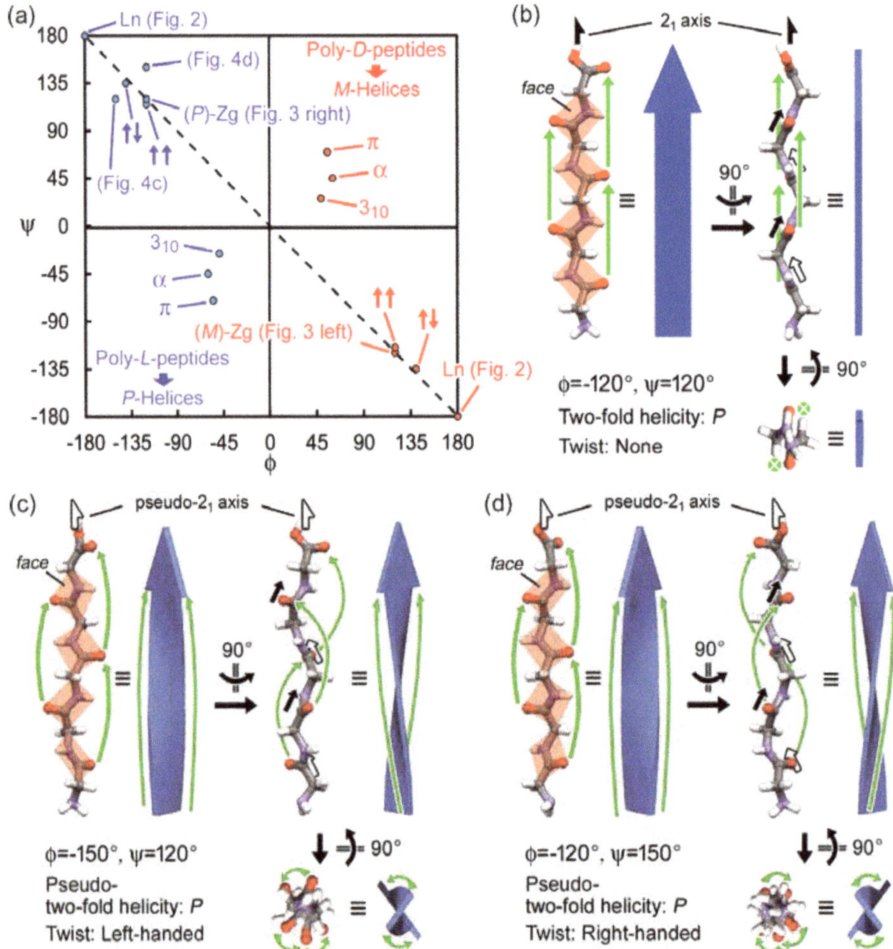

Figure 4. (**a**) The Ramachandran plot indicating φ and ψ values of the linear polypeptide (Ln, φ = 180° and ψ = −180° or φ = −180° and ψ = 180°), the zig-zag β-strands ((*M*)-Zg, φ = 120° and ψ = −120°; (*P*)-Zg, φ = −120° and ψ = 120°), the representative secondary structures: β-strands in a parallel β-sheet (↑↑, φ = −120° and ψ = 115°), antiparallel β-sheet (↑↓, φ = −140° and ψ = 135°), α-helix (α, φ = −60° and ψ = −45°), 3_{10}-helix (3_{10}, φ = −49° and ψ = −26°), and π-helix (π, φ = −55° and ψ = −70°). The linear pentaglycine (Figure 2) and (*M*)- or (*P*)-zig-zag pentaglycine (Figure 3) correspond with Ln and (*M*)- or (*P*)-Zg, respectively. Blue and red circles are in mirror-imaged relation to each other. (**b–d**) Relationship among dihedral angles (φ, ψ), (pseudo-)two-fold helicity, and twist: (**b**) (*P*)-Zg (φ = −120° and ψ = 120°) with no twist; (**c**) a β-strand with *P* pseudo-two-fold helicity and left-handed twist (φ = −150° and ψ = 120°); (**d**) a β-strand with *P* pseudo-two-fold helicity and right-handed twist (φ = −120° and ψ = 150°).

4.2. Correlation Between Molecular Chirality and Chirality of High Dimensional Structures

In characterization of chiral properties, it is important to consider not only molecular chirality (MC) but also high dimensional chiral structures, which we call high dimensional chirality (HDC) in the present study, because both of the chirality affect chiral functions. Previously, we succeeded in elucidating the linkage between MC and supramolecular chirality (SMC) [40]. At the same time, dependence of chiral characteristics on MC as well as SMC was confirmed by vibrational circular dichroism spectroscopy, which has been used in characterization of chirality in biopolymers [41–43]. This success demonstrated that geometrical characterization of MC and HDC leads to prediction and control of HDC. In the same way with the previous study, we investigated the correlation between MC of each amino acid residue and HDC, helicity in the present case, of β-strands and α-helices.

Structures of pentamers comprised of achiral glycine and chiral alanine were compared with each other to clarify effects of chiral centers on chiral secondary structures. In the case of pentaglycine ($RCH(NH_2)COOH$, R = H), *M* and *P* two-fold helices are formed almost equally without external chiral sources. On the other hand, there is clear selectivity between *M* and *P* two-fold helices in the other polypeptides ($RCH(NH_2)COOH$, R≠H). The selectivity is attributable to steric hindrance between the oxygen atom in a peptide bond and neighboring substituent R (Figure 5a). For example, linear penta-*L*-alanine (Figure 5a(i)) tends to form a zig-zag β-strand with *P* two-fold helicity (φ = −120° and ψ = 120°) (Figure 5a(ii)) by letting a methyl group be away from the oxygen atom of a neighboring peptide bond rather than that with *M* two-fold helicity (φ = 120° and ψ = −120°) (Figure 5a(iii)) by making a methyl group be close to the oxygen atom of a neighboring peptide bond. This fact suggests a correlation between MC and HDC, i.e., absolute structures of amino acids and helicity in polypeptides.

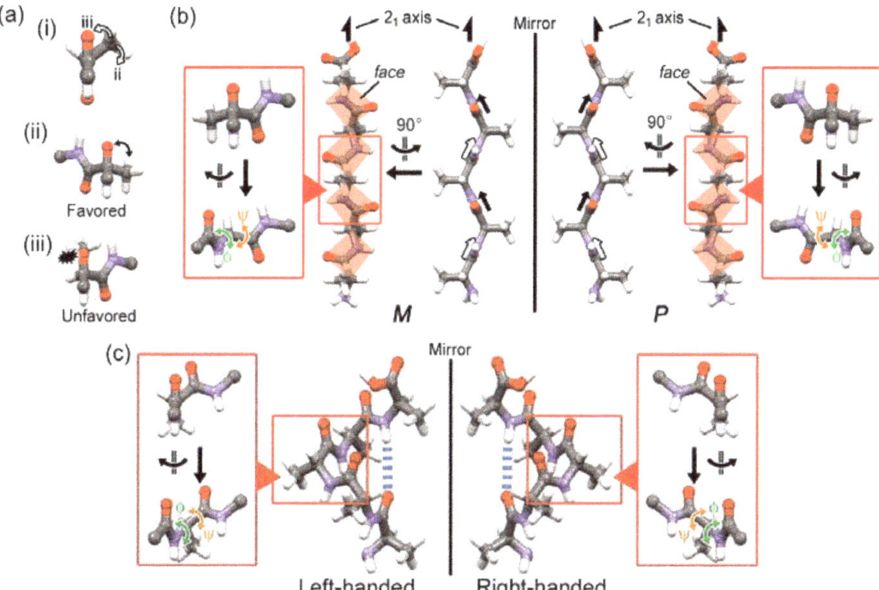

Figure 5. (**a**) Selectivity in rotation direction of dihedral angles for transformation of poly-*L*-alanine: (i) an extended linear structure to (ii) a favored and (iii) unfavored zig-zag structure (β-strand); (**b**) Two-fold helicity of zig-zag pentaalanine β-strands: left-handed (*M*, φ = 120° and ψ = −120°) and right-handed two-fold helices (*P*, φ = −120° and ψ = 120°) comprised of *D*-alanine and *L*-alanine, respectively; (**c**) Left-handed (*M*, φ = 60° and ψ = 45°) and right-handed (*P*, φ = −60° and ψ = −45°) α-helices of penta-*D*-alanine and penta-*L*-alanine, respectively.

In the same way, handedness selectivity of α-helices is explained. A zig-zag β-strand with *P* two-fold helicity can be formed from an extended linear poly-*L*-amino acid by rotating dihedral angles, φ = −180°→−120° (+60°) and ψ = 180°→120° (−60°), in which rotation directions are determined by chirality of component amino acid residues as mentioned above. The β-strand then becomes a right-handed α-helix by further rotation of the dihedral angles: φ = −120°→−60° (+60°) and ψ = 120°→−45° (−165°). The rotation directions, clockwise and counterclockwise in φ and ψ, respectively, are the same as those in the formation of the β-strand from an extended linear poly-*L*-amino acid, even though the value of rotation angle ψ is relatively large. This coincidence in the rotation directions suggests a structural relationship, or possibility of an inter-structural transition, between a β-strand and α-helix. In fact, there is a region, sometimes called the bridge region, which bridges the regions of β-strand (β-sheet) and α-helix in the Ramachandran plot [39].

5. Conclusions

Two-fold helicity as hidden chirality in a β-strand, which is an important structural motif in proteins, was successfully clarified and characterized by applying the concepts of tilt-chirality and multi-point approximation. We suggest that two-fold helicity is fundamental chirality of a β-strand. The geometric viewpoint brings us a new perspective to explain linkages between handedness of two-fold helicity and that of molecular chirality of amino acid residues in polypeptides. Helicity of an α-helix, twists in a β-strand and β-sheet, as well as central chirality of component amino acids are chiral structures in general polypeptides. The two-fold helicity in a β-strand is also a chiral structure of great importance observed in most polypeptides. Furthermore, the concepts are also applicable to chiral structures of other biopolymers, e.g., polysaccharides [44] and nucleic acids [45], and synthetic crystalline polymers [46] of which chirality is non-negligible. Our findings bring important knowledge for elucidating origins of chirality and homochirality in nature and chiral properties of proteins including amyloid fibrils [47] and also give new insights into the transition from prebiotic chemistry to protobiology [48–50].

Author Contributions: Conceptualization, T.S. and M.M.; Investigation, T.S. and M.M.; Supervision, M.M.; Writing—original draft, T.S.; Writing—review and editing, M.M.

Funding: This research received no external funding.

Conflicts of Interest: The authors declare no conflict of interest.

References

1. Mason, S.F. Origins of Biomolecular Handedness. *Nature* **1984**, *311*, 19–23. [CrossRef] [PubMed]
2. Feringa, B.L.; van Delden, R.A. Absolute Asymmetric Synthesis: the Origin, Control, and Amplification of Chirality. *Angew. Chem. Int. Ed.* **1999**, *38*, 3418–3438. [CrossRef]
3. Okano, K.; Taguchi, M.; Fujiki, M.; Yamashita, T. Circularly Polarized Luminescence of Rhodamine B in a Supramolecular Chiral Medium Formed by a Vortex Flow. *Angew. Chem. Int. Ed.* **2011**, *50*, 12474–12477. [CrossRef]
4. Inoue, Y. Asymmetric Photochemical Reactions in Solution. *Chem. Rev.* **1992**, *92*, 741–770. [CrossRef]
5. Lowry, T.M. *Optical Rotatory Power*; Dover Publications: New York, NY, USA; London, UK, 1964.
6. Jacques, J.; Collet, A.; Wilen, S.H. *Enantiomers, Racemates, and Resolutions*; Krieger: Malabar, India, 1991.
7. Matsuura, T.; Koshima, H. Introduction to Chiral Crystallization of Achiral Organic Compounds Spontaneous Generation of Chirality. *J. Photochem. Photobiol. C* **2005**, *6*, 7–24. [CrossRef]
8. Vogl, O. Chiral Crystallization and the Origin of Chiral Life on Earth. *J. Polym. Sci. Part A Polym. Chem.* **2011**, *49*, 1299–1308. [CrossRef]
9. Weissbuch, I.; Leiserowitz, L.; Lahav, M. Stochastic "Mirror Symmetry Breaking" via Self-Assembly, Reactivity and Amplification of Chirality: Relevance to Abiotic Conditions. *Top. Curr. Chem.* **2005**, *259*, 123–165.
10. Soai, K.; Shibata, T.; Morioka, H.; Choji, K. Asymmetric Autocatalysis and Amplification of Enantiomeric Excess of a Chiral Molecule. *Nature* **1995**, *378*, 767–768. [CrossRef]
11. Viedma, C. Chiral Symmetry Breaking During Crystallization: Complete Chiral Purity Induced by Nonlinear Autocatalysis and Recycling. *Phys. Rev. Lett.* **2005**, *94*, 065504. [CrossRef] [PubMed]
12. Green, B.S.; Lahav, M.; Rabinovich, D. Asymmetric Synthesis via Reactions in Chiral Crystals. *Acc. Chem. Res.* **1979**, *12*, 191–197. [CrossRef]
13. Hazen, R.M.; Sholl, D.S. Chiral Selection on Inorganic Crystalline Surfaces. *Nat. Mater.* **2003**, *2*, 367–374. [CrossRef] [PubMed]

14. Azumaya, I.; Yamaguchi, K.; Okamoto, I.; Kagechika, H.; Shudo, K. Total Asymmetric Transformation of an N-Methylbenzamide. *J. Am. Chem. Soc.* **1995**, *117*, 9083–9084. [CrossRef]
15. Iitaka, Y. The Crystal Structure of β-Glycine. *Acta Cryst.* **1960**, *13*, 35–45. [CrossRef]
16. Iitaka, Y. The Crystal Structure of γ-Glycine. *Acta Cryst.* **1961**, *14*, 1–10. [CrossRef]
17. Sasaki, T.; Ida, Y.; Tanaka, A.; Hisaki, I.; Tohnai, N.; Miyata, M. Chiral Crystallization by Non-Parallel Face Contacts on the Basis of Three-Axially Asymmetric Twofold Helices. *Cryst. Eng. Comm.* **2013**, *15*, 8237–8240. [CrossRef]
18. Sasaki, T.; Miyata, M.; Tsuzuki, S.; Sato, H. Experimental and Theoretical Analysis of Two-fold Helix-Based Chiral Crystallization by Confined Interhelical CH/π Contacts. *Cryst. Growth Des.* **2019**, *19*, 1411–1417. [CrossRef]
19. Kitaigorodskii, A.I. *Molecular Crystals and Molecules*; Academic Press: London, UK, 1973.
20. Groom, C.R.; Bruno, I.J.; Lightfoot, M.P.; Ward, S.C. The Cambridge Structural Database. *Acta Cryst.* **2016**, *72*, 171–179. [CrossRef] [PubMed]
21. International Tables for Crystallography Volume A: Space-Group Symmetry. 2016. Available online: https://it.iucr.org/A/ (accessed on 27 March 2019).
22. Hisaki, I.; Sasaki, T.; Sakaguchi, K.; Liu, W.-L.; Tohnai, N.; Miyata, M. Right- and Left-Handedness of 2_1 Symmetrical Herringbone Assemblies of Benzene. *Chem. Commun.* **2012**, *48*, 2219–2221. [CrossRef] [PubMed]
23. Hisaki, I.; Sasaki, T.; Tohnai, N.; Miyata, M. Supramolecular-Tilt-Chirality on Twofold Helical Assemblies. *Chem. Eur. J.* **2012**, *18*, 10066–10073. [CrossRef] [PubMed]
24. Miyata, M.; Tohnai, N.; Hisaki, I.; Sasaki, T. Generation of Supramolecular Chirality around Twofold Rotational or Helical Axes in Crystalline Assemblies of Achiral Components. *Symmetry* **2015**, *7*, 1914–1928. [CrossRef]
25. Miyata, M.; Tohnai, N.; Hisaki, I. Crystalline Host–Guest Assemblies of Steroidal and Related Molecules: Diversity, Hierarchy, and Supramolecular Chirality. *Acc. Chem. Res.* **2007**, *40*, 694–702. [CrossRef] [PubMed]
26. Miyata, M.; Hisaki, I. Twofold Helical Molecular Assemblies in Organic Crystals: Chirality Generation and Handedness Determination. In *Advances in Organic Crystal Chemistry: Comprehensive Reviews*; Tamura, R., Miyata, M., Eds.; Springer: Tokyo, Japan, 2015; pp. 371–390.
27. Sasaki, T.; Miyata, M.; Sato, H. Helicity and Topological Chirality in Hydrogen-Bonded Supermolecules Characterized by Advanced Graph Set Analysis and Solid-State Vibrational Circular Dichroism Spectroscopy. *Cryst. Growth Des.* **2018**, *18*, 4621–4627. [CrossRef]
28. Kawasaki, T.; Hakoda, Y.; Mineki, H.; Suzuki, k.; Soai, K. Generation of Absolute Controlled Crystal Chirality by the Removal of Crystal Water from Achiral Crystal of Nucleobase Cytosine. *J. Am. Chem. Soc.* **2010**, *132*, 2874–2875. [CrossRef] [PubMed]
29. Sasaki, T.; Ida, Y.; Hisaki, I.; Yuge, T.; Uchida, Y.; Tohnai, N.; Miyata, M. Characterization of Supramolecular Hidden Chirality of Hydrogen-Bonded Networks by Advanced Graph Set Analysis. *Chem. Eur. J.* **2014**, *20*, 2478–2487. [CrossRef]
30. Chothia, C. Conformation of Twisted β-Pleated Sheets in Proteins. *J. Mol. Biol.* **1973**, *75*, 295–302. [CrossRef]
31. Weatherford, D.W.; Salemme, F.R. Conformations of Twisted Parallel β-Sheets and the Origin of Chirality in Protein Structures. *Proc. Natl. Acad. Sci. USA* **1979**, *76*, 19–23. [CrossRef]
32. Ho, B.K.; Curmi, P.M.G. Twist and Shear in β-Sheets and β-Ribbons. *J. Mol. Biol.* **2002**, *317*, 291–308. [CrossRef]
33. Salemme, F.R. Structural Properties of Protein β-Sheets. *Prog. Biophys. Mol. Biol.* **1983**, *42*, 95–133. [CrossRef]
34. Ramachandran, G.N.; Ramakrishnan, C.; Sasisekharan, V. Stereochemistry of Polypeptide Chain Configurations. *J. Mol. Biol.* **1963**, *7*, 95–99. [CrossRef]
35. Dennington, R.D., II; Keith, T.A.; Millam, J.M. *Gauss View, Version 5*; Semichem Inc.: Shawnee Mission, KS, USA, 2009.
36. Macrae, C.F.; Bruno, I.J.; Chisholm, J.A.; Edgington, P.R.; McCabe, P.; Pidcock, E.; Rodriguez-Monge, L.; Taylor, R.; van de Streek, J.; Wood, P.A. Mercury CSD 2.0—New Features for the Visualization and Investigation of Crystal Structures. *J. Appl. Cryst.* **2008**, *41*, 466–470. [CrossRef]
37. Pauling, L.; Corey, R.B.; Branson, H.R. The Structure of Proteins: Two Hydrogen-Bonded Helical Configurations of the Polypeptide Chain. *Proc. Natl. Acad. Sci. USA* **1951**, *37*, 205–211. [CrossRef] [PubMed]

38. Pauling, L.; Corey, R.B. Configurations of Polypeptide Chains with Favored Orientations Around Single Bonds: Two New Pleated Sheets. *Proc. Natl. Acad. Sci. USA* **1951**, *37*, 729–740. [CrossRef] [PubMed]
39. Hollingsworth, S.A.; Karplus, A. A Fresh Look at the Ramachandran Plot and the Occurrence of Standard Structures in Proteins. *Biomol. Concepts* **2010**, *1*, 271–283. [CrossRef] [PubMed]
40. Sasaki, T.; Hisaki, I.; Miyano, T.; Tohnai, N.; Morimoto, K.; Sato, H.; Tsuzuki, S.; Miyata, M. Linkage Control between Molecular and Supramolecular Chirality in 2_1−Helical Hydrogen-Bonded Networks using Achiral Components. *Nat. Commun.* **2013**, *4*, 1787. [CrossRef]
41. Kurouski, D. Advances of Vibrational Circular Dichroism (VCD) in Bioanalytical Chemistry. A Review. *Anal. Chim. Acta* **2017**, *990*, 54–66. [CrossRef] [PubMed]
42. Nafie, L.A.; Freedman, T.A. Biological and Pharmaceutical Applications of Vibrational Optical Activity. *Pract. Spectrosc.* **2001**, *24*, 15–54.
43. Keiderling, T.A. Vibrational Circular Dichroism of Peptides and Proteins. Survey of Techniques, Qualitative and Quantitative Analyses, and Applications. *Pract. Spectrosc.* **2001**, *24*, 55–100.
44. Moon, R.J.; Martini, A.; Nairn, J.; Simonsen, J.; Youngblood, J. Cellulose Nanomaterials Review: Structure, Properties and Nanocomposites. *Chem. Soc. Rev.* **2011**, *40*, 3941–3994. [CrossRef]
45. Saenger, W. *Principles of Nucleic Acid Structure*; Springer: New York, NY, USA, 1984.
46. Rosa, C.D.; Auriemma, F. Structure and Physical Properties of Syndiotactic Polypropylene: A Highly Crystalline Thermoplastic Elastomer. *Prog. Polym. Sci.* **2006**, *31*, 145–237.
47. Sipe, J.D.; Cohen, A.S. Review: History of the Amyloid Fibril. *J. Struct. Biol.* **2000**, *130*, 88–98. [CrossRef]
48. Orgel, L.E. Prebiotic Chemistry and the Origin of the RNA World. *Crit. Rev. Biochem. Mol. Biol.* **2004**, *39*, 99–123.
49. Ruiz-Mirazo, K.; Briones, C.; de la Escosura, A. Prebiotic Systems Chemistry: New Perspectives for the Origins of Life. *Chem. Rev.* **2014**, *114*, 285–366. [CrossRef]
50. Krishnamurthy, R. Giving Rise to Life: Transition from Prebiotic Chemistry to Protobiology. *Acc. Chem. Res.* **2017**, *50*, 455–459. [CrossRef]

Article

Chiral Proportions of Nepheline Originating from Low-Viscosity Alkaline Melts. A Pilot Study

Ewald Hejl [1,*] and Friedrich Finger [2,*]

[1] Fachbereich für Geographie und Geologie der Universität Salzburg, Hellbrunnerstraße 34/III, A-5020 Salzburg, Austria

[2] Fachbereich für Chemie und Physik der Materialien, Universität Salzburg, Jakob Haringer Straße 2, A-5020 Salzburg, Austria

* Correspondence: Ewald.Hejl@sbg.ac.at (E.H.); Friedrich.Finger@sbg.ac.at (F.F.); Tel.: +43-662-8044-5437

Received: 14 August 2018; Accepted: 11 September 2018; Published: 18 September 2018

Abstract: Chromatographic interaction between infiltrating solutions of racemic mixtures of enantiomers and enantiomorphic minerals with chiral excess has been proposed as a scenario for the emergence of biomolecular homochirality. Enantiomer separation is supposed to be produced by different partition coefficients of both enantiomers with regard to crystal faces or walls of capillary tubes in the enantiomorphic mineral. Besides quartz, nepheline is the only common magmatic mineral with enantiomorphic symmetry. It crystallizes from SiO_2-undersaturated melts with low viscosity and is a promising candidate for chiral enrichment by autocatalytic secondary nucleation. Under liquidus conditions, the dynamic viscosity of silicate melts is mainly a function of polymerization. Melts with low concentrations of SiO_2 (<55 wt%) and rather high concentrations of Na_2O (>7 wt%) are only slightly polymerized and hence are characterized by low viscosities. Such melts can ascend, intrude or extrude by turbulent flow. Fourteen volcanic and subvolcanic samples from alkaline provinces in Africa and Sweden were chemically analyzed. Polished thin sections containing fresh nepheline phenocrysts were etched with 1% hydrofluoric acid at 20 °C for 15 to 25 min. Nepheline crystals suitable for a statistical evaluation of their etch figures were found in four samples. Crystals with chiral etch figures are mainly not twinned. Their chiral proportions in grain percentages of single crystals are close to parity in three samples. Only one sample shows a slight chiral excess (41.67% L-type vs. 58.33% D-type) but at a low level of significance (15 vs. 21 crystals, respectively).

Keywords: chirogenesis; enantiomorphism; nepheline; magmatic flow; etch figures

---

1. Introduction

Extraterrestrial scenarios for chiral enrichment [1,2] have recently been favoured by the scientific community because of experimental evidence for enantiomeric excesses induced by circularly polarized light [3], and because of the fascinating challenge of the cometary mission Rosetta and its enantiomer-separating Cometary Sampling and Composition (COSAC) experiment onboard the Philae lander [4–8]. Nevertheless, chiral enrichment of enantiomers by natural chromatographic processes on the early surface of planet Earth is still a matter of debate. One of us (E. Hejl) [9] has proposed a new hypothesis for enantiomer separation in the course of chemical etching of nuclear particle tracks originating from spontaneous or induced fission of ^{238}U or ^{235}U, respectively. The hypothesis relies on the principle of liquid chromatography, i.e., on partition of dissolved molecular species between the walls of capillary tubes and a liquid mobile phase. When such tubes or pores occur in an enantiomorphic crystal they are expected to produce a slight separation of dissolved enantiomers. This process can be amplified in the course of downhill infiltration of seepage water.

Chromatographic enantiomer separation requires different partition coefficients of both enantiomers relative to an enantiopure or chirally enriched stationary phase. In the scenario proposed by HEJL [9], enantiomorphic minerals in the subsoil of an early Precambrian mainland form the stationary phase for infiltrating fluids. Chromatographic separation of dissolved enantiomers might occur either by selective partition on grain boundaries, cracks and cleavage planes, or in tubular cavities produced by natural etching of nuclear particle tracks. This process necessitates a preexisting chiral enrichment of enantiomorphic minerals in a certain volume of rock or soil.

On Earth, only two enantiomorphic mineral species are quite common components of magmatic rocks: quartz and nepheline. These two minerals may be completed by the SiO_2-modification tridymite which could have been more frequent on the early Earth, and was recently discovered on planet Mars by the rover Curiosity of the Mars Science Laboratory [10]. Furthermore, the most frequent enantiomorphic mineral besides quartz and tridymite is nepheline. It belongs to the hexagonal system (hexagonal-pyramidal class, with a polar axis of hexagonal symmetry) and to the space-group $P6_3$ [11,12].

Nepheline has been ignored for enantioselective crystallization in general, and for the emergence of biomolecular homochirality in particular. More than 130 years after the discovery of its enantiomorphism [13], nothing is known about the chiral proportions of nepheline in magmatic rocks or on a global scale. This lack of information might be due to the fact that nepheline is optically not active [14], and that both enantiomorphs cannot be distinguished under polarized light. In this context it is important to notice that nepheline crystallizes in a single space group ($P6_3$)—in contrast to quartz which belongs to a pair of space groups, $P3_121$ (right-handed screw) and $P3_221$ (left-handed screw), i.e., space groups with screw axes of opposite handedness.

Determination of the absolute structure of enantiomorphic crystals (including handedness) is still a great challenge for experimental crystallography. Among other methods, X-ray diffraction with dispersion correction has been applied to this problem, and also resonant X-ray diffraction techniques have made some progress during the last decades [15]. TANAKA et al. [16] have studied structural chirality by use of circularly polarized resonant X-ray diffraction. They could demonstrate that the measurement of only one space-group forbidden reflection is sufficient to determine the chirality of α-quartz or berlinite ($AlPO_4$). The term "absolute structural chirality" refers to space groups having screw axes labeled right-handed or left-handed. This concept does not apply to nepheline whose two enantiomorphs belong to the same space group ($P6_3$).

Based on crystal morphology and optical activity, many investigations were dedicated to chiral proportions of quartz in nature, as for example in graphic granite [17], but strong chiral excess of quartz was neither found on a worldwide nor on a regional scale [18,19]. Because of its occurrence in alkaline rocks having crystallized from SiO_2-undersatureated low-viscosity melts, nepheline is a more promising candidate for chiral enrichment by autocatalytic secondary nucleation (seeding) than quartz which mainly crystallizes from melts with higher viscosity.

Spontaneous crystallization of almost enantiopure crystals was indeed observed under laboratory conditions [20,21]. Dissolved sodium chlorate $NaClO_3$ is not chiral, but its crystals are optically active and occur in two enantiomorphs. When such crystals precipitate from an aqueous solution of sodium chlorate, they can be either left-handed or right-handed, with corresponding opposite optical activities. In case of precipitation from a stationary—not flowing—solution, almost equal numbers of crystals of both enantiomorphic configurations are formed [20]. On the other hand, when the solution is stirred with a magnetic stirrer during crystallization, it was found that more than 99% of the crystals had the same handedness. The direction of enantiomorphic excess (either L- or D-crystals) was unpredictable and obviously not controlled by the direction of stirring. KONDEPOUDI et al. [20] have argued that the enantiomorphic excess is due to autocatalytic secondary nucleation and by the suppression of nuclei of the opposite handedness in the course of a competitive nucleation. Analogous seeding effects were observed for stirred solutions of $NaBrO_3$, and for the crystallization of chiral hydrocarbons from a melt [21].

If autocatalytic secondary nucleation [20,21] also occurs in magmatic crystallization, we can expect that a turbulent magmatic flow of low viscosity has a better chance to produce chiral excess than a gentle flowing magma with high viscosity. Nepheline, which crystallizes from SiO_2- undersaturated magmas with rather low viscosity, is a potential candidate for enantioselective crystallization. Transition from laminar to turbulent flow is more easily achieved in alkaline or carbonatitic melts than in silica-enriched magmas with high viscosity. The present investigation is a first test for an eventual chiral enrichment of nepheline in alkaline magmatic rocks.

2. Viscosities, Ascent Rates and Flow Patterns of Magma under Natural Conditions

Flow patterns of magma depend on viscosity and density as substantial properties of the magma itself, on magma's velocity or ascent rate relative to a solid frame of reference, and on the geometry of the walls that confine the magmatic flow. In contrast to magma densities, which are usually between 2000 and 3000 kg/m^3, viscosities of magma can vary by several orders of magnitude. Magmatic viscosity depends on temperature, the amount of fractional crystallization within the magmatic flow, and on the chemical composition of the molten fraction. The dynamic viscosity (η) connects the shear stress (τ) to the strain rate of a fluid ($\delta v/\delta y$ = velocity gradient perpendicular to planes of equal flow velocities). Silicate melts mainly behave as Newtonian fluids [22], i.e., they show a linear relationship between shear stress and shear rate for any given temperature and pressure ($\delta v/\delta y = \tau/\eta$). Above liquidus boundary conditions, the dynamic viscosity of a silicate melt is mainly a function of its polymerization [23,24]. In silicate minerals as well as in silicate melts, the Si^{4+} ions occur in tetrahedral coordination with oxygen over a wide range of temperature and pressure. Each of these oxygen atoms has the potential of bonding to another Si^{4+} ion which can result in chains, sheets or three-dimensional networks of connected SiO_4 tetrahedrons. Such linking of silicate tetrahedrons has been referred to as polymerization [23,24]. The equilibrium between various types of oxygen atoms in a silicate melt can be described by the following reaction:

$$Si\text{-}O\text{-}Si + M\text{-}O\text{-}M = 2(Si\text{-}O\text{-}M) \tag{1}$$

where M is a cation other than Si^{4+}, and the oxygens of the three terms of the equation are called bridging, free and non-bridging oxygens. This equilibrium between oxygen types can be shortly written as

$$O^0 + O^{2-} = 2O^- \tag{2}$$

Equilibrium conditions can be associated to the equilibrium constant K

$$K = \frac{(O^-)^2}{(O^0)(O^{2-})} \tag{3}$$

where the terms in brackets refer to activities or, in the case of an ideal solution, to molar concentrations of the oxygen species in the melt. A melt with $K = 0$ would be one with no reaction between molten SiO_2 and molten metal oxide to form non-bridging oxygen. On the other hand, a melt with $K = \infty$ is given when in reaction (1) all the silica or all the metal oxide is consumed to form non-bridging oxygen. Silicate magmas with a higher equilibrium constant K contain less bridging oxygen, are less polymerized and less viscous than melts with a small K. For a given pressure, temperature, and composition, the polymerization equilibrium of a silicate melt is realized when the distribution of oxygen species (O^0, O^{2-}, O^-) minimizes the free energy of the solution [24].

Such estimations were calculated for binary systems of SiO_2 on the one hand and FeO, MnO, CaO, and Na_2O on the other hand [24,25]. All curves (Gibbs free energy vs. molar concentration of SiO_2) show minimum free energies in the vicinity of about 40% molar concentration of SiO_2. For any given concentration of SiO_2, the free energy is highest for the system FeO-SiO_2, and becomes smaller in the order of MnO-, CaO-, and Na_2O. Consequently, a Na_2O-SiO_2 melt is always less polymerized and less viscous than a FeO-SiO_2 melt with the same SiO_2 concentration. In general, melts with cations of high ionization potential are expected to be more polymerized than melts with cations of low ionization potential, as for example Na^+ or K^+. This theoretical implication is well tested by X-ray diffraction studies and NMR spectroscopy of silica glass [26]. At higher pressures (>6 GPa), the degree of polymerization is further complicated by the formation of new oxygen clusters, including 5- and 6-coordinated Si and Al which result in a decrease of non-bridging oxygen [27,28].

Dynamic viscosities of fresh lava flows can be measured with a portable, motor-driven, rotating shear vane that records torque and rotation rate. Alternatively, open magma channels can be used as natural viscometers. Viscosities are calculated from the channel's slope and depth, as well as from flow velocity and density of magma. Both methods were successfully applied to alkali carbonatite lavas of the 1988 eruption of Oldoinyo Lengai, Tanzania [29].

Flow velocities of effusive lava can be determined by observations on the spot [29]. Emplacement velocities of magma injections into fractures or dykes, as well as magma transport rates in volcanic ascent channels can be derived from seismic data. The rate of stress release depends on the viscoelastic properties of rocks and on the rate of magma flow compensating extensional failure in sills or dykes. Seismic model calculations [30] have shown that at a differential pressure of 1000 bars, basaltic magma must be injected at 1000 m/s in a 1 m thick extensional dyke in order to produce a magnitude 5 earthquake. Magmatic flow velocities between 1 and 100 m/s are quite common when the injected volume and the dimension of extensional failure are smaller.

Transition from laminar to turbulent flow depends on the fluid's density and its dynamic viscosity, as well as on the flow velocity and a characteristic linear dimension of the flow (for example the diameter of a pipe or the depth of a flow with an open, unconstrained surface). The dimensionless Reynolds number (*Re*) helps to predict flow patterns for different flow situations:

$$Re = \rho.v.L/\eta \tag{4}$$

with ρ = density of the fluid (kg/m^3)
v = flow velocity (m/s)
L = hydraulic diameter (m)
η = dynamic viscosity (Pa.s)

Figure 1 is a double-logarithmic cross plot of dynamic viscosity vs. hydraulic diameter, with characteristic Reynolds numbers of 500 and 2500 for various flow velocities. Transition from laminar to turbulent flow usually occurs at a threshold between these *Re* values. The hydraulic diameter is either the diameter of a cylindrical pipe or the thickness of a dyke. Figure 1 is valid for a magma density of 2800 kg/m^3 (basalt and nephelinite).

By consideration of published dynamic viscosities of various types of magma above liquidus conditions [22,31], and of reasonable dimensions of magmatic channels, critical flow velocities for the transition from laminar to turbulent flow can be predicted in a semi-quantitative way (Figure 1). Low-viscosity melts of carbonatite or nephelinite ($\eta < 10^{-1}$ Pa.s) can exhibit turbulent flow even at very low flow velocities of less than 1 cm/s in a dyke or vent with hydraulic diameters above 10 m. Basaltic melts ($\eta > 10$ Pa.s) have a lesser probability of turbulent flow, which may only occur at higher flow velocities and/or in very large volcanic vents. Dry SiO_2-rich melts (rhyolite with $\eta > 10^8$ Pa.s) can never reach turbulence under realistic flow conditions. These predictions are supported by volcanologic field observations [29,32].

Chiral enrichment by seeding [20] requires not only turbulence but also unconstrained dispersion of crystalline nuclei within the entire magmatic volume. This is only possible when the enantiomorphic crystalline species precipitates early from the melt (Figure 2a), and not late under eutectic conditions (Figure 2b). In the latter case, any seeding will be restricted to the interstitial space of individual liquid pores between already crystallized minerals. The same is the case for crystals in individual magma droplets suspended in a gaseous phase of a rapidly expanding explosive eruption (Figure 2c). Mutual catalytic influence between liquid interstices or droplets is impossible.

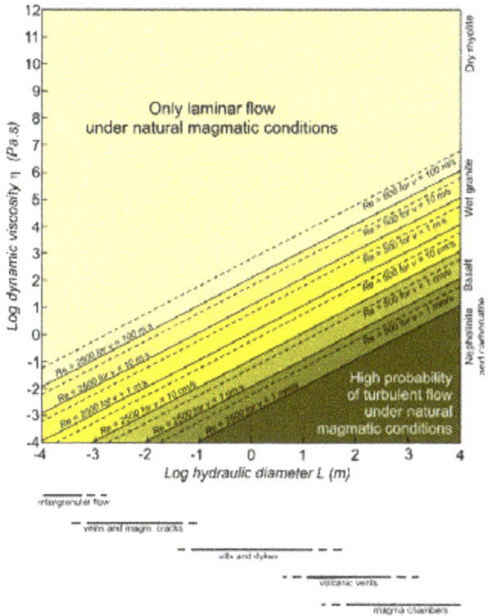

Figure 1. Crossplot of dynamic viscosity (η) vs. hydraulic diameter (L) with corresponding Reynold's numbers (both 500 and 2500) for a melt density (ρ) of 2800 kg/m^3, and for various flow velocities (v) between 1 mm/s and 100 m/s. Dynamic viscosities on the right-hand side according to [22,29,31].

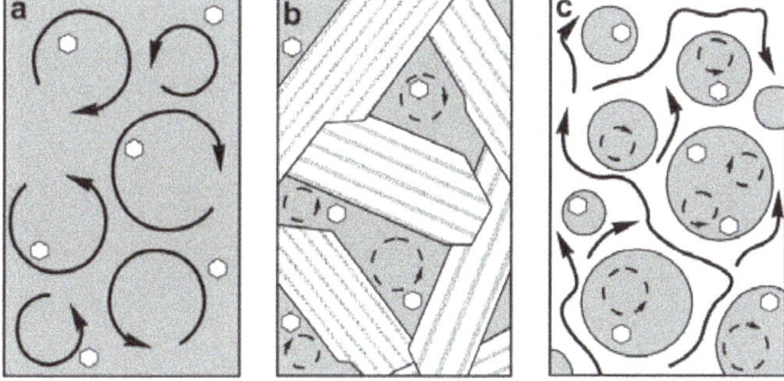

Figure 2. Schematic representation of crystallization under different magmatic conditions. (**a**) Early unconstrained crystallization in a mainly liquid environment. (**b**) Late interstitial crystallization in the pores between earlier crystallized phenocrysts. (**c**) Crystallization in liquid droplets of an expanding current of hot volcanic gas (ignimbrite or pyroclastic flow). Bold circular arrows symbolize turbulent mixing; dashed circular arrows symbolize low probability of turbulence.

Phase relations and the shape of the liquidus surface in the compositional triangle SiO_2-$NaAlSiO_4$-$KAlSiO_4$ [33–36] help to predict the crystallization behavior of alkaline melts. Figure 3a shows the liquidus temperatures (isothermal lines) for various compositions at 1 bar (=10^5 Pa) water-vapor pressure. Early and unconstrained crystallization of nepheline from the melt can only occur in a defined compositional field between the binary eutectic lines in the central part of the diagram and the $NaAlSiO_4$ corner—without the carnegieite field (Figure 3b). Melts with bulk compositions in the nepheline field at some distance from the binary lines will first crystallize nepheline, and by fractional crystallization the composition of the remaining melt will approach the binary eutectic lines. Alkaline rocks with bulk compositions in the quoted nepheline field should have porphyritic phenocrysts of nepheline in a finer-grained matrix with a composition that is closer to eutectic conditions.

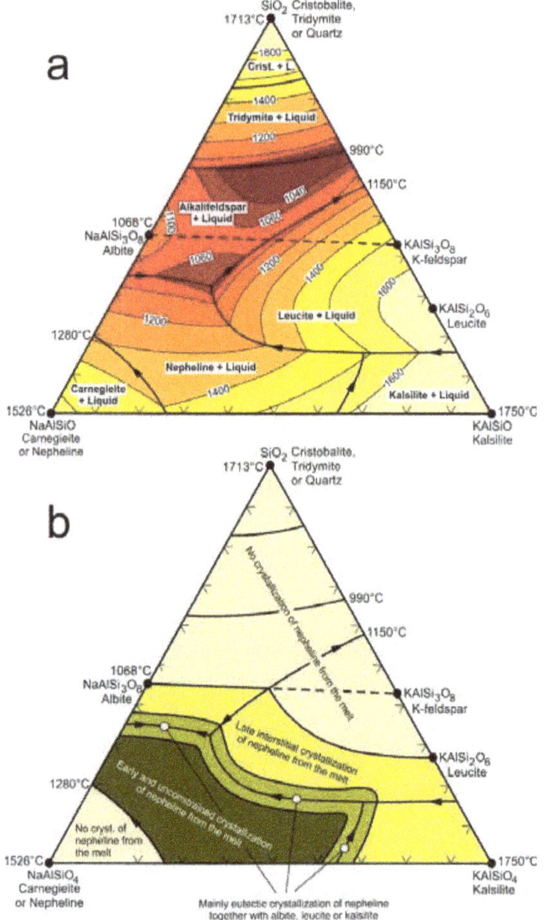

Figure 3. Phase relations in the system $NaAlSiO_4$-$KAlSiO_4$-SiO_2. The relative proportions of components represent mass concentrations (wt%). (**a**) Phase relationships with corresponding liquidus temperatures at 1 bar water-vapour pressure according to [34–36]. (**b**) Compositional areas with specific crystallization of nepheline (i.e., early, eutectic, or interstitial).

3. Occurrence, Etching Behavior and Chiral Proportions of Nepheline in Alkaline Igneous Rocks

Several SiO_2-undersaturated magmatic rocks from various alkaline provinces in Africa and northern Europe were chosen for the present investigation. Twelve volcanic rock specimens together with thin sections were provided by curator Epifanio Vaccaro from the Natural History Museum (NHM) of London, another two samples from Namibia were provided by courtesy of Robert Trumbull from the GFZ Potsdam. The absolute structure of nepheline (including handedness) cannot be determined by X-ray diffraction because both enantiomorphs belong to the same space group. For the present study, we decided to determine chiral proportions of polycrystalline nepheline by a statistical evaluation of etch figures on intersecting planes subparallel to the c-axis.

The 14 samples are mainly of volcanic or sub-volcanic origin, two are plutonic rocks, and one is a nepheline-bearing carbonatite (Table 1). Sample no. 1 (BM.1953, 133(2)) is a phonolite from Homa Montain (Kenya) which consists of a central area of concentric carbonatite dykes and breccia, and some subordinate plugs of phonolite to nephelinite. K-Ar ages of whole rocks and biotite range from 1.3 to

12 Ma [37]. Sample no. 2 (BM.1965, P19 (6)) is a phonolitic nephelinite from the Dorowa complex in Zimbabwe. This complex is intruded into Archaean granitic gneiss and is mainly composed of syenitic fenite with minor plutonic to subvolcanic intrusions of foyaite, Ijolite, and carbonatite [38]. Sample no. 3 (BM.1968, P37 (401)) is a phonolite from the Namangali hill in southern Malawi. It is composed of phonolitic and feldspathic breccia which forms vents in fenitized gneisses and in Precambrian basement rocks. Both fenites and vent rocks are cut by phonolitic dykes and small carbonatite veins [39]. Samples no. 4, 5, and 6 (BM.1980, P31 (2, 24, 27)) are alkaline volcanic rocks from the Tundulu complex in southern Malawi. Its central subvolcanic area, which comprises carbonatite, agglomerate, trachyte, nephelinite, and phonolite, is surrounded by a broad aureole of fenitized Precambrian basement [40]. Sample no. 7 (BM.1981, P14 (440)) is a nepheline carbonatite from the Alnö complex at the eastern coast of central Sweden. It intruded the Precambrian basement in Late Ediacarian times [41]. Sample no. 8 (BM.1981, P3) is a phonolitic nephelinite from Mt. Etinde in the vicinity of Limbe (Cameroon). Etinde is a steep-sided Late Cenozoic volcano composed of various kinds of nephelinites and nephelinitic tuffs. K-Ar ages of Etinde's volcanic rocks range between 0.065 and 6.3 Ma [42,43]. Samples no. 9 and 10 (BM.1995, P6 (40, 43)) are phonolitic nephelinites from Kerimasi in northern Tanzania. Kerimasi is a Quaternary volcanic cone that rises approximately 1000 m above the Serengeti Plain. It is mainly built up by nephelinites, corresponding tuffs and agglomerates [44]. Samples no. 11 and 12 (BM.2004, P12 (28, 75)) are from Oldoinyo Lengai which is situated immediately north of Kerimasi. Because of its natrocarbonatite lavas of extremely low viscosity, Oldoinyo Lengai is one of the most prominent carbonatite volcanoes of the world. Besides carbonatite, it is composed by nephelinites, agglomerates, and tuffs [43,45–47]. It has a Quaternary age and is still active. Samples no. 13 and 14 (KF 85 and KF 88) are nepheline syenites from the Kalkfeld ring complex in northwestern central Namibia. This bimodal carbonatite-alkali silicate complex belongs to the Damaraland Igneous Province which formed in the Early Cretaceous, between 137 and 124 Ma [48]. The Kalkfeld ring complex is hosted in granitic and metasedimentary rocks of the Pan-African Damara System. BÜHN & TRUMBULL [48] have found that the Kalkfeld silicate magma fractionated alkali-feldspar and nepheline in a CO_2-dominated, F- and Ca-poor system, and that euhedral nepheline phenocrysts are rare compared to predominant alkali-feldspar.

Table 1. Rock types and locations of investigated samples.

No.	Sample Code	Rock Type	Location	Coordinates	Ref.
1	BM.1953, 133 (22)	Phonolite	Homa Montain, Kenya	0°23′ S; 34°30′ E	[37]
2	BM.1965, P19 (6)	Phonolitic nephelinite	Dorowa compl., Zimbabwe	19°04′ S; 31°45′ E	[38]
3	BM.1968, P37 (401)	Phonolite	Namangali, Malawi	15°49′ S; 35°35′ E	[39]
4	BM.1980, P31 (2)	Aegirine biotite phonol.	Tundulu complex, Malawi	15°32′ S; 35°48′ E	[40]
5	BM.1980, P31 (24)	Nephelinite	Tundulu complex, Malawi	15°32′ S; 35°48′ E	[40]
6	BM.1980, P31 (27)	Phonolite	Tundulu complex, Malawi	15°32′ S; 35°48′ E	[40]
7	BM.1981, P14 (440)	Nepheline carbonatite	Alnö, Sweden	62°24′N; 17°28′ E	[41]
8	BM.1981, P3	Phonolitic nephelinite	Limbe (Victoria), Cameroon	4°04°N; 9°08′ E	[42,43]
9	BM.1995, P6 (40)	Phonolitic nephelinite	Kerimasi, Tanzania	2°52′ S; 35°57′ E	[44]
10	BM.1995, P6 (43)	Phonolitic nephelinite	Kerimasi, Tanzania	2°52′ S; 35°57′ E	[44]
11	BM.2004, P12 (28)	Phonolitic nephelinite	Oldoinyo Lengai, Tanzania	2°46′ S; 35°55′ E	[45–47]
12	BM.2004, P12 (75)	Phonolitic nephelinite	Oldoinyo Lengai, Tanzania	2°46′ S; 35°55′ E	[45–47]
13	KF 85	Nepheline syenite	Kalkfeld, Namibia	20°48′ S; 16°07′ E	[48]
14	KF 88	Nepheline syenite	Kalkfeld, Namibia	20°48′ S; 16°07′ E	[48]

Aliquots of samples no. 1 to 12 were ground to a fine powder in an agate mill. Chemical analyses were conducted by classical XRF methods on lithium tetraborate glass beads and pressed powder pellets using a Bruker Pioneer S4 crystal spectrometer at the Department for Chemistry and Physics of Materials, University of Salzburg. Obtained net count rates on single X-ray lines were recast into concentration data (wt% and ppm) based on an in-house calibration routine that involves measurements of ~30 international geostandards (USGS and GSJ). The calibration relies on the Bruker AXS software SPECTRAplus FQUANT (v1.7) and corrects absorption, fluorescence and line overlap effects. In addition, a monitor standard (GSJ Granodiorite JG-1a) was measured together with the samples. Analytical results of Table 2 include information on detection limits and typical analytical uncertainties for single elements. Reported errors are conservative and refer not only to the XRF counting statistics, but consider also the uncertainty of the linear fit of the calibrations. Loss on ignition (LOI) was determined gravimetrically after heating the dried samples to 1050 °C for two hours. Samples no. 13 and 14 had been already analyzed by BÜHN & TRUMBULL [48].

Table 2. Chemical compositions of investigated samples. L.O.I. = loss on ignition; bdl. = below detection limits. Total iron is given as Fe_2O_3. The CIPW Norm (cf. text) was calculated with an assumed ratio of Fe^{3+}/(total iron) = 0.5.

No.	1	2	3	4	5	6	7	8	9	10	11	12
Major elements (wt%)												
SiO_2	52.04	45.71	53.82	51.45	34.99	46.87	18.75	45.25	47.65	46.49	48.32	45.67
TiO_2	0.52	0.65	0.75	0.42	3.10	1.22	1.31	1.11	1.21	1.48	1.06	1.02
Al_2O_3	19.95	16.03	19.89	19.06	10.32	19.24	8.89	19.05	15.92	17.51	17.17	17.05
Fe_2O_3	6.28	7.46	3.81	5.71	11.05	5.93	9.17	6.74	9.68	8.82	7.52	7.27
MnO	0.30	0.15	0.19	0.44	0.19	0.21	0.30	0.37	0.28	0.24	0.22	0.21
MgO	0.33	3.60	0.88	0.16	14.72	1.29	1.30	1.23	1.07	1.63	0.42	0.42
CaO	1.97	7.64	2.16	1.63	12.18	3.82	33.46	6.32	8.07	6.93	3.97	5.26
Na_2O	6.40	11.55	9.96	7.51	3.00	8.68	2.95	8.10	8.54	9.29	10.05	9.51
K_2O	7.19	1.62	5.72	7.97	3.21	7.43	1.56	6.09	5.31	4.99	5.88	5.71
P_2O_5	0.08	1.21	0.20	0.17	1.12	0.25	1.89	0.16	0.34	0.46	0.14	0.61
SO_3	0.12	0.44	0.25	0.10	0.39	0.11	0.21	0.29	0.05	0.11	0.22	0.17
F	0.17	0.20	0.20	0.22	0.43	0.31	0.18	0.32	0.18	0.21	0.27	1.25
L.O.I.	4.77	3.64	1.66	4.35	4.83	3.70	18.91	3.51	1.16	1.26	4.04	5.06
Total	100.12	99.90	99.49	99.19	99.53	99.06	98.88	98.54	99.46	99.42	99.28	99.21
Trace elements (ppm)												
Ba	2272	980	705	286	2986	1942	1055	3370	1760	1333	1716	1923
Ce	248	39	273	375	232	185	1284	212	270	246	249	115
Cl	626	146	1739	90	173	886	71	2619	231	1184	1491	319
Co	5	20	8	4	57	8	17	8	9	15	9	8
Cr	8	32	28	8	384	28	26	14	22	70	10	3
Ga	40	20	25	42	13	26	10	26	32	28	30	33
La	122	20	169	167	123	112	647	175	161	143	156	98
Nb	278	55	193	749	99	202	528	343	190	156	159	132
Nd	41	11	51	bdl.	84	22	525	4	69	83	65	15
Ni	9	26	13	10	399	20	13	9	13	14	8	8
Pb	37	24	20	30	5	6	bdl.	bdl.	29	22	25	8
Rb	186	36	165	222	104	160	14	192	131	95	125	97
Sr	777	837	511	871	2267	2847	6205	6184	1539	1660	2201	1862
Th	89	bdl.	33	107	5	9	38	bdl.	29	26	28	7
U	13	bdl.	bdl.	47	bdl.	15	bdl.	10	5	bdl.	7	8
V	87	101	42	31	258	91	241	242	189	146	179	172
W	11	16	17	21	7	19	12	17	14	11	19	13
Y	29	12	30	45	25	23	59	26	33	36	31	45
Zn	246	114	153	338	101	145	91	210	203	166	187	198
Zr	985	168	882	4512	287	815	312	875	587	408	494	516

Except the carbonatite (no. 7, BM.1981, P14 (440)), all samples have low to intermediate SiO_2 contents between 34.99 and 53.82%. The nephelinite of sample no. 5 (34.99% SiO_2) and the carbonatite of sample no. 7 (18.75% SiO_2) are ultrabasic by definition (<45% SiO_2). All samples are undersaturated in silica (without normative quartz). With regard to molar proportions, 10 samples (except no. 1, 5, 7, and 14) have Al_2O_3 < (Na_2O + K_2O), and can be classified as subaluminous to peralkaline in composition—depending on their mafic minerals. In the classification scheme of COX et al. [49], all samples—except of the carbonatite (no. 7)—are in the compositional range of phonolite, phonolitic nephelinite and nephelinite. Normative nepheline contents range between 12.46% (no. 5) and 36.84

(no. 2). Samples no. 5, 6, 7, 8, 9, 10, 11, and 12 exhibit normative leucite (up to 23.55% in sample no. 6). Twelve samples (except of no. 5 and 7) have normative orthoclase (up to 50.65% in sample no. 14). Eight samples (except of no. 6, 8, 9, 10, 11, and 12) have normative plagioclase (up to 23.31% in sample no. 1). Except no. 14, all samples have normative diopside (up to 23.99% in sample no. 2). Most samples (except of no. 1, 5, 7, and 14) have normative aegirine (up to 14.06 in sample no. 9). The higher normative aegirine contents (>8%) coincide quite well with observed aegirine augite in the thin sections. Thus, a certain amount of Na_2O has been consumed by the growth of pyroxene and cannot be considered for the weight proportions in the system $NaAlSiO_4$-$KAlSiO_4$-SiO_2 (phase relationships of Figure 3). In this context it is important to note that powdered aliquots of sample no. 3, 9, 10, 11, and 12 melted during the determination of the loss on ignition. Thus, the bulk composition of the felsic components of sample no. 3, 9, 10, 11, and 12 must be very close to the ternary eutectic in the system SiO_2-$NaAlSiO_4$-$KAlSiO_4$ (cf. Figure 3a).

After careful examination under a petrographic microscope (cf. Figure 4), polished thin sections of 10 samples were chosen for chemical etching. Sample no. 3 (phonolite) was rejected because it contains no nepheline phenocrysts; sample no. 7 (carbonatite) was rejected because its nepheline is altered and not idiomorphic; samples 13 and 14 were rejected because they either contain no nepheline phenocrysts (no. 13) or only few interstitial phenocrysts of nepheline (no. 14).

Figure 4. Thin sections of nephelinite samples in transmitted light. (**a**) Strongly altered nepheline crystals (Ne) in nephelinite from the Tundulu Complex in Malawi; sample no. 6; BM.1980, P31 (27). (**b**) Porphyric nepheline (Ne) and green aegirine augite (Aeg) in a dark cryptocrystalline matrix; Kerimasi, Tanzania; sample no. 9; BM.1995, P6 (40). (**c**) Porphyric nepheline (Ne) and green aegirine augite (Aeg) together with fine-grained sanidine in a dark cryptocrystalline matrix; Kerimasi, Tanzania; sample no. 10; BM.1995, P6 (40). (**d**) Porphyric nepheline (Ne) and green aegirine augite (Aeg) in fine-grained to cryptocrystalline matrix (mainly sanidine); Oldoinyo Lengai, Tanzania; sample no. 11; BM.2004, P12 (28). Scale bars correspond to 500 μm.

Etch figures on prism or pyramidal faces of nepheline are highly asymmetric but they have the same orientation as other etch figures of the same crystallographic face [9,13,14,50,51]. Such etch figures can be produced with strongly diluted hydrofluoric acid at room temperature and are often depicted in mineralogical textbooks as an example for enantiomorphism (Figure 5). Nepheline can occur as compound twins, the twinning planes being the base and/or a second order prism [14]. In the course of etching, the shape of the etch figures evolves in a characteristic manner. Initial etch figures are asymmetric triangles. They become arcuate with progressive etching, and finally exhibit a drop-shaped form with strong asymmetry. The evolution schema of Figure 6 is a compilation from BAUMHAUER [13,50], TRAUBE [14], and observations of HEJL [9,51].

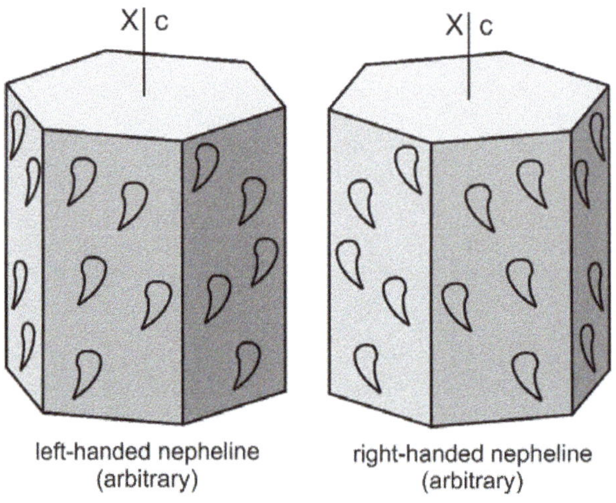

Figure 5. Idealized enantiomorphic single crystals of nepheline with asymmetric etch figures according to [13,14,50,51]. The hexagonal crystallographic c-axis corresponds to the X-axis of the optical indicatrix.

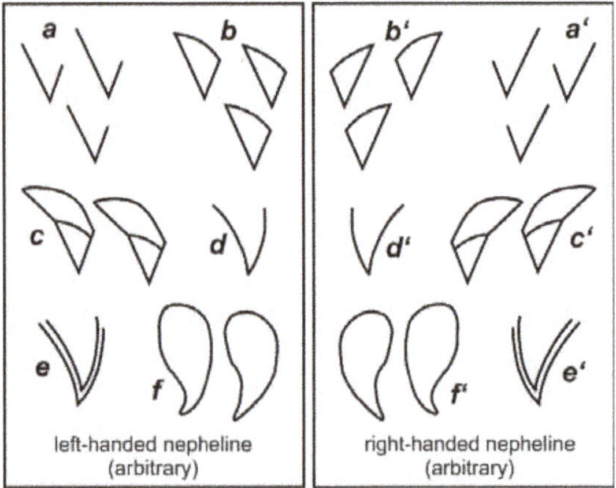

Figure 6. Typical shapes of asymmetric etch figures on prism faces of nepheline, depending on the handedness of the crystal [13,14,50,51].

Thin sections of 10 samples (i.e., all samples of Tables 1–3, except no. 3, 7, 13, and 14) were step-etched with 1 wt% HF at 20 °C for 20 to 35 min. Etch figures of nepheline in the thin sections no. 9, 10, 11, and 12 became visible after 15 min and were almost perfect after 25 min of etching. Etching of thin section 1 was terminated after 20 min because no typical etch figures became visible on the fine grained and partly altered nepheline. Etching of thin section no. 2 did not produce any etch figures after 35 min. Etching of thin Sections no. 4 and 5 was terminated after 20 min because only indistinct etch figures became visible. Etching of thin section no. 6 was etched for 35 min and did not exhibit any etch figures. Thin Section no. 8 was etched for 20 min; it exhibits few and mainly indistinct etch figures. Thus, characteristic asymmetric etch figures could only be produced in the thin sections of four samples (Table 3). Such etch figures at various stages of development are shown in Figure 7. They can be easily identified as either left-handed or right-handed by comparison with the development sequences of Figure 6.

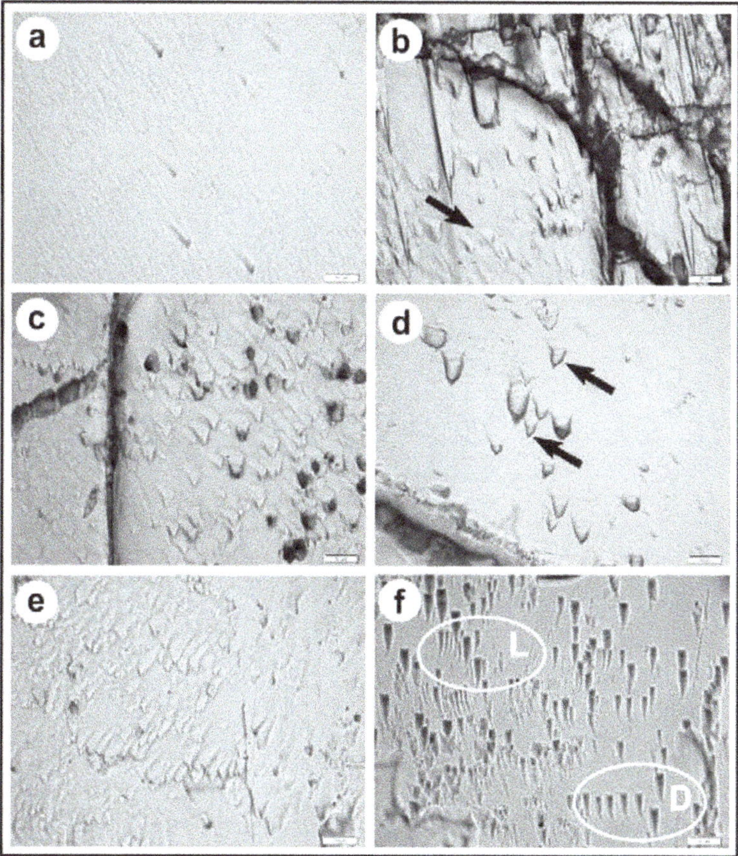

Figure 7. Etch figures on nepheline crystals from East African nephelinite samples. (**a**) Left-handed etch figures (type a) in sample no. 9, BM.1995, P6 (40). (**b**) Left-handed etch figures (arrow: type b) in sample no. 9. (**c**) Left-handed etch figures (mainly type c) in sample no. 9. (**d**) Left-handed etch figures (arrow: type d) in sample no. 11, BM.2004, P12 (28). (**e**) Right-handed etch figures (mainly type c') in sample no. 9. (**f**) left-handed (L) and right-handed (D) etch figures (type f and f', respectively) in a twinned crystal of sample 10, BM.1995, P6 (43). Scale bars correspond to lengths of 10 μm (**a**,**c**–**f**), and 20 μm (**b**).

Table 3. Chiral proportions of nepheline crystals in investigated samples.

No.	Remarks	Counting Statistics									
		Indistinct		Twins		L-Type		D-Type		Total	
		n	%	n	%	n	%	n	%	n	%
1	fine-grained matrix with few altered phenocrysts of nepheline and sanidine	—	—	—	—	—	—	—	—	—	—
2	fine-grained matrix with some small nepheline phenocrysts; no etch figures	—	—	—	—	—	—	—	—	—	—
3	sanidine phenocrysts (>5 mm) in fine-grained matrix; no neph. phenocrysts	—	—	—	—	—	—	—	—	—	—
4	altered phenocrysts of sanidine and aegirine in a fine-grained matrix	—	—	—	—	—	—	—	—	—	—
5	few large nepheline phenocrysts (>5 mm) in a cryptocrystalline matrix	—	—	—	—	—	—	—	—	—	—
6	cryptocryst. matrix with altered pheno-crysts of leucite and few nepheline	—	—	—	—	—	—	—	—	—	—
7	medium-grained fabric of carbonate, aegirine, and altered nepheline	—	—	—	—	—	—	—	—	—	—
8	dark cryptocrystalline matrix with leucite, nosean, and few nepheline	—	—	—	—	—	—	—	—	—	—
9	many fresh nepheline (>1 mm) and some aegirine in cryptocryst. matrix	53	44.20	13	10.80	28	**23.30**	26	**21.70**	120	100
10	many fresh nepheline (>1 mm) and some aegirine in cryptocryst. matrix	37	46.25	7	8.75	15	**18.75**	21	**26.25**	80	100
11	fresh and big nepheline (1–7 mm) and some aegirine in fine-grained matrix	23	46.00	6	12.00	11	**22.00**	10	**20.00**	50	100
12	slightly alterd nepheline (1–5 mm) and aegirine in a dark cryptocryst. matrix	25	50.00	5	10.00	9	**18.00**	11	**22.00**	50	100
13	porphyric K-feldspar in a groundmass of medium-grained feldspar and biotite	—	—	—	—	—	—	—	—	—	—
14	equigranular fabric of perthitic feldspar with few interstitial nepheline	—	—	—	—	—	—	—	—	—	—

Statistical proportions of left-handed and right-handed nepheline were determined in the following manner. After etching, the whole area of the thin sections was screened under a petrographic microscope at a magnification of 1250× for the identification of nepheline. It was decided to determine grain percentages instead of volumetric or mass percentages. Therefore, every discernible nepheline crystal with suitable orientation was recorded one time, regardless of its size. The counted crystals should fulfill the following criteria: they should be idiomorphic, larger than 0.3 mm, and their crystallographic c-axis should be parallel or subparallel to the plane of section (inclination < 20°). The latter criterion was tested under the petrographic microscope by the shape of the crystal boundaries and by the interference colors when a gypsum plate is inserted in addition position (thickness of the sections is about 30 μm). The crystals were divided into four classes: crystals without chiral etch figures, twinned crystals with chiral etch figures, single crystals with right-handed etch figures and single crystals with left-handed etch figures. Counting results are given in Table 3.

In exceptional cases a twinning plane can be parallel to the thin section without crosscutting it and would be erroneously counted as single L- or D-type. The twinning plane in Figure 7f crosscuts the thin section horizontally in the middle of the photograph; the symmetry relations of corresponding etch figures (both L and D) indicate that the twinning plane must be parallel to the c-axis, i.e., in an oblique (acute-angled) position to the section plane.

In sample no. 9 (BM.1995, P6 (40)) a total of 120 nepheline crystals fulfilling the required conditions was found. Fifty-three of them had only indistinct etch figures, 13 were found to be twins with etch figures of opposite handedness, and 54 had only etch figures of a single chiral type (either only L-type or only D-type). Among the latter, 28 are left-handed and 26 are right handed. Thus, 23.3% of the crystals are left-handed, 21.7 are right-handed, and 55% are indistinct or twinned. Among those crystals with distinct chirality, 51.85% are left-handed and 48.15% are right-handed.

In sample no. 10 (BM.1995, P6 (43)) a total of 80 nepheline crystals fulfilling the required conditions was found. Thirty-seven of them had only indistinct etch figures, 7 were obviously twinned, and 36 had only etch figures of a single chiral type (either only L-type or only D-type). Among the latter, 15 are left-handed and 21 are right handed. Thus, 18.75% of the crystals are left-handed, 26.25% are right-handed, and 55% are indistinct or twinned. Among those crystals with distinct chirality, 41.67% are left-handed and 58.33% are right handed.

It was difficult to find many suitable nepheline crystals in the samples no. 11 and 12 (BM.2004, P12 (28 and 75)). Therefore, only 50 crystals were evaluated in each of them. As in the above-mentioned samples, more than 50% of the crystals are indistinct or twinned (cf. Table 3). Among the well-defined chiral individual crystals, the proportions are close to parity: 11 L-Type (52.38%) vs. 10 D-type (47.62%) in sample no. 11; 9 L-type (45%) vs. 11 D-type (55%) in sample no. 12.

The level of significance of these statistical proportions can be estimated by the p-value (probability value) which indicates the probability that, when the null hypothesis is true, the deviation from the expected result is the same or greater than that of the actual observed result. The null hypothesis for chiral proportions of nepheline is that both L-type and D-type occur in similar amounts (probability = 0.5) and that for a number n of well-defined chiral crystals, the expected number of L-type and D-type crystals is $n/2$. The calculated p-values for samples no. 9, 10, 11, and 12 are 0.45, 0.20, 0.50, and 0.41, respectively. Each of these p-values is compatible with chiral parity, but the chiral proportions of sample 9 have the highest probability for a significant chiral excess, because the observed proportions have only a 20% probability to be found when the null hypothesis is true.

The counting results of nepheline crystals show that most crystals are not twinned and can be assigned as either left-handed or right handed. Measured chiral proportions in three samples are well compatible with chiral parity (no chiral excess). One sample (no. 10, BM.1995, P6 (43)) indicates a slight chiral excess of the D-type at a low level of significance. These findings do not necessarily exclude the possibility that under very special circumstances magmatic nepheline can exhibit a stronger chiral enrichment by autocatalytic secondary nucleation in a turbulent magmatic flow, but they show that this is not the general case.

None of the investigated samples has an Archean age and the sampled alkaline provinces in Africa and Sweden are certainly much younger than the appearance of biomolecular homochirality on planet Earth. Only little alkaline magmatism is known from Precambrian terrains. The oldest well-documented alkaline igneous rocks (leucite trachytes and phonolites) have an age of 2.7 Ga and occur in the Kirkland Lake area of the Superior Province, Canada [52]. To date, no alkaline rocks were reported from terrains older than 2.7 Ga. Thus, the time gap between the earliest evidence for life on Earth at 3.7 Ga [53] and the earliest known alkaline rocks is 1.0 Ga. Incomplete preservation of alkaline rocks could be due to preferential destruction by erosion of higher continental settings (volcanoes) or to almost complete subduction of alkaline volcanic islands. On the other hand, alkaline magmatic activity could have been rare because the Archean mantle was significantly hotter than today and has produced more extensive partial melting with only very small portions of low-degree melts.

4. Conclusions and Outlook

This pilot study has shown that nepheline enantiomorphs in unaltered volcanic rocks can be often identified by the shape of etch figures on section planes subparallel to the crystallographic c-axis. With regard to chiral proportions of nephelines originating from low-viscosity alkaline melts, the following conclusions can be drawn:

1. Not all the nepheline crystals exhibit chiral etch figures on such section planes. Up to more than 50% of the crystals with suitable orientation in a thin section do not develop chiral etch figures when they are treated with diluted hydrofluoric acid (1% HF aqu. at 20 °C).
2. Twinning occurs in magmatic nepheline but is not ubiquitous. In the investigated samples, most nepheline with chiral etch figures is not twinned.
3. The investigated samples do not exhibit a strong chiral excess of nepheline. Counting statistics of three of four evaluated samples are well compatible with chiral parity; only one sample shows a slight chiral excess (41.67% L-type vs. 58.33% D-type) but at a rather low level of significance (p-value = 0.20) because of the paucity of countable crystals (15 vs. 21, respectively).

When examples of significant chiral excess of nepheline are found in future research, two important issues need to be addressed:

1. Enantiomer separation by liquid chromatographic interaction between infiltrating molecular solutions and nepheline with chiral excess has not yet been tested in the laboratory.
2. Nepheline bearing rocks are not (yet) known from early Archean terrains. The eldest well-documented nepheline occurrences are about 1 Ga younger than the earliest evidence of life. At the present state of knowledge, this time gap is an obstacle for the validity of the outlined liquid chromatographic hypothesis with nepheline as a stationary phase.

Author Contributions: Conceptualization, E.H.; Methodology, E.H.; Validation, E.H. and F.F.; Investigation, E.H. and F.F.; Resources, E.H. and F.F.; Writing-Original Draft Preparation, E.H. and F.F.; Writing-Review & Editing, E.H.; Visualization, E.H.; Project Administration, E.H.; Funding Acquisition, E.H.

Funding: This investigation was funded by the Austrian Science Foundation (FWF—Der Wissenschaftsfonds, Fonds zur Förderung der Wissenschaftlichen Forschung) grant number P 30444-N28.

Acknowledgments: Samples no. 1–12 (BM code 1953 to 2004) were kindly provided by the Natural History Museum (NHM) of London (Cromwell Road, South Kensington, SW7 5BD London, United Kingdom). Special thanks to curator Epifanio Vaccaro for his prompt and competent support. Samples no. 13 and 14 (KF 85 and KF 88) were kindly provided by Robert Trumbull (Helmholz Centre Potsdam, GFZ German Research Centre for Geosciences, Telegrafenberg B125, 14473 Potsdam, Germany). Critical suggestions of two anonymous reviewers have helped to eliminate some shortcomings and to improve the overall argumentation.

Conflicts of Interest: The authors declare no conflict of interest.

References

1. Bailey, J.; Chrysostomou, A.; Hough, J.H.; Gledhill, T.M.; McCall, A.; Clark, S.; Ménard, F.; Tamura, M. Circular polarization in star-forming region: Implications for biomolecular homochirality. *Science* **1998**, *281*, 672–674. [CrossRef] [PubMed]
2. Bailey, J. Astronomical sources of circularly polarized light and the origin of homochirality. *Orig. Life Evol. Biosph.* **2000**, *31*, 167–183. [CrossRef]
3. Modica, P.; Meinert, C.; de Marcellus, P.; Nahon, L.; Meierhenrich, U.J.; le Sergeant d'Hendecourt, L. Enantiomeric excesses induced in amino acids by ultraviolet circularly polarized light irradiation of extraterrestrial ice analogs: A possible source of asymmetry for prebiotic chemistry. *Astrophys. J.* **2014**, *787*, 1–11. [CrossRef]
4. Bibring, J.-P.; Rosenbauer, H.; Boenhardt, H.; Ulamec, S.; Biele, J.; Espinasse, S.; Feuerbacher, B.; Gaudon, P.; Hemmerich, B.; Kletzkine, P.; et al. The Rosetta Lander ("Philae") investigations. *Space Sci. Rev.* **2007**, *128*, 205–220. [CrossRef]
5. Goesmann, F.; Rosenbauer, H.; Roll, R.; Szopa, C.; Raulin, F.; Sternberg, R.; Israel, G.; Meierhenrich, U.; Thiemann, W.; Munoz Caro, G.M. COSAC, the cometary sampling and composition experiment on Philae. *Space Sci. Rev.* **2007**, *128*, 257–280. [CrossRef]
6. Goesmann, F.; Rosenbauer, H.; Bredehöft, J.H.; Cabane, M.; Ehrenfreund, P.; Gautier, Th.; Giri, Ch.; Krüger, H.; leRoy, L.; MacDermott, A.J.; et al. Organic compounds on comet 67P/Churyumov-Gerasimenko revealed by COSAC mass spectrometry. *Science* **2015**, *349*. [CrossRef] [PubMed]

7. Meierhenrich, U. *Amino Acids and the Asymmetry of Life*; Springer: Berlin/Heidelberg, Germany, 2008; Volume XII, p. 241. ISBN 978-3-540-76885-2.
8. Meierhenrich, U. *Comets and Their Origin. The Tool to Decipher a Comet*; Wiley-VCH: Weinheim, Germany, 2015; Volume XXX, p. 352. ISBN 978-3-527-41281-5.
9. Hejl, E. Are fission tracks in enantiomorphic minerals a key to the emergence of homochirality? *J. Mineral. Geochem.* **2017**, *194*, 97–106. [CrossRef]
10. Morris, R.V.; Vaniman, D.T.; Blake, D.F.; Gellert, R.; Chipera, S.J.; Rampe, E.B.; Ming, D.W.; Morrison, S.M.; Downs, R.T.; Treiman, A.H.; et al. Silicic volcanism on Mars evidenced by tridymite in high-SiO_2 sedimentary rock at Gale crater. *Proc. Natl. Acad. Sci. USA* **2016**, *113*, 7071–7076. [CrossRef] [PubMed]
11. Bannister, F.A. A chemical, optical, and X-ray study of nepheline and kaliophilite. *Mineral. Mag.* **1931**, *22*, 569–608. [CrossRef]
12. Buerger, M.J.; Klein, G.E.; Donnay, G. Determination of the crystal structure of Nepheline. *Am. Mineral.* **1954**, *39*, 805–818.
13. Baumhauer, H. Ueber den Nephelin. *Zeitschrift für Kristallographie—Cryst. Mater.* **1882**, *6*, 209–216. (In German)
14. Traube, H. Beiträge zur Kenntnis des Nephelins und des Davyns. *N. Jb. Min. Geol. Palaeont.* **1895**, *IX*, 466–479. (In German)
15. Dmitrienko, V.E.; Ishida, K.; Kirfel, A.; Ovchinnikova, E.N. Polarization anisotropy of X-ray atomic factors and 'forbidden' resonant reflections. *Acta Crystallogr. A* **2005**, *61*, 481–493. [CrossRef] [PubMed]
16. Tanaka, Y.; Kojima, T.; Takata, Y.; Chainani, A.; Lovesey, S.W.; Knight, K.S.; Takeuchi, T.; Oura, M.; Senba, Y.; Ohashi, H.; Shin, S. Determination of structural chirality of berlinite and quartz using resonant X-ray diffraction with circularly polarized X-rays. *Phys. Rev. B* **2010**, *81*, 144104. [CrossRef]
17. Heritsch, H. Die Verteilung von Rechts- und Linksquarzen in Schriftgraniten [in german]. Tschermaks Miner. *Petrogr. Mitt.* **1953**, *3*, 115–125. [CrossRef]
18. Palache, C.; Bermann, G.B.; Frondel, C. Relative frequencies of left and right quartz. In *The System of Mineralogy*; Frondel, C., Ed.; Wiley: New York, NY, USA, 1962; p. 17.
19. Frondel, C. Characters of quartz fibers. *Am. Mineral.* **1978**, *63*, 17–27.
20. Kondepudi, D.K.; Kaufmann, R.J.; Singh, N. Chiral symmetry breaking in sodium chlorate crystallization. *Science* **1990**, *250*, 275–276. [CrossRef] [PubMed]
21. Kondepudi, D.K.; Asakura, K. Chiral autocatalysis, spontaneous symmetry breaking, and stochastic behavior. *Acc. Chem. Res.* **2001**, *34*, 946–954. [CrossRef] [PubMed]
22. Lesher, Ch.E.; Spera, F.J. Thermodynamic and transport properties of silicate melts and magma. In *The Encyclopedia of Volcanoes*, 2nd ed.; Sigurdson, H., Ed.; Elsevier: Amsterdam, The Netherlands, 2015; pp. 114–141.
23. Hess, P.C. Polymer model of silicate melts. *Geochim. Cosmochim. Acta* **1971**, *35*, 289–306. [CrossRef]
24. Hess, P.C. Polymerization model for silicate melts. In *Physics of Magmatic Properties*; Hargraves, R.B., Ed.; Princeton University Press: Princeton, NJ, USA, 1980; pp. 3–48.
25. Charles, R.J. The origin of immiscibility in silicate solutions. *Phys. Chem. Glasses* **1967**, *10*, 169–178.
26. Jones, A.R.; Winter, R.; Greaves, G.N.; Smith, I.H. MAS NMR study of soda-lime-silicate glasses with variable degree of polymerisation. *J. Non Cryst. Solids* **2001**, *293–295*, 87–92. [CrossRef]
27. Lee, S.K.; Cody, G.D.; Fei, Y.; Mysen, B.O. Nature of polymerization and properties of silicate melts and glasses at high pressures. *Geochim. Cosmochim. Acta* **2004**, *68*, 4189–4200. [CrossRef]
28. Scarfe, C.M.; Mysen, B.O.; Virgo, D. Pressure dependence of the viscosity of silicate melts. In *Magmatic Processes: Physicochemical Principles*; Mysen, B.O., Ed.; Special Publication no.1; Geochemical Society: Boston, MA, USA, 1987; pp. 59–67.
29. Dawson, J.B.; Pinkerton, H.; Norton, G.E.; Pyle, D.M. Physicochemical properties of alkali carbonatite lavas. Data from the 1988 eruption of Oldoinyo Lengai, Tanzania. *Geology* **1990**, *18*, 260–263. [CrossRef]
30. Shaw, H.R. The fracture mechanisms of magma transport from the mantle to the surface. In *Physics of Magmatic Processes*; Hargraves, R.B., Ed.; Princeton University Press: Princeton, NJ, USA, 1980; pp. 201–264.
31. Giordano, D.; Russel, J.K.; Dingwell, D.B. Viscosity of magmatic liquids: A model. *Earth Planet. Sci. Lett.* **2008**, *271*, 123–134. [CrossRef]
32. Stasiuk, M.V.; Jaupart, C. Lava flow shapes and dimensions as reflections of magma system conditions. *J. Volcanol. Geotherm. Res.* **1997**, *78*, 31–50. [CrossRef]

33. Schairer, J.F.; Bowen, N.L. Preliminary report on equilibrium-relations between feldspathoids, alkali-feldspars, and silica. *Trans. Am. Geophys. Union* **1935**, *16*, 325–328. [CrossRef]
34. Schairer, J.F. The alkali-feldspar joint in the system NaAlSiO4-KAlSiO4-SiO2. *J. Geol.* **1950**, *58*, 512–517. [CrossRef]
35. Hamilton, D.L.; MacKenzie, W.S. Phase equilibrium studies in the system NaAlSiO4 (nepheline)-KAlSiO4 (kalsilite). *Mineral. Mag.* **1965**, *34*, 214–231. [CrossRef]
36. Winter, J.D. *An Introduction to Igneous and Metamorphic Petrology*; Prentice Hall: Upper Saddle River, NJ, USA, 2001; p. 699. ISBN 978-0132403429.
37. Le Bas, M.J. *Carbonatite-Nephelinite Volcanism*; An African Case History; John Wiley: London, UK, 1977; p. 362. ISBN 978-0471994227.
38. Johnson, R.L. The Shawa and Dorowa carbonatite complexes, Rhodesia. In *Carbonatites*; Tuttle, O.F., Gittins, J., Eds.; John Wiley: New York, NJ, USA, 1966; pp. 205–224.
39. Garson, M.S. *Carbonatites in Southern Malawi*; Ministry of Natural Resources Geological Survey Department: Lilongwe, Malawi, 1965.
40. Garson, M.S. The Tundulu carbonatite ring complex in southern Nyasaland. *Mem. Nyasaland Geol. Surv.* **1962**, *2*, 248.
41. Kresten, P.; Troll, V.R. *The Alnö Carbonatite Complex, Central Sweden*; Springer: Berlin/Heidelberg, Germany, 2018; Volume XXXI, p. 196. ISBN 978-3-319-9022-1.
42. Nkoumbou, C.; Déruelle, B.; Velde, D. Petrology of Mt. Etinde nephelinite series. *J. Petrol.* **1995**, *36*, 373–395. [CrossRef]
43. Woolley, A.R. *Alkaline Rocks and Carbonatites of the World. Part 3: Africa*; London Geological Society: London, UK, 2001; p. 372. ISBN 1-86239-083-5.
44. Paslick, C.; Halliday, A.N.; James, D.; Dawson, J.B. Enrichment of the continental lithosphere by OIB melts: Isotopic evidence from the volcanic province of northern Tanzania. *Earth Planet. Sci. Lett.* **1995**, *130*, 109–126. [CrossRef]
45. Dawson, J.B. Sodium carbonatite intrusions from Oldoinyo Lengai, Tanzania: Implications for carbonatite complex genesis. In *Carbonatites Genesis and Evolution*; Bell, K., Ed.; Unwin Hyman: London, UK, 1989; pp. 255–277.
46. Klaudius, J.; Keller, J. Peralkaline silicate lavas at Oldoinyo Lengai, Tanzania. *Lithos* **2001**, *91*, 173–190. [CrossRef]
47. Maarten de Moor, J.; Fischer, T.P.; King, P.L.; Botcharnikov, R.E.; Hervig, R.L.; Hilton, D.R.; Barry, P.H.; Mangasini, F.; Ramirez, C. Volatile-rich silicate melts from Oldoinyo Lengai volcano (Tanzania): Implications for carbonatite genesis and eruptive behavior. *Earth Planet. Sci. Lett.* **2013**, *361*, 379–390. [CrossRef]
48. Bühn, B.; Trumbull, R.B. Comparison of petrogenetic signatures between mantle-derived alkali silicate intrusives with and without associated carbonatite, Namibia. *Lithos* **2003**, *66*, 201–221. [CrossRef]
49. Cox, K.G.; Bell, J.D.; Pankhurst, R.J. *Interpretation of Igneous Rocks*; George Allen & Unwin: London, UK, 1979; p. 450.
50. Baumhauer, H. Ueber die Krystallisation des Nephelin. *Zeitschrift für Kristallographie—Cryst. Mater.* **1891**, *18*, 611–618, in German.
51. Hejl, E. First observation of etched uranium fission tracks in nepheline by Hermann Traube (1895)? *Mitt. Österr. Miner. Ges.* **2017**, *162*, 83–90.
52. Blichert-Toft, J.; Arndt, N.T.; Ludden, J.N. Precambrian alkaline magmatism. *Lithos* **1996**, *37*, 97–111. [CrossRef]
53. Ohtomo, Y.; Kakegawa, T.; Ishida, A.; Nagase, T.; Rosing, M.T. Evidence for biogenic graphite in early Archean Isua metasedimentary rocks. *Nat. Geosci.* **2014**, *7*, 25–28. [CrossRef]

Article

Symmetry Breaking in Self-Assembled Nanoassemblies

Yutao Sang [1,2] and **Minghua Liu** [1,2,3,*]

1 CAS Key Laboratory of Colloid, Interface and Chemical Thermodynamics, Institute of Chemistry, Chinese Academy of Sciences, Zhongguancun North First Street 2, Beijing 100190, China
2 University of Chinese Academy of Sciences, Beijing 100049, China
3 Collaborative Innovation Centre of Chemical Science and Engineering, Tianjin 300072, China
* Correspondence: liumh@iccas.ac.cn

Received: 26 June 2019; Accepted: 18 July 2019; Published: 25 July 2019

Abstract: The origin of biological homochirality, e.g., life selects the *L*-amino acids and *D*-sugar as molecular component, still remains a big mystery. It is suggested that mirror symmetry breaking plays an important role. Recent researches show that symmetry breaking can also occur at a supramolecular level, where the non-covalent bond was crucial. In these systems, equal or unequal amount of the enantiomeric nanoassemblies could be formed from achiral molecules. In this paper, we presented a brief overview regarding the symmetry breaking from dispersed system to gels, solids, and at interfaces. Then we discuss the rational manipulation of supramolecular chirality on how to induce and control the homochirality in the self-assembly system. Those physical control methods, such as Viedma ripening, hydrodynamic macro- and micro-vortex, superchiral light, and the combination of these technologies, are specifically discussed. It is hoped that the symmetry breaking at a supramolecular level could provide useful insights into the understanding of natural homochirality and further designing as well as controlling of functional chiral materials.

Keywords: symmetry breaking; assemblies; supramolecular chirality; homochirality; self-assembly; vortex

1. Introduction

Chirality is one of the fundamental properties found in nature and also vital in many fields, including chemistry, physics, biochemistry, pharmacy, and materials science [1–5]. The chirality of molecular components has dramatic consequences in life systems, where biological homochirality, e.g., *L*-amino acids and *D*-sugar are solely selected, still remains a big mystery [6,7]. It has a profound effect across a spectrum of disciplines in both industrial and academic researches [8,9]. Beyond the molecule level, the chirality is extended down to subatomic and up to supramolecular and higher hierarchical levels [10]. The biological DNA double helixes and helical proteins are the typical consequence of molecular chirality and formed via self-assembly at a supramolecular level. In these chiral nanostructures, the structural complexity, information storage, and the realization of complicated functions significantly related to the chirality. At a supramolecular level, many chiral or achiral molecules prefer to self-assemble into an asymmetric packing mode, thus we could detect supramolecular chirality from various combinations of chiral and achiral molecules [11,12]. Generally, the chirality can be expressed from the chiral molecules to the self-assembled assemblies via non-covalent bonds. For the mixed system including chiral and achiral building blocks, the chirality can be transferred from the chiral components to achiral components, and thus endowing the achiral molecules with supramolecular chirality. Apart from that, even exclusively achiral molecules can form chiral aggregates due to the symmetry breaking [13–17]. While in most of the cases, an equal amount of the right and left-handed assemblies appeared from achiral molecules, predominant chiral aggregates have also been found. However, while these systems are interesting, a great challenge remains in controlling the supramolecular chirality or achieving the homochiral assemblies from the stochastic distributions.

It is generally believed that homochirality is attained in consecutive steps: starting from a minute chiral bias, and a subsequent chiral amplification process for enantiomeric enrichment and chiral transmission from one set of molecules to another [18,19]. For the first stage, the chiral bias is usually generated by the symmetry breaking. For the second stage, amplifying chirality by using self-assembly is a well-known strategy in supramolecular assemblies [11]. Therefore, exploring the symmetry breaking in self-assembled nanoassemblies is crucial and inspired for the understanding the of biomolecular homochirality during the evolution of life [19–23].

In this review, we will firstly present a brief synopsis of symmetry breaking in self-assembled systems, including crystals, liquid crystals, air/solution and solution/solid interface, and supramolecular gels. After that, we pay more attention to the manipulation of supramolecular chirality in the self-assembled systems by physical or chemical ways such as adding chiral dopants, applying Viedma ripening, hydrodynamic macro- and micro-vortex, superchiral light, and the combination of these technologies. Those physical control ways are mostly discussed.

2. Symmetry Breaking in Self-Assembly Systems

Over the past two decades, spontaneous symmetry breaking has been found in various self-assembly systems including crystals [24–26], Langmuir monolayers [27–29], liquid/solid interface [30–32], liquid crystals [33–37], dye aggregates [14,16,38,39], amphiphilic assemblies [40,41], and supramolecular gels [42–47]. Figure 1 illustrates some typical examples in these field. One chiral crystal that has been investigated in detail is tartrate [48]. During crystallization process, two kinds of chiral crystals with almost identical physical properties can be formed. However, the plane of polarization rotates are different when linearly polarized light (LPL) passes through these chiral crystals. It should be noticed that their optical activity maintains when these crystals are dissolved in water. However, another well-studied crystal with chirality in solid, named as sodium chlorate ($NaClO_3$), behaves quite differently, as it displays optical activity in crystal state (Figure 1A), but the optical properties disappears when the crystal totally dissolves in solution [49,50]. It is known that a crystal exists in a molecular or ion state in solution. Therefore, the above phenomenon suggests that ions of tartrate molecules are chiral. Conversely, the ions of the $NaClO_3$ molecule are achiral. This is one typical example of symmetry breaking in crystal systems, and thereafter many achiral molecules have been found to be chiral in crystals [51–55].

Apart from a few reports of symmetry breaking in solution [16,38,56–59], the introduction of an interface is effective to control the intermolecular interactions, thereby affecting their self-assembly properties. The symmetry also can be broken at a confined two-dimensional (2D) surface/interface during the self-assembly of achiral building blocks [60]. In fact, the formation of symmetry breaking at a surface/interface is more common than chiral crystallization in three dimensions [61]. However, compared with the solution state, the analysis of assemblies at surfaces/interface can be more complicated. For example, the chiral domains are usually hard to distinguish and the artefacts can confound the optical spectroscopy measurements. Even so, the study of symmetry breaking on surfaces and at interfaces not only assists our understanding of three-dimensional crystallization and self-assembly process, but also provides insight into how to fabricate otherwise complicated chiral materials, such as chiral graphene nanoribbons [62,63].

Among many kinds of 2D materials, Langmuir–Blodgett (LB) technology is an effective method to orderly arrange molecules on the interface [64,65]. Figure 1B gives a novel type of chiral assemblies fabricated from achiral amphiphilic molecules, a derivative of barbituric acid [28]. This molecule could form a chiral LB films at the air/water interface. As shown in the atomic force microscope (AFM) images, this LB film was consisted of spiral nanoarchitectures. More interestingly, these spirals were found to wind in both an anticlockwise (CCW) and a clockwise (CW) direction, and a careful investigation indicated this morphology was closely depended on the surface pressure. It was noted that the H-bonding between barbituric acid derivatives themselves is crucial for the self-assembly. Specifically, the carbonyls in 4-and 6-carbonyl of the pyrimidinetrione could form H-bonds with the hydrogen in the 1- and 3-N-H of the neighbouring pyrimidinetrione. Owing to the large aromatic rings and the directionality nature of hydrogen bonds between the amphiphilic molecules, the neighbouring molecules would incline from the same direction. Such aggregates gradually grow up to chiral fiber-like nanostructure, and then further curve in a fixed direction to form spirals [28].

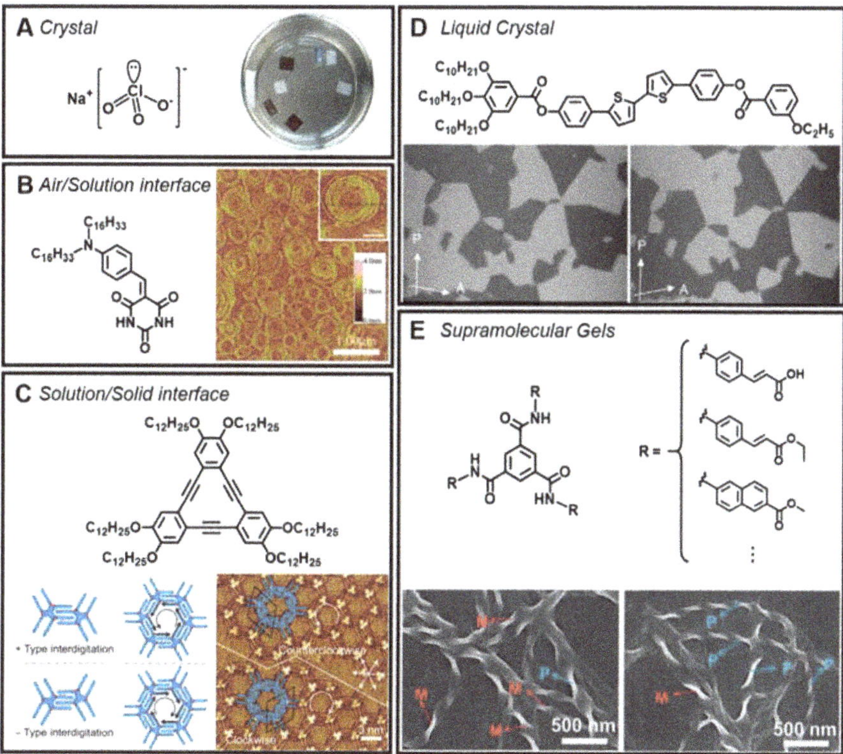

Figure 1. Symmetry breaking in different self-assembly systems. (**A**) Molecular structure of sodium chlorate and its crystal image observed through a pair of polarizers. Reprinted with permission from ref. [24]. (**B**) Atomic force microscope (AFM) image of one-layer Langmuir–Blodgett (LB) films deposited at 7 mN/m pressures after inflection point. Reprinted with permission from ref. [28]. (**C**) Chemical structures of an achiral alkoxylated dihydrobenzo [12] annulene derivative. Molecular models and scanning tunneling microscopy (STM) image of the honeycomb structure at the 1-phenyloctane/graphite interface. Reprinted with permission from ref. [66]. (**D**) Photomicrographs of chiral domains (dark/bright) fabricated from achiral liquid crystal molecule. Reprinted with permission from ref. [34]. (**E**) Chemical structures of some achiral gelators and the SEM images of optically active supramolecular gels. Reprinted with permission from ref. [43,46,67].

Figure 1C shows another kind of 2D supramolecular chirality, which is observed at the liquid–solid interface [66]. With the support of scanning tunnelling microscopy (STM), the chirality properties of 2D molecular organization can be clearly observed with high resolution under extreme conditions [30,68,69]. Surface-confined nanoporous assemblies, such as a periodic mesh of nanowells shown in Figure 1C, represent one special kinds of monolayers. The chirality of this monolayer is distinguished according to the molecular models of interdigitation motifs. Figure 1C gives the typical STM image of the honeycomb architectures from achiral molecules at the 1-phenyloctane/graphite interface. Clearly, the CW and CCW rotation of nanowells (indicated by white line) could be observed. The alkyl chain interdigitation interactions between neighbouring molecules at the liquid–solid interface were found to be crucial for the fabrication of the porous network structures [66].

In liquid crystal systems, there are many pioneering works that leading the research of symmetry breaking [33,35,70,71]. This is because liquid crystal is one unique phase that lies in between the crystals (or solid) and diluted solution. Different from the crystals, the weak intermolecular interactions in liquid crystal indicates that the driving force for chirality (long-range helical order) is relative weaker. On the other hand, the short-range chirality in liquid crystals is retained compared with the solution state. This combination leads to many distinct features, such as stimuli-responsive, self-healing, and adaptive behavior, which makes liquid crystals as the quintessential materials for self-assembly and symmetry breaking [72–74]. As shown in Figure 1D, when the crossed analyser and polarizer were slightly rotated, two kinds of domains are observed in some of the cubic phases [34]. Remarkably, when the rotation of the analyzer is reversed, the darker and brighter domains could exchange their contrast. However, the contrast does not change when the sample is rotated between the fixed polarizers, indicating that these domains belongs to a chiral structure with opposite handedness [34].

Recently, many achiral molecules, particularly the C_3-symmetric structure molecules, are found to form helical assemblies in supramolecular gel systems [45,75]. For example, as shown in Figure 1E, chiral symmetry breaking phenomena are observed in the supramolecular gel of achiral benzene-1,3,5-tricarboxamide/tricarboxylate-based molecules [43,46,67]. The directional H-bonds between amide groups and the $\pi-\pi$ stacking of the benzene rings are the main driven force during self-assembly process. When three non-chiral ethyl cinnamates were connected to the central benzene-1,3,5-tricarboxamide, the molecule was found to form instant gels with unequal number of right (*P*) and left (*M*)-handed twists, as observed from the scanning electron microscope (SEM) images in Figure 1E [43]. If the sample has more *M*-type twists than *P*-type twists, a negative Cotton effect was observed from the circular dichroism spectra; this was reversed if there were a relatively larger amount of *P*-type twists. Therefore, the uneven symmetry leads to the bulk macroscopic chirality of the supramolecular gels.

Based on the experimental data and the molecular dynamics simulation, it shows that the aggregation of these achiral C_3 molecules could initially form predominantly *P*-type or *M*-type aggregates with a random distribution. Thereafter, due to the steric hindrance caused by the closing-molecular packing, the subsequently growth of those small helical aggregates will follow the original chirality to form one-dimensional helical aggregates. Finally, such unequal numbers of *P*-type and *M*-type helical aggregates further twisted into twisted ribbons or larger helical fibers and intertwined with each other to hold the solvent [43].

It should be noted that appropriate manipulation of different noncovalent interactions can fabricate chiral ordered structures with various dimensions and complexities, which might be comparable with that found in nature systems. Among these noncovalent interactions, the hydrogen bond (H-bond) is crucial due to its directionality, strength, specificity of the interaction, and biological relevance [76–80]. For example, as shown in Figure 2A, 2,4-Diamino-6-phenyl-1,3,5-triazines with a single oligo (ethylene oxide) chain could form an optically isotropic mesophase [81]. This achiral molecule first formed a primarily double-hydrogen-bonded dimeric aggregates, and these aggregates paralleled side-by-side, leading to a highly ordered and hydrogen-bonded aromatic bilayer structures. Interestingly, the ethylene oxide chain at both ends prohibits the parallel alignment of the hydrogen-bonded cores, and thus induces a small helical twist between them. After that, the helical deformation of the bilayer ribbons is formed. The results shown that hydrogen bonding leads to chiral aggregates that undergo long-range chirality synchronization in the isotropic bulk state [81].

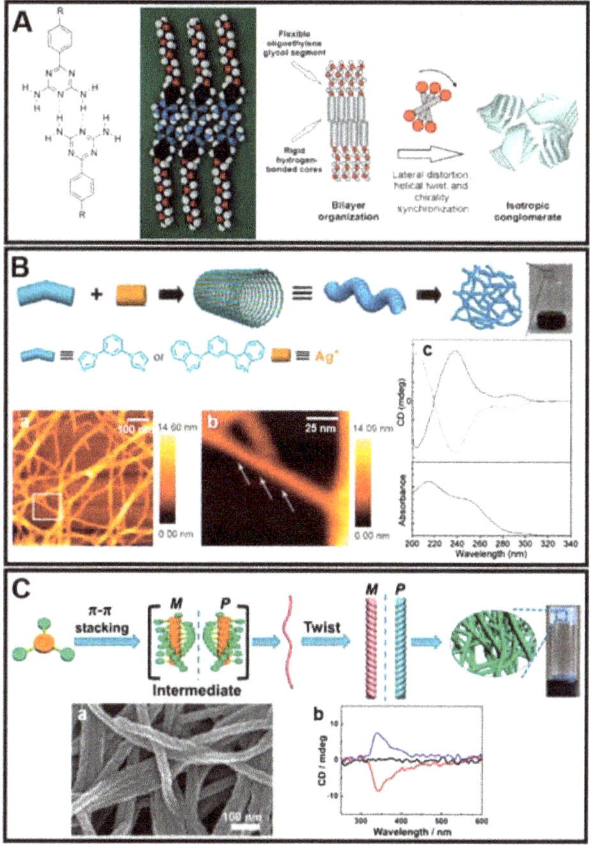

Figure 2. (**A**) Double hydrogen bonding, Corey-Pauling-Koltun (CPK) model and the scheme illustration of the chirality synchronization of hydrogen-bonded complexes of achiral N-heterocycles. Reprinted with permission from ref. [81]. (**B**) Schematic representation of the self-assembly process of the coordination polymers. (**a**) Tapping-mode atomic force microscope (AFM) height image and (**b**) a zoomed-in image of the area marked in (**a**). (**c**) Corresponding CD and UV-Vis spectra. Reprinted with permission from ref. [17]. (**C**) The possible gelation process (**a**) SEM image and (**b**) CD spectra of the chiral gels. Reprinted with permission from ref. [44].

Besides the hydrogen bond, there are a number of examples of dative-bond (coordination bond) driven symmetry breaking [82–86]. Figure 2B shows a novel coordination polymer gelators that stemmed from the coordination of Ag(I) and the achiral imidazole derivative [17]. Optically transparent gels could be formed when a methanol solution of the achiral monomer to an aqueous solution of silver nitrate at a 1:1 ratio of ligand: $AgNO_3$. The AFM measurements further revealed a well-developed network structure composed of fibrous aggregates. In addition, the zoomed-in image indicated that this thicker fiber consisted of a bundle of helical tubes. The CD spectra also exhibited mirror-imaged signals from different batches, indicating that chiral symmetry breaking is occurred during the coordination process. Due to the strong directional interactions between the rigid bent bridging ligands and Ag(I), the initial metal–ligand complexes with accidental excess of one helical direction were formed. Thereafter, the new aggregates would follow the same handedness to form a secondary helical structure. On the other hand, the formation of opposite helical aggregates is suppressed. Therefore, the eventual macroscopic chirality is observed [17].

As shown in Figure 2C, symmetry breaking that driven exclusively by weak $\pi-\pi$ interaction is studied in supramolecular assemblies [44]. An achiral C_3-symmetric gelator was found to form organogels in cyclohexane. Interestingly, the supramolecular gels were optically active with the helical nanofibers with predominant handedness. Since there are no any other noncovalent interactions in this system, the experiment results proved that purely $\pi-\pi$ stacking can also drive the symmetry breaking in the supramolecular gel system. In this case, $\pi-\pi$ interaction between benzene rings and cinnamate substituents were strong enough to form an overcrowded molecular packing. It is supposed that the achiral molecules could initially generate two kinds of helical conformer by chance, and subsequently grow up to form longer one-dimensional helical aggregates by following the original chiral conformation. At last, several helical aggregates further twisted into helical fibers [44].

In fact, combined noncovalent interactions including hydrogen bonding, electrostatic interaction, and $\pi-\pi$ stacking are generally necessary in most of the symmetry breaking systems. For example, both the hydrogen bonding and electrostatic interaction is essential in the fabrication of chiral tetraphenylporphyrin sulfonate (TPPS) aggregates [14]. Another example is the self-assembled achiral partially fluorinated benzene-1,3,5-tricarboxamides in solution, in which the 3-fold hydrogen bonding and dipole–dipole interaction play important roles [42].

3. Selection and Control of Supramolecular Chirality during Symmetry Breaking: Towards the Homochirality in Nanoassemblies

Even though the origin of chirality and homochirality in biological systems is still controversial, the quest to unravel this mystery has led to an intense research to select and control the supramolecular chirality during symmetry breaking. Actually, the manipulation of chirality in exclusive achiral systems can be much more useful, since it could not only provide a better understanding of the complicated biosystems, but also offer guidance on how to rational design of biomimetics as well as advanced chiral materials [87,88]. To date, various strategies such as changing pH [89], electroweak interaction [90,91] and microfluidic conditions [92,93], catalysis at prochiral crystal surfaces [94,95], adding chiral additive [96–99], applying circularly polarized light [100–102], rotational and magnetic force [103], and vortex and stirring motion [14,104–108], etc. are known to control the emerging chirality during symmetry breaking.

Generally, based on sergeant and soldier rule, adding chiral substance into the symmetry breaking systems is quite efficient and usual to control the macroscopic chirality [99,109,110]. For example, in the C_3 supramolecular gel system, a suitable amount of (*R*)-1-cyclohexyl ethylamine could result in the formation of *M*-type twists with a strong negative CD signal [43]. On the contrary, (*S*)-1-cyclohexyl ethylamine led to the *P*-type twists and a mirror-imaged CD spectrum. Owing to the interaction between amines and ethyl cinnamate, the added chiral amines achieved the chirality control through the ester–amide exchange reaction. In this case, the chiral amines could not completely remove from the system [43]. However, if the chiral dopants were replaced to chiral solvent, such as limonene and terpinen-4-ol, the induced chirality of the supramolecular gels could be maintained even after completely removing the chiral solvents [44].

In the solid state such as crystal systems, it is easier to achieve the chiral discrimination. Particularly, this applies to the crystallization process when the same enantiomers have a stronger interaction than that of the opposite enantiomer. Such kind of crystal is generally called as conglomerate crystals, representing that each crystal is homochiral (Figure 3A). In the words, the conglomerate can be regarded as a physical mixture of enantiomerically pure crystals of two kinds of enantiomer. However, only approximately 5–10% of chiral crystalline molecules belong to this case [9]. For the rest of approximately 90–95%, both enantiomers exist in one crystallographic unit cell, and the solid is called a racemic compound, racemate crystals, or true racemate. This is because the opposite enantiomers have a greater affinity than the same enantiomer in this case.

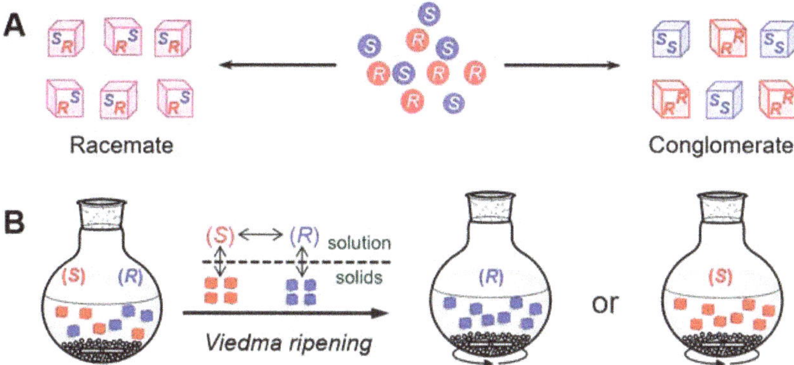

Figure 3. (**A**) Formation of conglomerate and racemate crystals during the crystallization of a racemic mixture. (**B**) Schematic representation of Viedma ripening.

It is worth noting that applying stirring during crystallization is usually effective to break the symmetry, leading to a high enantiomeric excess of the resulting crystal [25,111,112]. In 2005, Viedma demonstrated that the initially racemic mixture of $NaClO_3$ crystals could completely transform into one chiral form over a period of several days (Figure 3B) [113]. Remarkably, the completely transformation means that the whole system achieves homochirality although the final handedness could not control. This transformation process, which is now called Viedma ripening, involves the deracemization of solid-to-solid and solution-to-solid [114,115]. The novelty of Viedma's experiment lies in the addition of glass beads and magnetic bar during deracemization. The glass beads enhanced the grind of crystals, resulting in numerous small fragments with the identified handedness. Therefore, the homochirality is achieved by applying stir and grind for a period of time. It should be noted that the chiral bias can be controlled by crystal enantiomeric excess [113]. For example, solutions with initial 5% *L*-crystal enantiomeric excess give rise to 100% *L*-type crystals, and vice versa. Since that, this method was successfully extended to many crystal systems, and even organic reaction which contains crystal products. Very recently, Viedma ripening found broad applicability [18,116–122]. Without grinding treatment, a temperature gradient which involves several cycles of rapid heating and slow cooling could also realize the deracemization [123,124].

The responses of achiral molecules and its chiral assemblies to external stimulation can be also used to manipulate the supramolecular chirality after symmetry breaking. The J-aggregates of achiral amphiphilic porphyrins, such as 4-sulfonatophenyl and arylmeso-substituted porphyrins, may be the most representative model for the study of hydrodynamic forces [106,125]. Figure 4 illustrates the ground-breaking work of Ribo et al. in which the chiral signal of porphyrin aggregates can be controlled by vortex motion during self-assembly process [14]. The self-assembly of the achiral monomeric species were promoted by the rotary evaporation of very diluted solutions of deprotonated porphyrins.

Clearly, the randomly distributed chiral signals indicates that the unstirred experiments belong to a pure symmetry breaking process (Figure 4A). However, as shown in Figure 4B, the chiral selection was dependent on the rotation direction when the samples were applied rotary evaporation treatment. The statistical distribution indicated that 85% of the chiral signals could be controlled by the rotation direction, suggesting a biased symmetry breaking (Figure 4B). Due to the existence of anionic sulfonato groups and the positively charged porphyrin rings, the J-aggregation was achieved by the intermolecular electrostatic and hydrogen bonding interactions. Therefore, the arrangements of achiral porphyrins with different angles are indeed possible (Figure 4C). The schematic illustration shows that the chirality may be transferred from the macroscopic chiral force to the electronic distribution. After this report, the acting chiral hydrodynamic shear force during rotatory evaporation were theoretically investigated, and similar conclusions has been obtained in other type of vortices [126–129].

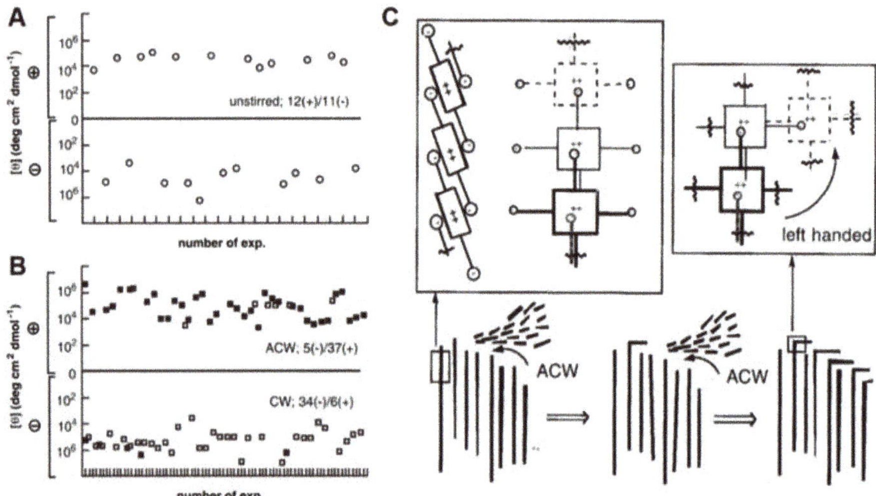

Figure 4. The chiral signals of solutions containing achiral amphiphilic porphyrins under (**A**) unstirred condition and (**B**) rotatory evaporation. The concentration was conducted by rotary evaporation from 500 mL to 20 mL for about 2 h. (**C**) Schematic illustration of the chiral selection during the self-assembly of substituted porphyrins. Reprinted with permission from ref. [14].

As one kind of the excited-state optical activity, circularly polarized light (CPL) has triggered intense research interests due to their potential applications in optical sensors [130,131], 3D displays [132,133], bioencoding [130,134,135], encrypted transmission, and storage of information [136], chiral catalysts [137–140], and photoelectric devices [141]. However, due to the small anisotropy factors ($<10^{-3}$), the obtained enantiomeric excess in most cases of asymmetric photolysis and photosynthesis is undesirable (<4%) [142,143]. When two counter-propagating CPL are interfered with opposite handedness, with different intensity but the same frequency, the generated light, which is called superchiral light (SCL), was demonstrated experimentally to enhance the enantioselective polymerization of achiral diacetylene monomer [144].

Figure 5A shows the experimental set-up for the generation of SCL from two CPL beams (325 nm) [144]. Clearly, compared with the conventional CPL, the achiral benzaldehyde-functionalized diacetylene (Figure 5B) LB films that irradiated by the SCL exhibited a stronger CD signal. In contrast, if the sample was polymerized by LPL, no signal was observed from the CD measurement (Figure 5C). As a result of the enhanced optical dissymmetry, the polydiacetylene films achieved in SCL field shows nearly 5-fold enhancement in dissymmetry factor. As illustrated in Figure 5D,E, compared with CPL irradiation, the enhanced optical dissymmetry in SCL may lead to more helical polydiacetylene (PDA) chains, thus resulting in an increased dissymmetry factor of PDA films. In addition, transmission electron microscopy (TEM) images also demonstrated that samples polymerized by SCL have more helical PDA chains than those polymerized with CPL [144].

Although all of the above approaches can be used alone to select the supramolecular chirality, not all of them are effective, due to their own limitations. In fact, a large number of reports employed more than one treatment to control the chirality of symmetry breaking [145]. For example, as shown in Figure 6A, Vlieg et al. demonstrated that the CPL irradiation caused symmetry breaking of an amino acid derivative could be amplified by a grinding process [146]. Specifically, a solid–liquid mixture was firstly irradiated with *L*- or *R*-CPL for 70 h (0.3 mW intensity). Thereafter, the slurry was grinded with a magnetic stirring bar and glass beads. In addition, organic base such as 1,8-diazabicyclo [5.4.0] undec-7-ene (DBU) was added in order to induce the racemization in solution. Five days later, the final chirality of enantiopure solid was found to be selected by the handedness of CPL. From the above description, it can be concluded that a non-racemizable chiral photoproduct may be formed under the effect of the CPL irradiation, while the subsequent grind treatment amplified the small initial chirality [146].

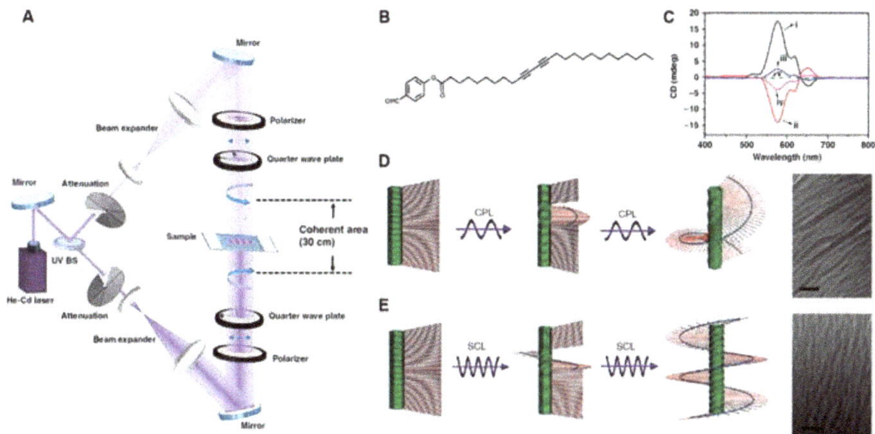

Figure 5. (**A**) Experimental set-up for the generation of superchiral light (SCL) field. (**B**) The molecular structure of achiral benzaldehyde-functionalized diacetylene. (**C**) CD spectra of polydiacetylene films polymerized by (i) left-handed or (ii) right-handed SCL; (iii) left-handed or (iv) right-handed circularly polarized light (CPL); (v) linearly polarized light (LPL), respectively. Schemes and corresponding transmission electron microscope (TEM) images for the helical PDA chains prepared by using (**D**) CPL and (**E**) SCL irradiation, respectively. Reprinted with permission from ref. [144].

Figure 6B gives an example of how the relative directions of rotation and effective gravity control the chirality of supramolecular assemblies constructed by achiral tris-(4-sulfonatophenyl) phenylporphyrin ($TPPS_3$) [103]. In order to simultaneously realize the manipulation of rotational, gravitational and orienting forces, an experiment set-up is designed and outlined in Figure 6B. To be brief, this set-up contains a tube with seven cylindrical vessels. Each of them was positioned at different positions and rotated for various time. Such a set-up ensures a solid-body rotation, and thus avoids the creation of pseudovortices. The rotation is characterized by two parameters, one is the angular momentum (L), which is set by CW or CCW rotation viewing from the top. The other factor is effective gravity (G_{eff}) that related to the magnetic levitation force. After rotation treatment, these samples were placed outside the magnet for three days before the CD measurement. The handedness of the aggregates is found to depended on the relative directions of the rotational and gravitational forces applied. For example, the antiparallel L and G_{eff} results in a positive CD signal, while parallel L and G_{eff} leads to the opposite signals. Further study revealed that the nucleation step was crucial to control the final handedness. On the basis of these experimental results, a possible schematic model was proposed. As a result of the electrostatic and π-stacking interactions, the $TPPS_3$ molecules could aggregate into small chiral nucleus. During this period, the hydrodynamic flow consistently applied. Meanwhile, the magnetic field arranges the nuclei along the rotation axis, which eliminated the randomizing Brownian motion and thus achieved the controlled chirality. After that, chiral nuclei worked as chiral seeds, and the subsequent growth followed the initial chirality even after the external stimulation was ceased [103].

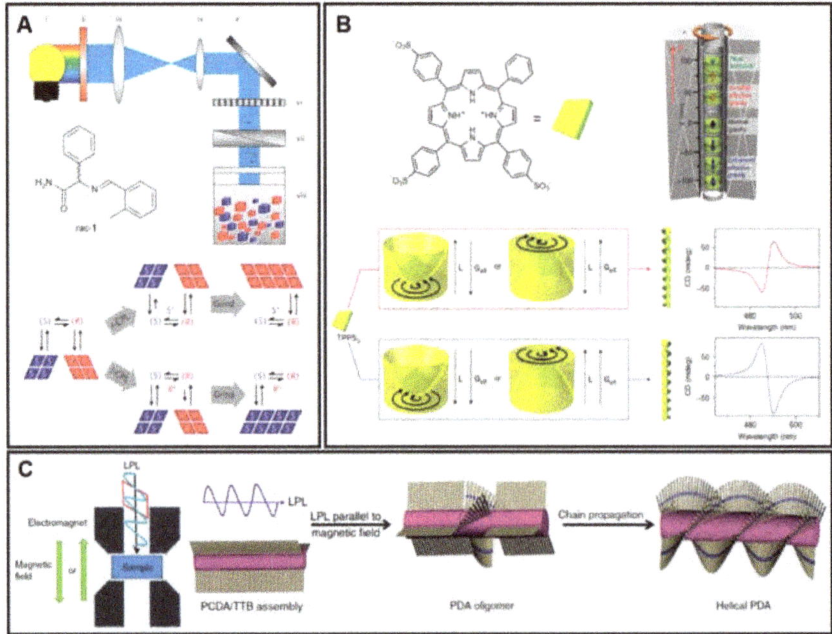

Figure 6. (**A**) Experimental set-up for circularly polarized light (CPL)-driven deracemization and the illustration of the cascade of events during this process. Reprinted with permission from ref. [146]. (**B**) Molecular formula of the achiral porphyrin tetraphenylporphyrin sulfonate ($TPPS_3$), experimental set-up and the relationship between the chirality and the applied physical forces. Reprinted with permission from ref. [103]. (**C**) Experimental set-up and the possible helix formation mechanism for the enantioselective synthesis of helical polydiacetylene by applying linearly polarized light (LPL) and magnetic field. Reprinted with permission from ref. [137].

Compared with CPL, the application of LPL irradiation is usually neglected. However, combined with other physical force, such as magnetic field, LPL is also effective to control the enantioselective polymerization of diacetylene derivative (Figure 6C) [137]. As a result of the magnetochiral dichroism effect, the achiral building blocks could selectively form one-handed helical oligomer chain. The dual effect of LPL and magnetic field might orient the helical oligomer chain in a chiral arrangement, which is benefit for the polymerization of closing monomer. Therefore, the final predominant helical chains can be directed by the relative orientation of LPL and magnetic field [137].

Very recently, inspired by natural rock micropores (Figure 7A), a microvortex generated by a microfluidic device was demonstrated to control the chirality after symmetry breaking either in gel or solution states [93]. Computational fluid dynamics (CFD) simulation suggested that the CCW and CW laminar vortices could be generated by the mismatched flow velocities between the main channel and the microchambers, and the microvortices in the left and right microchamber are predominantly *P* and *M* chirality, respectively. In addition, the high-speed microscopic observation showed that the highest rotation speed could be 4×10^4 rpm in this device. When the achiral building blocks (BTAC and $TPPS_4$) were injected into the microfluidic device, microvortex-induced symmetry breaking of these achiral molecules leads to the formation of supramolecular gels or $TPPS_4$ nuclei. Samples that obtained from two outlets always exhibited mirror-imaged CD signals. The unique feature of microvortices is the strong shear gradient, which allows the chiral alignment and formation of the supramolecular nuclei against the Brownian regime during the mirror symmetry breaking process. The microvortice controlled nuclei could be subsequently amplified into supramolecular aggregates with a certain chiral bias. As a result, the chirality distribution suggested the microvortices maintained 96% chirality control [93].

Compared with crystals, the homochirality after symmetry breaking is more difficult to achieve in soft matter and solution due to the dynamic features. In addition, the good stability of chiral assemblies is also essential. For example, the exchange between the monomers and helical aggregates is usually fast in diluted solution [99]. Figure 7B illustrated a novel strategy for obtaining almost homochiral supramolecular assemblies with controlled handedness in a totally achiral system [67]. Due to the hydrogen bond and π–π stacking, the achiral C_3 molecules could form instant gels in a mixed solvent of DMF/H_2O. However, the common gelation process only achieved racemic gels. Interestingly, applying vortex mixing during self-assembly process could significantly amplified the supramolecular chirality, leading to near-unity homochiral assemblies. In this case, the chiral signals were random distributed. The real-time monitor of CD intensity and the aggregation suggested that applying vortex mixing during the nucleation stage is sufficient. Due to the competition caused by vortex mixing, one kinds of helical nuclei occasionally dominated the system, and the chiral bias could be further amplified in the following growth. More importantly, by using a small amount of assemblies obtained as chiral seeds, a supramolecular ripening process could transform the racemic gels to the homochiral state with the seeds. Therefore, no additional chiral substances are required to obtain both chirality controlled and homochiral assemblies [67].

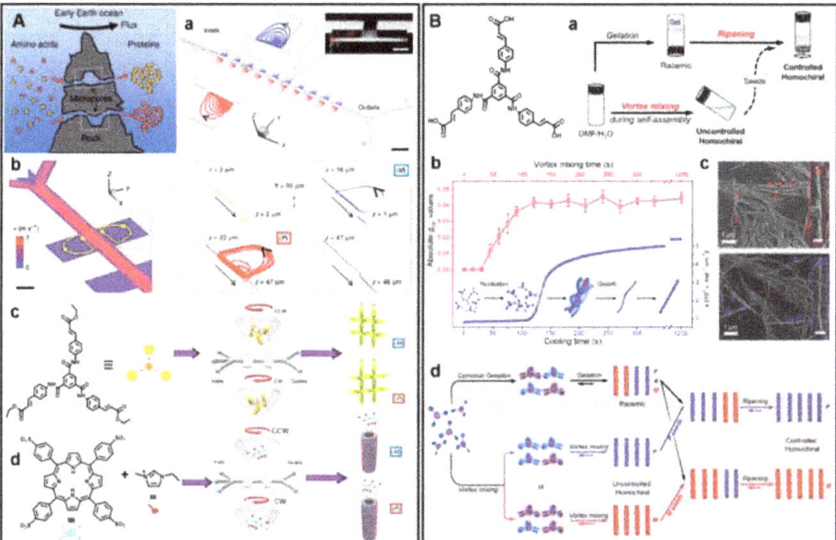

Figure 7. (**A**) Schematic hypothesis of the origin of supramolecular chirality in nature. (**a**) The imitated microvortices that generated by the microfluidic device. (**b**) computational fluid dynamics (CFD) simulation of the chiral microvortices. Formation of chiral supramolecular assemblies of achiral (**c**) (tris (ethyl cinnamate) benzene-1,3,5-tricarboxamide (BTAC) and (**d**) tetraphenylporphyrin (TPPS4) and C_2mim^+ ionic stabilizer within the microvortices. Reprinted with permission from ref. [93]. (**B**) Vortex mixing-accompanied self-assembly induced homochiral supramolecular assemblies from exclusively achiral molecules. (**a**) Schematic illustration of the procedures towards homochirality. (**b**) Red curve: the absolute dissymmetry factors (g_{CD}) of the samples prepared with different vortex times; blue curve: the correlation between the absorption data and cooling time. (**c**) SEM images of the helical structures after vortex mixing. (**d**) Schematic illustration of the mechanism towards homochirality. Reprinted with permission from ref. [67].

4. Conclusions

Symmetry breaking in self-assembled systems is an interesting phenomenon leading to an understanding of the homochirality in a biological system and providing an important method towards the construction of functional chiral materials, which significantly extend the potential applications of achiral building blocks. This review provides a summary and discussion of symmetry breaking based on the diverse systems from solution, interfaces to solids and gels. Moreover, we present a brief synopsis on the selection and control of supramolecular chirality in the self-assembled systems from achiral molecules. A number of novel physical manipulations, such as superchiral light field, microfluidic device induced microvortex, and vortex mixing have been successfully used to achieve the chiral selection/synthesis and even homochirality of nanoassemblies.

Despite these achieved developments in symmetry breaking, advanced techniques are still needed to follow the symmetry breaking process and the amplification of the supramolecular chirality from completely achiral molecules. For most of the systems, even though we could observe the microscopic chirality from the CD or CPL spectra and morphology, two enantiomers coexisted after symmetry breaking. Therefore, it is also necessary to quantitatively analyze the enantiomeric excess of the supramolecular assemblies although it is dynamic in many cases.

So far, various possibilities have been proposed to explain the emergency of the initial chiral bias in biomolecules [101,147–149]. However, these results are controversial [19,150]. On the other hand, attaining homochirality is still a challenge in exclusively achiral systems. In general, asymmetric environments or elements are necessary for the chirality control of symmetry breaking. Macroscopic chiral force caused by stirring or vortex mixing may transfer the chirality into the asymmetry molecule aggregation. Although many methods were discovered to control the chirality, the combination of two or more processes is more effective to the selection and subsequent amplification the supramolecular chirality during symmetry breaking towards the homochiral assemblies, which may enlighten the understanding the of biomolecular homochirality during the evolution of life.

Author Contributions: Both authors contributed to the writing and editing of this review.

Funding: This research was funded by Strategic Priority Research Program of the Chinese Academy of Sciences (XDB12020200), Key Research Program of Frontier Sciences, CAS (QYZDJ-SSWSLH044), and National Natural Science Foundation of China (21890734).

Acknowledgments: The authors gratefully acknowledge their collaborators whose names appear in the literature cited and the insightful comments from three reviewers.

Conflicts of Interest: The authors declare no conflict of interest.

References

1. Chelucci, G.; Thummel, R.P. Chiral 2,2′-bipyridines, 1,10-phenanthrolines, and 2,2′: 6′,2′′-terpyridines: Syntheses and applications in asymmetric homogeneous catalysis. *Chem. Rev.* **2002**, *102*, 3129–3170. [CrossRef] [PubMed]
2. Chin, J.; Lee, S.S.; Lee, K.J.; Park, S.; Kim, D.H. A metal complex that binds alpha-amino acids with high and predictable stereospecificity. *Nature* **1999**, *401*, 254–257. [CrossRef] [PubMed]
3. Engelkamp, H.; Middelbeek, S.; Nolte, R.J.M. Self-assembly of disk-shaped molecules to coiled-coil aggregates with tunable helicity. *Science* **1999**, *284*, 785–788. [CrossRef] [PubMed]
4. Lehn, J.M. Supramolecular chemistry. *Science* **1993**, *260*, 1762–1763. [CrossRef] [PubMed]
5. Prins, L.J.; Huskens, J.; de Jong, F.; Timmerman, P.; Reinhoudt, D.N. Complete asymmetric induction of supramolecular chirality in a hydrogen-bonded assembly. *Nature* **1999**, *398*, 498–502. [CrossRef]
6. Bada, J.L. Biomolecules - origins of homochirality. *Nature* **1995**, *374*, 594–595. [CrossRef]
7. Mason, S.F. Origins of biomolecular handedness. *Nature* **1984**, *311*, 19–23. [CrossRef]
8. Collins, A.N.; Sheldrake, G.; Crosby, J. *Chirality in Industry ii: Developments in the Commercial Manufacture and Applications of Optically Active Compounds*; John Wiley & Sons: Hoboken, NJ, USA, 1997; Volume 2.
9. Eliel, E.L.; Wilen, S.H. *Stereochemistry of Organic Compounds*; John Wiley & Sons: Hoboken, NJ, USA, 2008.
10. Hegstrom, R.A.; Kondepudi, D.K. The handedness of the universe. *Sci. Am.* **1990**, *262*, 108–115. [CrossRef]

11. Liu, M.; Zhang, L.; Wang, T. Supramolecular chirality in self-assembled systems. *Chem. Rev.* **2015**, *115*, 7304–7397. [CrossRef]
12. Duan, P.; Cao, H.; Zhang, L.; Liu, M. Gelation induced supramolecular chirality: Chirality transfer, amplification and application. *Soft Matter* **2014**, *10*, 5428–5448. [CrossRef]
13. Yamaguchi, T.; Kimura, T.; Matsuda, H.; Aida, T. Macroscopic spinning chirality memorized in spin-coated films of spatially designed dendritic zinc porphyrin j-aggregates. *Angew. Chem. Int. Ed.* **2004**, *43*, 6350–6355. [CrossRef] [PubMed]
14. Ribo, J.M.; Crusats, J.; Sagues, F.; Claret, J.; Rubires, R. Chiral sign induction by vortices during the formation of mesophases in stirred solutions. *Science* **2001**, *292*, 2063–2066. [CrossRef] [PubMed]
15. Azeroual, S.; Surprenant, J.; Lazzara, T.D.; Kocun, M.; Tao, Y.; Cuccia, L.A.; Lehn, J.-M. Mirror symmetry breaking and chiral amplification in foldamer-based supramolecular helical aggregates. *Chem. Commun.* **2012**, *48*, 2292–2294. [CrossRef] [PubMed]
16. DeRossi, U.; Dahne, S.; Meskers, S.C.J.; Dekkers, H. Spontaneous formation of chirality in j-aggregates showing davydov splitting. *Angew. Chem. Int. Ed.* **1996**, *35*, 760–763. [CrossRef]
17. Zhang, S.; Yang, S.; Lan, J.; Yang, S.; You, J. Helical nonracemic tubular coordination polymer gelators from simple achiral molecules. *Chem. Commun.* **2008**, 6170–6172. [CrossRef] [PubMed]
18. Saito, Y.; Hyuga, H. Colloquium: Homochirality: Symmetry breaking in systems driven far from equilibrium. *Rev. Mod. Phys.* **2013**, *85*, 603–621. [CrossRef]
19. Bonner, W.A. The origin and amplification of biomolecular chirality. *Orig. Life Evol. Biosph.* **1991**, *21*, 59–111. [CrossRef]
20. Luisi, P.L. *The Emergence of Life: From Chemical Origins to Synthetic Biology*; Cambridge University Press: Cambridge, UK, 2016.
21. Barron, L.D. Symmetry and molecular chirality. *Chem. Soc. Rev.* **1986**, *15*, 189–223. [CrossRef]
22. Cintas, P.; Viedma, C. On the physical basis of asymmetry and homochirality. *Chirality* **2012**, *24*, 894–908. [CrossRef]
23. Karunakaran, S.C.; Cafferty, B.J.; Weigert-Munoz, A.; Schuster, G.B.; Hud, N.V. Spontaneous symmetry breaking in the formation of supramolecular polymers: Implications for the origin of biological homochirality. *Angew. Chem. Int. Ed.* **2019**, *58*, 1453–1457. [CrossRef]
24. Alexander, A.J. Crystallization of sodium chlorate with d-glucose co-solute is not enantioselective. *Cryst. Growth Des.* **2008**, *8*, 2630–2632. [CrossRef]
25. Kondepudi, D.K.; Kaufman, R.J.; Singh, N. Chiral symmetry-breaking in sodium-chlorate crystallization. *Science* **1990**, *250*, 975–976. [CrossRef] [PubMed]
26. Kipping, F.S.; Pope, W.J. Lxiii.—Enantiomorphism. *J. Am. Chem. Soc.* **1898**, *73*, 606–617. [CrossRef]
27. Yuan, J.; Liu, M. Chiral molecular assemblies from a novel achiral amphiphilic 2-(heptadecyl) naphtha[2,3]imidazole through interfacial coordination. *J. Am. Chem. Soc.* **2003**, *125*, 5051–5056. [CrossRef] [PubMed]
28. Huang, X.; Li, C.; Jiang, S.; Wang, X.; Zhang, B.; Liu, M. Self-assembled spiral nanoarchitecture and supramolecular chirality in langmuir–blodgett films of an achiral amphiphilic barbituric acid. *J. Am. Chem. Soc.* **2004**, *126*, 1322–1323. [CrossRef] [PubMed]
29. Guo, P.; Zhang, L.; Liu, M. A supramolecular chiroptical switch exclusively from an achiral amphiphile. *Adv. Mater.* **2006**, *18*, 177–180. [CrossRef]
30. Elemans, J.A.A.W.; De Cat, I.; Xu, H.; De Feyter, S. Two-dimensional chirality at liquid-solid interfaces. *Chem. Soc. Rev.* **2009**, *38*, 722–736. [CrossRef] [PubMed]
31. Ernst, K. Amplification of chirality in two-dimensional molecular lattices. *Curr. Opin. Colloid Interface Sci.* **2008**, *13*, 54–59. [CrossRef]
32. Foster, J.S.; Frommer, J.E. Imaging of liquid-crystals using a tunnelling microscope. *Nature* **1988**, *333*, 542–545. [CrossRef]
33. Hough, L.E.; Spannuth, M.; Nakata, M.; Coleman, D.A.; Jones, C.D.; Dantlgraber, G.; Tschierske, C.; Watanabe, J.; Koerblova, E.; Walba, D.M.; et al. Chiral isotropic liquids from achiral molecules. *Science* **2009**, *325*, 452–456. [CrossRef]
34. Dressel, C.; Liu, F.; Prehm, M.; Zeng, X.; Ungar, G.; Tschierske, C. Dynamic mirror-symmetry breaking in bicontinuous cubic phases. *Angew. Chem. Int. Ed.* **2014**, *53*, 13115–13120. [CrossRef] [PubMed]

35. Tschierske, C.; Ungar, G. Mirror symmetry breaking by chirality synchronisation in liquids and liquid crystals of achiral molecules. *Chemphyschem* **2016**, *17*, 9–26. [CrossRef] [PubMed]
36. Keith, C.; Reddy, R.A.; Hauser, A.; Baumeister, U.; Tschierske, C. Silicon-containing polyphilic bent-core molecules: The importance of nanosegregation for the development of chirality and polar order in liquid crystalline phases formed by achiral molecules. *J. Am. Chem. Soc.* **2006**, *128*, 3051–3066. [CrossRef] [PubMed]
37. Young, W.R.; Aviram, A.; Cox, R.J. Stilbene derivatives—New class of room-temperature nematic liquids. *J. Am. Chem. Soc.* **1972**, *94*, 3976–3981. [CrossRef]
38. Qiu, Y.; Chen, P.; Liu, M. Evolution of various porphyrin nanostructures via an oil/aqueous medium: Controlled self-assembly, further organization, and supramolecular chirality. *J. Am. Chem. Soc.* **2010**, *132*, 9644–9652. [CrossRef] [PubMed]
39. Okano, K.; Taguchi, M.; Fujiki, M.; Yamashita, T. Circularly polarized luminescence of rhodamine b in a supramolecular chiral medium formed by a vortex flow. *Angew. Chem. Int. Ed.* **2011**, *50*, 12474–12477. [CrossRef] [PubMed]
40. Li, Y.; Wong, K.M.-C.; Wong, H.-L.; Yam, V.W.-W. Helical self-assembly and photopolymerization properties of achiral amphiphilic platinum(ii) diacetylene complexes of tridentate 2,6bis(1-alkylpyrazol-3-yl)pyridines. *ACS Appl. Mater. Inter.* **2016**, *8*, 17445–17453. [CrossRef]
41. Song, B.; Liu, B.; Jin, Y.; He, X.; Tang, D.; Wu, G.; Yin, S. Controlled self-assembly of helical nano-ribbons formed by achiral amphiphiles. *Nanoscale* **2015**, *7*, 930–935. [CrossRef]
42. Stals, P.J.M.; Korevaar, P.A.; Gillissen, M.A.J.; de Greef, T.F.A.; Fitie, C.F.C.; Sijbesma, R.P.; Palmans, A.R.A.; Meijer, E.W. Symmetry breaking in the self-assembly of partially fluorinated benzene-1,3,5-tricarboxamides. *Angew. Chem. Int. Ed.* **2012**, *51*, 11297–11301. [CrossRef]
43. Shen, Z.; Wang, T.; Liu, M. Macroscopic chirality of supramolecular gels formed from achiral tris(ethyl cinnamate) benzene-1,3,5-tricarboxamides. *Angew. Chem. Int. Ed.* **2014**, *53*, 13424–13428. [CrossRef]
44. Shen, Z.; Jiang, Y.; Wang, T.; Liu, M. Symmetry breaking in the supramolecular gels of an achiral gelator exclusively driven by pi-pi stacking. *J. Am. Chem. Soc.* **2015**, *137*, 16109–16115. [CrossRef] [PubMed]
45. Maity, A.; Gangopadhyay, M.; Basu, A.; Aute, S.; Babu, S.S.; Das, A. Counter anion driven homochiral assembly of a cationic achiral c3-symmetric gelator through ion-pair assisted hydrogen bond. *J. Am. Chem. Soc.* **2016**, *138*, 11113–11116. [CrossRef] [PubMed]
46. Sang, Y.; Duan, P.; Liu, M. Nanotrumpets and circularly polarized luminescent nanotwists hierarchically self-assembled from an achiral c3-symmetric ester. *Chem. Commun.* **2018**, *54*, 4025–4028. [CrossRef] [PubMed]
47. Kimura, M.; Hatanaka, T.; Nomoto, H.; Takizawa, J.; Fukawa, T.; Tatewaki, Y.; Shirai, H. Self-assembled helical nanofibers made of achiral molecular disks having molecular adapter. *Chem. Mater.* **2010**, *22*, 5732–5738. [CrossRef]
48. Pasteur, M.L. Recherches sur les relations qui peuvent exister entre la forme cristalline: La composition chimique et les sens de la polarisation rotatoire. *Ann. Chim. Phys.* **1848**, 442–459.
49. Pagni, R.M.; Compton, R.N. Asymmetric synthesis of optically active sodium chlorate and bromate crystals. *Cryst. Growth Des.* **2002**, *2*, 249–253. [CrossRef]
50. Kipping, F.S.; Pope, W.J. Stereochemistry and vitalism. *Nature* **1898**, *59*, 53. [CrossRef]
51. Koby, L.; Ningappa, J.B.; Dakessian, M.; Cuccia, L.A. Chiral crystallization of ethylenediamine sulfate. *J. Chem. Educ.* **2005**, *82*, 1043–1045.
52. Saito, Y.; Hyuga, H. Chirality selection in crystallization. *J. Phys. Soc. Jpn.* **2005**, *74*, 535–537. [CrossRef]
53. Ziach, K.; Jurczak, J. Mirror symmetry breaking upon spontaneous crystallization from a dynamic combinatorial library of macrocyclic imines. *Chem. Commun.* **2015**, *51*, 4306–4309. [CrossRef]
54. Chen, S.-C.; Zhang, J.; Yu, R.-M.; Wu, X.-Y.; Xie, Y.-M.; Wang, F.; Lu, C.-Z. Spontaneous asymmetrical crystallization of a three-dimensional diamondoid framework material from achiral precursors. *Chem. Commun.* **2010**, *46*, 1449–1451. [CrossRef] [PubMed]
55. Shu, C.-Y.; Huang, F.-P.; Yu, Q.; Yao, P.-F.; Bian, H.-D.; Lan, R.-Q.; Wei, B.-L. Ph-dependent co(ii) assemblies from achiral 2-benzothiazolylthioacetic acid: Crystal structures, symmetry breaking, and magnetic properties. *J. Coord. Chem.* **2015**, *68*, 2107–2120. [CrossRef]
56. Mineo, P.; Villari, V.; Scamporrino, E.; Micali, N. Supramolecular chirality induced by a weak thermal force. *Soft Matter* **2014**, *10*, 44–47. [CrossRef] [PubMed]

57. Wang, Y.; Zhou, D.; Li, H.; Li, R.; Zhong, Y.; Sun, X.; Sun, X. Hydrogen-bonded supercoil self-assembly from achiral molecular components with light-driven supramolecular chirality. *J. Mater. Chem. C* **2014**, *2*, 6402–6409. [CrossRef]
58. Hu, Q.; Wang, Y.; Jia, J.; Wang, C.; Feng, L.; Dong, R.; Sun, X.; Hao, J. Photoresponsive chiral nanotubes of achiral amphiphilic azobenzene. *Soft Matter* **2012**, *8*, 11492–11498. [CrossRef]
59. Romeo, A.; Castriciano, M.A.; Occhiuto, I.; Zagami, R.; Pasternack, R.F.; Scolaro, L.M. Kinetic control of chirality in porphyrin j-aggregates. *J. Am. Chem. Soc.* **2014**, *136*, 40–43. [CrossRef] [PubMed]
60. Katsonis, N.; Lacaze, E.; Feringa, B.L. Molecular chirality at fluid/solid interfaces: Expression of asymmetry in self-organised monolayers. *J. Mater. Chem. C* **2008**, *18*, 2065–2073. [CrossRef]
61. Perez-Garcia, L.; Amabilino, D.B. Spontaneous resolution, whence and whither: From enantiomorphic solids to chiral liquid crystals, monolayers and macro-and supra-molecular polymers and assemblies. *Chem. Soc. Rev.* **2007**, *36*, 941–967. [CrossRef]
62. Parschau, M.; Ernst, K.-H. Disappearing enantiomorphs: Single handedness in racemate crystals. *Angew. Chem. Int. Ed.* **2015**, *54*, 14422–14426. [CrossRef]
63. Sakaguchi, H.; Song, S.; Kojima, T.; Nakae, T. Homochiral polymerization-driven selective growth of graphene nanoribbons. *Nat. Chem.* **2017**, *9*, 57–63. [CrossRef]
64. George, L.; Gaines, J. *Insoluble Monolayers at Liquid-Gas Interfaces*; Interscience Publishers: New York, NY, USA, 1966.
65. Ulman, A. *An Introduction to Ultrathin Organic Films: From Langmuir-Blodgett to Self-Assembly*; Academic Press: Cambridge, MA, USA, 2013.
66. Tahara, K.; Yamaga, H.; Ghijsens, E.; Inukai, K.; Adisoejoso, J.; Blunt, M.O.; De Feyter, S.; Tobe, Y. Control and induction of surface-confined homochiral porous molecular networks. *Nat. Chem.* **2011**, *3*, 714–719. [CrossRef]
67. Sang, Y.; Yang, D.; Duan, P.; Liu, M. Towards homochiral supramolecular entities from achiral molecules by vortex mixing-accompanied self-assembly. *Chem. Sci.* **2019**, *10*, 2718–2724. [CrossRef] [PubMed]
68. Barth, J.V. Molecular architectonic on metal surfaces. *Annu. Rev. Phys. Chem.* **2007**, *58*, 375–407. [CrossRef] [PubMed]
69. Bartels, L. Tailoring molecular layers at metal surfaces. *Nat. Chem.* **2010**, *2*, 87–95. [CrossRef]
70. Link, D.R.; Natale, G.; Shao, R.; Maclennan, J.E.; Clark, N.A.; Korblova, E.; Walba, D.M. Spontaneous formation of macroscopic chiral domains in a fluid smectic phase of achiral molecules. *Science* **1997**, *278*, 1924–1927. [CrossRef] [PubMed]
71. Tschierske, C. Development of structural complexity by liquid-crystal self-assembly. *Angew. Chem. Int. Ed.* **2013**, *52*, 8828–8878. [CrossRef]
72. Stoddart, J.F. Thither supramolecular chemistry? *Nat. Chem.* **2009**, *1*, 14–15. [CrossRef]
73. Tschierske, C. Liquid crystals materials design and self-assembly preface. In *Liquid Crystals: Materials Design and Self-Assembly*; Tschierske, C., Ed.; Springer Science & Business Media: Berlin, Germany, 2012; Volume 318, pp. IX–X.
74. Goodby, J.W. The nanoscale engineering of nematic liquid crystals for displays. *Liq. Cryst.* **2011**, *38*, 1363–1387. [CrossRef]
75. Cantekin, S.; de Greef, T.F.; Palmans, A.R. Benzene-1, 3, 5-tricarboxamide: A versatile ordering moiety for supramolecular chemistry. *Chem. Soc. Rev.* **2012**, *41*, 6125–6137. [CrossRef]
76. Desiraju, G.R.; Steiner, T. *The Weak Hydrogen Bond: In Structural Chemistry and Biology*; International Union of Crystal: Weinheim, Germany, 2001; Volume 9.
77. Kimizuka, N.; Kawasaki, T.; Hirata, K.; Kunitake, T. Tube-like nanostructures composed of networks of complementary hydrogen bonds. *J. Am. Chem. Soc.* **1995**, *117*, 6360–6361. [CrossRef]
78. George, S.J.; Tomović, Ž.; Smulders, M.M.; de Greef, T.F.; Leclère, P.E.; Meijer, E.W.; Schenning, A.P. Helicity induction and amplification in an oligo (p-phenylenevinylene) assembly through hydrogen-bonded chiral acids. *Angew. Chem. Int. Ed.* **2007**, *46*, 8206–8211. [CrossRef] [PubMed]
79. Yan, X.; Li, S.; Pollock, J.B.; Cook, T.R.; Chen, J.; Zhang, Y.; Ji, X.; Yu, Y.; Huang, F.; Stang, P.J. Supramolecular polymers with tunable topologies via hierarchical coordination-driven self-assembly and hydrogen bonding interfaces. *Proc. Natl. Acad. Sci. USA* **2013**, *110*, 15585–15590. [CrossRef] [PubMed]

80. Ji, X.; Shi, B.; Wang, H.; Xia, D.; Jie, K.; Wu, Z.L.; Huang, F. Supramolecular construction of multifluorescent gels: Interfacial assembly of discrete fluorescent gels through multiple hydrogen bonding. *Adv. Mater.* **2015**, *27*, 8062–8066. [CrossRef] [PubMed]
81. Buchs, J.; Vogel, L.; Janietz, D.; Prehm, M.; Tschierske, C. Chirality synchronization of hydrogen-bonded complexes of achiral n-heterocycles. *Angew. Chem. Int. Ed.* **2017**, *56*, 280–284. [CrossRef] [PubMed]
82. Ding, S.; Gao, Y.; Ji, Y.; Wang, Y.; Liu, Z. Homochiral crystallization of single-stranded helical coordination polymers: Generated by the structure of auxiliary ligands or spontaneous symmetry breaking. *CrystEngComm* **2013**, *15*, 5598–5601. [CrossRef]
83. Wu, S.T.; Wu, Y.R.; Kang, Q.Q.; Zhang, H.; Long, L.S.; Zheng, Z.; Huang, R.B.; Zheng, L.S. Chiral symmetry breaking by chemically manipulating statistical fluctuation in crystallization. *Angew. Chem. Int. Ed.* **2007**, *46*, 8475–8479. [CrossRef] [PubMed]
84. Meng, W.; Ronson, T.K.; Nitschke, J.R. Symmetry breaking in self-assembled m4l6 cage complexes. *Proc. Natl. Acad. Sci. USA* **2013**, *110*, 10531–10535. [CrossRef]
85. Zhou, T.-H.; Zhang, J.; Zhang, H.-X.; Feng, R.; Mao, J.-G. A ligand-conformation driving chiral generation and symmetry-breaking crystallization of a zinc (ii) organoarsonate. *Chem. Commun.* **2011**, *47*, 8862–8864. [CrossRef] [PubMed]
86. Mamula, O.; von Zelewsky, A. Supramolecular coordination compounds with chiral pyridine and polypyridine ligands derived from terpenes. *Coord. Chem. Rev.* **2003**, *242*, 87–95. [CrossRef]
87. Zhang, L.; Qin, L.; Wang, X.; Cao, H.; Liu, M. Supramolecular chirality in self-assembled soft materials: Regulation of chiral nanostructures and chiral functions. *Adv. Mater.* **2014**, *26*, 6959–6964. [CrossRef] [PubMed]
88. Pérez-García, L.; Amabilino, D.B. Spontaneous resolution under supramolecular control. *Chem. Soc. Rev.* **2002**, *31*, 342–356. [CrossRef] [PubMed]
89. Janssen, P.G.A.; Ruiz-Carretero, A.; Gonzalez-Rodriguez, D.; Meijer, E.W.; Schenning, A.P. Ph-switchable helicity of DNA-templated assemblies. *Angew. Chem. Int. Ed.* **2009**, *48*, 8103–8106. [CrossRef] [PubMed]
90. Kondepudi, D.K.; Nelson, G.W. Weak neutral currents and the origin of biomolecular chirality. *Nature* **1985**, *314*, 438–441. [CrossRef]
91. Berger, R.; Quack, M. Electroweak quantum chemistry of alanine: Parity violation in gas and condensed phases. *Chemphyschem* **2000**, *1*, 57–60. [CrossRef]
92. Sorrenti, A.; Rodriguez-Trujillo, R.; Amabilino, D.B.; Puigmarti-Luis, J. Milliseconds make the difference in the far-from-equilibrium self-assembly of supramolecular chiral nanostructures. *J. Am. Chem. Soc.* **2016**, *138*, 6920–6923. [CrossRef]
93. Sun, J.; Li, Y.; Yan, F.; Liu, C.; Sang, Y.; Tian, F.; Feng, Q.; Duan, P.; Zhang, L.; Shi, X.; et al. Control over the emerging chirality in supramolecular gels and solutions by chiral microvortices in milliseconds. *Nat. Commun.* **2018**, *9*, 2599. [CrossRef] [PubMed]
94. Kuhn, A.; Fischer, P. Absolute asymmetric reduction based on the relative orientation of achiral reactants. *Angew. Chem. Int. Ed.* **2009**, *48*, 6857–6860. [CrossRef]
95. Kawasaki, T.; Kamimura, S.; Amihara, A.; Suzuki, K.; Soai, K. Enantioselective c-c bond formation as a result of the oriented prochirality of an achiral aldehyde at the single-crystal face upon treatment with a dialkyl zinc vapor. *Angew. Chem. Int. Ed.* **2011**, *50*, 6796–6798. [CrossRef]
96. Fang, Y.; Ghijsens, E.; Ivasenko, O.; Cao, H.; Noguchi, A.; Mali, K.S.; Tahara, K.; Tobe, Y.; De Feyter, S. Dynamic control over supramolecular handedness by selecting chiral induction pathways at the solution-solid interface. *Nat. Chem.* **2016**, *8*, 711–717. [CrossRef]
97. Wilson, A.J.; Masuda, M.; Sijbesma, R.P.; Meijer, E.W. Chiral amplification in the transcription of supramolecular helicity into a polymer backbone. *Angew. Chem. Int. Ed.* **2005**, *44*, 2275–2279. [CrossRef]
98. Palmans, A.R.; Meijer, E.E.W. Amplification of chirality in dynamic supramolecular aggregates. *Angew. Chem. Int. Ed.* **2007**, *46*, 8948–8968. [CrossRef] [PubMed]
99. Smulders, M.M.J.; Schenning, A.P.H.J.; Meijer, E.W. Insight into the mechanisms of cooperative self-assembly: The "sergeants-and-soldiers" principle of chiral and achiral c-3-symmetrical discotic triamides. *J. Am. Chem. Soc.* **2008**, *130*, 606–611. [CrossRef] [PubMed]
100. Flores, J.J.; Bonner, W.A.; Massey, G.A. Asymmetric photolysis of (rs)-leucine with circularly polarized uv light. *J. Am. Chem. Soc.* **1977**, *99*, 3622–3625. [CrossRef] [PubMed]

101. Bailey, J.; Chrysostomou, A.; Hough, J.H.; Gledhill, T.M.; McCall, A.; Clark, S.; Menard, F.; Tamura, M. Circular polarization in star-formation regions: Implications for biomolecular homochirality. *Science* **1998**, *281*, 672–674. [CrossRef] [PubMed]
102. Kim, J.; Lee, J.; Kim, W.Y.; Kim, H.; Lee, S.; Lee, H.C.; Lee, Y.S.; Seo, M.; Kim, S.Y. Induction and control of supramolecular chirality by light in self-assembled helical nanostructures. *Nat. Commun.* **2015**, *6*, 6959. [CrossRef] [PubMed]
103. Micali, N.; Engelkamp, H.; van Rhee, P.G.; Christianen, P.C.M.; Scolaro, L.M.; Maan, J.C. Selection of supramolecular chirality by application of rotational and magnetic forces. *Nat. Chem.* **2012**, *4*, 201–207. [CrossRef] [PubMed]
104. D'Urso, A.; Randazzo, R.; Lo Faro, L.; Purrello, R. Vortexes and nanoscale chirality. *Angew. Chem. Int. Ed.* **2010**, *49*, 108–112. [CrossRef] [PubMed]
105. Escudero, C.; Crusats, J.; Díez-Pérez, I.; El-Hachemi, Z.; Ribó, J.M. Folding and hydrodynamic forces in j-aggregates of 5-phenyl-10, 15, 20-tris (4-sulfophenyl) porphyrin. *Angew. Chem. Int. Ed.* **2006**, *118*, 8200–8203. [CrossRef]
106. Ribo, J.M.; El-Hachemi, Z.; Arteaga, O.; Canillas, A.; Crusats, J. Hydrodynamic effects in soft-matter self-assembly: The case of j-aggregates of amphiphilic porphyrins. *Chem. Rec.* **2017**, *17*, 713–724. [CrossRef] [PubMed]
107. Crusats, J.; El-Hachemi, Z.; Ribo, J.M. Hydrodynamic effects on chiral induction. *Chem. Soc. Rev.* **2010**, *39*, 569–577. [CrossRef]
108. Aquilanti, V.; Maciel, G.S. Observed molecular alignment in gaseous streams and possible chiral effects in vortices and in surface scattering. *Orig. Life Evol. Biosph.* **2006**, *36*, 435–441. [CrossRef] [PubMed]
109. Jalilah, A.J.; Asanoma, F.; Fujiki, M. Unveiling controlled breaking of the mirror symmetry of eu(fod)(3) with alpha-/beta-pinene and binap by circularly polarised luminescence (cpl), cpl excitation, and f-19-/p-31{h-1}-nmr spectra and mulliken charges. *Inorg. Chem. Front.* **2018**, *5*, 2718–2733. [CrossRef]
110. Fujiki, M. Supramolecular chirality: Solvent chirality transfer in molecular chemistry and polymer chemistry. *Symmetry-Basel* **2014**, *6*, 677–703. [CrossRef]
111. McBride, J.M.; Carter, R.L. Spontaneous resolution by stirred crystallization. *Angew. Chem. Int. Ed.* **1991**, *30*, 293–295. [CrossRef]
112. Kondepudi, D.K.; Bullock, K.L.; Digits, J.A.; Yarborough, P.D. Stirring rate as a critical parameter in chiral-symmetry breaking crystallization. *J. Am. Chem. Soc.* **1995**, *117*, 401–404. [CrossRef]
113. Viedma, C. Chiral symmetry breaking during crystallization: Complete chiral purity induced by nonlinear autocatalysis and recycling. *Phys. Rev. Lett.* **2005**, *94*, 065504. [CrossRef] [PubMed]
114. Sogutoglu, L.C.; Steendam, R.R.; Meekes, H.; Vlieg, E.; Rutjes, F.P. Viedma ripening: A reliable crystallisation method to reach single chirality. *Chem. Soc. Rev.* **2015**, *44*, 6723–6732. [CrossRef] [PubMed]
115. Palmans, A.R.A. Deracemisations under kinetic and thermodynamic control. *Mol. Syst. Des. Eng.* **2017**, *2*, 34–46. [CrossRef]
116. Steendam, R.R.; Verkade, J.M.; van Benthem, T.J.; Meekes, H.; van Enckevort, W.J.; Raap, J.; Rutjes, F.P.; Vlieg, E. Emergence of single-molecular chirality from achiral reactants. *Nat. Commun.* **2014**, *5*, 5543. [CrossRef] [PubMed]
117. Engwerda, A.H.J.; Koning, N.; Tinnemans, P.; Meekes, H.; Bickelhaupt, F.M.; Rutjes, F.P.J.T.; Vlieg, E. Deracemization of a racemic allylic sulfoxide using viedma ripening. *Cryst. Growth Des.* **2017**, *17*, 4454–4457. [CrossRef] [PubMed]
118. Noorduin, W.L.; Meekes, H.; van Enckevort, W.J.P.; Millemaggi, A.; Leeman, M.; Kaptein, B.; Kellogg, R.M.; Vlieg, E. Complete deracemization by attrition-enhanced ostwald ripening elucidated. *Angew. Chem. Int. Ed.* **2008**, *47*, 6445–6447. [CrossRef] [PubMed]
119. Hein, J.E.; Cao, B.H.; Viedma, C.; Kellogg, R.M.; Blackmond, D.G. Pasteur's tweezers revisited: On the mechanism of attrition-enhanced deracemization and resolution of chiral conglomerate solids. *J. Am. Chem. Soc.* **2012**, *134*, 12629–12636. [CrossRef] [PubMed]
120. Noorduin, W.L.; Izumi, T.; Millemaggi, A.; Leeman, M.; Meekes, H.; Van Enckevort, W.J.P.; Kellogg, R.M.; Kaptein, B.; Vlieg, E.; Blackmond, D.G. Emergence of a single solid chiral state from a nearly racemic amino acid derivative. *J. Am. Chem. Soc.* **2008**, *130*, 1158–1159. [CrossRef] [PubMed]
121. Viedma, C.; Ortiz, J.E.; de Torres, T.; Izumi, T.; Blackmond, D.G. Evolution of solid phase homochirality for a proteinogenic amino acid. *J. Am. Chem. Soc.* **2008**, *130*, 15274–15275. [CrossRef] [PubMed]

122. Tsogoeva, S.B.; Wei, S.; Freund, M.; Mauksch, M. Generation of highly enantioenriched crystalline products in reversible asymmetric reactions with racemic or achiral catalysts. *Angew. Chem. Int. Ed.* **2009**, *48*, 590–594. [CrossRef] [PubMed]
123. Viedma, C.; Cintas, P. Homochirality beyond grinding: Deracemizing chiral crystals by temperature gradient under boiling. *Chem. Commun.* **2011**, *47*, 12786–12788. [CrossRef] [PubMed]
124. Suwannasang, K.; Flood, A.E.; Rougeot, C.; Coquerel, G. Use of programmed damped temperature cycles for the deracemization of a racemic suspension of a conglomerate forming system. *Org. Process Res. Dev.* **2017**, *21*, 623–630. [CrossRef]
125. Ribó, J.M.; Hochberg, D.; Crusats, J.; El-Hachemi, Z.; Moyano, A. Spontaneous mirror symmetry breaking and origin of biological homochirality. *J. R. Soc. Interface* **2017**, *14*, 20170699. [CrossRef]
126. Rubires, R.; Farrera, J.A.; Ribo, J.M. Stirring effects on the spontaneous formation of chirality in the homoassociation of diprotonated meso-tetraphenylsulfonato porphyrins. *Chem. Eur. J.* **2001**, *7*, 436–446. [CrossRef]
127. Raudino, A.; Pannuzzo, M. Hydrodynamic-induced enantiomeric enrichment of self-assemblies: Role of the solid-liquid interface in chiral nucleation and seeding. *J. Chem. Phys.* **2012**, *137*, 134902. [CrossRef]
128. Kitagawa, Y.; Segawa, H.; Ishii, K. Magneto-chiral dichroism of organic compounds. *Angew. Chem. Int. Ed.* **2011**, *50*, 9133–9136. [CrossRef] [PubMed]
129. Hamba, F.; Niimura, K.; Kitagawa, Y.; Ishii, K. Helicity transfer in rotary evaporator flow. *Phys. Fluids* **2014**, *26*, 017101. [CrossRef]
130. Heffern, M.C.; Matosziuk, L.M.; Meade, T.J. Lanthanide probes for bioresponsive imaging. *Chem. Rev.* **2014**, *114*, 4496–4539. [CrossRef] [PubMed]
131. Carr, R.; Evans, N.H.; Parker, D. Lanthanide complexes as chiral probes exploiting circularly polarized luminescence. *Chem. Soc. Rev.* **2012**, *41*, 7673–7686. [CrossRef] [PubMed]
132. Schadt, M. Liquid crystal materials and liquid crystal displays. *Annu. Rev. Mater. Sci.* **1997**, *27*, 305–379. [CrossRef]
133. Kim, D.-Y.; Kim, D.-Y. Potential application of spintronic light-emitting diode to binocular vision for three-dimensional display technology. *J. Korean Phys. Soc.* **2006**, *49*, 505–508.
134. Zinna, F.; Di Bari, L. Lanthanide circularly polarized luminescence: Bases and applications. *Chirality* **2015**, *27*, 1–13. [CrossRef]
135. Muller, G. Luminescent chiral lanthanide(iii) complexes as potential molecular probes. *Dalton Trans.* **2009**, 9692–9707. [CrossRef]
136. Wagenknecht, C.; Li, C.-M.; Reingruber, A.; Bao, X.-H.; Goebel, A.; Chen, Y.-A.; Zhang, Q.; Chen, K.; Pan, J.-W. Experimental demonstration of a heralded entanglement source. *Nat. Photonics* **2010**, *4*, 549–552. [CrossRef]
137. Xu, Y.; Yang, G.; Xia, H.; Zou, G.; Zhang, Q.; Gao, J. Enantioselective synthesis of helical polydiacetylene by application of linearly polarized light and magnetic field. *Nat. Commun.* **2014**, *5*, 5050. [CrossRef]
138. Tang, Y.; Cohen, A.E. Enhanced enantioselectivity in excitation of chiral molecules by superchiral light. *Science* **2011**, *332*, 333–336. [CrossRef] [PubMed]
139. Sato, I.; Sugie, R.; Matsueda, Y.; Furumura, Y.; Soai, K. Asymmetric synthesis utilizing circularly polarized light mediated by the photoequilibrium of chiral olefins in conjunction with asymmetric autocatalysis. *Angew. Chem. Int. Ed.* **2004**, *43*, 4490–4492. [CrossRef] [PubMed]
140. Kawasaki, T.; Sato, M.; Ishiguro, S.; Saito, T.; Morishita, Y.; Sato, I.; Nishino, H.; Inoue, Y.; Soai, K. Enantioselective synthesis of near enantiopure compound by asymmetric autocatalysis triggered by asymmetric photolysis with circularly polarized light. *J. Am. Chem. Soc.* **2005**, *127*, 3274–3275. [CrossRef] [PubMed]
141. Yang, Y.; da Costa, R.C.; Fuchter, M.J.; Campbell, A.J. Circularly polarized light detection by a chiral organic semiconductor transistor. *Nat. Photonics* **2013**, *7*, 634–638. [CrossRef]
142. Meinert, C.; Bredehoeft, J.H.; Filippi, J.-J.; Baraud, Y.; Nahon, L.; Wien, F.; Jones, N.C.; Hoffmann, S.V.; Meierhenrich, U.J. Anisotropy spectra of amino acids. *Angew. Chem. Int. Ed.* **2012**, *51*, 4484–4487. [CrossRef] [PubMed]
143. Meinert, C.; Hoffmann, S.V.; Cassam-Chenai, P.; Evans, A.C.; Giri, C.; Nahon, L.; Meierhenrich, U.J. Photonenergy-controlled symmetry breaking with circularly polarized light. *Angew. Chem. Int. Ed.* **2014**, *53*, 210–214. [CrossRef] [PubMed]

144. He, C.; Yang, G.; Kuai, Y.; Shan, S.; Yang, L.; Hu, J.; Zhang, D.; Zhang, Q.; Zou, G. Dissymmetry enhancement in enantioselective synthesis of helical polydiacetylene by application of superchiral light. *Nat. Commun.* **2018**, *9*, 5117. [CrossRef]
145. Fujiki, M.; Kawagoe, Y.; Nakano, Y.; Nakao, A. Mirror-symmetry-breaking in poly (9,9-di-n-octylfluorenyl-2, 7-diyl)-alt-biphenyl (pf8p2) is susceptible to terpene chirality, achiral solvents, and mechanical stirring. *Molecules* **2013**, *18*, 7035–7057. [CrossRef]
146. Noorduin, W.L.; Bode, A.A.C.; van der Meijden, M.; Meekes, H.; van Etteger, A.F.; van Enckevort, W.J.P.; Christianen, P.C.M.; Kaptein, B.; Kellogg, R.M.; Rasing, T.; et al. Complete chiral symmetry breaking of an amino acid derivative directed by circularly polarized light. *Nat. Chem.* **2009**, *1*, 729–732. [CrossRef]
147. Japp, F.R. Asymmetry and vitalism. *Nature* **1898**, *58*, 616–618. [CrossRef]
148. Feringa, B.L.; Van Delden, R.A. Absolute asymmetric synthesis: The origin, control, and amplification of chirality. *Angew. Chem. Int. Ed.* **1999**, *38*, 3418–3438. [CrossRef]
149. Yamagata, Y. A hypothesis for the asymmetric appearance of biomolecules on earth. *J. Theor. Biol.* **1966**, *11*, 495. [CrossRef]
150. Avalos, M.N.; Babiano, R.; Cintas, P.; Jiménez, J.L.; Palacios, J.C. From parity to chirality: Chemical implications revisited. *Tetrahedron Asymmetry* **2000**, *11*, 2845–2874. [CrossRef]

Article

A General Phenomenon of Spontaneous Amplification of Optical Purity under Achiral Chromatographic Conditions

K. Michał Pietrusiewicz [1], Mariusz Borkowski [1,2], Dorota Strzelecka [1], Katarzyna Kielar [1], Wioleta Kicińska [1], Sergei Karevych [1], Radomir Jasiński [3] and Oleg M. Demchuk [1,4,*]

1. Faculty of Chemistry, Maria Curie-Skłodowska University, 2 M. Sklodowskiej-Curie Square, 20-031 Lublin, Poland; Kazimierz.Pietrusiewicz@poczta.umcs.lublin.pl (K.M.P.); ncborkow@cyf-kr.edu.pl (M.B.); dorin538@gmail.com (D.S.); kasiak1005@teln.pl (K.K.); kicinskawioleta@gmail.com (W.K.); kackac16@gmail.com (S.K.)
2. Institute of Catalysis and Surface Chemistry, Polish Academy of Sciences, 8 Niezapominajek Str., 31-155 Cracow, Poland
3. Institute of Organic Chemistry and Technology, Cracow University of Technology, 24 Warszawska Str., 31-155 Cracow, Poland; radomir@indy.chemia.pk.edu.pl
4. Pharmaceutical Research Institute, 8 Rydygiera Str., 01-793 Warsaw, Poland

* Correspondence: O.Demchuk@IFarm.eu

Received: 2 April 2019; Accepted: 15 May 2019; Published: 17 May 2019

Abstract: This work explores the behavior of chiral compound mixtures enriched in one of the enantiomers whilst a typical chromatography on the achiral stationary phase is employed. The influence of several factors, such as the eluent composition, ratio of the compound to the stationary phase, and the initial enatiomeric purity of the compound used on the distribution of the enantiomers in the collected chromatographic fraction, was studied. The obtained results indicate that the phenomenon of Self Disproportionation of Enantiomer (SDE) occurred in all cases, and some of the collected fractions got higher optical purities than the initial one. Thus, achiral column chromatography could be applied in some cases as the simplest approach for chiral purification. Based on the experimental results and DFT calculations, an alternative concept explaining the SDE phenomenon was proposed. Due to its generality and simplicity, SDE may also be responsible for the formation of the first chiral non-racemic compounds on the early Earth.

Keywords: enantiomer self-disproportionation; SDE; achiral stationary phase; homochiral and heterochiral aggregates; chiral separation; chirality; genesis of life chirality

---

1. Introduction

The mystery of life on Earth is closely connected with the phenomenon of the chiral nature of all living organisms (human beings, animals, plants, insects, etc.). Chirality, as a fundamental property of 3-dimensional objects (including many molecules and ions) to be non-superinposable on their mirror images, is imprinted in such basic building blocks of life as amino acids and sugars; it is further reflected by the chirality of more complex biochemical objects such as DNA and eventually by the chirality of entire bodies [1,2]. Interestingly, amino acids [3] and sugars [4] derived from natural sources are non-racemic and are constituted of single stereoisomers. At the same time, according to the fundamental chemical principles the formation of non-racemic chemical compounds is impossible when the chiral component of the chemical reaction (substrate, catalyst, medium) is absent [5]. Despite its crucial value for the understanding of the origin of life, the origins of the chirality of molecules of life are still unknown [6]. Apart from the creationist approach stating that "the only possible way for unique homochirality to exist in the chiral biochemical molecules found in living organisms is for those biomolecules to have been created with unique homochirality when that organism was first created. Just as a fingerprint identifies its creator" [7], there are several other theories trying to address this issue [8]: partial photochemical decomposition of the racemic mixtures of chiral compounds in the interstellar clouds caused by polarized cosmic irradiation [9,10]; photochemical reactions in the primordial soup influenced by the Sun's light polarized by passing though the Earth atmosphere [11,12]; the effect of the parity-symmetry violation of the weak force [13]; the stereochemical outcome of the chemical reactions influenced by a magnetic field [14]; asymmetric catalysis mediated with chiral minerals e.g., single quartz crystals [15,16]; stereoselective crystallization of a single enantiomer from the racemic mixture [17,18]. Since the formation of both enantiomers is statistically equal, all the variety of the chiral compounds on Earth should have originated from almost a single act of creation. Moreover, in all those rare cases, mixtures with only an extremely low excess of one of the enantiomers could be formed. Thus, the major question of the chirality genesis is how—from an almost racemic mixture of chiral probiotic compounds formed accidentally, once or few times only—an optically pure substance could have been spontaneously created. The process of the spontaneous enrichment of the enantiomer mixture and the generality of this phenomenon could be studied on the example of several typical benchmark asymmetric syntheses where some enantio-enriched mixture of the chiral compounds is formed.

As a result of the significant progress made in the last few decades, the asymmetric catalysis has become the basic tool of synthetic organic chemistry used to obtain modern medicines [19], and cosmetics [20], plant protecting substances [21], and variety of functional materials [22,23]. Valuable features of those kinds of transformations include: the highest atom economy (thus, only one of two possible enantiomers are usually desirable); a short synthetic pathway because of the elimination of chiral derivatisation and racemic mixture separation steps; mild reaction conditions; wide substitution pattern toleration; and many others—all of the above made these transformations highly attractive economically. Therefore, they draw the attention of both academics and the industry. The importance of this field has also entailed the development of methods for the determination of the enantiomeric composition of the obtained products directly or after derivatisation by chiral reagents. There are the classical measurement of optical rotation power [24,25], chiral chromatographic approaches [26,27], NMR methodologies utilising chiral shifting reagents [28–31] and chiral derivatisation reagents [32–34].

In the majority of cases, the determination of the enantiomeric composition of the mixtures of isomeric products obtained in the reaction is performed after preliminary purification of the sample studied. Distillation is only marginally useful in the case of polysubstituted organic materials. Such routine purification techniques as crystallisation and sublimation cannot ensure the invariability of the enantiomeric composition in a rough reaction mixture and isolated substance because of a widely observed *enantiomer self-disproportionation* phenomenon (SDE). SDE was defined in 2006 as a transformation of an enantiomerically enriched system resulting in the formation of fractions with different, in comparison to the initial enantiomeric distribution [35]. The SDE phenomenon is attributed to two distinct models of the intermolecular interaction between the enantiomers of a chiral compound. These are homochiral (*R*,*R* or *S*,*S*) and heterochiral (*R*,*S*)-associations (Figure 1) which are present in a not completely racemic mixture and are largely responsible for the nonlinear behavior of the optical rotation [35,36].

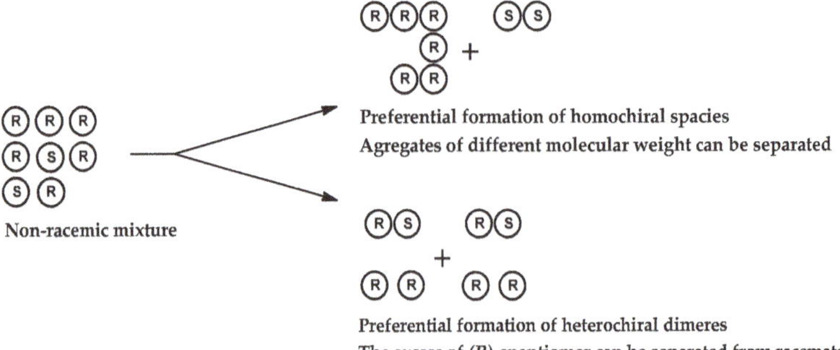

Figure 1. Preferential formation of homochiral or heterochiral aggregates in the solution of optically enriched compounds.

Thus, column chromatography is usually the method of choice for the purification of reaction mixtures in laboratory practice and it is extensively used for the preparation of analytical samples from different asymmetric reactions. For a long time, the chemical community accepted the assumption that, with very few exceptions [37], the ratio of enantiomers would not be significantly changed during chromatographic purification, unless some other chiral players such as chiral solvents or additives or a chiral stationary phase are used. Recent studies indicate that this rule could be broken in the case of chiral fluorinated compounds because of notable chiral self-aggregation promoted by a strong dispersive interaction of fluorous tails [35,38]. Rare literature examples also suggest that a partial separation of the racemate from the enantiomer could be observed for some other enantiomericaly enriched compounds subjected to achiral chromatographic purification via regular gravity driven chromatography on silica gel [39–42], or HPLC [43], MPLC [37], flash [44], or other chromatographic purification techniques [45]. However, this possibility is still usually neglected in routine chromatographic purification on regular achiral stationary phases.

Herein, we would like to demonstrate that SDE is a common phenomenon observed during the achiral low-pressure column chromatography of the chiral compounds obtained in benchmark asymmetric reactions such as allylic substitution, [46,47] cross-coupling [48–52] and organocatalytic aldol condensation [53,54] as well as other important asymmetric transformations. Thus, such processes may play a significant role in the course of the chiral, non-racemic, pro-biotic compound formation.

2. Materials and Methods

All the chemicals used in this research were purchased from Sigma-Aldrich, Avantor Performance Materials Poland S.A. (formerly POCH S.A) and Merck. Analytical thin-layer chromatography (TLC) was performed using silica gel 60 F_{254} precoated plates (0.25 mm thickness). Visualization of TLC plates was performed by means of UV light. NMR spectra were recorded on Bruker Avance 500 MHz spectrometers, and chemical shifts are reported in ppm, and calibrated to residual solvent peaks at 7.27 ppm and 77.00 ppm for ^{1}H and ^{13}C in $CDCl_3$ or TMS as internal reference compounds. The enantiomeric compositions of the obtained fractions were determined by a chromatographic method (Shimadzu LCMS IT-TOF spectrometer) using chiral chromatographic columns Chiralpak® and Chiralcel® with the following dimensions: 150 × 2.1 mm, or by ^{1}H NMR and ^{13}C NMR techniques using an appropriate chiral differentiating factor and solvents. The purification process and the study of the SDE phenomenon were conducted on a flash low pressure BUCHI chromatograph using 12 × 150 mm PE columns packed by authors with 10 g spherical silica gel 60 with the particle size of 230–400 mesh as the stationary phase. The studied compounds were dissolved in 5 mL of an appropriate solvent and mixed with 1g of silica gel. Next, the solvent was evaporated and the sample supported on silica was placed at the top of the column prepacked with silica. After that, the column was connected to the MPLC to be subjected to SDE studies. In the case of compounds **1** and **6**, instead of supporting them on silica, the solutions of the studied compounds in 0.6 mL of mobile phase were injected onto the column prepacked with 11 g of silica and prewashed with the mobile phase. The fractions were automatically collected by 18 mL volume. The elution progress was monitored by TLC. All DFT calculations were performed using the "Prometheus" cluster in the "Cyfronet" computational center in Cracow. A new generation M062x [55]; [56] functional, implemented in the Gaussian 09 package [57], was used. This functional has been recently used by us to solve several similar problems [58–60]. All stationary structures have been optimized using the advanced 6-311++G(d,p) basis set and were characterized by only positive eigenvalues in their diagonalized Hessian matrices. For the optimized structures, thermochemical data for the temperature T = 298 K and pressure p = 1 atm were computed using vibrational analysis data. For the simulation of the solvent presence, the PCM model has been applied, similar to in previous our works [59,60].

3. Results and Discussion.

In order to investigate the impact of a routine chromatographic purification process on the distribution of enantiomers in collected fractions, non-racemic mixtures of the studied compounds were obtained in the usual benchmark reactions.

3.1. Study on the SDE of Dicarboxylic Acid Esters: (R)-Dimethyl [(2E)-1,3-Diphenyl-2-Propen-1-yl] Malonate (1)

The asymmetric allylic substitution reaction leading to (*R*)-dimethyl [(2*E*)-1,3-diphenyl-2-propen-1-yl] malonate (**1**) is a standard process used to assess the efficiency and compare the chiral transition metal complex based catalysts [59]. The product of this benchmark reaction was evaluated first. The reaction was carried out using (*R*)-(-)-*BisNap*-Phos [48] as a ligand and furnished crude product **1** (Scheme 1), which was subsequently subjected to the enantiomeric purity determination by an HPLC technique. The enantiomeric composition and (*R*)-absolute configuration of the product were determined by the peak integration and elution order from chiral HPLC using a Chiralcel® OD-H column [61]. The enantiomeric excess of the crude product was 49.0%.

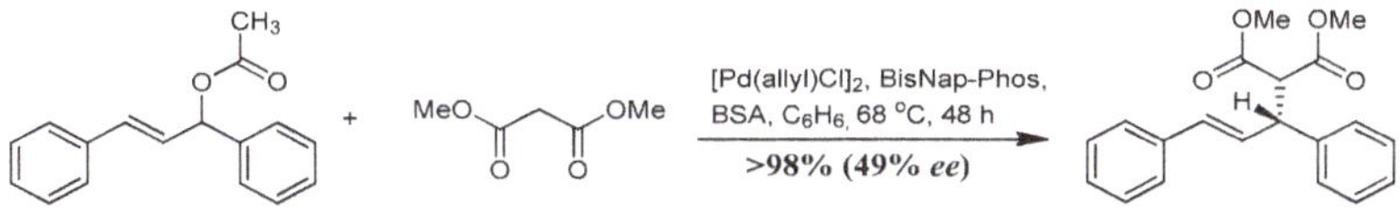

CH_3 + MeO OMe; $[Pd(allyl)Cl]_2$, BisNap-Phos, BSA, C_6H_6, 68 °C, 48 h; >98% (49% *ee*); OMe OMe, H, **1**

Scheme 1. Synthesis of (*R*)-dimethyl [(2E)-1,3-diphenyl-2-propen-1-yl] malonate (**1**).

Next, the product was chromatographically purified on the achiral stationary phase. The eluent system selection was based on the R_f values (~0.2) and hence the mixtures of hexane/acetone in the ratios of (99/1) and (99.7/0.3) were selected as mobile phases. 50.0 mg of compound **1** was subjected to the purification process on an achiral 12 × 150 mm column filled with 11.0 g silica gel. The separations were run at the eluent flow rate of 18 mL/min and the collected fractions had the volume of 18 mL. The fractions were collected from the moment when the eluted compound was detected by TLC. The enantiomeric compositions of consecutive fractions were determined by means of chiral HPLC with the application of a Chiralpak® AS-RH column and the mobile phase H_2O/CH_3CN = 55/45, and 0.4 mL/min flow (Figure S1). The Chiralpak® OJ-RH and Chiralcel® OD-H columns could also be used for this purpose. The elution profiles for both eluent systems are depicted in Figure 2 and Table S1.

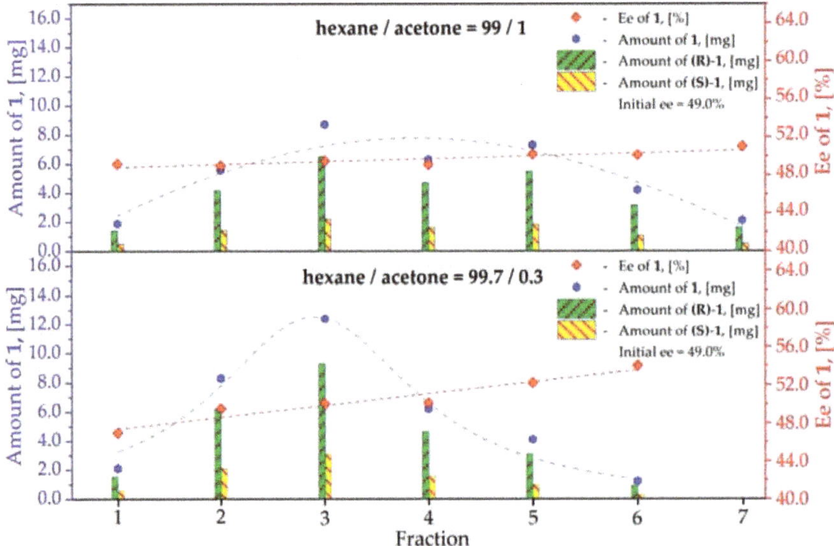

Figure 2. SDE of compound **1** in achiral MPLC conditions: the elution profiles. The fitting equation for hexane/acetone = 99/1 is y = 0.018x + 49.429 (R^2 = 0.851) and for hexane/acetone = 99.7/0.3 is y = 2.099x + 40.4 (R^2 = 0.768).

The mobile phases were selected by the optimization process using the TLC method. Several possible solvent mixtures and pure solvents were tested in this process. In the next step, those mobile phases, for which the separation on the plates gave promising results were selected; these mobile phases were then tested in the column separation process. In those experiments, several mobile phases were examined and finally those for which the separation result was good enough were obtained and results obtained in the context of the SDE phenomenon. This procedure of mobile phase optimization was used in a course of all further separations. In this case, in the optimization process, hexane/acetone with ratios of 99:1 and 99.7:0.3 was selected as mobile phase. It can be seen that in the course of the purification, the *ee* values of the collected fractions increased slightly from the initial level of 49.0% for hexane/acetone = 99:1 and more significantly for hexane/acetone in the ratio of 99.7: 0.3 (to 51.0% and 54.0%, respectively). The values of the SDE magnitude (Δ*ee*) parameter (defined as the difference between the highest and the lowest value of the enantiomeric excess of the collected fractions) equaled 1.8% and 7.0% for hexane/acetone = 99:1 and hexane/acetone = 99.7:0.3, respectively. These results indicate that the eluent used in the first series of the experiment was characterized by a polarity too high to allow effectively the occurrence of SDE. In contrast to the first series, it is noticeable that the utilization of a less polar eluent leads to obtaining a compound with a slightly higher *ee* value in the latter fractions. These results immediately allowed us to state that the purification process of the crude reaction mixture on achiral silica gel may affect the *ee* value. Secondly, the SDE of compound **1** depends on the eluent composition used in the purification process; the less polar eluent facilitates the SDE phenomenon.

3.2. *Study on the SDE of Atropisomeric Biaryl: 2,2′-Dimethoxy-1,1′-Binaphthyl (**2**)*

Inspired by the initial results, we decided to investigate whether the SDE phenomenon may be applicable to the enantiopurification of 2,2′-dimethoxy-1,1′-binaphthyl (**2**) enriched in the (*R*)-isomer (Scheme 2). Compound **2** was obtained in a benchmark Suzuki-Miyaura reaction which is usually used to assess the efficiency of a palladium catalyst in cross-coupling reactions. The reaction that was run in the presence of 0.3% SDS and Na_2CO_3 in an aqueous medium at ambient temperature utilized the (*S*)-*BisNap*-Phos ligand (1 mol%) and $Pd(C_6H_5CN)_2Cl_2$ pre-catalyst (0.5 mol%). The enantiomeric excess of **2** was determined by an HPLC technique using a Chiralpak® AS-RH column, and the mobile phase H_2O/CH_3CN = 50/50 (Figure S2). The initial enantiomeric excess was assigned as 41.0% *ee*.

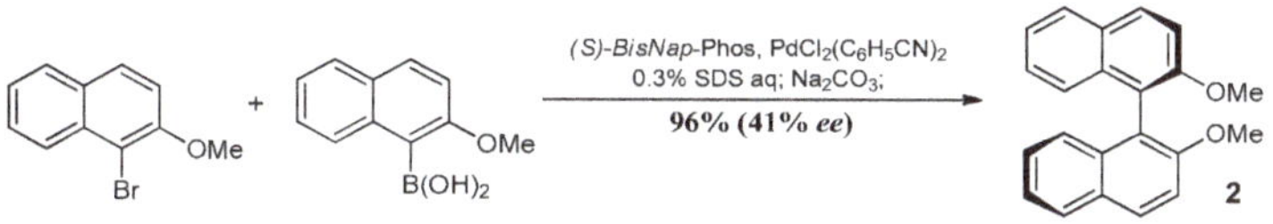

Scheme 2. Synthesis of (*R*)-2,2′-dimethoxy-1,1′-binaphthalene (**2**).

Further, 100.0 mg of compound **2** was fractionated on a 12 × 150 mm column filled with 10.0 g silica gel, using a mixture of hexane/ethyl acetate = 97:3 as an eluent. Considering the low solubility of compound 2 in the eluent, the compound was applied on the gel and loaded onto the column in this form. The separations were run at the eluent flow rate of 18 mL/min and the collected fractions had the volume of 18 mL. The SDE profile of compound **2** is shown in Figure 3 (the numeric data are given in Table S2). As one can see, the enantiomeric excess of the examined compound increased with the elution volume from 21.0% *ee* in fraction 32 (in which compound **3** was first detected by TLC) to 63.0% *ee* in the last fraction (fraction 52). The SDE magnitude (Δee) in the studied cases equaled 42.0%. The chiral HPLC chromatograms of the first and last fractions are depicted in Figure S2. The final fraction (fraction 52) is 22.0% more enantiomerically enriched in comparison to the initial sample of 41.0% *ee*. Such highly significant value of Δee could be caused by the π-π stacking interaction between the naphthyl moieties [62] as it has been previously observed for other types of substituted naphthalenes [63]. On the other hand, 2,2′-dimethoxy-1,1′-binaphthyl could easily undergo crystallization, and it is also theoretically possible that a multiple micro recrystallization process may have taken place on the column and eventually influenced the enantiomeric distribution in the collected fractions. In any case, the obtained results indicate a possibility of the enantiomeric enrichment of potentially crystalline compounds (or even a separation of the enantiomeric fraction from the racemate) using the methodology based on the achiral chromatographic technique.

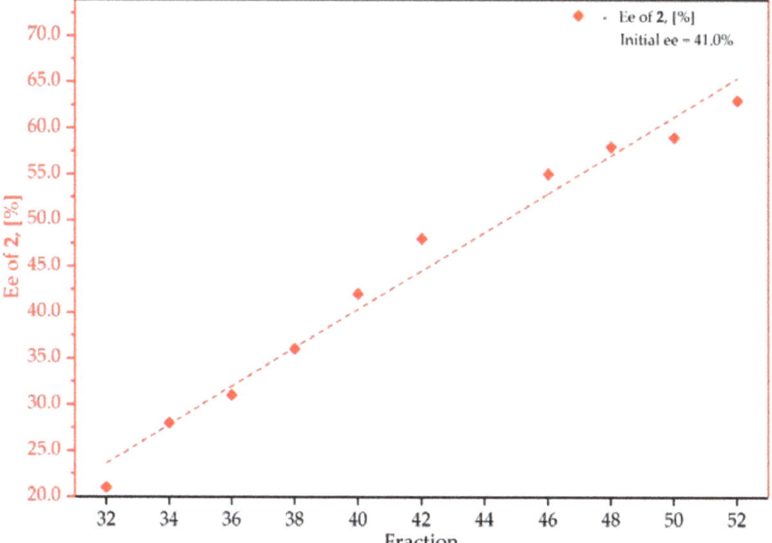

Figure 3. SDE of compound **2** in achiral MPLC conditions: the elution profile. The fitting equation is y = 2.089x -46.242 (R^2 = 0.979).

3.3. *Study on the SDE of β-Hydroxy Ketone: (4R)-(4-nitrophenyl)-4-Hydroxy-2-Butanone* (**3**)

Taking into account a possible strong effect of the π-π forces in the crystal net, we turned our attention to another class of organic compounds. The next chiral model compound (4*R*)-(4-nitrophenyl)-4-hydroxy-2-butanone (**3**) was synthesized under aldol condensation conditions (Scheme 3) [64] in a reaction between 4-nitrobenzaldehyde and technical grade acetone in the presence of 20 mol% of *L*-proline. In this benchmark organocatalytic reaction the chiral organic molecule plays the role of a catalyst, which is more inherent to biomimetic processes. As a result, the reaction products are not contaminated with toxic metal impurities, the catalyst does not undergo decomposition, and in general, the process itself is compatible with the requirements of green chemistry, as well as chemical processes occurring on the early Earth. The optical purity of synthesized compound **3** was determined by an HPLC technique using a Chiralpak® AD-RH column, and the CH_3CN/H_2O = 30/70 mobile phase [65] (Figure S3).

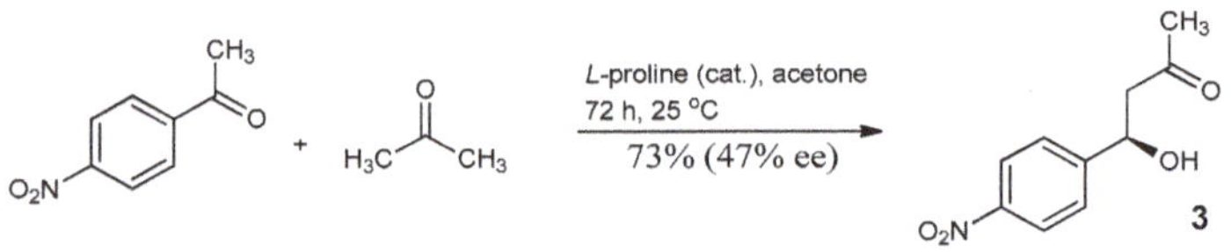

L-proline (cat.), acetone
72 h, 25 °C
73% (47% ee)

Scheme 3. Synthesis of (*R*)-4-hydroxy-4(nitrophenyl)butan-2-one (**3**).

The ratio of the amount of the studied compound to the amount of the stationary phase used may affect the efficiency of SDE. To explore this dependence three different quantities of **3** (50.0 mg, 100.0 mg and 150.0 mg supported on 1.0 g of silica) were eluted with hexane/MTBE = 85:15 in flash chromatography conditions on a 12 × 150 mm column filled with 10.0 g silica gel. On the basis of the obtained results (Table S3) the enantiomeric distribution profiles are presented in Figure 4. In all cases a clear decreasing dependence of the enantiomeric excess was observed. First fractions exhibited a higher enantiomeric enrichment in the (*R*)-isomer than the latter ones. The values of Δ*ee* were 23.6%, 17.2% and 16.8% for 50.0 mg, 100.0 mg and 150.0 mg of the examined compound sample size, respectively. In the case of the sample with the smallest quantity of the compound the highest *ee* value of 55.6% occurred in fraction 37 (first fraction in which compound **3** was detected by TLC), while the lowest in fraction 70 and equaled 32.0% *ee*. In comparison with the initial value of 47.0% *ee*, the first fractions were enantiomerically enriched in (*R*)-isomer by 8.6% (for a 50.0 mg sample), 5.5% (for a 100.0 mg sample), 1.8% (for a 150.0 mg sample), while the last fractions were enantiomerically depleted by 15% (for the 50.0 mg and 150.0 mg sample), 10.6% (for a 50.0 mg sample). As it can be seen in Figure 4, for 100.0 mg and 150.0 mg of the compound in the sample the SDE phenomenon occurred, yet to a lesser degree. In consequence, these outcomes indicate that the magnitude of the SDE phenomenon depends on the ratio of the compound amount to the silica gel amount in the chromatographic purification process. The observed relationship between the SDE and the amount of the separated compound in comparison to the achiral stationary phase may be due to the fact that a smaller quantity has a larger number of unsaturated adsorption centers on a gel, which leads to the intensification of the SDE process. However, when the separation is performed using a larger compound to stationary phase ratio, an increase in gel loading and therefore saturation of the adsorption centers could be observed, which may result in the slowdown of the SDE process.

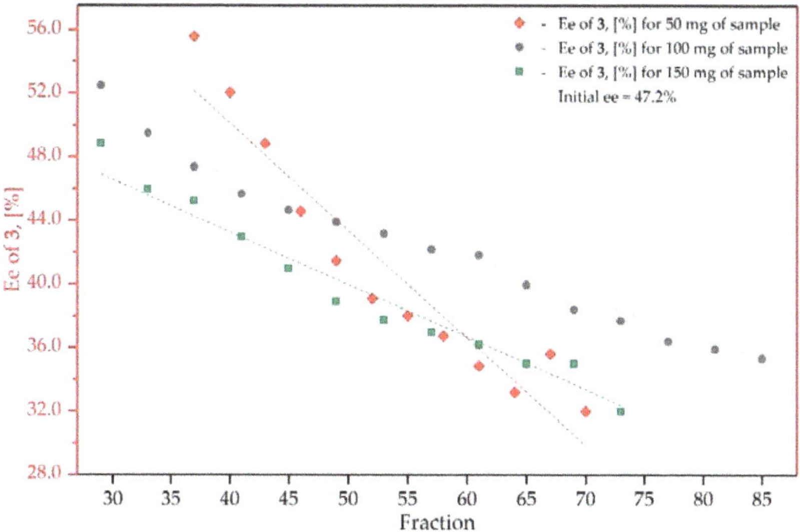

Figure 4. SDE of compound **3** in achiral MPLC conditions: the elution profile. The fitting equation for 50.0 mg of sample is y = −0.676x + 77.190 (R^2 = 0.909) and for 100.0 mg of sample is y = −0.352x + 57.612 (R^2 = 0.966) and for 150.0 mg of sample is y = −0.283x + 58.408 (R^2 = 0.972).

3.4. Study on the SDE of P-Chiral Compounds: tert-Butylphenylphosphinothioic acid (**4**)

The influence of the starting material's enantiomeric composition on the magnitude of SDE was examined with the *tert*-butylphenylphosphinothioic acid (**4**) (Figure 5). The chosen model compound was earlier synthesized in accordance with the literature procedure [41]. This compound has already found its application as a chiral solvating agent (CSA) for the enantiomeric excess determination by an NMR technique [66]. It is worth noting that the enantiomeric purity of **4** could also be determined by an NMR technique since **4** may simultaneously play a role of the analyzed compound and CSA [67] (Figure S4). The reason for the observed chiral recognition lays in the ability of **4** to act as a donor and an acceptor of hydrogen bonds and hence the formation of homochiral and heterochiral dimeric species (Figure 5). Taking into account the properties of the selected compound, we thought that it would prove an ideal candidate for investigating the dependence of the SDE phenomenon on the initial enantiomeric excess of the tested samples.

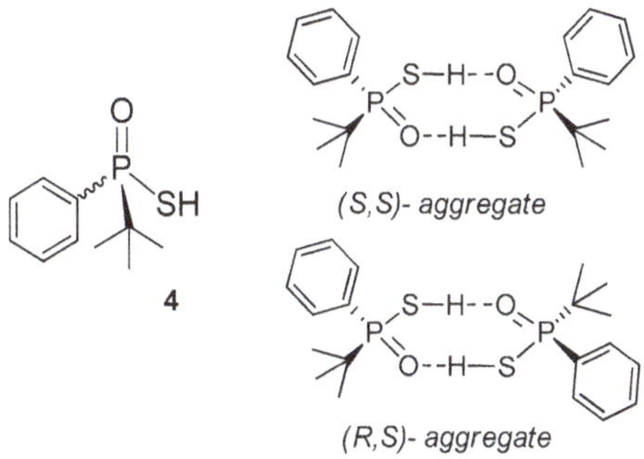

4

(*S,S*)- aggregate

(*R,S*)- aggregate

Figure 5. (*R*)-*tert*-butylphenylphosphinothioic acid (**4**): chemical structures on the homochiral and heterochiral aggregates.

For the purpose of this study, three samples of (*R*)-**4** in the amount of 100.0 mg and with the optical purity of 25.0% *ee*, 60.0% *ee* and 75.0% *ee* were prepared, supported on 1.0 g of silica, and then subjected to the purification process driven by flash chromatography on a 12 × 150 mm column filled with 10.0 g achiral silica gel.

The samples were then eluted in gradient conditions using a mixture of cyclohexane and *tert*-butyl methyl ether in the ratio from 100/0 to 0/100. The separations were run at the eluent flow rate of 18 mL/min and the collected fractions had the volume of 18 mL. To our disappointment, the resulting Δ*ee* was not significant and reached only about 3.0% (Figure 6). The first fractions in which compound 4 was detected by TLC were fractions 7, 6, 9 of 25.0, 60.0, and 75.0% ee, respectively. Compound 4 obtained in the course of the purification exhibited a slightly lower enantiomeric purity in the last fraction (Figures 6 and 7, Table S4).

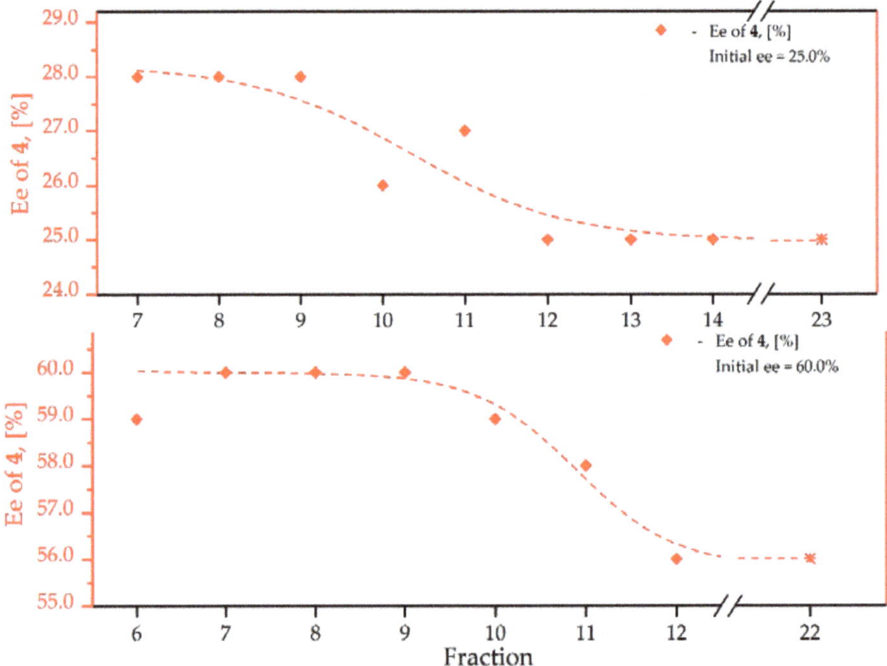

Figure 6. SDE of compound **4** of 25.0% and 60.0% *ee* in achiral MPLC conditions: the elution profiles. The model of the fitting is polynomial.

Since the concentration of the compounds in later fractions was low, those fractions (14–23 for 25.0% *ee* sample, 12–22 for 60.0% *ee* sample, and 12–24 for 75.0% *ee* sample) were combined before the *ee* determination. The distribution of the enantiomer in consequent fractions was determined in the experiment where sample of **4** with 75.0% *ee* was used. As it could be seen from Figure 7 and Table S4 the majority of the sample of slightly higher *ee* was eluted in fractions 9–10, while a less enantioenriched compound was eluted with the next 12 fractions. Such behavior of the compounds may indicate that part of the sample was adsorbed on stationary phase and slowly liberated upon elution.

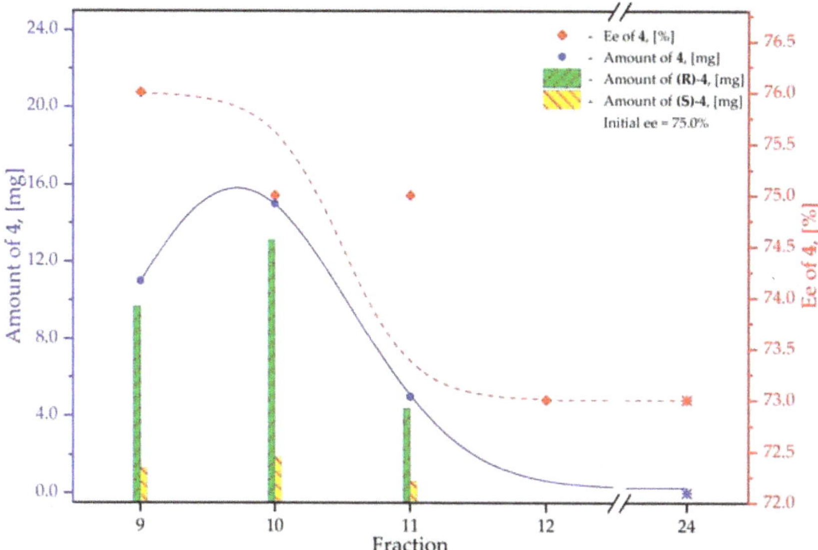

Figure 7. SDE of compound **4** of 75.0% *ee* in achiral MPLC conditions: the elution profile. The model of the fitting is polynomial.

An analysis of the collected data showed that in the case of compound **4**, even at a low polarity of the eluent, the magnitude of SDE is insignificantly influenced by the enantiomeric purity of the starting material. Additionally, it was observed that the amount of the compound collected after the elution constitutes only around 64.0% of the weight loaded onto the column. This is caused by strong interactions between the examined compound and the stationary phase and therefore its irreversible adsorption.

In the cases of compounds **1–3**, where the separation process was carried out with the isocratic elution, the phenomenon of SDE occurs in a linear manner in time. This may mean that the process takes place at the same speed during the separation procedure. The gradient of a straight line informs us about whether the enrichment or depletion *ee* of the fraction is observed, and its value represents the speed of these changes. The linear effect of the SDE overtime changes can be explained by the fact that during the isocratic elution, the constants of adsorption and desorption do not change. In the case of gradient elution conditions, as it had been presented for the compounds 4, the constants of adsorption are change during the elution process and therefore no linear correlation was observed.

3.5. *Study on the SDE of Amino Acid Derivatives: N-Acetyl-Phenylalanine* (**5**)

Biologically important small molecules are of especial interest in terms of the SDE studies. Therefore, the possibility of enantioseparation of the amino acid derivatives has been studied. *N*-acetyl-phenylalanine (**5**) was selected as a model since it is readily available from natural and synthetic amino acid [68], **5** is also a product of widely used benchmark asymmetric hydrogenation reaction [69]. In theory, strong intermolecular hydrogen bonding could be responsible for the enantio recognition in the case of amino acids. The simplest possible homochiral (**A**) and heterochiral (**B**) dimeric aggregates formed for *N*-acetyl-phenylalanine are depicted in Figure 8. At the same time, the crystallographic analysis indicates that in the solid state, *N*-acetyl-amino acids form a more complicated net of hydrogen bonds in which the moieties of amino acids are usually groups in tetrameric cycles (**C**) [70] [CSD Refcodes: COQHAR, DADGUK01, ACNVAL]. Interestingly, in the presence of protic solvents, the latter could replace one (or more) of the amino acid derivative molecules to form simpler aggregates such as e.g., trimers (D) [CSD Refcode: ACDAHO]. For comparison, similar studies on the

SDE of *N*-acetylated 1-phenylethylamine [71], and some other amides [72–74], which show a simpler mode of coordination, have already been reported.

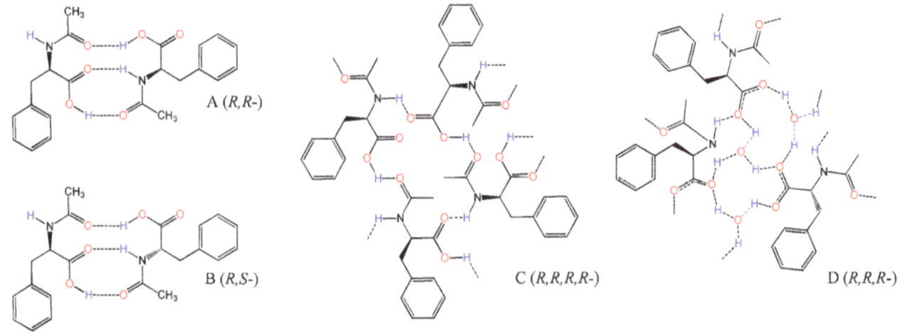

Figure 8. Chemical structures of the possible aggregates of *N*-acetyl-phenylalanine (**5**).

For the purposes of this study, the samples of **5** with the initial value of *ee* 50.0% were evaluated in different eluent systems. First elution was carried out using hexane/i-PrOH = 92/8 and the second series employed dichloroethene as the mobile phase. Since compound **5** is not soluble in the mobile phase, 100.0 mg of it was initially dissolved in methanol, and the solvent was evaporated in the presence of about 1.0 g of silica, then the compounds supported on the stationary phase were loaded at the top of a 12 × 150 mm column filled with additional 10.0 g silica gel. The separations were run at the eluent flow rate of 18 mL/min and the collected fractions had the volume of 18 mL. In both cases 26 fractions were collected, compound **5** first appeared in fraction 9 (hexane/^{i}-PrOH) and in fraction 12 (ClC_2H_4Cl). The enantiomeric composition of the consecutive fractions was determined be means of a chiral HPLC Chiralpak® OJ-RH column, with the mobile phase using H_2O/CH_3CN = 85/15 at the flow rate equal to 0.3 mL/min (Figure S5).

The amount of compound **5** and its enantiomeric composition in the consecutive fractions were evaluated and are presented in Table S5. The SDE profiles are presented in Figure 9. Whenever the mixture hexane/^{i}-PrOH = 92/8 was used as the eluent, the value Δ*ee* was 5.9% and the enantiomeric excess of the most enriched fraction 9 exceeded the initial mixture by 3.5%. Using dichloroethane allowed us to reach a greater magnitude of the SDE phenomenon with the 7.5% value of the Δ*ee* parameter. The enantiomeric excess of fraction 12 deviated the initial by 7.2%. The use of an aprotic solvent strongly affected the SDE, and (in contrast to where hexane/^{i}-PrOH was applied) furnished an elution with the opposite profile, where the initial fractions were enantiomerically depleted and the last ones were enantiomerically enriched.

The total amounts of compound **5** recovered after the purification were 65.7% (in the case of hexane/*i*-PrOH), and only 47.0% (in the case of dichloroethane) of the weight loaded onto the column. The results presented in Table S5 clearly indicate that an irreversible sorption of 5 on the stationary phase was much bigger when the aprotic solvent was used. It also facilitated the formation of heterochiral aggregates (*R*,*S*)- in the solution which (being symmetric and less polar—see Figure 10) are eluting faster to form first fractions of a lower *ee*.

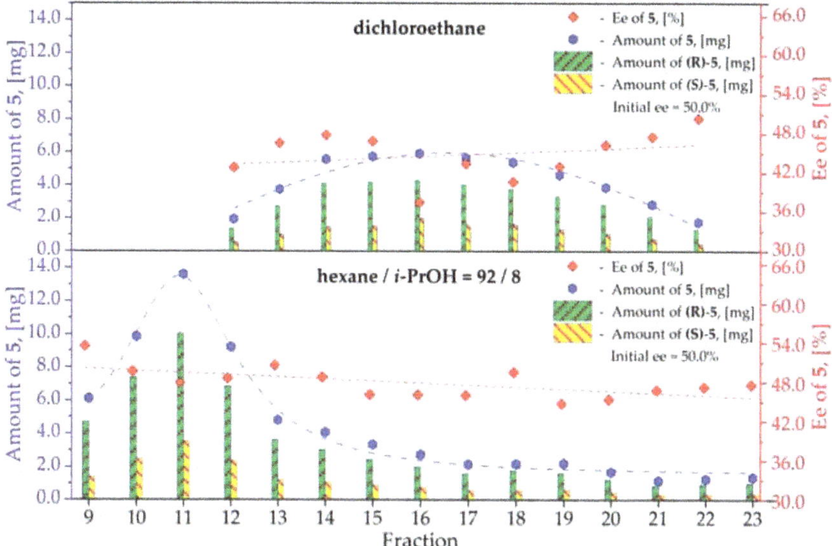

Figure 9. SDE of compound **5** in achiral MPLC conditions: the elution profiles. Fitting equation for dichloroethane is y = 0.730x + 32.462 (R^2 = 0.261), and for hexane/i-PrOH is y = −0.557x + 56.961 (R^2 = 0.428).

The experimental observations correlate well with the thermodynamic analysis of the potential equilibrium between possible stereoisomeric aggregates of 5 (Figure 10). It was found that the bimolecular (*R*,*S*)-aggregate is more stable than the isomeric (*S*,*S*)- or (*R*,*R*)-aggregates. In particular, an advanced M062x/6-311++G(d,p)(PCM) computational study showed that the differences between the values of Gibbs free energies of the formation of both molecular systems equals 4.0 and 2.5 kcal/mol respectively (Table 1). Additionally, we analyzed key distances between the substructures within the considered aggregates. It was found that these distances are practically identical in both cases. So, the difference between thermodynamic stabilities is a not a consequence of the power of local intermolecular interactions, but rather a more favorable arrangement of the structural elements in space, which provides more favorable electrostatic interactions and/or lower steric repulsions. Two aggregates are also characterized with different polarities where heterochiral aggregate has a lower dipole moment (Figure 10).

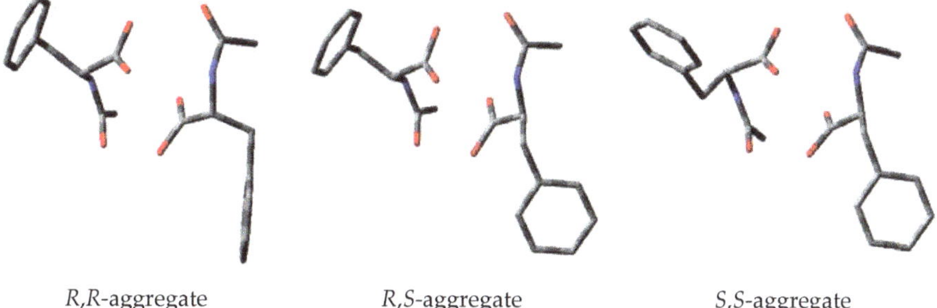

Figure 10. Views of the homo- and heterochiral aggregates of compound **5** in the DCM solution optimized at the M062x/6-311++G(d,p)(PCM) level of theory.

Table 1. Differences between the heterochiral and homochiral aggregates of compound **5** in the DCM solution optimized at the M062x/6-311++G(d,p)(PCM) level of theory.

Transition	ΔG [kcal/mol]
(*R*,*S*) → (*S*,*S*)	4.0
(*R*,*S*) → (*R*,*R*)	2.5

The limited solubility of compound **5**, and its tendency to form diverse and intricate crystalline forms, make the SDE process less efficient and make its study very complicated. Taking that into consideration, a simpler liquid derivative of α-amino acids should be selected to eliminate the possible separation via micro crystallization on the stationary phase, and facilitate enantioseparation by limiting the number of possible aggregation models which may be in an equilibrium and have an opposite effect on the SDE efficiency.

3.6. Study on the SDE of Aamino Acid Derivatives: Methyl Phenylalaninate (**6**)

Thus, methyl phenylalaninate (**6**) was chosen to be examined. The compound was obtained by esterification [75] of *ee* = 36.0% (*S*)-phenylalanine in standard conditions for this sequence of the reactions (Scheme 4).

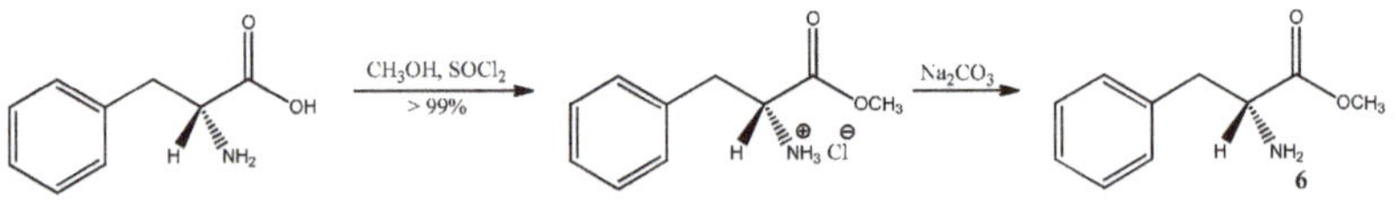

Scheme 4. Synthesis of methyl phenylalaninate (**6**).

Since **6** is a liquid well soluble in the majority of organic solvents there was no need to support the compound on the stationary phase before the separation. The solution of **6** in 0.5 mL of the mobile phase with the (S)- enantiomer of *ee* = 36.0% in the amount of 100.0 mg was subjected to the separation process on an achiral 12 × 150 mm column filled with additional 11.0 g silica gel and prewashed with the mobile phase. The separations were carried out at the eluent flow rate of 18 mL/min and the collected fractions had the volume of 18 mL. In this step, diethyl ether (R_f = 0,43) was used as the mobile phase. The presence of the compound in the obtained fraction was verified on the basis of the TLC technique. At first compound **6** appeared in fraction 12. The optical purity of the fractions was determined based on the ^{1}H and ^{13}C NMR techniques using, as CSA, 100 mol% of enantiomericaly pure **4** (Figure S6).

In the case of ^{13}C NMR, two sets of signals corresponding to the OCH_3 (51.41/51.45 ppm) and NCH (54.29/54.34 ppm) carbon atoms were analyzed, while in the case of ^{1}H NMR only signals corresponding to NCH proton at 3.95/4.01 (t, *J* = 6.5 Hz) ppm were used for the *ee* calculation (Figure S6). The individual optical purities, measured by NMR, and their average values are shown in Table S6.

As a result of this separation (Figure 11, and Table S6), 11 fractions were collected, in which the total amount of the compound was 68.1 mg (where: the mass of the (*R*)-enantiomer was 21.8 mg, and the mass of the *S* enantiomer was 46.3 mg). This means that during the separation 31.0 mg of compound **6** was permanently adsorbed on the stationary phase, including 10.2 mg of the (*R*)-enantiomer and 21.7 mg of the (*S*)-enantiomer. The Δ*ee* value for this separation was 13.8% *ee* (*S*). The differentially calculated enantiomeric composition of compound **6** irreversibly adsorbed on the stationary phase is the same as for the initial sample and constitutes 36.0% *ee* (*S*), which means that the adsorption process is not selective and the amount of the adsorbed enantiomer depends only on the initial composition of the starting mixture.

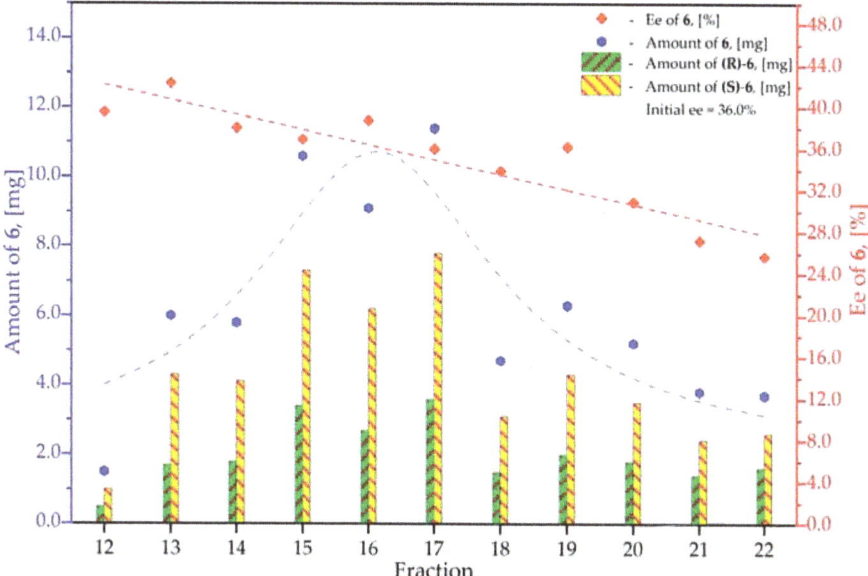

Figure 11. SDE of compound **6** using diethyl ether: the elution profile. The fitting equation is y = −1.426x + 59.275 (R^2 = 0.836).

Since the initial compound is not racemic, its irreversible nonstereoselective adsorption leads to a change in the column's character from achiral to chiral. The enantiomeric purity of the chiral selector on a newly-formed chiral surface depends on the enantiomeric composition of the mixture during loading and on the adsorptive capacity of the stationary phase. Such a newly-created chiral column differentiates the enantiomers of the chromatographed compound and causes their natural separation. This could be a good alternative explanation of the SDE phenomenon under chromatographic condition.

In order to confirm this hypothesis and explain the mechanism of the SDE process, the following experiments were carried out. Compound **6** of the same initial enantiocomposition (36.0% *ee*, (*S*) was eluted by *tert*-butyl methyl ether. In the first separation experiment, 100.0 mg of **6** with an initial purity of 36.0% *ee* (*S*) in 0.5 mL of hexane was injected into a 12 × 150 mm column packed with 11.0 g of spherical achiral silica gel, washed with the mobile phase. The separation was performed using *tert*-butyl methyl ether (MTBE) as the mobile phase (R_f = 0,4) and the eluent flow rate of 18 mL/min, 18 mL of the fractions were collected. As a result of this separation, 8 fractions containing **6** were obtained (fractions from 7 to 14). The presence of the compound in the fractions was verified based on the TLC technique. Then, when no more compound was found in the subsequent fractions collected, the column was used for the second separation experiment, where the same portion of compound **6** (as in the case of the first separation) was injected. The second separation experiment was carried out using the same mobile phase MTBE, as in the first separation process. As a result of this separation, 10 fractions were collected (fractions from 36 to 45). The optical purity of the fractions collected in the first and second separation experiments was determined based on the ^{1}H ^{13}C NMR spectroscopic technique with CSA **4**. The data collected in the course of both experiments was analyzed, and is presented in Table S7, while the elution profiles are shown in Figure 12.

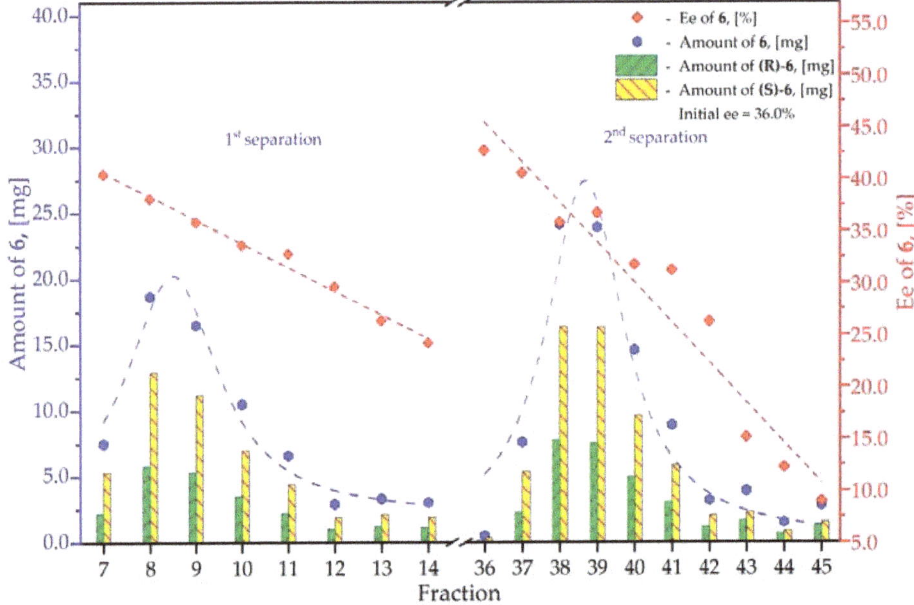

Figure 12. SDE of compound **6** by *tert*-butyl methyl ether: the elution profile. The fitting equation for 1st separation is y = −2.284x + 56.615 (R^2 = 0.988) and for 2nd separation is y = −3.863x + 184.578 (R^2 = 0.936).

As a result of the first separation experiment, 8 fractions were obtained in which the total mass of the compound was 68.0 mg (68.0% of initial), where the calculations were as follows: the amount of the (*R*)-enantiomer was 22.3 mg and the amount of the (*S*)-enantiomer was 46.7 mg (of 35.1% *ee* (*S*), calculated on the basis of each fraction). The value of the Δ*ee* for the first separation experiment was 16.0%, and the most enriched fraction 7 *ee* equaled 40.5%, while the less enriched fraction 14 equaled 24.2%. This means that during the first separation 31 mg (31.0%) of compound **6** (of 37.0% *ee* (S)) was permanently adsorbed on the stationary phase, including 9.7 mg of the (*R*)-enantiomer and 21.3 mg of the (*S*)-enantiomer. In the first separation experiment, similarly as it had been observed in the case of the elution with diethyl ether, during the chromatographic process part of the eluted compounds (31.0 mg) was adsorbed on the stationary phase not enantio-selectively with the ratio *R/S* depending only on the initial composition of the mixture.

Notably, in the case of the second separation experiment, the same column was used, but with already adsorbed compound **6.** The total amount of **6** collected after the separation was much higher and equal to 91.0 mg (91.0%). The weighted arithmetic mean ratio of the (*R*)-enantiomer to the (*S*)-enantiomer was 30.7 to 60.3 mg (33.0% *ee* (S)). The value of the Δ*ee* for this separation calculated based on the *ee* of fraction 36 (42.7%) and fraction 45 (9.0%) equaled 33.7%. Thus, only 9.0 mg (9.0%) of compound **6** of 71.0% *ee* (S) was lost on the stationary phase.

The efficiency of the second separation experiment was therefore notably higher than of the first one. The simple rationalization of this phenomenon could be as follows: the effect can be explained by the formation of a new interaction between the enantiomers and the modified stationary phase which formed after part of the chiral compound was adsorbed on the stationary phase surface. The homochiral and heterochiral aggregates are formed during the separation between the free compound in the mobile phase and the chiral selectors adsorbed on the stationary phase. Such process follows the rule of three points (Pirkle rule) [76,77]. According to this principle, for the chiral separation of enantiomers, it is necessary to have a minimum of three interactions between the enantiomers and the chiral selector, as shown in Figure 13, while at least one of these interactions depends on the configuration of the chiral centers of the selector and the selectand. The separation of the enantiomers by HPLC with the chiral stationary phases depends on the formation of transient diastereoisomeric linkages (in our case they are homochiral and heterochiral aggregates). Different stabilities of these complexes result in the possibility of the chiral separation of enantiomers.

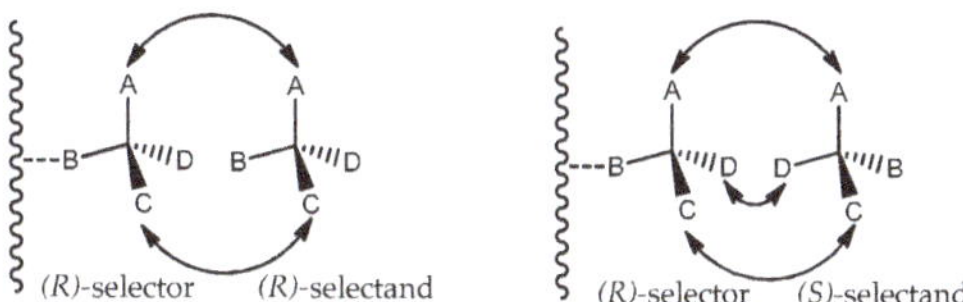

Figure 13. The model of the interactions between selectand and chiral selector.

The efficiency of the enentiomer separation can be explained based on the quantumchemical calculations, as in the case of **6**. In particular, the M062x/6-311++G(d,p) computational study showed that the homochiral aggregate is more stable than the isomeric heterochiral one (Table 2). The model of coordination was based on crystallographic data [70] [CSD Refcodes: IWOMIS] where hydrogen bonds NH—O=C with the bond length of 2.28 Å were responsible for the crystal net formation. Next, the quantumchemical study confirmed that the less polar homochiral aggregates (Figure 14) are eluting faster than the heterochiral ones. The DFT calculation also indicated that the thermodynamic stabilities of the formed aggregates are dependent on the arrangement of their structural elements in space.

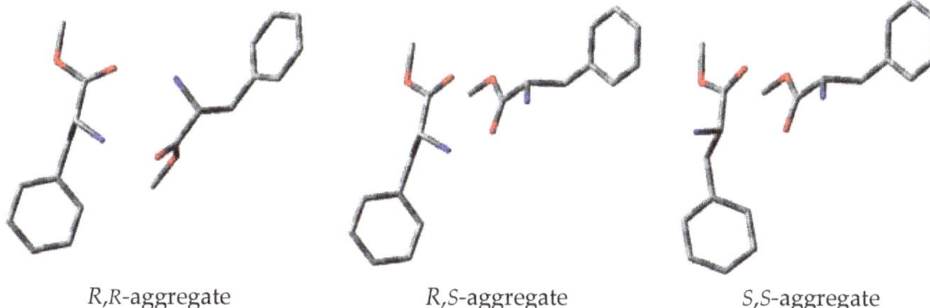

Figure 14. Views of the homo- and heterochiral aggregates of compound **6** in the DCM solution optimized at the M062x/6-311++G(d,p)(PCM) level of theory.

Table 2. Differences between heterochiral and homochiral aggregates of compound **6** in the DCM solution optimized at the M062x/6-311++G(d,p)(PCM) level of theory.

Transition	ΔG [kcal/mol]
(*R*,*R*) → (*R*,*S*)	2.2
(*S*,*S*) → (*R*,*S*)	3.3

4. Conclusions

The data presented above indicate that the optical purification approach based on SDE has a potential and in some cases could be applied as an alternative to the conventional procedures of chiral separations. The SDE phenomenon of six non-racemic compounds on achiral silica gel was explored, and some correlation between the eluent used, ratio of the compounds to the stationary phases and compound structures was found to influence the magnitude of the enantiomer self-disproportionation process. An explanation of a possible mechanism of this phenomenon, which involves the in situ formation of the chiral surface on the achiral stationary phase, was also provided. The presented results may mean that the SDE phenomenon should not only be explained by the formation of the aggregates between the enantiomers in a mobile phase, but also by the change of the column character from achiral to chiral. Since crystallization and sublimation are purification methods applicable only for enantiomerically enriched crystalline samples, SDE-based chromatographic methodology could be applied as a general tool for practical enantiopurification of organic compounds enriched in one of the enantiomers. Such a self enantio-enriching process may take place in nature, and could be responsible for the formation of the first chiral non-racemic compounds on the early Earth.

Additionally, it worth mentioning that, taking into consideration the fact that in many publications on asymmetric synthesis, the optical purity of the furnished products exceeds 99% *ee*, it is important to measure the enantiomeric purity before any purification of the product obtained in an asymmetric reaction to achieve reliable results of the measurement of the level of an asymmetric induction.

From a practical point of view, to facilitate a SDS effect, is may be recommend to use an aprotic solvents of low polarity and higher ratio of the stationary phase to the separated compound. The mobile phases should have a R_f value in the range of 0.2–0.4 and the resulting TLC spot should be wider rather than narrow. Probably, better separations may be obtained in cases of substances forming weak intermolecular hydrogen bonds.

Supplementary Materials: Supplementary materials can be accessed at: http://www.mdpi.com/2073-8994/11/5/680/s1.

Author Contributions: Conceptualization, O.M.D.; methodology, O.M.D. and M.B.; investigation, O.M.D., M.B., K.K., R.J., W.K., S.K.; writing—original draft preparation, O.M.D., M.B., D.S.; writing—review and editing, O.M.D.; visualization, M.B., O.M.D.; supervision, O.M.D.; inspiration and discussions: K.M.P.

Funding: This research received no external funding. The research was carried out in part using the PLGrid ('Prometheus' cluster) infrastructure (ACK 'Cyfronet' in Cracov), and the equipment purchased thanks to the financial support of the European Regional Development Fund under the framework of the Operational Program Development of Eastern Poland 2007–2013 (Contract No. POPW.01.03.00-06-009/11-00, equipping the laboratories of the Faculties of Biology and Biotechnology, Mathematics, Physics and Informatics, and Chemistry for studies of biologically active substances and environmental samples) as well as the Polish National Science Centre research grant (2012/05/B/ST5/00362).

Conflicts of Interest: The authors declare no conflict of interest.

References

1. Carreira, E.M.; Yamamoto, H. *Comprehensive Chirality*, 1st ed.; Elsevier: Amsterdam, The Netherlands, 2012; ISBN 978-008-095-167-6.
2. Hegstrom, R.A.; Kondepudi, D.K. The Handedness of the Universe. *Sci. Am.* **1990**, *262*, 108–115. [CrossRef]
3. Lubec, G.; Rosenthal, G.A. *Amino Acids, Chemistry, Biology and Medicine*; ESCOM: Leiden, The Netherlands, 1990; ISBN 9072199049.
4. Ernst, B.; Hart, G.W.; Sinay, P. *Biosynthesis and Degradation of Glycoconjugates*; Wiley-VCH: Chichester, UK, 2000; ISBN 352-729-511-9.
5. Gawley, R.E.; Aubé, J. *Principles of Asymmetric Synthesis*, 2nd ed.; Elsevier: Oxford, UK, 2012; ISBN 978-008-044-860-2.
6. Cohen, J. Getting all turned around over the origins of life on earth. *Science* **1995**, *267*, 1265–1266. [CrossRef]
7. McCombs, C. The Extraterrestrial Search for the Origin of Homochirality. *Creat. Res. Soc. Q.* **2014**, *51*, 5–13.
8. Rauchfuss, H.; Mitchell, T.N. *Chemical Evolution and the Origin of Life by Horst Rauchfuss*; Springer: Berlin/Heidelberg, Germany, 2008; ISBN 978-354-078-823-2.
9. McGuire, B.A.; Carroll, P.B.; Loomis, R.A.; Finneran, I.A.; Jewell, P.R.; Remijan, A.J.; Blake, G.A. Discovery of the interstellar chiral molecule propylene oxide (CH3CHCH2O). *Science* **2016**, *352*, 1449–1452. [CrossRef]
10. Cooper, G.; Reed, C.; Nguyen, D.; Carter, M.; Wang, Y. Detection and formation scenario of citric acid, pyruvic acid, and other possible metabolism precursors in carbonaceous meteorites. *Proc. Natl. Acad. Sci. USA* **2011**, *108*, 14015–14020. [CrossRef] [PubMed]
11. Angel, J.R.P.; Illing, R.; Martin, P.G. Circular polarization of twilight. *Nature* **1972**, *238*, 389–390. [CrossRef]
12. Wolstencroft, R.D. Terrestiral and Astronomical Sources of Circular Polarisation: A fresh look at the origin of Homochirality on Earth. *Bioastron. Life Stars* **2004**, *213*, 154–158.

13. Garay, A.S.; Ahigren-Beckendorf, J.A. Differential interaction of chiral β-particles with enantiomers. *Nature* **1990**, *346*, 451–453. [CrossRef]
14. Rikken, G.L.J.A.; Raupach, E. Enantioselective magnetochiral photochemistry. *Nature* **2000**, *405*, 932–935. [CrossRef]
15. Soai, K.; Osanai, S.; Kadowaki, K.; Yonekubo, S.; Shibata, T.; Sato, I. d- and l-Quartz-Promoted Highly Enantioselective Synthesis of a Chiral Organic Compound. *J. Am. Chem. Soc.* **1999**, *121*, 11235–11236. [CrossRef]
16. Tang, L.; Shi, L.; Bonneau, C.; Sun, J.; Yue, H.; Ojuva, A.; Lee, B.L.; Kritikos, M.; Bell, R.G.; Bacsik, Z.; et al. A zeolite family with chiral and achiral structures built from the same building layer. *Nat. Mater.* **2008**, *7*, 381–385. [CrossRef] [PubMed]
17. Nógrádi, M.; Fogassy, E.; Pálovics, E.; Schindler, J. Resolution of Enantiomers by Non-Conventional Methods. *Synthesis* **2005**, *10*, 1555–1568.
18. Faigl, F.; Fogassy, E.; Nógrádi, M.; Pálovics, E.; Schindler, J. Strategies in optical resolution: A practical guide. *Tetrahedron Asymmetry* **2008**, *19*, 519–536. [CrossRef]
19. Nag, A. *Asymmetric Synthesis of Drugs and Natural Products*; CRC Press: Boca Raton, FL, USA, 2018; ISBN 978-113-803-361-0.
20. Lough, W.J.; Wainer, I.W. *Chirality in Natural and Applied Science*; CRC Press: Boca Raton, FL, USA, 2002; ISBN 084-932-434-3.
21. Kurihara, N.; Miyamoto, J. *Chirality in Agrochemicals*; Wiley: New York, NY, USA, 1998; ISBN 047-198-121-4.
22. Kitzerow, H.-S.; Bahr, C. *Chirality in Liquid Crystals*; Springer: New York, NY, USA, 2001; ISBN 038-798-679-0.
23. Li, Q. *Photoactive Functional Soft Materials: Preparation, Properties, and Applications*; Wiley: Blackwell, UK, 2019; ISBN 978-352-781-676-7.
24. Mason, S.F. Optical rotatory power. *Q. Rev. Chem. Soc.* **1963**, *17*, 20–66. [CrossRef]
25. Schellman, J.A. Circular dichroism and optical rotation. *Chem. Rev.* **1975**, *75*, 323–331. [CrossRef]
26. Lloyd, D.K.; Goodall, D.M. Polarimetric detection in high-performance liquid chromatography. *Chirality* **1989**, *1*, 251–264. [CrossRef]
27. Roussel, C.; Del Rio, A.; Pierrot-Sanders, J.; Piras, P.; Vanthuyne, N.C. Chiral liquid chromatography contribution to the determination of the absolute configuration of enantiomers. *J. Chromatogr. A* **2004**, *1037*, 311–328. [CrossRef]
28. Pakulski, Z.; Demchuk, O.M.; Kwiatosz, R.; Osiński, P.W.; Świerczyńska, W.; Pietrusiewicz, K.M. The classical Kagan's amides are still practical NMR chiral shift reagents: Determination of enantiomeric purity of P-chirogenic phospholene oxides. *Tetrahedron Asymmetry* **2003**, *14*, 1459–1462. [CrossRef]
29. Demchuk, O.M.; Świerczynska, W.; Pietrusiewicz, K.M.; Woźnica, M.; Wójcik, D.; Frelek, J. A convenient application of the NMR and CD methodologies for the determination of enantiomeric ratio and absolute configuration of chiral atropoisomeric phosphine oxides. *Tetrahedron Asymmetry* **2008**, *19*, 2339–2345. [CrossRef]
30. Goering, H.L.; Eikenberry, J.N.; Koermer, G.S. Tris[3-(trifluoromethylhydroxymethylene)-d-camphorato]europium(III) a chiral shift reagent for direct determination of enantiomeric compositions. *J. Am. Chem. Soc.* **1971**, *3*, 5913–5914. [CrossRef]
31. McCreary, M.D.; Lewis, D.W.; Wernick, D.L.; Whitesides, G.M. Determination of enantiomeric purity using chiral lanthanide shift reagents. *J. Am. Chem. Soc.* **1974**, *96*, 1038–1054. [CrossRef]
32. Kolodiazhnyi, O.I.; Demchuk, O.M.; Gerschkovich, A.A. Application of the dimenthyl chlorophosphite for the chiral analysis of amines, amino acids and peptides. *Tetrahedron Asymmetry* **1999**, *10*, 1729–1732. [CrossRef]
33. Seco, J.M.; Latypov, S.; Quiñoá, E.; Riguera, R. New chirality recognizing reagents for the determination of absolute stereochemistry and enantiomeric purity by NMR. *Tetrahedron Lett.* **1994**, *18*, 2921–2924. [CrossRef]
34. Wenzel, T.J.; Wilcox, J.D. Chiral reagents for the determination of enantiomeric excess and absolute configuration using NMR spectroscopy. *Chirality* **2003**, *15*, 256–270. [CrossRef] [PubMed]
35. Soloshonok, V.A. Remarkable Amplification of the self-disproportionation of enantiomers on achiral-phase chromatography columns. *Angew. Chem. Int. Ed.* **2006**, *45*, 766–769. [CrossRef]
36. Soloshonok, V.A.; Berbasov, D.O. Self-disproportionation of enantiomers of (*R*)-ethyl 3-(3,5-dinitrobenzamido)-4,4,4-trifluorobutanoate on achiral silica gel stationary phase. *J. Fluorine Chem.* **2006**, *127*, 597–603. [CrossRef]

37. Cundy, K.C.; Crooks, P.A. Unexpected phenomenon in the high-performance liquid chromatographic analysis of racemic 14C-labelled nicotine: Separation of enantiomers in a totally achiral system. *J. Chromatogr. A* **1983**, *281*, 17–33. [CrossRef]
38. Soloshonok, V.; Sorochinsky, A.; Aceña, J. Self-Disproportionation of Enantiomers of Chiral, Non-Racemic Fluoroorganic Compounds: Role of Fluorine as Enabling Element. *Synthesis* **2012**, *45*, 141–152. [CrossRef]
39. Faigl, F.; Fogassy, E.; Nógradi, M.; Palovics, E.; Schindler, J. Separation of non-racemic mixtures of enantiomers: An essential part of optical resolution. *Org. Biomol. Chem.* **2010**, *8*, 947–959. [CrossRef]
40. Garin, D.L.; Greco, D.J.C.; Kelley, L. Enhancement of optical activity by fractional sublimation. An alternative to fractional crystallization and a warning. *J. Org. Chem.* **1977**, *42*, 1249–1251. [CrossRef]
41. Omelańczuk, J.; Mikołajczyk, M. Chiral t-butylphenylphosphinothioic acid: A useful chiral solvating agent for direct determination of enantiomeric purity of alcohols, thiols, amines, diols, aminoalcohols and related compounds. *Tetrahedron Asymmetry* **1996**, *7*, 2687–2694. [CrossRef]
42. Tsai, T.L.; Herman, K.; Hug, E.; Rohde, B.; Dreiding, A.S. Enantiomer-Differentiation Induced by an Enantiomeric Excess during Chromatography with Achiral Phases. *Helv. Chem. Acta* **1985**, *68*, 2238–2243. [CrossRef]
43. Viedma, C.; Noorduin, W.L.; Ortiz, J.E.; de Torres, T.; Cintas, P. Asymmetric amplification in amino acid sublimation involving racemic compound to conglomerate conversion. *Chem. Commun.* **2011**, *47*, 671–673. [CrossRef] [PubMed]
44. Charles, R.; Gil-Av, E. Self-amplification of optical activity by chromatography on an achiral adsorbent. *J. Chromatogr. A* **1984**, *298*, 516–520. [CrossRef]
45. Dobashi, A.; Motoyama, Y.; Kinoshita, K.; Hara, S.; Fukasaku, N. Self-induced chiral recognition in the association of enantiomeric mixtures on silica gel chromatography. *Anal. Chem.* **1987**, *59*, 2209–2211. [CrossRef]
46. Frost, C.G.; Howarth, J.; Williams, J.M.J. Selectivity in palladium catalyzed allylic substitution. *Tetrahedron Asymmetry* **1992**, *3*, 1089–1122. [CrossRef]
47. Dawson, G.J.; Frost, C.G.; Williams, J.M.J. Asymmetric palladium catalysed allylic substitution using phosphorus containing oxazoline ligands. *Tetrahedron Lett.* **1993**, *34*, 3149–3150. [CrossRef]
48. Demchuk, O.M.; Kielar, K.; Pietrusiewicz, K.M. Rational design of novel ligands for environmentally benign cross-coupling reactions. *Pure Appl. Chem.* **2011**, *3*, 633–644. [CrossRef]
49. Demchuk, O.M.; Kapłon, K.; Kącka, A.; Pietrusiewicz, K.M. The utilization of chiral phosphorus ligands in atroposelective cross-coupling reactions. *Phosphorus Sulfur Silicon Relat. Elem.* **2016**, *191*, 180–200. [CrossRef]
50. Jasiński, R.; Demchuk, O.M.; Babyuk, D. A Quantum-Chemical DFT Approach to Elucidation of the Chirality Transfer Mechanism of the Enantioselective Suzuki–Miyaura Cross-Coupling Reaction. *J. Chem.* **2017**, *2017*, 3617527. [CrossRef]
51. Zhang, D.; Wang, Q. Palladium catalyzed asymmetric Suzuki–Miyaura coupling reactions to axially chiral biaryl compounds: Chiral ligands and recent advances. *Coord. Chem. Rev.* **2015**, *286*, 1–16. [CrossRef]
52. Sawai, K.; Tatumi, R.; Nakahodo, T.; Fijihara, H. Asymmetric Suzuki–Miyaura coupling reactions catalyzed by chiral palladium nanoparticles at room temperature. *Angew. Chem.* **2008**, *120*, 7023–7025. [CrossRef]
53. Braun, M.; Devant, R. (*R*)- and (*S*)-2-acetoxy-1,1,2-triphenylethanol—Effective synthetic equivalents of a chiral acetate enolate. *Tetrahedron Lett.* **1984**, *25*, 5031–5034. [CrossRef]
54. Tempkin, O.; Abel, S.; Chen, C.P.; Underwood, R.; Prasad, K.; Chen, K.M.; Repic, O.; Blacklock, T.J. Asymmetric synthesis of 3,5-dihydroxy-6(E)-heptenoate-containing HMG-CoA reductase inhibitors. *Tetrahedron* **1997**, *53*, 10659–10670. [CrossRef]
55. Zhao, Y.; Truhlar, D.G. Density Functionals with Broad Applicability in Chemistry. *Acc. Chem. Res.* **2008**, *41*, 157–167. [CrossRef]
56. Zhao, Y.; Truhlar, D.G. The M06 suite of density functionals for main group thermochemistry, thermochemical kinetics, noncovalent interactions, excited states, and transition elements: Two new functionals and systematic testing of four M06-class functionals and 12 other functionals. *Theor. Chem. Acc.* **2007**, *120*, 215–241.
57. Frisch, M.J.; Trucks, G.W.; Schlegel, H.B.; Scuseria, G.E.; Robb, M.A.; Cheeseman, J.R.; Scalmani, G.; Barone, V.; Mennucci, B.; Petersson, G.A. *Gaussian 09, Revision B.01*; Gaussian Inc.: Wallingford, UK, 2010.
58. Demchuk, O.M.; Jasiński, R.; Pietrusiewicz, K.M. New Insights into the Mechanism of Reduction of Tertiary Phosphine Oxides by Means of Phenylsilane. *Heteroatom Chem.* **2015**, *26*, 441–448. [CrossRef]

59. Pietrusiewicz, K.; Szwaczko, K.; Mirosław, B.; Dybała, I.; Jasiński, R.; Demchuk, O.M. New Rigid Polycyclic Bis(phosphane) for Asymmetric Catalysis. *Molecules* **2019**, *24*, 571. [CrossRef]
60. Mirosław, B.; Babyuk, D.; Łapczuk-Krygier, A.; Kącka-Zych, A.; Demchuk, O.M.; Jasiński, R. Regiospecific formation of the nitromethyl-substituted 3-phenyl-4,5-dihydroisoxazole via [3 + 2]cycloaddition. *Monatshefte für Chemie Chem. Mon.* **2018**, *149*, 1877–1884. [CrossRef]
61. Jin, M.-J.; Takale, V.B.; Sarkar, M.S.; Kim, Y.-M. Highly enantioselective Pd-catalyzed allylic alkylation using new chiral ferrocenylphosphinoimidazolidine ligands. *Chem. Commun.* **2006**, *6*, 663–664. [CrossRef]
62. Martinez, C.R.; Iverson, B.L. Rethinking the term "pi-stacking". *Chem. Sci.* **2012**, *3*, 2191–2201. [CrossRef]
63. Demchuk, O.M.; Justyniak, I.; Miroslaw, B.; Jasinski, R. 2-Methoxynaphthylnaphthoquinone and its solvate: Synthesis and structure-properties relationship. *J. Phys. Organ. Chem.* **2014**, *27*, 66–73. [CrossRef]
64. Nakashima, E.; Yamamoto, H. Asymmetric Aldol Synthesis: Choice of Organocatalyst and Conditions. *Chem. Asian J.* **2017**, *12*, 41–44. [CrossRef] [PubMed]
65. Sakthivel, K.; Notz, W.; Bui, T.; Barbas, C.F. Amino Acid Catalyzed Direct Asymmetric Aldol Reactions: A Bioorganic Approach to Catalytic Asymmetric Carbon−Carbon Bond-Forming Reactions. *J. Am. Chem. Soc.* **2001**, *123*, 5260–5267. [CrossRef]
66. Chrzanowski, J.; Krasowska, D.; Urbaniak, M.; Sieroń, L.; Pokora-Sobczak, P.; Demchuk, O.M.; Drabowicz, J. Synthesis of Enantioenriched Aryl-tert-Butylphenylphosphine Oxides via Cross-Coupling Reactions of tert-Butylphenylphosphine Oxide with Aryl Halides. *Eur. J. Organ. Chem.* **2018**, *2018*, 4614–4627. [CrossRef]
67. Harger, M.J.P. Chemical shift non-equivalence of enantiomers in the proton magnetic resonance spectra of partly resolved phosphinothioic acids. *J. Chem. Soc. Perkin II* **1978**, *4*, 326–331. [CrossRef]
68. Dressen, M.H.C.L.; van de Kruijs, B.H.P.; Meuldijk, J.; Vekemans, J.A.J.M.; Hulshof, L.A. From Batch to Flow Processing: Racemization of N-Acetylamino Acids under Microwave Heating. *Organ. Process Res. Dev.* **2009**, *13*, 888–895. [CrossRef]
69. Demchuk, O.M.; Arlt, D.; Jasiński, R.; Pietrusiewicz, K.M. Relationship between structure and efficiency of atropisomeric phosphine ligands in homogeneous catalytic asymmetric hydrogenation. *J. Phys. Organ. Chem.* **2012**, *25*, 1006–1011. [CrossRef]
70. Groom, C.R.; Bruno, I.J.; Lightfoot, M.P.; Ward, S.C. The Cambridge Structural Database. *Acta Crystallogr. Sect. B Struct. Sci. Cryst. Eng. Mater.* **2016**, *72*, 171–179. [CrossRef]
71. Wzorek, A.; Sato, A.; Drabowicz, J.; Soloshonok, V.A. Self-disproportionation of enantiomers via achiral gravity-driven column chromatography: A case study of N-acyl-α-phenylethylamines. *J. Chromatogr. A* **2016**, *1467*, 270–278. [CrossRef] [PubMed]
72. Wzorek, A.; Kamizela, A.; Sato, A.; Soloshonok, V.A. Self-Disproportionation of Enantiomers (SDE) via achiral gravity-driven column chromatography of N-fluoroacyl-1-phenylethylamines. *J. Fluorine Chem.* **2017**, *196*, 37–43. [CrossRef]
73. Suzuki, Y.; Han, J.; Kitagawa, O.; AceÇa, J.L.; Klika, K.D.; Soloshonok, V.A. Comprehensive examination of the self-disproportionation of enantiomers (SDE) of chiral amides via achiral, laboratory-routine, gravity-driven column chromatography. *RCS Adv.* **2015**, *5*, 2988–2993. [CrossRef]
74. Wzorek, A.; Sato, A.; Drabowicz, J.; Soloshonok, V.A.; Klika, K.D. Enantiomeric enrichments via the self-disproportionation of enantiomers (SDE) by achiral, gravity-driven column chromatography: A case study using *N*-(1-Phenylethyl)acetamide for optimizing the enantiomerically pure yield and magnitude of the SDE. *Helv. Chem. Acta* **2015**, *98*, 1147–1159. [CrossRef]
75. Mori, M.; Deodato, D.; Kasula, M.; Ferraris, D.M.; Sanna, A.; De Logu, A.; Rizzi, M.; Botta, M. Design, synthesis, SAR and biological investigation of 3-(carboxymethyl)rhodanine and aminothiazole inhibitors of Mycobacterium tuberculosis Zmp1. *Bioorg. Med. Chem. Lett.* **2018**, *28*, 637–641. [CrossRef]
76. Pirkle, W.H.; Pochapsky, T.C. Considerations of chiral recognition relevant to the liquid chromatography separation of enantiomers. *Chem. Rev.* **1989**, *89*, 347–362. [CrossRef]
77. Polavarapu, P.L. *Chiral Analysis—Advances in Spectroscopy, Chromatography and Emerging Methods*, 2nd ed.; Elsevier Science: Atlanta, GA, USA, 2018; ISBN 978-044-464-028-4.

Concept Paper

Chemical Basis of Biological Homochirality during the Abiotic Evolution Stages on Earth

Josep M. Ribó [1,*] **and David Hochberg** [2]

1 Department of Inorganic and Organic Chemistry, University of Barcelona, Institute of Cosmos Science (UB-EEC), c. Martí i Franquès 1, 08028-Barcelona, Catalonia, Spain
2 Centro de Astrobiología (CSIC-INTA), Department of Molecular Evolution, Carretera Ajalvir Kilómetro 4, 28850 Torrejón de Ardoz, Madrid, Spain; hochbergd@cab.inta-csic.es
* Correspondence: jmribo@ub.edu

Received: 31 May 2019; Accepted: 18 June 2019; Published: 20 June 2019

Abstract: Spontaneous mirror symmetry breaking (SMSB), a phenomenon leading to non-equilibrium stationary states (NESS) that exhibits biases away from the racemic composition is discussed here in the framework of dissipative reaction networks. Such networks may lead to a metastable racemic non-equilibrium stationary state that transforms into one of two degenerate but stable enantiomeric NESSs. In such a bifurcation scenario, the type of the reaction network, as well the boundary conditions, are similar to those characterizing the currently accepted stages of emergence of replicators and autocatalytic systems. Simple asymmetric inductions by physical chiral forces during previous stages of chemical evolution, for example in astrophysical scenarios, must involve unavoidable racemization processes during the time scales associated with the different stages of chemical evolution. However, residual enantiomeric excesses of such asymmetric inductions suffice to drive the SMSB stochastic distribution of chiral signs into a deterministic distribution. According to these features, we propose that a basic model of the chiral machinery of proto-life would emerge during the formation of proto-cell systems by the convergence of the former enantioselective scenarios.

Keywords: biological homochirality; enantioselective reaction; autocatalysis; origin of life; replicators

---

1. Introduction

The origin of biological homochirality (BH) [1–4] is often described as a scientific mystery [5] and the lack of a scientific explanation of the BH phenomenon is often used as proof against the theory of evolution [6]. However, it is possible nowadays to give a rigorous basic physico-chemical justification for the origin of BH. There are indisputable experimental reports on spontaneous mirror symmetry breaking (SMSB) [7–9] as well as many theoretical reports. SMSB can be interpreted within the framework of entropy production and dissipative processes [10,11]. This means that BH belongs to the same physico-chemical scenarios as those of open chemical networks proposed for the description of minimal self-reproducing proto-cells, primordial biochemical cycles, and the chemical thermodynamics of the reaction networks that maintain the phenomenon of life [12–26]. The paradox is that all of these reports eschew the enantioselective character of biological replicators, organocatalysis and autocatalytic reaction networks. Here we report that, in spite of the current mainstream ideas that dominate the possible explanations of BH, there are currently well-established theoretical and experimental chemical reports supporting chemical abiotic scenarios, which show a resilience to racemization and a preference for homochiral reaction outcomes with respect to the racemic ones, i.e. these are the very trends defining BH. We discuss these points with the aim to close the gap in the understanding of SMSB and of its place within the differentiated and hierarchical stages of chemical evolution.

The advances made in asymmetric induction during the 20th Century dominate the mainstream opinion on chiral methods in chemistry. Consequently, historically the emergence of homochirality has been justified mainly on the basis of the asymmetric induction of enantioselective reactions by physical chiral forces [2,27,28], in conjunction with non-linear asymmetric inductions leading to chiral amplifications [29–31] and the kinetic trapping of the final chiral species. To assume this as the basis for BH implies that during the evolutionary stages (see Figure 1) corresponding to the condensation reactions leading to the homochiral functional polymers and chiral replicators, homochirality is achieved by starting from enantiopure pools of homochiral building blocks (amino acids and carbohydrates). From a chemical point of view, this is an unlikely assumption that places SMSB at one of the earlier stages of chemical evolution as a singular event. In this regard, the low probability of such a scenario has been previously reported by some authors. Root-Bernstein proposed [32] the simultaneous origin of BH and the genetic code. Moreover, the relationship between the emergence of autocatalytic replication and homochirality has been reported, as well as the cooperative effect of external chiral polarizations [33]. However, despite their seminal significance, in our opinion, these proposals lack a description of their physico-chemical basis. Without this, the acceptance of such models by applied chemists is not possible. Furthermore, the design and experimental research in the field of artificial life cannot progress without the assumption of the chemical relationship between homochirality and autocatalysis. In this report, we describe: a) What type of enantioselective reaction networks may or may not lead to racemic biases; b) the general physico-chemical basis of how such enantioselective autocatalytic reaction networks may or may not lead to racemic biases; and c) how these SMSB systems (reaction networks plus open dissipative systems) depend on external chiral polarizations for the deterministic emergence in chemical evolution of one of the two chiral signs.

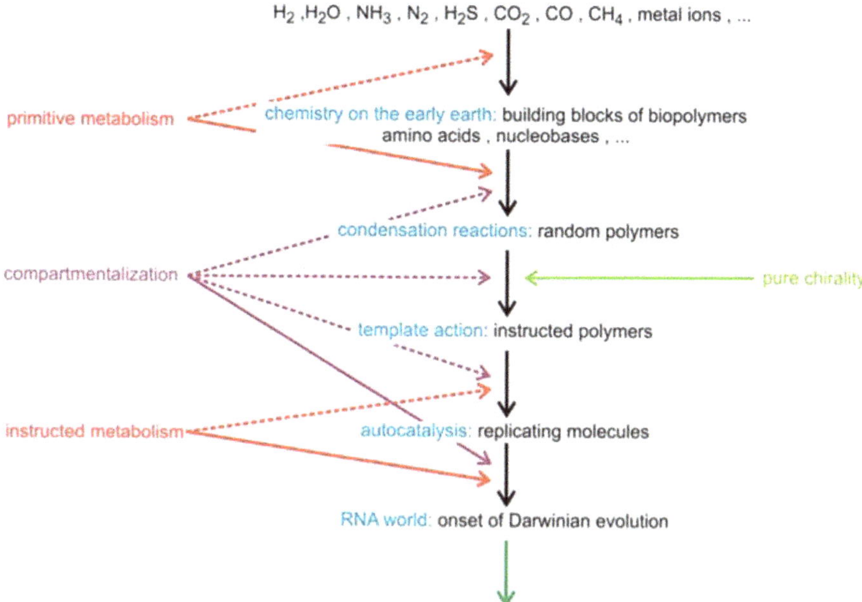

Figure 1. Presently, most accepted stages of a scheme for a reasonable hierarchical order of the increase of complexity during chemical evolution. The scheme assumes the unreasonable existence of enantiopure pools of chemical building blocks for the formation of the chiral polymers of life. Reproduced with author's permission from reference [34].

2. Racemization and Racemic Mixtures

Asymmetric synthesis of simple chiral organic compounds taking place under the effect of chiral forces and during the early stages of chemical evolution, i.e. in astrophysical scenarios, is an hypothesis that is being experimentally confirmed [35,36]. In such physical scenarios, strong chiral polarization forces can be expected to be operative. The chiral compounds obtained in such asymmetric synthesis must undergo racemization processes, which in spite of their low rates due to the low temperatures of interstellar space, would most likely be significant on long time scales. Therefore, in the currently and widely accepted hypothesis of the bombardment of Earth by planetesimal objects containing organic compounds [37], the enantiomeric excess (*ee*) values could be expected to be on the order of a few percent. Furthermore, on Earth, the increase of complexity towards supra and macromolecular compounds could only begin at the latter stages of the Hadean Era, when the Earth's temperature decreased low enough so as to not decompose organic compounds. This means the complexity increase could not have occurred before the beginning of the zircon chronology. This also means that the starting temperatures of chemical evolution on Earth should be high enough to lead to significant racemization rates. Furthermore, condensation polymer synthesis probably requires experimental conditions favoring racemization processes [38]. In summary, it is chemically reasonable to expect an $ee \neq 0$ during the stage of the formation of condensation polymers, but certainly much lower than the homochirality required for obtaining homochiral polymers (Figure 1) from pure enantiomeric pools of monomers. Note also that when assuming the first formation of polymers was not the result of a simple direct synthesis yet occurred under heterogeneous catalysis as mediated for example by inorganic materials such as clays [39], the existence of non-racemizing enantiopure pools of amino acids and sugars is highly unlikely.

A key point for starting a discussion of SMSB is the chemical meaning of a racemic mixture. Mislow discussed this concept in a Socratic-like method by proposing the question of whether a racemic mixture is composed by an exact number of molecules of both enantiomers [40]. In fact, older reports [41] had already established that there is an unavoidable statistical deviation from the ideal racemic composition as expressed by:

$$ee(100\%) = 67.43 \times \mathrm{N}^{0.5} \quad (1)$$

where N is the number of molecules. Notice that this statistical *ee* deviation (1) from the ideal racemic composition increases when the concentration decreases, so that for extremely dilute solutions there are low but statistically significant $ee \neq 0$ values. Extremely low concentrations of racemic mixtures occur, for example, when we consider the chirality arising by isotopic element substitution according to natural isotopic compositions [42], or in the case of polymers of non-identical building blocks [43], such as those of the RNA-world and proteins, where random condensation leads to a sufficiently high number of isomers such that their concentrations are low enough to show detectable *ee* values [44]. However, chiral amplifications of such stochastically distributed initial *ee's* in reversible reactions cannot take place via asymmetric induction reactions. This is because both at thermodynamic equilibrium and in non-equilibrium stationary states (NESS) in the linear regime of non-equilibrium thermodynamics, the racemic composition represents a stable potential well even in the presence of the statistical fluctuations about the ideal racemic composition.

Chemical evolution is not supported by irreversible reaction pathways and the existence of certain racemization rates must be assumed to occur. Therefore, despite the fluctuations about the definition of the ideal/mathematical racemic mixture, the real meaning of the racemic mixture is the existence of a thermodynamic potential minimum for the ideal racemic composition. This means that for any compositional bias from the ideal racemic composition, such as those associated with the statistical *ee* values (1), the system will tend to return to an $ee = 0$. Obviously, there is a statistical chiral fluctuation or chiral noise about the potential minimum of the pure 1:1 enantiomeric mixture, but the system tends to return to $ee = 0$. This is a consequence of the energy degeneracy between the enantiomers and of the law of large numbers acting in the thermodynamic limit [45]. In summary, chiral statistical fluctuations around a stable racemic configuration cannot be amplified.

In contrast to asymmetric synthesis, absolute asymmetric synthesis (AAS), in the absence of any chiral polarization other than that between the very enantiomeric species of the reaction network, is possible for some reaction networks having non-linear kinetics of the enantiomer concentrations. When the system is maintained far from thermodynamic equilibrium, it is due to the imposed boundary conditions. The final reaction states, provided the boundary conditions are maintained, are non-equilibrium stationary states (NESS). These types of reactions, or simply the deracemization of racemic NESS to scalemic or homochiral NESS, are called spontaneous mirror symmetry breaking (SMSB). It is worth noting that in a SMSB scenario the chiral statistical fluctuations about the ideal racemic composition are now necessary to expose the metastability of the racemic NESS, which does not correspond to a potential well topology but to that of a saddle point. It is important to stress that the role of the fluctuations is not that of an activation energy and that their effect at the SMSB event is not at all an amplification of chirality.

3. Stable and Unstable Non-Equilibrium Stationary States (NESS) in Enantioselective Reactions

Reaction networks that are able to lead to SMSB have focused the research interest on the topic [3,46]. The fact that such reaction networks may exhibit permanent SMSB as non-equilibrium stationary states (NESS) and only when operative in open or some special closed systems is assumed [47], but they are rarely analyzed within the thermodynamic framework of dissipative reaction systems. Moreover, previous seminal reports on dissipative reaction networks actually hamper, rather than facilitate, the chemical understanding of such systems. This is because the erroneous use of the kinetic approximation of clamped concentrations for the species exchanged with the environment overlooked the role of the boundary conditions of open systems for the correct description of the NESSs. This error originates from the first seminal reports on dissipative chemical reactions (e.g. [10]) and was most likely a consequence, at that time, of the mathematical difficulties of studying the sets of coupled non-linear differential equation describing reaction networks. Nowadays, this simplification cannot be excused when simple and powerful computational tools for solving such equations are readily available. The use of the clamped concentration approximation has even led to the refutation [48] of Prigogine's theorem on minimum entropy production for the linear thermodynamic regime and to cast doubts on the validity of the General Evolution Criterion (GEC) valid for the non-linear thermodynamic regime. Further discussion concerning the correct thermodynamic description of models for the simulation of SMSB as NESS can be found in references [49,50].

3.1. Entropy Production and Balance in Open Systems

The boundary conditions (either systems open to matter exchange or closed systems unable to equilibrate energy with their surroundings) can keep the reaction network operating far from thermodynamic equilibrium. By increasing the "distance" from thermodynamic equilibrium, the internal entropy production of the reaction network increases [51]: $dS_i \geq 0$, where the equality corresponds to the thermodynamic equilibrium state. Depending how far the reaction network is driven from equilibrium, a NESS can result, forming the so-called thermodynamic branch (for example the black trace of Figure 2). However, in a real chemical system, above a critical value of the entropy production, instability of the NESS can occur. This is because taking the system very far from equilibrium leads to an increase of the entropy production, which is obtained by increasing the chemical affinities (forces) and the absolute reaction rates (fluxs/currents). This in turn corresponds to high concentrations, which can lead to increased viscosities, and so to a breakdown of the mean field approximation of chemical kinetics. Therefore, the reaction rates become diffusion-controlled and inhomogeneous distributions of the reaction species and the temperature takes place. This may lead to either chaotic distributions or to organized structures (oscillatory or stationary). The ordered structures arising as a consequence of the increased entropy production are called "dissipative" because they are maintained thanks to the entropy dissipated by the reaction network. However, for some autocatalytic reaction networks, and before such a spatial instability occurs, the thermodynamic branch can become unstable for homogeneous distributions of matter and energy that yield other types of NESS, or even oscillatory compositional states, distinct from the single NESS of the thermodynamic branch. It was recognized early on that this could occur only in the case of autocatalytic reactions [11]. However, such chemical reactions have received much less attention than the study of oscillating reactions or the formation of macroscopic dissipative spatial structures. Surprisingly, the enantioselective character of the autocatalytic reaction networks significant in the chemical machinery of life and in consequence the possible relationship with SMSB processes and BH has been mostly overlooked: only some reports considerer SMSB in the framework of entropy production and dissipative systems ([52–55]).

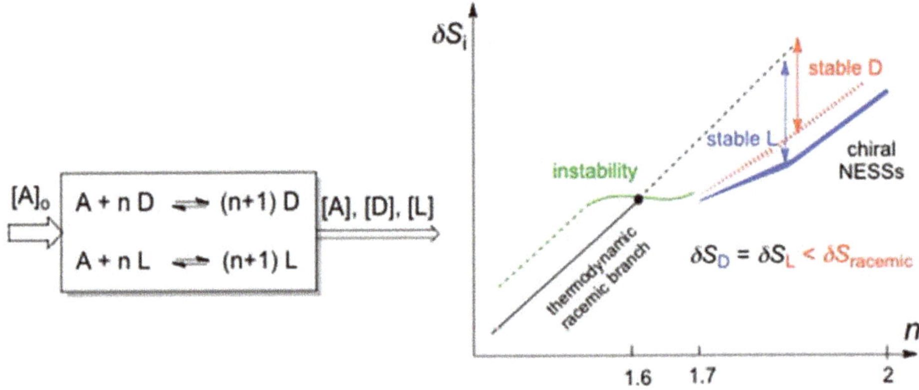

Figure 2. Spontaneous mirror symmetry breaking (SMSB) emerging in an open system for selective enantiocatalysis, where A is the achiral resource and D and L are enantiomers (initial concentrations $[D]_o$ and $[L]_o \neq 0$). SMSB does not take place for n = 1. Metastability of the racemic thermodynamic branch can only occur for values of n > 1. For reasonable reaction rate constants and reaction parameters (not specified here) in this example instability of the racemic branch occurs for n > 1.6 and SMSB for n ≥ 1.7. Non-integer values of n correspond to the simplification of complex autocatalytic networks [56], which leads to the same dynamic change of the solution species.

The entropy production (dS_i) of a chemical reaction is always positive definite:

$$dSi \geq 0 \tag{2}$$

where the equality corresponds to thermodynamic equilibrium. In an open system at any stationary state the balance between the internal (reaction) entropy production (dS_i) and the exchange entropy production (dS_e) must be zero:

$$dS = dSe + dSi = 0 \tag{3}$$

At thermodynamic equilibrium dS = 0 and $dS_i = 0$, hence there is no exchange entropy with the exterior $dS_e = 0$. At a NESS $dS = 0$, this means that the exchange entropy production is negative, allowing a more ordered internal configuration, than that corresponding to thermodynamic equilibrium: as occurs for example in SMSB. Non-equilibrium states of the system fulfilling equation (2) may be unstable, metastable or stable. States that do not have positive time derivative of the second variation of entropy may become unstable, but it is not a sufficient condition for instability [10,11,50]. Jacobian linear stability analysis distinguishes metastable from the true stable states [53]. Notice that there are non-stationary states with positive entropy production $dS_i > 0$ obeying dS = 0. Non-stationary states with dS = 0 and $d^2S \neq 0$ correspond to states where the local equilibrium concept can be applied. These states may form continuous pathways between NESS and, therefore, represent the existence of thermodynamic reversible routes between different NESS, i.e. those NESSs that are not located in kinetic traps.

SMSB, and in the case of one asymmetric center, leads to a racemic NESS of saddle point topology (similar to that of transition state), which is able to evolve through reversible paths (while maintaining

$dS = 0$) to one of two degenerate enantiomeric NESS located in an entropy production well $dS_i > 0$ (the entropy production is minimized). More details about this will be published elsewhere. A schematic outline of this is shown in Figure 2, wherein the open system consists of a linear flow reactor where the achiral resources enter with a zero-order rate constant (as a molecular pump) in a well-mixed solution that has the same volume exiting the reactor, i.e. the exit of all species under the law of mass action. This open system at the final NESSs have constant volume and the same chemical mass ($[A]_O$) for both the racemic and the homochiral NESS, which simplifies the evaluation of the associated entropy productions.

It is worth noting that the NESSs are thermodynamically defined by equality (3), referring to the entropy production balance between the internal reactions and the imposed boundary conditions, which is entirely overlooked under the clamped concentration approximation of SMSB commonly used to describe reaction networks in open systems.

The calculation of the entropy production (equal to the product of the forces and the currents that these forces lead to) is a chemical reaction that is relatively straightforward [52]. Furthermore, the way the contribution to the entropy production is calculated from the fluxes exchanging matter with surroundings and on the basis of the chemical potentials of the species can be found, for example, in reference [49]. From the point of view of chemical evolution, SMSB is subject to thermodynamic constraints and boundary conditions similar to those of the evolutionary stages of chemical networks.

3.2. *Potential Reaction Networks Able to Yield SMSB.*

The direct synthesis

$$A \leftrightarrows D,\ A \leftrightarrows L \tag{4}$$

fulfils the entropy balance (3) along the thermodynamic branch of racemic stationary states. However, enantioselective autocatalysis may also fulfill condition (3) for several different compositions.

Autocatalytic reaction networks capable of yielding species/enantiomer selection have been previously reviewed [3,46,47,56–58]. In addition to the networks cited/discussed therein, we must include the recently reported enantioselective hypercycle [50,59], which is highly significant due to its coincidence with the replicators of the nucleic acid and protein domain [60].

These reaction networks for SMSB contain a central mechanism in the first order (quadratic) enantioselective catalysis:

$$A + D \leftrightarrows 2\,D,\ A + L \leftrightarrows 2\,L, \tag{5}$$

where A is achiral and D and L are enantiomers. Reaction (4) may also represent a simplification for the formation of chiral polymers from achiral building blocks. Reaction (5) can form homochiral NESS in the case that the initial conditions contain only one of the enantiomers. However, in any model concerning chemical evolution, the model should include the direct synthesis (4), albeit for a much slower reaction rate than for the enantioselective autocatalysis (5). This is because the autocatalytic functionality of (5) can only emerge if reaction (4) is operative. Reaction (4) can start for initial conditions without any D and L.

The first order autocatalysis (5), also called quadratic autocatalysis, cannot lead to species selection [56] (between D and L in SMSB), i.e., the entropy production needed to achieve the instability of the racemic branch is not high enough. This is because the kinetic/dynamic sigmoidal curve of growth implicit in autocatalytic processes ($n > 0$; see the meaning of n in Figure 2) is needed, as well as a super-exponential growth ($n > 1$) is necessary for species selection. Of course, this is not the case of cubic autocatalysis ($n = 2$), but in this case, high concentrations are necessary for the reaction progress in open system scenarios. Furthermore, to our knowledge, no natural replicators (autocatalytic systems) have been detected having a reaction order higher than the quadratic. Therefore, the reaction networks proposed for SMSB are based on first order autocatalysis, which through the coupling of other enantioselective reactions (Figure 3) yields, autocatalytic kinetic/dynamic signatures such as those expected for values of $n > 1$ necessary for achieving the instability of the thermodynamic branch and for certain reaction parameters and boundary conditions.

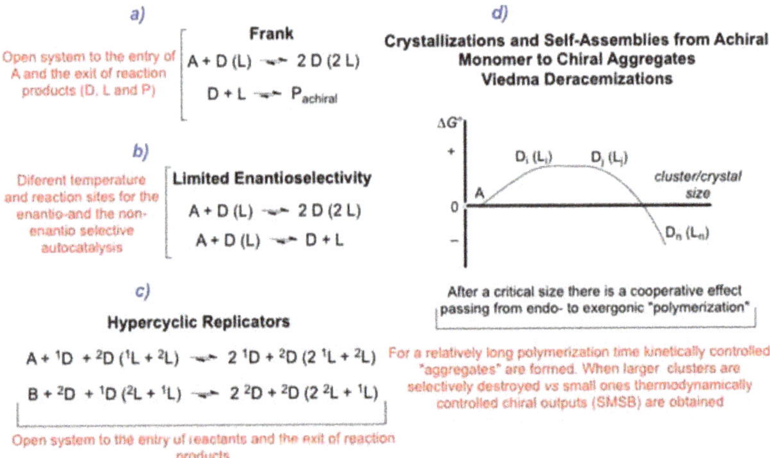

Figure 3. Reaction networks (simplified) that may lead to SMSB processes, both by enantioselective transformations from achiral to chiral compounds a), b), and c), or through the deracemization of racemic mixtures d).

3.2.1. Frank-like Models

The first theoretical proposal of a reaction network able to lead to SMSB was the so-called Frank model (1953) [61] (Figure 3a). This model has its historical precedent in the Lotka-Volterra model for the competitive exclusion between two species [59], and when we take into account that enantiomerism is the limiting case of species distinguishability. Frank-like models (Figure 3a) contain the homochiral reaction between enantiomers as the additional chiral recognition step that allows for the increase in the dynamic of growth allowing enantiomer selection. All evidence indicates that the experimental SMSB of the Soai reaction [7,8] is a complex reaction network that basically fits into a Frank-like reaction network [62–64].

3.2.2. Limited Enantioselective Model

The enantioselectivity model (Figure 3b) proposed by Goldanskii and Avetisov [65] is based on the reasonable assumption of the presence of the corresponding non-enantioselective autocatalysis. However, thermodynamic constraints [66] determine that LES only makes thermodynamic sense when the enantioselective and the non-enantioselective autocatalytic steps are located at different points in the reaction domain and held at quite different temperatures [67–69]. There are no experimental examples for this model, but it adapts well to SMSB for the abiotic scenario of deep ocean thermal vents [68].

3.2.3. Enantioselective Hypercyclic Replicators

SMSB in hypercyclic replicators [59] is a consequence of the increase of non-linearity of the quadratic autocatalysis when cross-catalysis appears. In fact, the hypercycle model [14] was developed to solve the problem of selection and Darwian evolution for low autocatalytic orders. Although the implications of this model cannot be an ultimate explanation in systems biology, its raison d'etre is during the chemical evolution stages of compartmentalization of the RNA-world and the nucleic acid/ protein domain. Notice that this implies that the formation of systems able to self-reproduce while transmitting functionalities at the same time could be the origin of the emergence of BH.

Notice that if autocatalysis is present, the emergence of cross catalysis for the same family of compounds and chemical processes, for example polymerization of oligonucleotides or peptides, is quite reasonable from a point of view of chemical reactivity and structure. It is quite reasonable that autocatalytic and cross-catalytic functionalities could emerge simultaneously.

3.2.4. SMSB in Enantioselective Autocatalytic Polymerization/Depolymerization

In our opinion, the Viedma deracemization of chiral racemic conglomerates [9] (Figure 3c) in the presence of achiral or racemizing building blocks gives unprecedented hints on reaction networks able to lead to SMSB in the formation of condensation polymers. There are still controversial interpretations of the Viedma deracemization. In the following, we will discuss this according to our experimental and theoretical reports. Viedma deracemization can be interpreted, both experimentally and theoretically, as a phenomenon taking place within the framework of dissipative systems [70] which we summarize as follows.

A far from equilibrium system is created by mechanical grinding or by establishing temperature gradients [71,72] in solutions of crystals of racemic conglomerates made up from of either achiral or racemic monomers. These constitute closed systems unable to achieve energy equilibrium (energy balance) with their surroundings because:

(a) In spite of being a closed system, the energy input is given selectively to only some of the species of the system. In other words, the mechanical grinding (the energy input) affects only the largest crystals.
(b) The higher solubility of the smaller crystals obtained by grinding creates supersaturated solutions for the larger crystals and in consequence, a constant and permanent cycle of solubilization and crystal growth is maintained.
(c) Homochiral cluster-to-cluster growth (this is equivalent—through its dynamic signature—to first order enantioselective autocatalysis). Notice that in saturated solutions monomer-to-cluster exchange between solution and crystal/cluster is not autocatalytic and therefore is unable to lead to any chiral amplification, nor to racemization, of the crystal mixture *ee* value.
(d) Additional growth dynamics to first order autocatalysis of the cluster-to-cluster growth in (c) that increase the dynamic growth signature are provided by a mutualistic effect in the growth of homochiral material; each cluster coming from the fragmentation of large crystals can react with crystals of many different sizes leading to the formation of larger cluster/crystals of many different sizes, but of the same chiral sign. Furthermore, the free energy profile of the polymer formation, which is endergonic during the first "polymerization" steps, with cooperative growth beyond a critical size as well the intermediate sizes of the clusters coming from the breakage of the large ones, are decisive features for the achievement of the growth dynamics leading to SMSB [70].

From a synthetic chemical point of view, it is paradoxical how the resulting homochiral mixture does not racemize when the experimental conditions a) and b) above are halted. This is simply due to the fact that the solid chiral species at saturated conditions do not show solid-to-solid interactions. Therefore, in the absence of chiral recognition between enantiomeric phases, they are thermodynamically identical (for a detailed discussion of this, see refs. [73,74]). Furthermore, there are results pointing to the absence of ripening mechanisms of the larger crystals at the expense of the smaller ones in Viedma deracemization [75]. This suggests that the former results on the homochiral biases reported earlier by Kondepudi, in the formation of strong racemic biases in such types of racemic conglomerates [76,77], could belong to this type of SMSB mechanism. Regarding a generalization of the Viedma deracemization, the increasing number of reports on the spontaneous formation of strong bias from the racemic composition in the self-assembly of achiral building blocks towards chiral aggregates or liquid crystals, is surely significant [78–80].

The lessons learned from Viedma deracemization, make SMSB scenarios reasonable for the formation of primordial homochiral biopolymers based on cluster-to-cluster growth, as for example by template mechanisms in conjunction with the catalyzed depolymerization of the larger polymer chains to intermediate oligomer sizes. This type of SMSB polymerization has been previously theoretically studied [81]. There are experimental reports on oligomer formation/depolymerization cycles of peptides [82] that open prospects on the translation of the former speculations into experimental research [47].

3.2.5. On the Detection of SMSB in Low Exergonic Reactions in Solution

Most of the examples on SMSB correspond to phase transitions: crystals, self-assembly aggregates, e.g. Viedma-like deracemizations (e.g. [9,76–80]), that is, cases for which when the boundary conditions disappear allowing the system to equilibrate with its surroundings, racemization cannot occur. The only indisputable example of SMSB in solution is the Soai reaction [7]. However, the Soai reaction corresponds to a high exergonic reaction (organozinc addition to a carbaldehyde) that allows a reaction workup withouth significant racemization. By contrast, prebiotic chemistry involves many low exergonic reactions [83].

The lack of experimental results on the emergence of natural optical activity in autocatalytic reaction networks in solution, for example in the formose reaction (see below) in the carbohydrate domain, is hampered by the lack of synthetic methods and techniques for the direct study of the final stationary states in open systems. SMSB in solution should be performed in open systems under strength-controlled parameters, in contrast to the usual synthetic experiments carried out in the flask. Furthermore, the emergence of natural optical activity in low exergonic reactions in these open systems should be performed in real time, because racemization may occur in the common reaction workups when the open system conditions disappear. However, open system reactors at the bench scale are today possible thanks to microfluidic instrumentation. In addition, detection of natural optical activity for specific compounds, or families of compounds, could be carried out by circular dichroism (CD) spectroscopy, direct observation of reactor content, or of its effluents.

4. BH Based on SMSB Requires Previous Asymmetric Synthetic Scenarios: Stochastic vs. Deterministic Chiral Signs in SMSB

We assume that the decisive stage in the formation of proto-cells is the compartmentalization [84] of similar autocatalytic sets in a myriad of abiotic reactors [12,13]. Therefore, a hypothesis on a BH based on SMSB for the autocatalytic sets reveals the paradox that the stochastic chiral sign distribution in the different compartments would lead to a set of racemic outcomes when averaged over all the compartments. The hypothesis of the emergence of two competing enantiomeric worlds is in part contradictory to the hypothesis that by the exchange of species showing similar functions between compartmentalized autocatalytic sets, life emerges as a cooperative phenomenon [12–19,85].

However, the change from a stochastic to a deterministic distribution of chiral signs in SMSB is a remarkable property of SMSB. Under the effect of very weak polarizations, sufficiently weak so that such polarizations have no chemical consequence in common asymmetric synthesis, the stochastic distribution of chiral signs in SMSB changes to a single deterministic chiral sign. There are simulations of this [86,87], but the Soai reaction show dramatic examples, such as the deterministic distribution of chiral signs between experiments by such weak polarizations as those due to cryptochiral isotopic enantiomers [8].

Asymmetric induction by the action of natural physical chiral forces, active during the first stages of chemical evolution, can be considered as proven, despite of the diverse possible origins and organic compounds. The unavoidable slow racemization times (Section 2) do not prevent these compounds from arriving to the stage of the formation of instructed polymers and compartmentalization, showing small residual *ee* values. These low *ee* values of the organic compounds, constituting the autocatalytic replicators, would convert the stochastic SMSB chiral sign into a deterministic one.

According to this, BH would belong to a process developed through all the stages of chemical evolution. The selection of a deterministic chiral sign in the formation of instructed/functional polymers would constitute a first Darwinian selection, namely that of the phenotypic replicator able to survive by adopting the boundary chiral polarizations (the entry of resources into the compartment).

An interesting question is that the emergence of order in the evolutionary stage of compartmentalization arises because of a cooperative effect between autocatalytic networks and quasi-species. This information exchange has been proposed to arise thanks to similar catalytic functionalities of the exchanged species. This overlooks the chirality of the exchanged species. However, the chirality sign of similar functionalized polymers shows a chemical recognition of chirality based on a digitized +/− type of signal.

5. Hypothesis on the Emergence of BH

The former sections correspond to an attempt to describe selected concepts from different scientific fields, which are essential for the understanding of the emergence of chirality in chemical evolution and for the description of the basic principles necessary for the understanding of BH. In our opinion, the former discussion allows for a reasonable speculation of how, and in which stages of chemical evolution, the present chiral machinery of BH has been formed.

5.1. Is Carbohydrate Synthesis the Third Leg of a Tripodal SMSB Scenario?

Cross-catalysis between peptide and oligonucleotide hypercyclic replicators justifies a SMSB driving force to one enantiomer class in nucleic acids and in protein polymers [59]. Although weak this would require the chiral induction, coming from the asymmetric inductions transformations at previous stages of chemical evolution. However, the formation of primordial nucleotide-like compounds requires the selection of the adequate chiral carbohydrate monomer to achieve the structural functionalities of the nucleic acid chains [44,88]. Therefore, chirality in the domain of carbohydrate synthesis should be an essential element along the path leading to BH. In this respect, assuming SMSB as the central part of the emergence of BH, the question arises if the selection of one enantiomer of the carbohydrate family of compounds occurs by enantioselective selection at the formation of the nucleotides, or if it originated from its own SMSB process during the prebiotic stage of carbohydrate synthesis.

Prebiotic carbohydrate formation is based on aldol-like reactions integrating formaldehyde units to carbohydrate chains [89–91]. The synthetic integration of formaldehyde is the expected chemical pathway towards pentoses and hexoses: This was shown already by Butlerov (1861) in the so-called formose reaction. This reaction leads to a huge diversity of compounds, but the significant point with respect to SMSB is that it is composed by a family of interconnected autocatalytic cycles [92]. This strongly suggests that the necessary, but not sufficient, condition of autocatalysis for SMSB may occur at the prebiotic carbohydrate synthesis. However, there are no experimental data on the detection of spontaneous natural optical activity in the integrative synthesis of sugars, but we believe that this a consequence of a lack of methods to detect SMSB in solutions in open systems as discussed above.

The difficulty for theoretical studies of SMSB, and for future research experimental work in this topic in the carbohydrate domain, compared to those described in Section 3.2., is that, and with the exception of the first chiral compound of the family (glyceraldehyde or glycerol-1-phosphate), carbohydrates have more than one asymmetric center. This means that because of epimerization processes, i.e, the existence of chiral diastereoisomers, when the instability of the thermodynamic branch occurs, oscillatory final states may arise instead of the simple NESSs (saddle-point racemic NESS and two chiral NESS at minimum well potentials) described above. In this respect, it is surely significant that oscillations are an experimental behavior of glycolysis [90], and that in theoretical models implying compounds with two asymmetric centers, oscillatory behavior between epimers has been reported [93–95].

The difficulty for experimental studies of formose-like reactions is that, for the synthetic chemist, the final outcome is a "messy" mixture of compounds. However, from the point of view of the origin of life topic, it leads to a "high diversity" of carbohydrates. Both points of view are true, but diversity is probably a characteristic of the first stages of chemical evolution (see for example the thousands of compounds detected in a carbonaceous meteorite [96]). The first stages of chemical evolution are characterized by the formation of a very high diversity of organic compounds of the structural families that will constitute the chemical support of the phenomenon of life. Later, the increase of complexity, by forming interconnected reaction networks, the diversity would be thinned out towards a lower number of structural families and compounds. Synthetic chemical research in the origin of life is needed to discern which transformations are possible and which reaction mechanisms and interactions would have be at work to discover the reaction pathways that justify chemical evolution. However, abiotic scenarios need to be based on "impure" mixtures of compounds of related structural families to be reliable [97].

Reaction networks that show enantioselective autocatalysis are rare in chemistry. However, in the protein and nucleic acid domain the presence of replicators showing template mechanisms and the mutual catalytic activities between the RNA-world and peptide synthesis justify the connections between SMSB in these two domains. Yet the connection of these two with a possible SMSB at the carbohydrate metabolisms is more speculative. In the carbohydrate domain the question of BH is not only related to the supply of a specific pentose to the pre-RNA- or RNA-world, but also to membrane formation. In this respect, such a lateral connection is shown in the different starting chiral sign of the glycerol-1-phosphate building block to form the membrane lipids either in archeobacteria or in bacteria: chiral configuration L (sn-glycerol-1-phosphate) for archeobacteria and the chiral configuration D (sn-glycerol-3-phosphate) for bacteria [98].

5.2. *Enantioselective Linear Reactions in the Chiral Machinery of Life*

Regarding the chemical connection between chiral nitrogen-containing compounds and chiral hydroxy organic compounds, i.e. between the simple compounds of the amino acid and the carbohydrate domain (for example in the prebiotic chemical cycles proposed by Eschenmoser [83]), chirality can only imply simple asymmetric inductions and chirality transfer in linear kinetic dependences between enantiomers.

Metabolic cycles are necessary for the activation of low exergonic reactions, for example by substrate phosphorylation. Without the energy provided by metabolic cycles it is not possible for species activation to undergo endergonic or low exergonic reactions with higher yields and rates. It is surely significant that all indicate a minor role of enantiomerism in the metabolic world proposals, for example, the citric acid cycle, independent of the enzymes (enantioslective catalysis) that currently catalyze the cycle, chirality appears for only a few compounds and apparently as a mere accident that does not have an effect on the energetic function of the cycle. Chirality in metabolic cycles probably had appeared because of its interconnection with other worlds where SMSB had emerged. A reasonable hypothesis on the formation of proto-cells is to avoid the term "first" for the proposed models on the origin of life and consider the proto-cell emergence as a consequence of the mutualistic effects originating in the compartmentalization of the RNA-, protein-, carbohydrate-, lipid-, and metabolic-worlds (Figure 4). There, homochirality would be preserved by the SMSB autocatalysis of the interconnected RNA/protein and carbohydrate domains, which additionally would drive the rest of the enantioselective reactions and chiral transfer transformations to a common chiral sign.

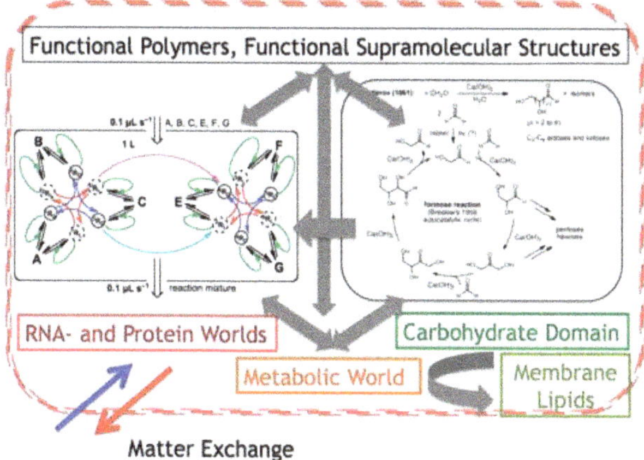

Figure 4. The compartmentalization of the different worlds proposed for the origin of life, would determine that the driving force of SMSB at the interconnected nucleic acid, protein and carbohydrate domain preserves biological homochirality (BH) and determines the chiral sign of all the rest of the enantioselective reactions and chiral transfer interactions.

6. Conclusions

(a) BH has its physico-chemical basis in enantioselective autocatalytic reaction networks operating in open dissipative systems. This thermodynamic scenario is similar to that proposed for the compartmentalization stages of chemical evolution where the question of chirality has been avoided up to the present. Moreover, the reaction networks proposed for the formation of pre-RNA worlds and peptides showing catalytic and autocatalytic functionalities and able to develop Darwinian evolution, also possess the ability for SMSB, i.e. of the selection between enantiomers.

(b) Asymmetric inductions originating through natural chiral forces during previous stages of chemical evolution, as for example in astrophysical scenarios, could provide the necessary chiral polarization to transform, at later stages of chemical evolution, the stochastic chiral sign outcome of the SMSB processes into a deterministic one. This is in spite of the unavoidable racemization processes acting during the long-time interval of chemical evolution.

(c) SMSB autocatalytic cycles provide the resilience against racemization characteristic of BH and would drive the rest of the enantioselective reactions towards a definite chiral sign.

(d) Such a general scenario of BH may be applied to all current models, which aim to find the links between prebiotic chemistry and biological chemistry [97]. The emergence of BH is concurrent with the emergence of autocatalytic sets, proto-cells, etc., and it probably represents an energetic advantage (lower entropy production) [51], with respect to the racemic outcome.

(e) Homochirality has been considered as an inherent property of matter [98]. From a chemical point of view, it is probably also a necessary condition for life because of the informational entropy advantage of asymmetry in molecular recognition, that is, in the emergence of catalytic and autocatalytic functionalities. Once the reasonable existence of SMSB in the decisive stages of chemical evolution is recognized, in our opinion, then the models concerning the emergence of autocatalytic sets, proto-cells, etc., cannot avoid the question of the enantioselectivity of the processes involved. Note that this also has direct consequences for the applied biotechnological fields of new metabolic cycles and of artificial cells.

Author Contributions: J.M.R. and D.H. contributed equally to the present work.

Funding: This research was funded by MINECO (Spain), coordinated grant numbers CTQ2017-87864-C2-1-P and –C2-2-P.

Acknowledgments: We have benefitted from fruitful discussions with J. Crusats, Z. El-Hachemi and A. Moyano on the topic of this report, and in the framework of the coordinated project cited above.

Conflicts of Interest: The authors declare no conflict of interest.

References

1. Guijarro, L.; Yus, M. *The Origin of Chirality in the Molecules of Life: A Revision from Awareness to the Current*; RSC Publishing: Cambridge, UK, 2009.
2. Gal, J.; Cintas, P. Early History of the Recognition of Molecular Biochirality. In *Biochirality*; Springer: Berlin/Heidelberg, Germany, 2013; Volume 333, pp. 1–40.
3. Blackmond, D.G. The origin of biological homochirality. *Cold Spring Harb Perspect. Biol.* **2010**, *2*, a002147. [CrossRef]
4. Davankov, V.A. Biological homochirality on the earth, or in the universe? A selective review. *Symmetry* **2018**, *10*, 749. [CrossRef]
5. An Evolutionary Mystery: Mirror Asymmetry in Life and in Space. Available online: https://www.acs.org/content/acs/en/acs-webinars/popular-chemistry/homochirality.html (accessed on 20 February 2019).
6. Origin of Life. The Chirality Problem. Available online: https://creation.com/origin-of-life-the-chirality-problem (accessed on 20 February 2019).
7. Soai, K.; Kawasaki, T. Asymmetric autocatalysis with amplification of chirality. In *Amplification of Chirality*; Springer: Berlin/Heidelberg, Germany, 2008; Volume 284, pp. 1–33.
8. Kawasaki, T.; Matsumura, Y.; Tsutsumi, T.; Suzuki, K.; Ito, M.; Soai, K. Asymmetric autocatalysis triggered by carbon Isotope ($^{13}C/^{12}C$) chirality. *Science* **2009**, *324*, 492–495. [CrossRef]
9. Viedma, C. Chiral symmetry breaking during crystallization: Complete chiral purity induced by non-linear autocatalysis. *Phys. Rev. Lett.* **2005**, *94*, 065504. [CrossRef]
10. Glansdorff, P.; Prigogine, I. *Thermodynamic Theory of Structure, Stability and Fluctuations*; Wiley-Interscience: London, UK, 1971.
11. Nicolis, G.; Prigogine, I. *Self-Organization in Nonequilibrium Systems*; Wiley: New York, NY, USA, 1977.
12. Ruiz-Mirazo, K.; Briones, C.; de la Escosura, A. Prebiotic systems chemistry: New perspectives for the origins of life. *Chem. Rev.* **2014**, *114*, 285–366. [CrossRef]
13. Kauffman, S. Autocatalytic sets of proteins. *J. Theor. Biol.* **1986**, *119*, 1–24. [CrossRef]
14. Eigen, M.; Schuster, P. *The Hypercycle: A Principle of Natural Self-Organization*; Springer: Berlin, Germany, 1979.
15. Hordijk, W.; Hein, J.; Steel, M. Autocatalytic sets and the origin of life. *Entropy* **2010**, *12*, 1733–1742. [CrossRef]
16. Lifson, S. On the crucial stages in the origin of animate matter. *J. Molec. Evol.* **1997**, *44*, 1–8. [CrossRef]
17. Schneider, E.D.; Kay, J.J. Life as manifestation of the second law of thermodynamics. *Math. Comp. Model.* **1994**, *19*, 25–48. [CrossRef]
18. Mavelli, F.; Ruiz-Mirazo, K. Stochastic simulations of minimal self-reproducting cellular systems. *Phil Trans. Royal Soc.: Biol. Sci.* **2007**, 362. [CrossRef]
19. Rasmussen, S.; Chen, L.; Deamer, D.; Krakauer, D.C.; Packard, N.H.; Stadler, P.F.; Bedau, M.A. Transitions from Nonliving to Living Matter. *Science* **2004**, *303*, 963–965. [CrossRef] [PubMed]
20. Qian, H.; Beard, D.A. Thermodynamic of stoichiometric biochemical networks in living system far from equilibrium. *Biophys. Chem.* **2005**, *114*, 213–220. [CrossRef]
21. Schmiedl, T.; Seifert, U. Stochastic thermodynamics of chemical reaction networks. *J. Chem. Phys.* **2007**, *126*, 044101. [CrossRef] [PubMed]
22. Michaelian, K. Thermodynamic dissipation theory for the origin of life. *Earth Syst. Dynam.* **2011**, *2*, 37–51. [CrossRef]
23. Ruiz-Mirazo, K.; Briones, C.; de la Escosura, A. Chemial roots of biological evolution: The origins of life as a process of development of autonomous functional systems. *Open Biol.* **2017**, *7*, 170050. [CrossRef] [PubMed]
24. Schuster, P.; Stadler, P.F. Networks in chemical evolution. *Complexity* **2001**, *8*, 34–42. [CrossRef]
25. Polettini, M.; Esposito, M. Irreversible thermodynamics of open chemical networks. *J. Chem. Phys.* **2014**, *141*, 024117. [CrossRef]
26. Eschenmoser, A.; Kisakürek, M.V. Chemistry and the Origin of Life. *Helv. Chim. Acta* **1996**, *79*, 1249–1259. [CrossRef]

27. Barron, L.D. Symmetry and molecular chirality. *Chem. Soc. Rev.* **1986**, *15*, 189–223. [CrossRef]
28. Avalos, M.; Babiano, R.; Cintas, P.; Jimenez, J.L.; Palacios, J.C.; Barron, L. Absolute asymmetric synthesis under physical fields: Facts and fictions. *Chem. Rev.* **1998**, *98*, 2391–2404. [CrossRef]
29. Feringa, B.L.; van Delden, R.A. Absolute asymmetric synthesis: The origin, control, and amplification of chirality. *Angew. Chem. Int. Ed.* **1999**, *38*, 3418–3438. [CrossRef]
30. Gu, H.; Nakamura, Y.; Sato, T.; Teramoto, A.; Green, M.M.; Jha, S.K.; Andreola, C.; Reidy, M.P.; Mark, H.F. Optical Rotation of Random Copolyisocyanates of Chiral and Achiral Monomers: Sergeant and Soldier Copolymers. *Macromolecules* **1998**, *31*, 6362–6368. [CrossRef]
31. Guillaneux, D.; Zhao, S.H.; Samuel, O.; Rainford, D.; Kagan, E.B. Nonlinear Effects in Asymmetric Catalysis. *J. Am. Chem. Soc.* **1994**, *116*, 9430–9439. [CrossRef]
32. Root-Bernstein, R. Simultaneous origin of homochirality, the genetic code and its directionality. *BioEssays* **2007**, *29*, 689–698. [CrossRef]
33. Wu, M.; Walker, S.I.; Higgs, P.G. Autocatalytic replication and homochirality in biopolymers: Is homochirality a requierment of life or a result of it? *Astrobiology* **2012**, *12*, 818–829. [CrossRef] [PubMed]
34. Schuster, P. Germany-Japan round table. Heidelberg. 2011. Available online: https://www.tbi.univie.ac.at/~{}pks/ (accessed on 30 May 2019).
35. Cronin, J.R.; Pizzarello, S. Enantiomeric excesses in meteoric amino acids. *Science* **1997**, *275*, 951–955. [CrossRef]
36. McGuire, A.B.; Carroll, P.B.; Loomis, R.A.; Finneran, I.A.; Jewell, P.R.; Remijan, A.J.; Blake, G.A. Discovery of the interstellar chiral molecule propylene oxide (CH_3CHCH_2O). *Science* **2016**, *352*, 1449–1452. [CrossRef]
37. Oró, J.; Mills, T.; Lazcano, A. Comets and the formation of biochemical compounds on the primitive Earth–A review. *Orig. Life Evol. Biosph.* **1991**, *21*, 267–277. [CrossRef]
38. Kempe, S.; Kazmierczak, J. Biogenesis and early life on Earth and Europa: Favored by an alkaline ocean? *Astrobiology* **2002**, *2*, 123–130. [CrossRef]
39. Brack, A. Clay minerals and the origin of life. In *Handbook of Clay Science, Developments in Clay Science*; Bergaya, F., Theng, B.K.G., Lagaly, G., Eds.; Elsevier: Oxford, UK, 2006; Volume 1, pp. 379–391.
40. Mislow, K. Absolute asymmetric synthesis: A commentary. *Collect. Czech. Chem. Commun.* **2003**, *68*, 849–864. [CrossRef]
41. Mills, W.H. Some aspects of stereochemsitry. *Chem. Ind. (Lond.)* **1932**, *51*, 750–759. [CrossRef]
42. Barabás, B.; Kurdi, R.; Pályi, G. Natural abundance isotopic chirality in the reagents of the soai reaction. *Symmetry* **2016**, *8*, 2. [CrossRef]
43. Bolli, M.; Micura, R.; Eschenmoser, A. Pyranosyl-RNA: Chiroselective self-assembly of base sequences by ligative oligomerization of tetranucleotide-2′,3′-cyclophosphates (with a commentary concerning the origin of biomolecular homochirality). *Chem. Biol.* **1997**, *4*, 309–320. [CrossRef]
44. Ben-Naim, B. *Entropy Demystified: The Second Law of Thermodynamics Reduced to Plain Common Sense*; World Scientific: Singapore, 2015.
45. Plasson, R.; Kondepudi, D.K.; Bersini, H.; Commeyras, A.; Asakura, K. Emergence of homochirality in far-from-equilibrium systems: Mechanisms and role in prebiotic chemistry. *Chirality* **2007**, *19*, 589–600. [CrossRef] [PubMed]
46. Ribo, J.M.; Blanco, C.; Crusats, J.; El-Hachemi, Z.; Hochberg, D.; Moyano, A. Absolute asymmetric synthesis in enantioselective autocatalytic reaction networks: Theoretical games, speculations on chemical evolution and perhaps a synthetic option. *Chem. Eur. J.* **2014**, *20*, 17250–17271. [CrossRef]
47. Ross, J.; Vlad, M. Exact solutions for the entropy production rate of several irreversible processes. *J. Phys. Chem. A* **2005**, *109*, 10607–10612. [CrossRef]
48. Hochberg, D.; Ribó, J.M. Stochiometric network analysis of entropy production in reaction networks. *Phys. Chem. Chem. Phys.* **2018**, *20*, 23726–23739. [CrossRef]
49. Hochberg, D.; Ribó, J.M. Entropic analysis of mirror symmetry breaking in chiral hypercycles. *Life* **2019**, *9*, 28. [CrossRef]
50. Werth, H. Über Irreversibilität, Naturprozesse und Zeitstrukture. In *Offene Systeme I*; Weizsäcker, E., Ed.; Klett-Cotta: Stuttgard, Germany, 1974; pp. 114–199.
51. Kondepudi, D.; Kapcha, K. Entropy production in chiral symmetry breaking transitions. *Chirality* **2008**, *20*, 524–528. [CrossRef]
52. Kondepudi, D.; Prigogine, I. *Modern Thermodynamics*, 2nd ed.; John Wiley & Sons: Chichester, UK, 2015.

53. Plasson, R.; Bersini, H. Energetic analysis of mirror symmetry breaking processes in a recycled microreversible chemical system. *J. Phys. Chem. B* **2008**, *113*, 3477–3490. [CrossRef]
54. Mauksch, M.; Tsogoeva, S.B. Spontaneous Emergence of Homochirality via Coherently Coupled Antagonistic and Reversible Reaction Cycles. *ChemPhysChem* **2008**, *9*, 2359–2371. [CrossRef] [PubMed]
55. Plasson, R.; Brandenburg, A.; Jullient, L.; Bersini, H. Autocatalyses. *J. Phys. Chem. A* **2011**, *115*, 8073–8085. [CrossRef]
56. Hochberg, D.; Bourdon García, R.D.; Ägreda Bastidas, J.A.; Ribó, J.M. Stoichiometric network analysis of spontaneous mirror symmetry breaking in chemical reactions. *Phys. Chem. Chem. Phys.* **2017**, *19*, 17618–17636. [CrossRef] [PubMed]
57. Gellman, A.J.; Ernst, K.-H. Chiral autocatalysis and mirror symmetry breaking. *Catal. Lett.* **2018**, *148*, 1610. [CrossRef]
58. Ribó, J.M.; Crusats, J.; El-Hachemi, Z.; Moyano, A.; Hochberg, D. Spontaneous mirror symmetry breaking in heterocatalytically coupled enantioselective replicators. *Chem. Sci.* **2017**, *8*, 763–769.
59. Szathmary, E. The origin of replicators and reproducers. *Philos. Trans. R. Soc. Lond. Ser. B* **2006**, *361*, 1761–1776. [CrossRef] [PubMed]
60. Frank, F.C. On spontaneous asymmetric synthesis. *Biochim. Biophys. Acta* **1953**, *11*, 459–463. [CrossRef]
61. Ribó, J.M.; Hochberg, D. Competitive exclusion principle in ecology and absolute asymmetric synthesis in chemistry. *Chirality* **2015**, *27*, 722–727. [CrossRef]
62. Rivera Islas, J.; Lavabre, D.; Grevy, J.-M.; Hernández Lamoneda, R.; Rojas Cabrera, K.; Micheau, J.C.; Buhse, T. Mirror-symmetry breaking in the Soai reaction: A kinetic understanding. *Proc. Natl. Acad. Sci. USA* **2005**, *102*, 13743–13748. [CrossRef]
63. Lavabre, D.; Micheau, J.-C.; Islas, J.R.; Buhse, T. Kinetic Insight into specific features of the autocatalytic Soai reaction. In *Amplification of Chirality*; Springer: Berlin, Germany, 2008; Volume 284, pp. 67–96.
64. Crusats, J.; Hochberg, D.; Moyano, A.; Ribó, J.M. Frank models and spontaneous emergence of chirality in closed systems. *ChemPhysChem* **2009**, *10*, 2123–2131. [CrossRef]
65. Avetisov, V.; Goldanskii, V. Mirror symmetry breaking at the molecular level. *Proc. Natl. Acad. Sci. USA* **1996**, *93*, 11435–11442. [CrossRef] [PubMed]
66. Ribó, J.M.; Hochberg, D. Stability of racemic and chiral steady states in open and closed chemical systems. *Phys. Lett. A* **2008**, *373*, 111–122. [CrossRef]
67. Blanco, C.; Crusats, J.; El-Hachemi, Z.; Moyano, A.; Hochberg, D.; Ribó, J.M. Spontaneous Emergence of Chirality in the Limited Enantioselectivity Model: Autocatalytic Cycle Driven by an External Reagent. *ChemPhysChem* **2013**, *14*, 2432–2440. [CrossRef] [PubMed]
68. Ribó, J.M.; Crusats, J.; El-Hachemi, Z.; Moyano, A.; Blanco, C.; Hochberg, D. Spontaneous Mirror Symmetry Breaking in the Limited Enantioselective Autocatalysis Model: Abyssal Hydrothermal Vents as Scenario for the Emergence of Chirality in Prebiotic Chemistry. *Astrobiology* **2013**, *13*, 132–142. [CrossRef]
69. Blanco, C.; Crusats, J.; El-Hachemi, Z.; Moyano, A.; Veintemillas-Verdaguer, S.; Hochberg, D.; Ribó, J.M. The Viedma Deracemization of Racemic Conglomerate Mixtures as a Paradigm of Spontaneous Mirror Symmetry Breaking in Aggregation and Polymerization. *ChemPhysChem* **2013**, *14*, 3982–3993. [CrossRef]
70. El-Hachemi, Z.; Crusats, Z.; Ribó, J.M.; Veintemillas-Verdaguer, S. Spontaneous transition toward chirality in the $NaClO_3$ crystallization in boiling solutions. *Cryst. Growth Des.* **2009**, *9*, 4802–4806. [CrossRef]
71. Cintas, P.; Viedma, C. Homochirality beyond grinding: Deracemizing chiral crystals by temperature gradient under boiling. *Chem. Commun.* **2011**, *47*, 12786–12788. [CrossRef]
72. Crusats, J.; Veintemillas-Verdaguer, S.; Ribó, J.M. Homochirality as a Consequence of Thermodynamic Equilibrium? *Chem. Eur. J.* **2006**, *12*, 7776–7781. [CrossRef]
73. El-Hachemi, Z.; Arteaga, O.; Canillas, A.; Crusats, J.; Sorrenti, A.; Veintemillas-Verdaguer, S. Achiral-to-chiral transition in benzil solidification: Analogies with racemic conglomerates systems showing deracemization. *Chirality* **2013**, *25*, 393–399. [CrossRef]
74. Viedma, C. Fighting fire with fire: Racemization drives deracemization. In Proceedings of the Chiral Symmetry Breaking at Molecular Level, Solvay Workshop, Brussels, Belgium, 28–30 November 2018.
75. Kondepudi, D.K.; Kauffman, R.; Singh, N. Chiral symmetry breaking in sodium chlorate crystallization. *Science* **1990**, *250*, 975–976. [CrossRef]
76. Kondepudi, D.K.; Asakura, K.; Laudadio, J. Chiral symmetry breaking in stirred crystallization of 1,1′-binaphthyl melt. *J. Am. Chem. Soc.* **1999**, *121*, 1448–1451. [CrossRef]

77. Ribó, J.M.; El-Hachemi, Z.; Arteaga, O.; Canillas, A.; Crusats, J. Hydrodynamic Effects in Soft-Matter Self-Assembly: The Case of J-Aggregates of Amphiphilic Porphyrins. *Chem. Record.* **2017**, *17*, 1–17. [CrossRef] [PubMed]
78. Li, P.; Lv, B.; Duan, P.; Liu, M.; Yin, M. Stoichiometry-controlled inversion of circulaly polarized luminiscence in co-assembly of chiral gelators with achiral tetraphenylethylene derivative. *Chem. Commun.* **2019**, *55*, 2194–2197. [CrossRef] [PubMed]
79. Karunakaran, S.C.; Cafferty, B.J.; Weigert-Muñoz, A.; Schuster, G.B.; Hud, N.V. Spontaneous Symmetry Breaking in the Formation of Supramolecular Polymers: Implications for the Origin of Biological Homochirality. *Angew. Chem. Int. Ed.* **2019**, *58*, 1453–1457. [CrossRef] [PubMed]
80. Blanco, C.; Stich, M.; Hochberg, D. Mechanically induced homochirality in nucleated enantioselective polymerization. *J. Phys. Chem. B* **2017**, *121*, 942–955. [CrossRef] [PubMed]
81. Saghatelian, A.; Yokobayashi, Y.; Soltaini, K.; Gadhiri, M.R. A chiroselective peptide replicator. *Nature* **2001**, *409*, 797–801. [CrossRef] [PubMed]
82. Eschenmoser, A. The search of the chemistry of life's origin. *Tetrahedron* **2001**, *63*, 12821–12844. [CrossRef]
83. Monnard, P.-A.; Walde, P. Current ideas about prebiological compartmentalization. *Life* **2015**, *5*, 1239–1262. [CrossRef]
84. Kauffman, S. *At Home in the Universe*; Oxford University Press: Oxford, UK, 1995.
85. Kondepudi, D.K.; Prigogine, I.; Nelson, G. Sensitivity of branch selection in nonequilibrium systems. *Phys. Lett. A* **1985**, *111*, 29–32. [CrossRef]
86. Avetisov, V.A.; Kuz'min, V.V.; Anikin, S.A. Sensitivity of chemical chiral systems to weak asymmetric factors. *Chem. Phys.* **1987**, *112*, 179–187. [CrossRef]
87. Pitsch, S.; Wendeborn, S.; Jaun, B.; Eschenmoser, A. Why Pentose- and Not Hexose-Nucleic Acids? *Helvetica Chim. Acta* **1993**, *76*, 2161–2183. [CrossRef]
88. Shapiro, R. Prebiotic ribose synthesis: A critical analysis. *Orig. Life Evol. Biosph.* **1988**, *18*, 71–85. [CrossRef] [PubMed]
89. Stern, R.; Jedrzejas, M.J. Carbohydrate polymers at the center of life's origin. *Chem. Rev.* **2008**, *108*, 50161–55085. [CrossRef] [PubMed]
90. Pestunova, O.P.; Simonov, A.N.; Suytnikov, V.N.; Parmon, V.N. Prebiotic carbohydrates and their derivatives. In *Biosphere Origin and Evolution*; Dobretsov, N., Kolchanov, N., Rozanov, A., Zavarzin, G., Eds.; Springer: New York, NY, USA, 2008; pp. 103–118.
91. Orgel, L.E. Self-organized metabolic cycles. *Proc. Natl. Acad. Sci. USA* **2000**, *97*, 12503–12507. [CrossRef] [PubMed]
92. Sel'kov, E.E. Self-oscillations in glycolysis. *Eur. J. Biochem.* **1968**, *4*, 79–86. [CrossRef] [PubMed]
93. Plasson, R.; Bersini, H.; Commeyras, A. Recycling Frank: Spontaneous emergence of homochirality in noncatalytic systems. *Proc. Natl. Acad. Sci. USA* **2004**, *101*, 16733–16738. [CrossRef] [PubMed]
94. Stich, M.; Blanco, C.; Hochberg, C. Chiral and chemical oscillations in a simple dimerization model. *Phys. Chem. Chem. Phys.* **2013**, *15*, 255–261. [CrossRef] [PubMed]
95. Schmitt-Kopplin, P.; Gabelica, Z.; Gougeon, R.D.; Fekete, A.; Kanawati, B.; Harir, M.; Gebefuegi, I.; Eckel, G.; Hertkorn, N. High molecular diversity of extraterrestrial organic matter in Murchison meteorite revealed 40 years after its fall. *Proc. Natl. Acad. Sci. USA* **2010**, *107*, 2763–2768. [CrossRef]
96. Krishnamurthy, R. Life's Biological Chemistry: A Destiny or destination starting from prebiotic chemistry? *Chem. Eur. J.* **2018**, *24*, 16708–16715. [CrossRef]
97. Peretó, J.; López-García, P.; Moreira, D. Ancestral lipid biosynthesis and early membrane evolution. *Trends Biochem. Sci.* **2004**, *29*, 469–477. [CrossRef]
98. Davankov, V. The chirality as an inherent general property of matter. *Chirality* **2006**, *18*, 459–461. [CrossRef] [PubMed]

MDPI
St. Alban-Anlage 66
4052 Basel
Switzerland
Tel. +41 61 683 77 34
Fax +41 61 302 89 18
www.mdpi.com

Symmetry Editorial Office
E-mail: symmetry@mdpi.com
www.mdpi.com/journal/symmetry

www.ingramcontent.com/pod-product-compliance
Lightning Source LLC
Chambersburg PA
CBHW051715210326
41597CB00032B/5489

* 9 7 8 3 0 3 9 2 1 7 2 2 9 *